# Glassy, Amorphous and Nano-Crystalline Materials

**Hot Topics in Thermal Analysis and Calorimetry**

VOLUME 8

*Series Editor*
**Judit Simon**, *Budapest University of Technology and Economics, Hungary*

For other titles published in this series, go to
http://www.springer.com/series/6056

Jaroslav Šesták • Jiří J. Mareš • Pavel Hubík
Editors

# Glassy, Amorphous and Nano-Crystalline Materials

## Thermal Physics, Analysis, Structure and Properties

Springer

*Editors*
Jaroslav Šesták
New Technologies –
Research Centre in the Westbohemian
Region
University of West Bohemia
Univerzitní 8
30614 Plzeň
Czech Republic
sestak@fzu.cz

Jiří J. Mareš
Institute of Physics, v.v.i.
Academy of Sciences
of the Czech Republic
Cukrovarnická 10
16200 Prague 6
Czech Republic
maresjj@fzu.cz

Pavel Hubík
Institute of Physics, v.v.i.
Academy of Sciences
of the Czech Republic
Cukrovarnická 10
16200 Prague 6
Czech Republic
hubik@fzu.cz



ISBN 978-90-481-2881-5 e-ISBN 978-90-481-2882-2
DOI 10.1007/978-90-481-2882-2
Springer Dordrecht London Heidelberg New York

Library of Congress Control Number: 2010938473



Printed on acid-free paper

Springer is part of Springer Science+Business Media (www.springer.com)

# Preface

## Early Research into Amorphous Semiconductors

Numerous aspects of physics and chemistry of non-crystalline solids and glassy state which are discussed in the present book can only hardly be prefaced without rude simplification. Therefore, let us make instead a short excursion to the prehistory of research into amorphous semiconductors, the topic on which the common epistemological features of the other subjects treated in this book may be demonstrated.

It is a matter of fact that in everyday life we encounter more frequently non-crystalline than crystalline solids. We can even with some exaggeration say that in the Nature the perfect crystals are as rare as diamonds. In spite of that, the existence of the class of non-crystalline materials has been recognized only recently. One of the reasons for such a state of the art is probably the fact that the positivistic continuous model of matter dominated till the end of the nineteenth century and that the fundamental conjectures of atomism were too closely bound up with the idea of the regular ordering of atoms; early atomic theories accounting for the regular shape of snow flakes [1] and for the anisotropy of optical properties of transparent crystals [2], namely, exploited the idea that such a regularity is due to the closest filling of the space by identical hard polyhedrons or spheres, atoms. Denying atomic order in solids would thus undermine the strongest intuitive argument in favour of atomism, namely, that just the satisfaction of geometrical constrains between neighbouring atoms and their close packing accounts for actually observed regular shape of crystals. Interestingly enough, in the scientific disputations about the structure of matter, the existence of glass, for a long time known amorphous, i.e. "shapeless" material par excellence, was tacitly ignored.

The serious attempts to treat the atomic structure of amorphous or glassy state are thus relatively recent, belonging to the first half of the twentieth century. The glass was at that time considered to be nothing but undercooled liquid i.e. a solid having essentially the atomic structure of original melt. Such a picture was an immediate consequence of phenomenological principle of continuity between liquid and solid state proposed by Frenkel [3], which was, among others, partially confirmed by means of X-ray analysis. Accordingly, namely, some characteristic structural patterns observed in the solid phase are, as a rule, observed in the liquid melt as well. From the broadening of X-ray diffraction patterns it was even possible to conclude that the glass comprises nanocrystals of typical size ~2 nm. (Today such a material would be classified rather as a nanocrystalline aggregate than glass.) Nevertheless, the phenomenological model provided neither the algorithm for the reconstruction of atomic lattice of glass nor the basis for the derivation of its physical properties from the structural ones. The first realistic model of glass lattice due to Zachariasen [4] was thus rather a result of physical reasoning than of a purely descriptive phenomenology. Crucial for the establishment of the model was the observation (from today's point of view not very exact!) that the mechanical properties of a glass, e.g. elastic bulk modulus or hardness, and of a corresponding crystal are similar. Consequently, the underlying building blocks have to have similar structural energies and must be matched together without further expense of energy. In order to satisfy these conditions it was suggested to identify the "underlying building blocks" with the nearest neighbourhood of each atom, the arrangement of which was only slightly changed with respect to that in a crystal. Cumulating small deviations of bond lengths (~1%) and tiny variations of bond angles (~1°) made it possible to match building blocks together and to construct a random amorphous lattice without appreciable increase of structural energy. It can be shown that such a random lattice being locally almost identical with the periodic lattice of the corresponding crystal should reveal up to the second or the third coordination sphere practically the same radial distribution of atoms. For the following coordination spheres, however, huge differences are expected. The overall structural properties of amorphous solid may then be characterized quite simply by saying that the short-range order of atoms is preserved while the long-range order is absent. Evidently, this is quite a new concept of disorder, differing essentially from that encountered e.g. in gases. Being once established, the model of locally ordered homogeneous random network based on the absence of long-range order started to play more and more important role in modern solid state physics and chemistry.

By admitting new structural model, the problem of amorphous state was by no means definitely solved but just on the contrary. The researchers had to struggle with qualitatively new difficulties which appeared by computing the mechanical, optical and electronic properties of amorphous solids. As the semiconductors are materials which are known to be most sensitive to the changes of structural and chemical disorder, they may serve as a good example illustrating the fascinating development of solid state physics which followed. The quantum band model of solids as established in the 1930s was tailor-made for crystals having perfectly periodic lattice disturbed by only a small number of imperfections. Accordingly to Wilson's classification [5], the semiconductors were materials characterized by a "not very large" energy gap (~1 eV) in their electronic band structure. The very existence of the electronic band structure of solids was at that time treated as a direct consequence of periodicity of crystal lattice. Such an opinion was basically due to the establishment of famous Bloch's theorem [6] which enabled one to solve electronic structure of periodically arranged atoms even analytically (see e.g. Kronig-Penney model [7]). According to Bloch's theorem, namely, the interaction of an electron with the periodic crystal lattice may be replaced by its movement in a certain periodic potential. The solution of Schrödinger equation in such a periodic field has a special form of harmonic wave with periodically modulated amplitude. In amorphous solids such a marvelous mathematical simplicity due to the perfect translational symmetry and long range order of crystals was lost forever and researchers have to learn how to do without Bloch's theorem. Moreover, the situation was further complicated by quite an astonishing experimental observation of Kolomiets [8] that chalcogenide glasses behave like intrinsic semiconductors which are essentially non-sensitive to the doping. Unexpectedly enough, theorists had to account for the surviving of the band model in the absence of lattice periodicity and for experimentalists quite a new field of research was opened, amorphous semiconductors. It should be stressed here that just these two circumstances had enormous impact on the further development of semiconductor science and technology.

The real boom of research into amorphous semiconductors was initialized by works of Ovshinski who reported about switching between a highly resistive and conductive state effected by an electric field and memory effects in chalcogenide glasses [9]. Immediately afterwards a plenty of new effects such as photodoping, reversible photostructural changes and optical memory effect having a huge application potential for imaging and electrophotography (e.g. Xerox process) were discovered. Hand in hand with the promise of further applications steeply increased interest not only in chalcogenide glasses but also in other types of amorphous semiconductors (e.g. tetrahedrally bounded semiconductors) and in non-crystalline materials in general. Quite naturally there appeared a demand for a new scientific journal (Journal of Non-Crystalline Solids, 1st issue 1969) covering all these hot topics.

What were actually the main achievements in amorphous semiconductors during the decade from1965 to 1975? First of all it was recognized that the band gap does exists also in amorphous semiconductors in spite of the absence of atomic long-range order. The band gap is, however, not empty but it contains an appreciable amount of localized states and its edges are no more sharp [10]. The position of the band gap edges and actual width of the band gap thus depends on the method of measurement. Transport band edges coincide with so called mobility edges where the mobility of carriers dramatically changes while the optical band gap (Tauc's optical gap [11, 12]) is determined by means of extrapolation of absorption curve. The localized states within the gap are, moreover, no passive entity. As a rule, at room and lower temperatures, they enable a special type of carrier transport via localized states, the so called hopping [13]. A new light on the character of localized states within the gap shed the path-breaking discovery of possibility of effective doping of amorphous silicon prepared by glow-discharge technique [14]. Passivation of localized gap states by hydrogen or other chemicals opened a new way to the tailoring of these materials. Besides the purely technological progress, the research into amorphous semiconductors stimulated development of qualitatively new methods of computing electronic structure of disordered solids and brought about changes in understanding to apparently closed topics [15].

We can thus claim that the research into amorphous semiconductors completely changed, within practically 1 decade, the gist of solid state physics as a whole. The concept of disorder became a corner stone of a lot of theories; the era of order was shared by the era where the concept of disorder dominated. And finally, in epistemological context provided the discovery of various aspects of disorder a valuable key to our understanding of a lot of natural systems and phenomena belonging not only to the scope of solid state physics and chemistry but also to biology, astrophysics and even to sociology.

May 2010

Jiří J. Mareš
Institute of Physics ASCR, v.v.i.
Prague, Czech Republic

Jan Tauc
Brown University
Providence, Rhode Island, USA

1. Kepler J (1987) Strena seu de Nive Sexangula. G. Tambach, Fracofurti ad Moenum (1611), German transl.: Vom Sechseckigen Schnee. Ostwald's Klassiker vol 273, Geest and Portig, Leipzig
2. Huyghens C (1903) Traité de la Lumiére. Pierre van der Aa, Leide (1690), German transl.: Abhandlung über das Licht. W. Engelmann, Leipzig
3. Frenkel YaI (1945) Kinetitsheskaya teoria zhidkostei. Izd. AN SSSR, Moscow; English transl.: Kinetic theory of liquids. Clarendon, Oxford (1946)
4. Zachariasen WH (1932) The atomic arrangement in glass. J Am Chem Soc 54:3841–3851
5. Wilson AH (1931) The theory of electronic semi-conductors. Proc R Soc London A 133:458–491; Proc R Soc London A 134:277–287
6. Bloch F (1929) Über die Quantenmechanik der Elektronen in Kristallgittern. Z Phys 52:555–600
7. Kronig R de L, Penney WG (1931) Quantum mechanics of electrons in crystal lattices. Proc R Soc London A 130:499–513
8. Kolomiets BT (1964) Vitreous semiconductors (I), (II). Phys Stat Sol 7:359–372, 713–731; Original report in: Proc IC on Semicon Phys Prague'60, Czechoslovak Acad Sci (1961), p 884
9. Ovshinski SR (1968) Reversible electrical switching phenomena in disordered structures. Phys Rev Lett 21:1450–1453
10. Cohen MH, Fritzsche H, Ovshinsky SR (1969) Simple band model for amorphous semiconducting alloys. Phys Rev Lett 20:1065–1068
11. Tauc J, Grigorovici R, Vancu A (1966) Optical properties and electronic structure of amorphous germanium. Phys Stat Sol 15:627–637
12. Tauc J (ed) (1974) Amorphous and liquid semiconductors. Plenum, London

13. Mott NF, Davies EA (1971) Electronic processes in non-crystalline materials. Clarendon, Oxford
14. Spear WE, LeComber PG (1975) Substitutional doping of amorphous silicon. Solid State Commun 17:1193–1196
15. Shklovskii BI, Efros AL (1979) Elektronnye svoistva legirovanykh poluprovodnikov (Electronic Properties of Doped Semiconductors). Nauka, Moscow

# About the Editors

**Prof. Jaroslav Šesták, MEng., Ph.D., DSc.**
Senior Scientist of the Institute of Physics, Academy of Sciences of the Czech Republic and New Technologies Research Centre, University of West Bohemia – specialised in thermodynamics (kinetics) and material science (particularly applied to inorganic glasses), 287 papers in impact journals, over 2,500 SCI citation responses, 15 books and book chapters, received degree of Doctor Honoris Causa of the University of Pardubice (January 2010).

**Dr. Jiří J. Mareš, Ph.D.**
Deputy Director of the Institute of Physics, Academy of Sciences of CR – specialised in condensed matter physics, deals with quantum properties of disordered systems and with fundamental problems of thermal physics and electrostatics, 123 papers in impact journals, over 230 citation responses, 3 books and book chapters.

**Dr. Pavel Hubík, Ph.D.**
Senior Scientist of the Institute of Physics, Academy of Sciences of CR – deals with defects in semiconductors and with electron and thermal properties of solid state, 51 papers in impact journals, over 110 citation responses, 2 books and book chapters.

# Acknowledgments

Prepared under the support of the following projects:

Institutional Research Plan No AV0Z10100521 of the Institute of Physics, v.v.i., Academy of Sciences of the Czech Republic

Grant Agency of the Academy of Sciences of the Czech Republic Project No IAA100100712

Czech Science Foundation Projects No P204/10/0212 and No P204/11/0964

Czech Industrial Grants (MPO) No 2A-1TP1/037 in the program 2A – Sustainable welfare and Nos FR-TI 1/335, FR-TI 1/369 and FR-TI 1/278 in the program TIP

Ministry of Education, Youth and Sport (MŠMT) Project No 1M06031

# Contents

# Chapter 1
# Introduction: Some Essential Attributes of Glassiness Regarding the Nature of Non-crystalline Solids

**Hiroshi Suga**

## 1.1 Character of the Amorphous Solids

Glass products have long been used from ancient times not only in our daily life but also in some laboratory experiments such as the U tube for the measurement of volume of a gas as a function of pressure. Liquefaction of the last "permanent gas helium" was done successfully with an apparatus entirely made of glass. Formerly the glasses have been produced by cooling the melts of silicate minerals without crystallization until they becomes hard and brittle solids. Later the glasses were found to exhibit hallo diffraction patterns similar to those of the liquids. Some important concepts are involved in this description. The first is the method of preparation. The melt-cooling was used in some of the modern definitions of glasses. The second is the starting materials of inorganic origin. Organic substances such as glycerol and synthetic polymers were found to behave similarly. Thus the term glasses can be extended to a wide range of substances that easily undercool to form amorphous solids. The third is the metastability of the undercooled liquids and glasses compared to the corresponding crystalline solids. If the cooling rate is adequately slow to induce nucleation, the melt becomes crystalline solid possessing regular lattice with lower Gibbs energy. Thus the formation of glass is a problem of bypassing or avoiding the crystallization. Although the main subjects of this book are the structures and properties of ordinary network glasses of inorganic origin, it will be instructive to start with the description of the general features of glassiness exhibited by various kinds of condensed matters in which the constituents are held together by interaction forces, such as the van der Waals, hydrogen bonding, ionic or covalent bonds.

In spite of the long history of practical usage of the ordinary glasses, the scientific research of the glassy state began in the twentieth century. It is surprising to realize

H. Suga (✉)
Osaka University, Toyonaka, Osaka 560-0043, Japan
e-mail: suga-hiroshi777@hcn.zaq.ne.jp

J. Šesták et al. (eds.), *Glassy, Amorphous and Nano-Crystalline Materials*,
Hot Topics in Thermal Analysis and Calorimetry 8, DOI 10.1007/978-90-481-2882-2_1,

that the first measurement of glass transition is more recent event than the discovery of the radioactivity. Pioneering work of the calorimetric measurement by Gibson and Giauque [1] has shown that glycerol glass exhibited a sudden increase in the heat capacity $C_p$ over a narrow temperature region, and possessed a residual entropy $S_0$ at 0 K. The temperature region is called the glass transition temperature $T_g$, and is an important keyword for the glass science and technology. Other thermodynamic quantities such as the volume $V$ and thermal expansivity $\alpha$ also undergo some changes at $T_g$. This experiment gave the first paramount example of the solid that deviated from the third law of thermodynamics. In Simon's word [2], the law is expressed as "The contribution to the entropy of a system by each aspect which is in internal equilibrium tends to zero at 0 K." The realization of the internal equilibrium in some systems becomes quite serious for a human being with a short life span. In this way, the glasses are the frozen-in non-equilibrium states with respect to some degrees of freedom in a metastable phase of the system. In modern terminology, the glasses or glassy states are designated as non-ergodic. Thus any liquid that satisfies the ergodicity hypothesis in statistical mechanics becomes out of equilibrium with some aspects of the system during a continuous cooling. The breaking of the ergodicity occurs at a small temperature interval $T_g$, which depends naturally on the experimental time-scale. In this way, the glass transition temperature, the residual entropy, the metastability, and the crystallization are the important keywords for the thermodynamic characterizations of the glassy systems.

Over the past 50 years, many amorphous solids have been prepared by various methods other than the traditional liquid-cooling. Amorphous semiconductor used for the solar cells is one example. Much of the intense research interest in amorphous solids is driven by the technological importance of these solids. Intellectual fascination with the amorphous solids arises from the basic understanding of the state of aggregation of the constituent entities in the solids. These amorphous solids combine liquid-like disordered structure and crystal-like rigidity, and display unique features that are generally absent in their counterparts in the crystalline solids. The exotic formation of amorphous solids can be done by either physical or chemical methods. Some methods do not pass at all their liquid state during the formation of amorphous solids. Glasses are just one example of the amorphous solids formed from the liquid states. What is the relation between glasses and amorphous solids? The question was raised by Secrist and Mackenzie [3]. The two terms had been used for some times without any clear distinction.

Molecular assembly is traditionally divided into four states of aggregation of the constituent molecules. The division of assembly into gas, liquid, and solid is based on the mechanical properties of the assembly, such as density and fluidity. Further division of the solid into crystalline and non-crystalline solids is obviously based on the structural point of view. The division of metastable liquids below their melting points into glassy and undercooled liquids is based on the non-equilibrium and the equilibrium nature of both states based on the thermodynamic reasoning. Thus criteria of the classification are multifarious in nature. The glasses are just one example of the amorphous solids with frozen-in disordered structure of liquids, but the reverse is not true necessarily without some experimental verification.

One method to answer this question is to detect in amorphous solids possible existence of either the glass transition or the residual entropy that have been observed to occur concomitantly in many glass-forming liquids. This problem is discussed first in the following section.

It is worthy to note here another metastable nature of the polymeric materials. Most of the synthetic polymers consist of the crystalline and amorphous parts. The ratio of both parts can be changed arbitrarily by keeping them at particular temperature range. This nature produces a wide variety of properties in one and the same polymer, and enhances the usefulness in the applications of the polymeric materials. According to the phase rule, however, only a single phase should appear generally in any single-component system at a particular $T$ and $p$ other than the melting line. The coexistence of two parts will be a consequence of a distribution of masses in the constituent molecules of the system. Any synthetic polymers, even composed of one kind of pure monomer, cannot be regarded as the "single-component system," for which the phase rule holds rigorously. One intriguing exception would be atactic polymers in which a trace of crystalline part cannot be observed to exist. It is not easy to imagine any conformation of the molecules in the atactic polymer that guarantees the translational invariance in the spatial arrangement of molecules placed on any regular lattice points. Strictly speaking, there exists no single-component system in pure substances if we take into consideration the existence of isotopes. The effect of isotope on the phase behaviour appears, however, only at very low temperatures. In fact, liquid helium, being a mixture of $He^3$ and $He^4$, exhibits a phase separation at a cryogenic temperature that depends on the composition. The phase separation is a kind of ordering process that reduces the entropy of the system. Most of the substances are believed to be in frozen-in disordered states with respect to their isotopic species.

## 1.2 Preparation and Characterizations of Amorphous Solids

A wide variety of amorphous solids can be prepared by physical and chemical methods. The former is (1) rapid cooling of liquid; (2) deposition of vapour onto a cold substrate; (3) mechanical milling of crystal; (4) particle bombardment onto crystal; (5) compression or depression of crystal, and so on. While the latter is (6) gelation; (7) precipitation by chemical reaction; (8) dehydration of hydrate crystal, and so on. In the former processes, the composition of substance does not change during the formation of amorphous solids [4]. The liquid-cooling (1) is the most familiar one. A hyper-quenching method, with a cooling rate of nearly $10^6$ K $s^{-1}$, was developed and applied to water or metallic liquids that had strong crystallization tendency. The other physical methods do not pass in principle their liquid state. Vapour deposition is one of the powerful methods in forming amorphous solids. The kinetic energies of vapour molecules are extracted efficiently during the condensation on to a substrate kept at a temperature far below the hypothetical $T_g$, so as to arrest the deposited molecules in a frozen disordered state. Crystals are used in the remaining physical methods. High energy in the form of compression, shear stress, or radiation is supplied to the crystal to destroy the neatly ordered arrangements of the constituents. Amorphization of ice, $SiO_2$ and others were done by this way. Mechanical milling was another energizing method and applied first to a metallic mixture forming an amorphous alloy. These non-equilibrium processes can be done without passing their liquids. Amorphization by any low-temperature routs is useful particularly for substances that are unstable at high temperatures. Thermal decomposition can be observed in some of the molecular crystals. The various methods for production of the amorphous solids are depicted schematically in Fig. 1.1.

Mechanical alloying is a non-equilibrium amorphization process and has been used in the development of new metallic systems. This can be done by milling of a mixture composed of two or more metals at room temperature. In contrast to the metallic systems, milling of molecular crystals produces amorphous solids even in a single system. For example, salicin, deoxycholic acid, trehalose, tri-*O*-methyl-β-cyclodextrin, etc. have been amorphized easily even in single system. Supply of mechanical energy to a crystalline substance beyond a critical level induces lattice instability of the crystal and freezes it in an energized state that has lost the original periodicity. Obviously the process must be carried out at temperatures below the "crystallization-dangerous" region of the resulting amorphous solids. Nature of the formed amorphous solid depends on the milling conditions. Both the $T_g$-value and the enthalpy of crystallization of the obtained solid are found to increase asymptotically with the milling time.

For some binary molecular crystals, milling of a mixture results in the molecular alloys with a single $T_g$ varying with the composition. This indicates that the component molecules mix uniformly in a molecular level to exhibit a single relaxation process over the whole composition range. Formation of solid solutions in the crystalline state is highly limited by many factors such as the shared crystal

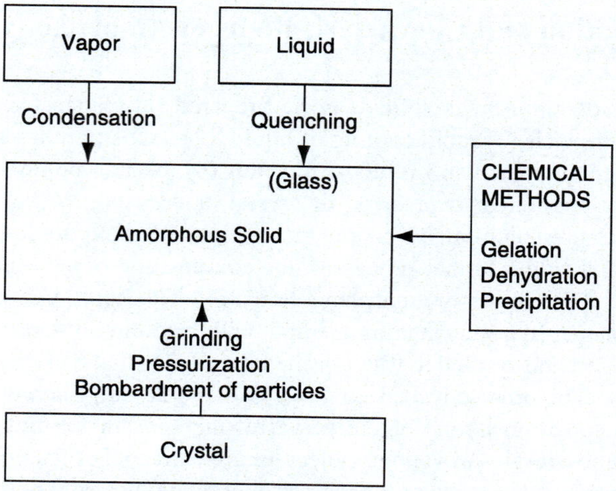


**Fig. 1.1** Various methods of preparation of amorphous solids

symmetry, the similar sizes of the unit cells, the similar molecular shapes, and so on. Alloying of otherwise immiscible substances in the solid state is possible only under non-equilibrium condition.

The chemical methods accompany necessarily some changes in chemical potentials of the system. In method (6), the sample in a sol is brought into a gel state and then removal of extra components forms corresponding amorphous solid. Some oxides were amorphized by this way in the form of fibres or lumps. The method (7) utilizes the fact that some chemical reactions in solution produce amorphous precipitates. Students in the course of analytical chemistry experience fine precipitates in the exercise when $H_2S$ gas is passed into an aqueous solution including $As^{3+}$ or $Sb^{3+}$ ions. The yellow precipitates are fine particles and are known to be amorphous from powder X-ray diffraction experiment. Some hydroxides are also amorphous when they are precipitated in solutions. While, water molecules are important ingredients in hydrate crystals. If the molecules are extracted by rapid evacuation, the resulting anhydride cannot keep anymore the crystalline lattice in some hydrate crystals. Magnesium acetate tetrahydrate, $Mg(CH_3COOH)_2{\cdot}4H_2O$, belongs to this category. By evacuation, the hydrate crystal changes into amorphous anhydride. In this way, there exist many methods for the formation of amorphous solids other than the traditional liquid cooling.

The chemical precipitates $As_2S_3$(am) was examined by thermal analysis, and found to exhibit a glass transition $T_g$ at about 477 K. The temperature was essentially the same as that of $As_2S_3$(am) prepared by the liquid-cooling of orpiment, being crystalline mineral $As_2S_3$(cr) occurring in nature. While, $Sb_2S_3$(am) can be prepared either by the chemical precipitation or by vapour deposition. Both the samples showed $T_g$ at essentially the same temperature of 489 K. These results [5, 6] are summarized in Table 1.1.

The $C_p$ values of the amorphous tri-*O*-methyl-β-cyclodextrin were essentially the same at temperatures below and above $T_g$ for both samples prepared by the mechanical milling and the liquid-quenching. Thus it turned out that the terms amorphous solids and glasses are synonymous. Obviously the extent of frozen-in disorder of any amorphous solids depends on the condition by which the amorphous solids were formed. In the vapour deposition experiments, it was revealed that the lower the substrate temperature, the larger the amount of residual entropy $S_0$. This indicates that the faster removing of the kinetic energy of vapour molecules results in larger extent of frozen-in disorder in the solid. The extent of frozen-in disorder in amorphous solid can be described also by the concept of fictive temperature $T_{\text{fic}}$, that was introduced by Tool as an extra variable necessary for the description of glass and was defined as the temperature at which a glass with a given enthalpy would be at equilibrium if it followed the iso-structural change as the temperature was changed [7]. The thermal motions in high-temperature liquids permit rapid and effective exploration of alternative molecular packing. But such structural change becomes increasingly sluggish as the temperature is lowered, and it largely ceases if further cooling passes through $T_g$. If a liquid is cooled at a constant rate, the equilibrium enthalpy deviates from the equilibrium curve at a particular temperature $T_{\text{fic}}$. The value becomes lower as the cooling rate is lowered. The value $T_{\text{fic}}$ will

**Table 1.1** Glass transition temperature $T_g$ of various amorphous solids

| Substance | $T_g$/K | Method | Substance | $T_g$/K | Method |
|---|---|---|---|---|---|
| Propane | 46 | VC | Methanol | 103 | VC |
| (Propane)$_x$(propene)$_{1-x}$ | 46 ~ 55 | VC | Vinyl acetate | 123 | LQ |
| Propene | 55 | VC | Isopropylbenzene | 126 | LQ |
| | 55 | LQ | $H_2O$ | 135 | VC |
| (Propene)$_x$(1-butene) $_{1-x}$ | 55 ~ 60 | VC | Ethylene glycol | 153 | LQ |
| 1-Butene | 60 | VC | 1,3-Propanediol | 166 | LQ |
| | 60 | LQ | *m*-Cresol | 200 | LQ |
| (Propene)$_x$(1-pentene)$_{1-x}$ | 55 ~ 72 | VC | Resorcinol | 250 | LQ |
| (Butane)$_x$(1-butene)$_{1-x}$ | 60 ~ 62 | VC | Salicin | 333 | MM |
| (1-Butene)$_x$(1-pentene)$_{1-x}$ | 60 ~ 72 | VC | Sucrose | 342 | MM |
| 1-Pentene | 72 | VC | Tri-*O*-methyl-β-cyclodextrin (TMCD) | 352 | LQ |
| $CCl_4$ | 60 | VC | | 352 | MM |
| Isopentane | 65 | LQ | Phenolphthalene | 362 | MM |
| $CHCl_3$ | 79 | VC | TMCD + benzoic acid | 337 | LQ |
| 3-Methylpentane (0.1 MPa) | 77 | LQ | | 337 | MM |
| (108 MPa) | 85 | LQ | Brucine | 359 | MM |
| (198 MPa) | 92 | LQ | Deoxycholic acid (DCA) | 363 | MM |
| Ethanol | 90 | LQ | Trehalose | 384 | MM |
| Cyclohexene | 92 | LQ | Magnesium acetate | 469 | DH |
| Dichloromethane | 93 | LQ | $As_2S_3$ | 477 | CR |
| 1-Propanol (0.1 MPa) | 97 | LQ | | 477 | LQ |
| (108 MPa) | 104 | LQ | $Sb_2S_3$ | 489 | VC |
| (198 MPa) | 109 | LQ | Polytetrahydrofuran | 185 | LQ |
| Butyronitrile | 97 | VC | Polyoxacyclobutane | 193 | LQ |
| | 97 | LQ | Polyoctamethyleneoxide | 255 | LQ |

LQ Liquid quenching, VC Vapour condensation, CR Chemical reaction, MM Mechanical milling of crystal, DH Dehydration of hydrate crystal

change with time in the glassy state, depending on the annealing conditions at temperatures below the initial $T_{fic}$. Thus the determination of $T_{fic}$ of a sample at the measurement stage requires the determinations of enthalpy of the actual glass and of the equilibrium liquid. In this way, calorimetric measurements give the quantitative data of $C_p$, $T_g$, $S_0$ and $T_{fic}$ at the same time. Figure 1.2 shows (a) the heat capacity and (b) the entropy of isopropylbenzene [5] plotted against temperature. The figure shows common features of the amorphous solids prepared by the rapid-cooling of liquid. The heat capacity of a liquid is generally larger than its crystal, primarily because the configurational degrees of freedom are excited in the liquid but not in the crystal. In addition to the primary or α glass transition $T_{g,\alpha}$, a secondary or β glass transition $T_{g,\beta}$ can be found by the measurement. The heat capacity jump associated with the β transition is so small that very careful and

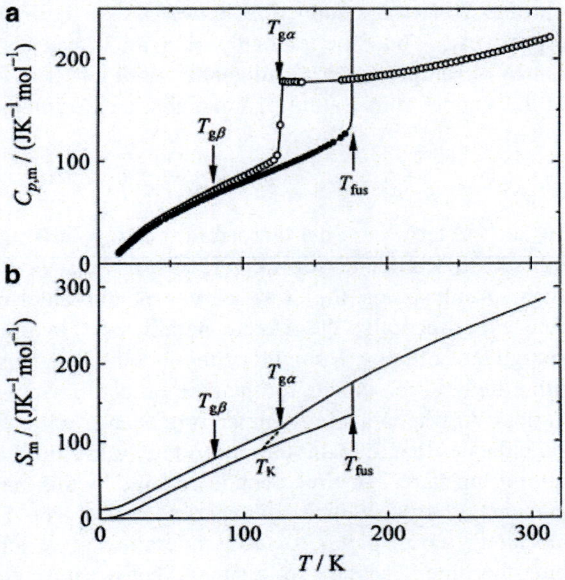


**Fig. 1.2** Heat capacity and entropy of isopropylbenzene

precise measurements are required for the detection. The quantity $T_K$ is designated as the Kauzmann temperature and is described later.

Many physical properties, say the enthalpy, of glass tend to relax toward the equilibrium values of the liquid at a constant temperature with a characteristic time. The relaxation process for the configurational enthalpy is well described by the Kohlrausch-Williams-Watts' (KWW) equation [7],

$$\Delta H_c(t) = \Delta H_c(0)\exp[-(t/\tau)^{\beta}], \tag{1.1}$$

where $H_c$ is the part of enthalpy related to the structure or configuration of the liquid, $\tau$ is the average relaxation time, $\beta$ ($0 \leqq \beta \leqq 1$) is the non-exponential parameter, $\Delta H_c(t)$ and $\Delta H_c(0)$ are the configurational enthalpy to be relaxed at the time $t = t$ and $t = 0$, respectively. The non-exponential relaxation is one of the characteristic phenomena observed in many glasses. The smaller is the $\beta$ value, the larger is the distribution of relaxation times. The remaining part of the enthalpy, called the vibrational enthalpy $H_{vib,}$ responds always quickly to the temperature variation. Since the total enthalpy of a sample, $H_{total} = H_{vib} + H_c$, can be kept constant under the adiabatic condition, it follows

$$\mathrm{d}\Delta H_c(t)/\mathrm{d}t = -[\mathrm{d}H_{vib}(t)/\mathrm{d}T][\mathrm{d}T/\mathrm{d}t] = -C_{vib}[\mathrm{d}T/\mathrm{d}t]. \tag{1.2}$$

Obviously the temperature $T$ means the vibrational temperature. The structural relaxation is, therefore, a process by which $T$ and $T_{fic}$ equalize. This is essentially a problem of heat conduction between two thermal reservoirs with $T$ and $T_{fic}$, respectively. Thus the enthalpy relaxation causes a spontaneous temperature change of sample under an adiabatic condition. The following equation is derived for the sample temperature in an adiabatic calorimeter as a function of time [8].

$$T(t) = a + bt + c\exp[-(t/\tau)^{\beta}]. \tag{1.3}$$

The second term $bt$ is the temperature change arising from residual heat leakage, and is of the order of several mK h$^{-1}$ in most of the experimental temperature range. Analysis of the $T \sim t$ curve of the calorimeter enables us to determine experimentally the kinetic parameters governing the structural relaxation through the enthalpy. It should be reminded that a constant temperature of a sample with time is a prerequisite for the thermal equilibrium (the zeroth law of thermodynamics). Any spontaneous temperature change with time under an adiabatic condition indicates that the sample is in a non-equilibrium state and is relaxing toward the equilibrium state. The time domain covered by this method is 0.1 ks ~ 1 Ms and can be extended by patience of experimentalist. The $T_g$ value determined by the calorimetry corresponds to the temperature at which becomes $\tau$ $10^3$ s or 1 ks, being the time necessary for a single heat-capacity determination. It is interesting to note that for 1-pentene prepared by vapour-deposition, $T_{fic} = 85$ K and $\beta = 0.45$, is quite different from those for the liquid-cooled sample, $T_{fic} = 72$ K and $\beta = 0.92$. This indicates the effectiveness of the formation of glass with high extent of disorder along with wider distribution of the relaxation times by the vapour-condensation process compared to an ordinary cooling process. In this way, the adiabatic calorimeter does work as an ultra-low frequency spectrometer in addition to the conventional tool.

The viscosity of liquid is one measure of the structural relaxation time for the undercooled fluid, and is an important quantity from a practical point of view. The viscosity $\eta$ can be measured routinely and it becomes a convenient kinetic parameter for the glass relaxation. Over a small temperature interval, the viscosity changes by several orders of magnitude from typically $10^{-1}$ Pa s in the liquid state to about $10^{12}$ Pa s in the glassy state. Since a value of $\eta$ in the order of $10^{12}$ Pa s is associated with the solid state, one can define $T_g$ newly as the temperature at which the viscosity reaches this particular value as the temperature is lowered. According to the Maxwell theory of viscoelasticity, the viscosity $\eta$ can be related to the relaxation time $\tau$ through the following equation [9].

$$\tau = G_{\infty}^{-1}\eta, \tag{1.4}$$

where $G_{\infty}$ is the instantaneous shear modulus. Numerical evaluation of this equation shows that the new definition, $\eta\ (T_g) = 10^{12}$ Pa s, corresponds to the relaxation time of several hundred seconds at $T_g$, being again the time scale for a single heat-capacity determination. For some network-forming glasses, such as ordinary silicate glasses, the viscosity changes with temperature $T$ in the Arrhenius fashion,

$\eta = \eta_0 \exp[A/T]$. Many molecular liquids, on the other hand, show pronounced deviation from the Arrhenius behaviour and can generally be better described by the Vogel-Tammann-Fulcher (VTF) equation.

$$\eta = \eta_0 \exp[A/(T - T_0)], \tag{1.5}$$

where $\eta_0$, $A$, and $T_0$ are the constants depending on the nature of the liquid. In this way the viscosity $\eta$, hence the relaxation time $\tau$ tends to diverge towards the temperature $T_0$.

Experiments of the enthalpy relaxation at a temperature below $T_g$ showed that the relaxation rate depended not only the amount of $\Delta H_c$ but also the initial sign of departure. The rate depends on whether the relaxation is structure-breaking or structure-forming process. The data for $\tau$ plotted against $1/T$ exhibit generally non-Arrhenius behaviour, and the relaxation time diverges towards a temperature $T_0$ as the temperature is lowered, as already mentioned in the case of viscosity. Experimental data showed that this temperature is quite near to the Kauzmann temperature $T_K$, at which the extrapolated entropy of liquid crosses that of crystal [9]. Since any liquid that possesses entropy less than that of crystal is hard to imagine, the extrapolated entropy of the liquid should bend at $T_K$, and follows that of the crystal in the temperature range below $T_K$. The seemingly second-order transition that the actual liquid exhibits at $T_g$ is considered to be a kinetic manifestation of the underlying equilibrium phase transition arising from the variation of the entropy at the Kauzmann temperature $T_K$. There exists other scenario that avoids the entropy crisis at $T_K$. Experimental verification of the validity of any models that should occur at $T_K$ in an equilibrium liquid is not possible owing to our short life span.

The deviation from the Arrhenius behaviour of $\tau$ is correlated phenomenologically with a fragility parameter $m$ as follows [10].

$$m = \mathrm{d}\log\tau/\mathrm{d}(T_g/T)|_{T=Tg} \tag{1.6}$$

For the Arrhenius behaviour, for which $m$ is assigned to be 17, the liquid is designated as strong liquid. The structure does not change much with temperature and the heat capacity jump $\Delta C_p$ at $T_g$ is rather small. The larger the $m$ value is, the structure becomes fragile with change in temperature. These are designated as fragile liquids, and exhibit rather large jump $\Delta C_p$ at $T_g$.

Adam and Gibbs [11] proposed a model for explaining the non-Arrhenius behaviour of the relaxation time. They expressed the relaxation time $\tau$ by the following equation.

$$\tau = A\exp[z^*\Delta\mu/kT] = A\exp[N_A\Delta\mu s_c{}^*/kTSc(T)]. \tag{1.7}$$

Here $Sc(T)$ is the macroscopic configurational entropy of liquid, $\Delta\mu$ the chemical potential per molecule hindering the cooperative rearrangement of a group of molecules, $z^*$ the number of molecules constituting the group (or cluster), $s_c{}^*$ the configurational entropy of the smallest group that can undergo the rearrangement, and $N_A$ is Avogadro constant. The configurational entropy is determined by subtracting the vibrational entropy from the experimentally determined entropy. The vibrational heat capacity of the glass expressed in terms of a combination of the Debye and the Einstein heat-capacity functions should be determined first by the least-square's fitting of the experimental data in the low-temperature range. The value $Sc(T)$ becomes zero at $T_K$ for the ideally equilibrated liquid, and this situation results in the divergence of $\tau$ towards $T_K$ as the temperature is lowered.. The non-Arrhenius behaviour can be considered to be due to the fact that $\tau$ is a function of $T$ as well as of $Sc$. Likely the number $z^*$ will not be constant but change with temperature. This entropy theory is supported by experiment [12], showing good correlation between $TSc(T)$ and $\log \tau$ for some glasses prepared under different conditions. In spite of the large differences in $T_g$ among the samples, the glass transition takes place at essentially the same $TSc$ value.

The number of molecules $z^*$ in a cooperatively rearranging unit, or a cluster, can be estimated by considering the following relation that can be derived from Eq. (1.7).

$$z^*(T_1)/z^*(T_2) = Sc(T_2)/Sc(T_1) \tag{1.8}$$

If the value $z^*$ be assumed to be 1 for the liquid at high temperatures where the unimolecular process is dominant, the equation makes possible to estimate the number of molecules $z^*$ in the undercooled and glassy states. This has been done for some systems [13].

## 1.3 Glass Transitions in Crystals: Glassy Crystals

Most of the molecular crystals fuse directly into isotropic liquids. Some crystals fuse in two steps, either through the liquid crystal phases or the orientationally disordered crystal phases. These mesophases have the entropy values close to those of the isotropic liquids, indicating a great deal of disorder in these mesophases. Most of the orientationally disordered crystals have face-centred cubic (fcc) or body-centred cubic (bcc) lattice with molecules possessing a great deal of orientational disorder. If such a phase is cooled rapidly by avoiding the transformation into an ordered crystal, the undercooled disordered phase undergoes freezing-in process at a temperature at which the relevant relaxation time $\tau$ crosses the experimental time, as in the case of ordinary glasses. Particularly the glassy states of crystals designated as "glassy crystals" are interesting. This is because glasses and crystals are two extremes in our concept of structural regularity in solids, and the seemingly contradicting concepts must be combined in this new state of aggregation of molecules [5].

Analogy between the dielectric and enthalpy relaxations will be helpful in understanding the concept of frozen-in disordered crystals. Dipolar liquids generally show the dielectric relaxation in the equilibrium or undercooled liquid phases depending on the frequency $f$ of an external electric field. The dielectric permittivity drops at a temperature at which the dielectric relaxation time $\tau_d$ becomes $(2\pi f)^{-1}$, indicating a freezing-in of the orientation polarization. The actual $T_g$ value of the liquid corresponds to the temperature at which the value $\tau_d$ becomes ~1 ks. The same thing can happen in some orientationally disordered crystals in which the centres-of-mass of the molecules form three-dimensional lattice but their orientations are random among several equi-energetical directions, say eight [111] directions of their fcc or bcc lattices. These crystals exhibit, therefore, the dielectric relaxation likely if the crystal is composed of polar molecules. The reorientational motion is dynamic at high temperatures but becomes quiet as the temperature is lowered until the motion becomes dormant. The heat capacity drops suddenly more or less in a narrow temperature range just in a way similar to the liquids. In spite of the paradoxical nature of the concepts involved in the nomenclature "glassy crystals," time is now giving it proof as an intriguing new state of aggregation of molecules.

Cyclohexanol forms an fcc lattice below $T_{fus}$. The high crystallographic symmetry arises entirely from a great deal of orientational disorder of the molecule that has only mirror symmetry. When the fcc phase is cooled by avoiding the transformation into an ordered low-temperature phase, the undercooled orientationally-disordered phase exhibits $T_g$ at about 150 K and has a definite amount of residual entropy. The behaviour of enthalpy relaxation is unable to discriminate in all aspects from those of ordinary glasses only except for the existence of a long-range positional order in cyclohexanol crystal. The heat-capacity tail below $T_g$ is noticeable. The long $C_p$ tail below $T_g$ causes the $S_c$ to decrease over a wide range of temperature. This phenomenon has been observed more or less in many glassy liquids and the effect on the residual entropy of cyclohexanol crystal was discussed in details [14]. The heat capacity along with the entropy of cyclohexanol crystal is shown in Fig. 1.3.

In the case of ethanol, two kinds of glass transitions were observed for its liquid and the orientational disordered phases, respectively. Some textbooks described that ethanol was a good glass-forming liquid based on the observation of $T_g$. The old observation of $T_g$ by calorimetric measurements turned out to be for its orientational disordered phase and not for the liquid. Glassy liquid of ethanol can be realized only by a rapid cooling of the liquid with a rate faster than 50 K min$^{-1}$. The glass transition for the liquid takes place at essentially the same temperature of 97 K. Two glass transitions can occur for different phases of one and the same substance of $C_2H_5OH$. This observation strongly suggests that the kinetic feature of the molecular reorientation in condensed phases is primarily responsible for the freezing process and that the reorientational motion governs the translation-rotation-coupled rearrangement of molecules in the liquid. Figure 1.4 shows the heat capacity and entropy of ethanol around their glass transitions, respectively [5].

Table 1.2 lists some examples of the glassy crystals along with their $T_g$ values. Likely, glassy state of undercooled liquid crystal can be realized by avoiding the transformation of the liquid crystalline phase into an ordered crystal [5]. In a liquid crystalline phase, molecules of an elongated shape have almost parallel orientational order, but positional disorder. Another name "anisotropic liquids" given to the liquid crystals comes from this fact. The next substance

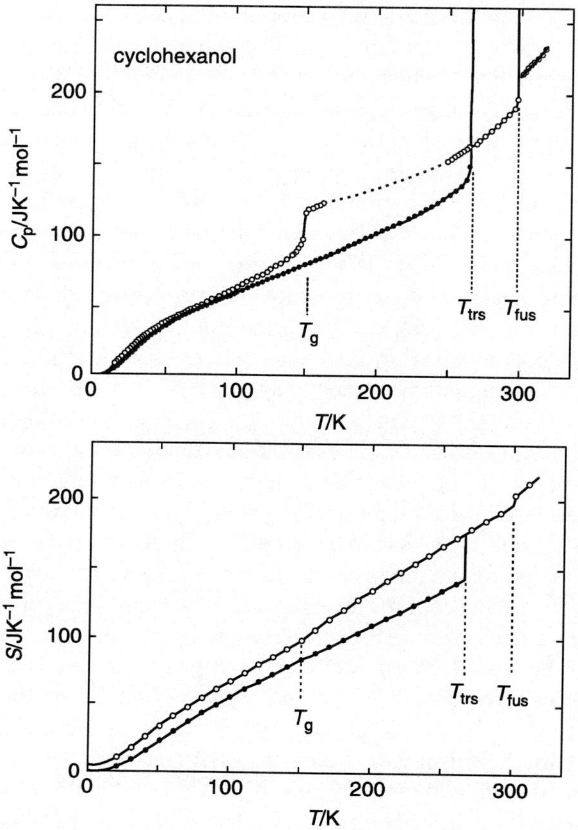


**Fig. 1.3** Heat capacity and entropy of cyclohexanol

*N*-*p*-*n*-hexyloxy-benzylidene-*p*′-*n*-butylaniline, $C_6H_{13}O - C_6H_4 - CH = N - C_6H_4 - C_4H_9$ provides one example of the liquid crystals. This substance shows the following phase sequence: Crystal – Smectic G – Smectic B – Smectic A – Nematic – Isotropic liquid (IL). The highly disordered nature of the Smectic G phase is quantified by the large entropy change $\Delta_{trs}S$ (Cr – $S_G$) amounting 75.98 J K$^{-1}$ mol$^{-1}$ that exceeds far the sum of the remaining entropies of transition: 2.53 ($S_G - S_B$) + 10.14 ($S_B - S_A$) + 9.37 ($S_A$–N) + 5.37 (N–IL). The smectic G phase possessing a great deal of disorder readily undercools and exhibits two-step glass transitions at temperatures around 200 K. The nature of double glass transitions is not clear at the moment, but one of them will be associated with the freezing of molecular modes characteristic of layered structure, either the undulation mode of a layer or anisotropic translational diffusions parallel and perpendicular to a smectic layer.

It turned out that the glass transitions were not characteristic property of liquids but of wide occurrence in a variety of substances irrespective of their translational invariance with respect to their centres-of-mass. The glassy crystals can be a model

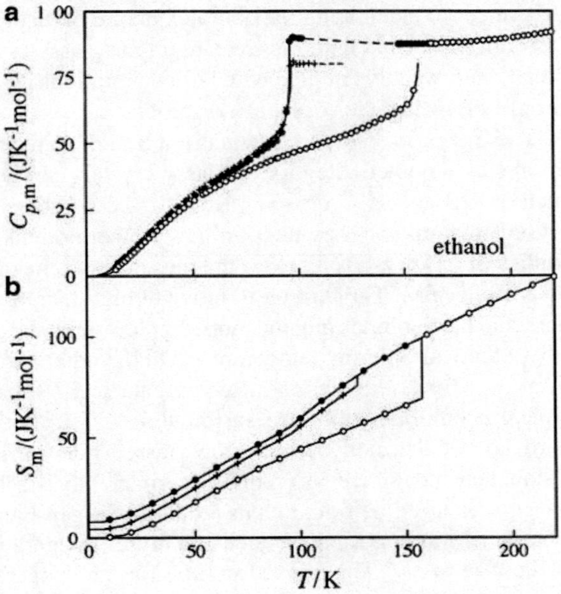


**Fig. 1.4** Heat capacity and en-tropy of ethanol

**Table 1.2** Glass transition temperature $T_g$ of various glassy crystals

| Metastable phase | $T_g$/K | Stable phase | $T_g$/K |
|---|---|---|---|
| Thiophene | 37 | Thiophene | 42 |
| 2,3-Dimethylbutane | 76 | Buckminsterfullerene $C_{60}$ | 87 |
| Isocyanocyclohexanol | 55 | β-Cyclodextrin·11$H_2O$ | 150 |
| | 130 | Ethylene oxide·6.86$H_2O$ | 85 |
| | 160 | Tetrahydrofuran·17$H_2O$ | 85 |
| $CFCl_2$–$CFCl_2$ | 60 | Acetone·17$H_2O$ | 90 |
| | 90 | CO | 18 |
| | 130 | RbCN | 30 |
| Ethanol | 90 | $CsNO_2$ | 42 |
| Cyclohexene I | 92 | Tl $NO_2$ | 60 |
| II | 93 | $SnCl_2$·2$H_2O$ | 150 |
| Cycloheptane I | 100 | $SnCl_2$·2$D_2O$ | 155 |
| II | 100 | $H_2O$ (hexagonal) | 105 |
| III | 93 | $H_2O$ (cubic) | 140 |
| Cycloheptatriene | 106 | $D_2O$ (hexagonal) | 115 |
| Cycloheptanol | 135 | Pinacol·6$H_2O$ | 155 |
| Cyclohexanol | 150 | $H_3BO_3$ | 290 |
| $Cs_{0.7}Tl_{0.3}NO_2$ | 48 | $D_3BO_3$ | 298 |
| $C_2Cl_6$ in thiourea adduct | 59 | Lysozyme[a] | *ca* 150 |
| | | Myoglobin[a] | *ca* 170 |

[a] $T_g$ value of the protein crystal depends strongly on the water content

substance for elucidating the complex nature of disordered solids, as the degrees of freedom associated with the freezing in the glassy crystals are almost pure orientational ones, weakly coupled with lattice vibrations [5, 14]. The situation will greatly simplify the character of frozen disorder.

The frozen-in disordered state is not limited to the metastable undercooled phase of crystal. Another category of glassy crystals can be found in the stable crystals, such as CO and $H_2O$ crystals. Both of the substances provide good examples of crystals that do not obey the third law of thermodynamics. Structural and dynamic studies of CO crystals disclose the presence of a head-to-tail motion of the weakly polar molecules. The motion is active at high temperatures but cannot overcome a potential barrier hindering the motion at low temperatures before the crystal reaches a hypothetical ordering temperature. On the other hand, the name "crystal" derives from the Greek "krustallos" which meant ice. Ice has long been considered as a typical crystalline substance surrounding us. Thus the deviation of ice from the third law of thermodynamics embarrassed many scientists since 1936 when the residual entropy of ice was found experimentally. Each water molecule in ice $I_h$ has six equi-energetical orientations in the hydrogen-bonded network, but their reorientational motions must be done in a highly cooperative way under the constraints of the "ice rules." The relaxation time for the cooperative motion becomes longer and longer as the temperature is lowered until the disorder becomes immobilized below a certain temperature. Whether the disorder should be described as static or dynamic depends *inter alia* on the height of the potential barrier, along with the temperature and the time-scale of the observation. The half-hydrogen model proposed by Pauling cannot be an equilibrium structure of ice crystal at 0 K, but absence of any anomalous heat capacity ascribable to an ordering transition has puzzled many scientists for a long time [15].

In fact, a glass transition was found to occur at about 15 K for CO and 105 K for ice $I_h$. Thus the freezing process is believed to occur during cooling before a hypothetical ordering temperature is reached for each crystal [5, 14]. The dielectric and enthalpy relaxation times of ice, $\tau_d$ and $\tau_H$, turned out to lie on the same straight line in an Arrhenius plot. Extrapolation of these data to 60 K, a hypothetical ordering temperature of ice $I_h$ proposed by Pitzer and Polissar [15], gave a value of $10^{13}$ s for $\tau$. It must be this geological time that hindered the crystal from realizing an ordered phase of ice $I_h$ in our laboratory time. Onsager [16] has suggested using some impure ice samples for inducing the expected ordered phase. The impurity might hopefully relax the severe constraints "ice rules" imposed to the cooperative motion of water molecules in the lattice and enhance the orientational mobility.

A minute amount of KOH doped into the ice lattice was found to accelerate dramatically the reorientational motion of water dipoles [17]. The hydroxide ion $OH^-$ creates necessarily a hydrogen bond without proton in the neighbourhood, and the creation of this L (leer in German) defect in the network will surely relax the severe constraints for the cooperative water motion. Thus, an ice sample doped with KOH in the mole fraction of $2 \times 10^{-5}$ and kept at 65 K for 2 days induced a first-order phase transition at 72 K. The transition removed a substantial fraction of the residual entropy and changed the lattice symmetry from $P6_3/mmc$ of $I_h$ to $Cmc2_1$ of the low temperature phase. The proton-ordered phase was designated as ice XI. Examination of the experimental data showed that the relaxation time of the doped ice was shortened by a factor of $10^8$ at 90 K. This is the reason why the suggested ordering transition has escaped notice since 1936 when the residual entropy was first observed, and now appeared by the catalytic action of the minute amount of dopant that relaxed greatly the serious constraints imposed to the rearrangements of the water molecules in ice $I_h$. The Kauzmann paradox does not occur in ice crystal because $I_h$ is the most stable modification of ice under atmospheric pressure. This experiment shows the wide validity of the third law of thermodynamics in one hand, and reveals a facet of a possible ordering process for the otherwise frozen-in disordered systems in other hand.

Essentially the same things were observed for the clathrate hydrates of the type II structure. The host lattice composed of water molecules forms two kinds of cages that accommodate some guest molecules of suitable size and shape. Two kinds of molecular disorder exist in the crystal; the first is the orientational disorder of the host water molecules similar to that of ice, and the second is the reorientational motion of the guest molecules inside the cages. This can be clearly indicated by the observation of two kinds of the dielectric relaxations, if the guest is dipolar molecule. The one that appears at higher temperature is for the water dipole, and the other occurring at cryogenic temperature is for the guest dipole. The latter exhibits a great deal of orientational disorder inside the cavity with almost spherical symmetry. The calorimetric measurement of the clathrate hydrate enclathrating tetrahydrofuran showed a tiny heat capacity anomaly with relaxational nature that is similar to the ice crystal. When the crystals are doped with a minute amount of KOH, a first-order phase transition appeared at 61.9 K. The transition temperature depended on the nature of the guest molecules: 46.6 K for acetone hydrate, and 34.5 K for trimethylene oxide hydrate. Dielectric measurements showed that both of the host and guest molecules in the tetrahydrofuran hydrate were ordered at temperatures below $T_{trs}$. Most probably, the ordering of the host water dipoles produced a strong electric field that forced the guest molecules to align along a preferred orientation.

Our experiments are always governed by the Deborah number $D$ [18], which is defined as the ratio of the relaxation time $\tau$ and observation time $t$; $D = \tau/t$. As far as $D \ll 1$, we can observe the whole shape of heat capacity anomaly associated with the possible ordering transition that is determined by the intermolecular interaction and cooperative nature of the interaction. For a system $D \gg 1$, we will miss a part or full of the relevant thermodynamic quantity owing to our limited experimental time. The relaxation time is determined by the disordered structure and barrier height hindering the rearrangement of the structure. A particular impurity doped into the system will modify the cooperative nature of the rearrangement and shorten the relaxation time as a whole. Only this process makes possible to observe a transition that has been concealed for a kinetic reason. Modification of chemical potential of a system by a particular impurity can hopefully result in a drastic change of dynamical situation in the system. We may call this new field "doping chemistry."

A particular kind of dopants acted on frozen-in disordered system as catalyst for releasing the immobilized state and recovering the equilibrium state in our observation time. In relation to this fact, one might raise an inquiry "What kind of physical or chemical impurity will release the frozen-in state of the liquid?" If a particular dopant were discovered luckily for this purpose, we will be blessed to observe in reality what will happen at the Kauzmann temperature of a liquid? As a matter of fact, it is not easy for us to imagine some disordered solids without any configurational entropy, designated as an "ideal glass."

## 1.4 Other Aspects of Glassy Solids

In spite of its simple molecular structure, a wide variety of states of aggregation of water molecules have been observed hitherto. At least 13 forms of ice designated as $I_h$, $I_c$, II, III, IV, ···, XI, as well as vapour-deposited amorphous ice have been found to exist under atmospheric and high pressures as of 1982. Amorphous ice is, galactically speaking, a very abundant material in space. The Comet' tail is strongly connected to the formation of amorphous ice. The idea that there is an amorphous solid form of water was first put forward in 1935 when water vapour was deposited onto a cold substrate [15]. A calorimetric measurement showed that the amorphous ice exhibited a glass transition at 135 K [19] and had a strong crystallization tendency. The crystallization starts to occur progressively at temperatures even below $T_g$, so that the heat capacity jump is masked to some extent by an exothermic effect arising from the crystallization. Hyper-quenching of the liquid water can also produce the amorphous ice and its $T_g$ value was determined to be 136 K by a DSC measurement [20]. These values are consistent with that obtained by extrapolation of unambiguous $T_g$ data from glass-forming binary solutions up to pure water. On the other hand, Angell has proposed that the $T_g$ of amorphous ice must be higher than 150 K [21]. The discussion is mainly based on the fact that if the heat capacity of the undercooled liquid follows the extrapolated line of the equilibrium water, the entropy of the amorphous ice becomes smaller than that of hexagonal ice below about 150 K. Heated discussions about this problem are still continued.

Pressurization of $I_h$ at low temperatures above 1.3 GPa turned out to be a new way for preparation of high-density amorphous (HDA) ice [22]. The HDA ice transforms to a low-density amorphous (LDA) ice by warming the HDA ice under atmospheric pressure. Both the amorphous ices change each other by an apparent first-order transition. The phenomenon designated as "polyamorphism" has opened a new realm in amorphous solids. Particularly interesting is the properties of LDA ice. It is becoming increasingly clear that the LDA ice is of extraordinary character among amorphous solids. What is unexpected is that the thermal conductivity increases sharply when the HDA ice transforms to LDA ice. Usually the denser forms of substances are the better thermal conductors. Even more unexpected is that the thermal conductivity of the LDA ice increases with decreasing temperature, the opposite behaviour to that of the HDA ice and of all other amorphous solids. The LDA behaviour is like that of crystals in which the scattering of phonons by other phonons decreases with decreasing temperature. This anomalous behaviour implies the existence of a special kind of order in the LDA ice [23]. This conclusion is consistent with the observation that LDA ice does not show boson peaks that are widely observed in glass-forming liquids. An X-ray scattering experiment showed that the dispersion relation of LDA ice for vibrational modes is similar to that of cubic ice $I_c$, the metastable form of ice at atmospheric pressure. It seems likely that in LDA ice we are closely approaching an ideal glassy state [24]. It is an embarrassment, however, that it is not certain that a characteristic $T_g$ can be assigned to this almost ideal glass. Recent discovery of very high-density amorphous (VHDA) ice makes the situation more complex [25]. Further works with novel idea is highly encouraged for an improved understanding of this intriguing and important substance.

Another interest is concerning with cryo-preservation of biological substances: cryogenic storage by amorphization of the water solution system within the biologically active cell. When the biological cell is kept at low temperature without formation of ice, the biological activity is preserved, independently of the storage period. The solution system is in a glassy state at low temperatures, with all the molecules and ions being in a frozen state. For successful vitrification, diverse treatments of the cell or issues are required to protect from ice formation. The optimum conditions for the successful preservation depend on the particular cells/tissues and added glass-forming substances.

It is interesting to describe here some analogies between proteins and glasses. Proteins are known to have a free-energy landscape possessing a large number of energy valleys, separated by potential barriers. This arises from the conformational degrees of freedom of the polypeptide chains. The energy valleys are called the conformational substates and correspond to slightly different higher-order structures in a protein molecule. At physiological temperatures, protein molecule fluctuates among many substates, as if they were liquid. This fluctuation is believed to play an important role in the biological activity such as the enzyme-substrate reactions. With lowering temperature, they are expected to freeze in a substate. In fact, protein crystals undergo the glass transition at low temperatures, depending strongly on the amount of water in the crystal. Water molecules are important ingredient of the protein crystal [26]. Lowering of the water content in the crystal causes a decrease in the mobility of the component molecules, resulting in an increase of $T_g$. Below a certain amount of water, the glass transition takes place at temperatures above 300 K. This means that conformational and orientational degrees of freedom of both molecules are in a frozen-in disordered state at our physiological temperatures. Any biological function cannot be expected to appear in such a frozen-in non-equilibrium state.

Figure 1.5 summarizes all the observed states of aggregation of molecules in the equilibrium and non-equilibrium states of molecular assembly, as revealed by calorimetric measurements. It can be concluded that the glass transitions observed exclusively in liquids so far are just one example of "transitions" that must be of wide occurrence in solids in relation to the freezing of some degrees of freedom, irrespectively of the translational invariance in the system. Molecular assembly can produce various states of aggregation that are much richer than our anticipation.

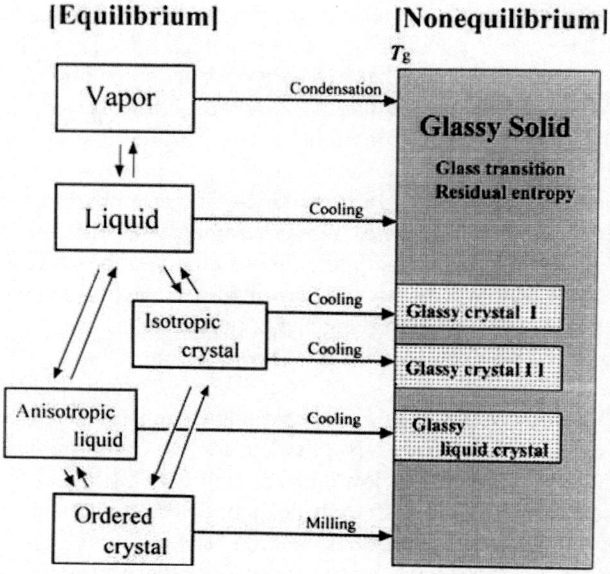


**Fig. 1.5** Various states of aggregation of molecules in equilibrium and nonequilibrium states

The subdivision of the glassy states will deepen our basic understanding of the substance and require a renewal of the concept of the glass transition. Freezing processes in seemingly dissimilar substances could be analyzed on the same thermodynamic basis, indicating the universality of the underlying thermodynamic principles. A new realm of amorphous solid-state science with many unsolved problems is confronting us and must be challenged.

**Acknowledgements** The author would like to express his sincere thanks to the late Professor I. Nitta and Professor Emeritus S. Seki for the continued encouragements throughout the present works. Thanks are extended to many collaborators for their hard works and useful discussions.

## References

1. Gibson GE, Giauque WF (1923) The third law of thermodynamics. Evidence from the specific heats of glycerol that the entropy of a glass exceeds that of a crystal at absolute zero. J Am Chem Soc 45:93–104
2. Wilks J (1961) The third law of thermodynamics. Oxford University Press, Oxford
3. Mackenzie JD (ed) (1964) Modern aspects of the vitreous state, vol 3, Chapter 6. Butterworth, London
4. Oguni M, Suga H (1999) Amorphous materials and their elucidation by adiabatic calorimetry. In: Letcher T (ed) Chemical thermodynamics. IUPAC, North Carolina, pp 227–237
5. Suga H, Seki S (1974) Thermodynamic investigation on glassy states of pure simple compounds. J Non-cryst Solids 16:171–194; Suga H, Seki S (1980) Frozen-in States of

Orientational and Positional Disorder in Molecular Solids. Faraday Discussion No 69, 221–240 (1980)
6. Suga H (2000) Frozen-in disorder and slow relaxation in crystals. J Chem Thermodyn 25:463–484; Suga H (2000) Prospects of material science; from crystalline to amorphous solids. J Thermal Anal Calor 60:957–969
7. Brawer S (1985) Relaxation in viscous liquids and glasses. In: Review of phenomenology, molecular dynamics simulations, and theoretical treatment. American Ceramic Society, Columbus
8. Suga H, Matsuo T (1989) Adiabatic calorimeter as an ultra-low frequency spectrometer. Pure Appl Chem 61:1123–1132
9. Kauzmann W (1948) The nature of the glassy states and the behavior of liquids at low temperature. Chem Rev 43:219–287
10. Angell CA (1991) Relaxation in liquids, polymers and plastic crystals. J Non-Cryst Solids 13:131–133
11. Adam G, Gibbs JH (1965) On the temperature dependence of cooperative relaxation properties in glass-forming liquids. J Chem Phys 43:139–165
12. Takahara S, Yamamuro O, Suga H (1994) Heat capacities and glass transitions of 1-propanol and 3-methylpentane under high pressure. New evidence for the entropy theory. J Non-Cryst Solids 171:259–270
13. Yamamuro O, Tsukushi I, Lindqvist A, Takahara S, Ishikawa M, Matsuo T (1998) Calorimetric study of toluene and ethylbenzene: thermodynamic approach to spatial heterogeneity in glass-forming liquids. J Phys Chem B 102:1605–1609
14. Suga H (2003) Calorimetric study of frozen-in disordered solids. J Phys Condens Matter 15: S775–S789; Suga H (2005) Frozen-in disorder in condensed phases. Russian J Phys Chem 77: S7–S16 (2003); Suga H (2005) Study of transition phenomena in molecular solids. J Thermal Anal Calor 80:49–55
15. Eisenberg D, Kauzmann W (1969) The structure and properties of water. Clarendon, Oxford; Petrenko VF, Whitworth RW (1999) Physics of ice. Oxford University Press, Oxford
16. Onsager L (1967) Ferroelectricity of ice? In: Weller E (ed) Ferroelectricity. Elsevier, Amsterdam, pp 16–19
17. Tajima Y, Matsuo T, Suga H (1982) Phase transition in KOH-doped ice. Nature 299:810–812; Suga H, Matsuo T, Yamamuro O (1992) Thermodynamic study of ice and Clathrate hydrates. Pure Appl Chem 64:17–26; Suga H (2005) Ultra-slow relaxation in ice and related substances. Proc Japan Acad B 81:349–362.
18. Reiner M (1964) The Deborah number. Phys Today 17:62
19. Sugisaki M, Suga H, Seki S (1968) Calorimetric study of glassy state V. Heat capacities of glassy water and cubic ice. Bull Chem Soc Jpn 41:2591–2599
20. Johari GP, Hallbrucker A, Mayer E (1987) The Glass-liquid transition of hyperquenched water. Nature 330:552–553
21. Yue Y, Angell CA (2004) Clarifying the glass transition behaviour of water by comparison with hyperquenched inorganic glasses. Nature 427:717–720
22. Mishima O, Calvert LD, Whalley E (1984) Melting of ice I at 77 K and 10 kbar. Nature 310:393–395
23. Andersson O, Suga H (2002) Thermal conductivity of amorphous ices. Phys Rev B 65:140201–140204
24. Angell CA (2004) Amorphous water. Annu Rev Phys Chem 55:559–583
25. Loerting T, Schustereder W, Winkel K, Salzmann CG, Kohl I, Mayer E (2006) Amorphous ice: stepwise formation of very high density amorphous ice from low density amorphous ice. Phys Rev Lett 96:025702(4)
26. Miyazaki Y, Matsuo T, Suga H (2000) Low temperature heat capacity and glassy behavior of lysozyme crystal. J Phys Chem B 104:8044–8052

# Chapter 2
# Heat Capacity and Entropy Functions in Strong and Fragile Glass-Formers, Relative to Those of Disordering Crystalline Materials

**C. Austen Angell**

## 2.1 Introduction

The glassy state problem is often separated into two major components [1, 2]. One of these concerns the reasons that glasses form in the first place, and deals with the circumstance that glasses are usually metastable with respect to crystals so that crystallization must be avoided. The second deals with the question of how liquids behave when crystals do not form, and it is with this component that we are concerned in this chapter. Here the central phenomenon with which we must deal, in seeking to understand vitrification, is the heat capacity function and the change in that function that accompanies the freezing in of the disordered state. This phenomenon is illustrated in Fig. 2.1 for a typical molecular liquid, 2-pentene vitrified by both liquid cooling and by vapor deposition [3].

The fairly abrupt change in heat capacity that is observed as the system falls out of equilibrium (essentially because the systems' molecular motions have become too sluggish to follow the cooling) irrespective of whether the substance under study is a liquid, a plastic crystal, or a disordered superlattice. It is the phenomenon that many would consider to be the most characteristic feature of glassformers and vitrification ... and yet, for the archetypal glassformer $SiO_2$ in its purest anhydrous state, and for glassy water –the most abundant form of water in the universe – this change in heat capacity on structural arrest during cooling is barely detectable [4]. It is not surprising that the vitrification of liquids is found to be the source of much confusion.

It is clear to any reader of the recent literature on glassforming liquids (now "glassformers") that investigators in the field have given most of their attention to the spectacular manner in which the majority of glassformers change their viscosity

C.A. Angell (✉)
Department of Chemistry and Biochemistry, Arizona State University,
Tempe, AZ 85287-1604, USA
e-mail: austenangell@gmail.com

J. Šesták et al. (eds.), *Glassy, Amorphous and Nano-Crystalline Materials*,
Hot Topics in Thermal Analysis and Calorimetry 8, DOI 10.1007/978-90-481-2882-2_2,

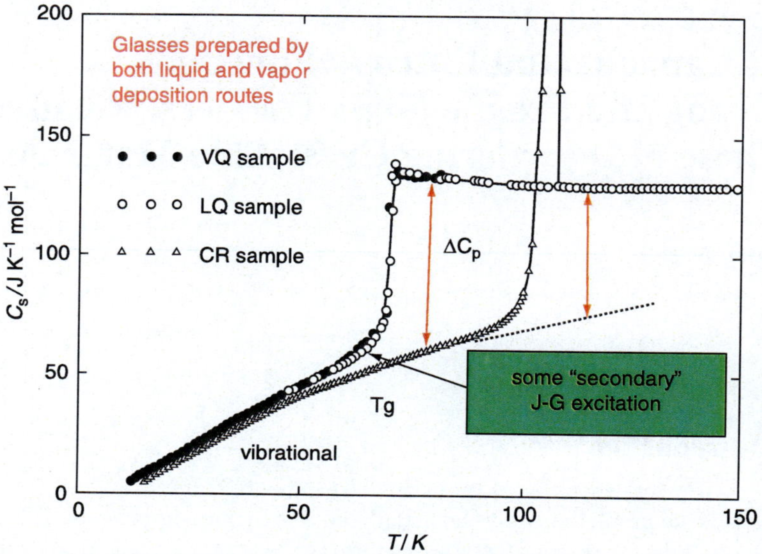


**Fig. 2.1** Heat capacity of 1-pentene prepared by both liquid cooling and by vapor deposition. The excess heat capacity is indicated by double arrows and is seen to increase markedly with decreasing temperature (From Takeda et al. Ref. [3], by permission)

(or relaxation times) on approach to the glass transition. In the most interesting cases now known as "fragile" liquids [5, 6], the temperature dependence of viscosity departs dramatically from the Arrhenius equation typical of most rate processes in condensed matter, accelerating as the glass transition is approached to such an extent that a temperature change of 3 K can induce an order of magnitude change in viscosity. The behavior is not as dramatic as the power law divergences observed in critical phenomena, but in some cases this behavior is approached. Indeed, as we will discuss further below, this is probably not an accident.

Most theories for the viscosity (or, more generally, for the relaxation time) connect viscosity temperature dependence to the temperature dependence of one or other thermodynamic property. In the 1959 free volume model of Cohen and Turnbull [7], the thermodynamic property was the unoccupied volume (integral of an expansivity component), in Adam-Gibbs theory [8] it was the configurational part of the entropy (integral of a heat capacity component), while the more recent "shoving model" of Dyre and colleagues [9] use the temperature dependence of a modulus, the "infinite frequency" shear modulus, to explain the super-Arrhenius behavior. So a proper appreciation of the thermodynamic properties of the glassformer would seem to be a prerequisite of a detailed understanding of the transport behavior.

Here we will consider the highly variable forms that the heat capacities of glass-forming liquids can take, and will then show that, in spite of this great variation, the integrals of the heat capacities, the entropies, $S = \int C_p dT$ – and particularly its excess over the entropy due to vibrations – fall into a pattern that mimics the well-known pattern made by the viscosities of glassformers when plotted in $T_g$-scaled Arrhenius form (the "strong/fragile" liquid pattern). While the origin of the "strong/fragile" pattern for viscosities (and relaxation times) remains more or less mysterious, the pattern for the entropies can be simply reproduced using a two-parameter model of configurational excitations [10], as will be demonstrated. Finally, we will interpret the "strong liquid" extreme of this pattern in a provocative way by showing the similarity that the heat capacity function for these liquids bears to a well-known heat capacity form for disordering excitations in solids, the lambda transition.

The entropy variations can be accounted for quite pleasingly with the help of a simple two-state excitations model [11], provided that apparently pathological systems like water and silicon, which are extremely poor glassformers by liquid routes, and the liquids classified as "strong" on the basis of their viscosity data, are excluded. Even the latter follow the expectations of two-state excitations when only very high temperature data, obtained by computer simulation, are considered [12]. This encourages consideration of the possibility that the much more challenging pattern offered by the heat capacities themselves might have a systematic explanation. As an aid to understanding this problem we include consideration of the relatively well-understood cases of disordering solids, the order–disorder transitions or lambda transitions.

Before launching into a review of the heat capacity functions, their integrals and their patterns of behavior, let us give some brief consideration to non-glassy systems in which structural arrest of ordering processes is commonly observed. Since the archetypal glassformer, silica, has barely any thermal signature of the glass transition, we cannot make a large jump in heat capacity on structural arrest a criterion for inclusion in the discussion. Thus we start with the simplest and most familiar case, the freezing in of a defect population during the cooling of a simple crystal, for instance potassium iodide.

We do not know of any heat capacity studies of this trapping of the configurational state of the system, but its generic relation to the glass transition phenomenology can be judged from a comparison of a quantity which, like the viscosity of silica, is exponentially sensitive to the freezing of the excitation population. This is the ionic conductivity, which is shown in Arrhenius form in Fig. 2.2, in the form preferred by many in the ionics field because it is suggested by theoretical treatments, viz., log ($\sigma$T) vs 1/T where $\sigma$ is the specific conductivity. Comparison is made with the variation of the diffusivities of different elemental components of various metallic glassformers through their $T_g$s [13].

Notwithstanding the similarity of Fig. 2.2a, b, the metallic glassformers of Figure 2b have quite marked heat capacity jumps at their $T_g$s. At higher temperatures it becomes clear, also, that their diffusivities follow a super-Arrhenius path in temperature, though in the temperature domain immediately above $T_g$ they are usually following the Arrhenius law with higher activation energies than below $T_g$.

Passing to systems that begin to resemble glassformers a little more closely, we show the relaxation times for some ionic crystal rotator phases, in which the rotation of a structural element causes little distortion of the crystal lattice. A series of these were studied by Fujimori and Oguni [14], and their relaxation time behavior is shown in Fig. 2.3a. In this case the thermal consequences of the structural arrest at the "glass

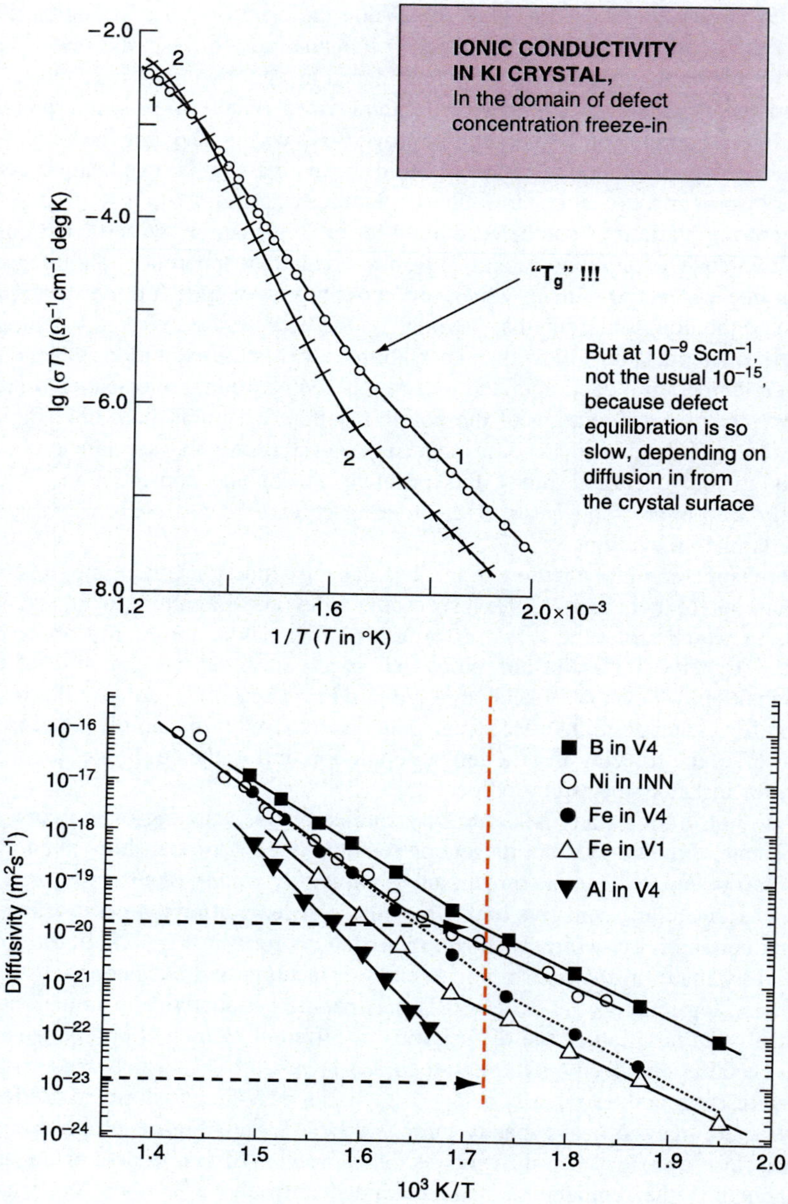


This figure will be printed in b/w

**Fig. 2.2** (**a**) Temperature dependence of ionic conductivity in single crystal of potassium iodide in the temperature range where the defect population freezes in for this crystal dimension. (**b**) Temperature dependence of metallic diffusivity of atomic components through Tg in different metallic glasses (From Faupel et al. Ref. [13], by permission)

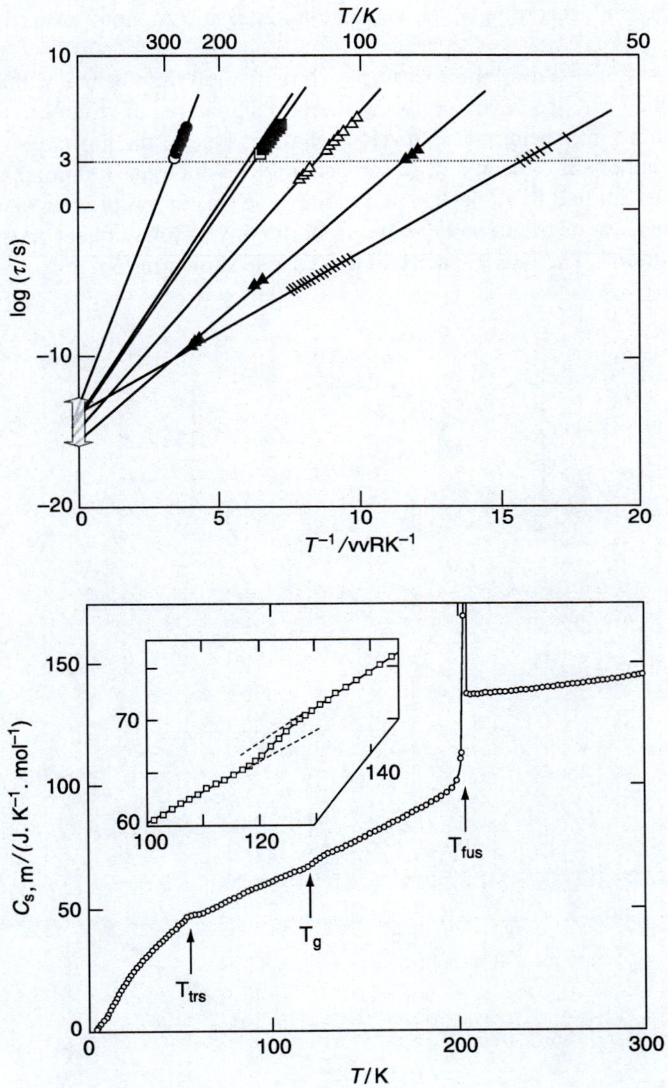


**Fig. 2.3** (**a**) Arrhenius plots of the dielectric (higher T points) and calorimetric relaxation times (lower T points) for several rotator phases (in order from *left*) $H_3BO_3$, $D_3BO_3$ (identical curve with $H_3BO_3$ one), $SnC1_2{\cdot}2H_20$, $SnC1_2{\cdot}2D_20$, $C_4H_3BrS$, Cm, $TINO_2$ (From Fujimori and Oguni Ref. [14]). (**b**) The weak jump in heat capacity that accompanies the glass transition in $C_4H_3BrS$, typical of the systems with the relaxation time behavior seen in Fig. 2.3a (From Fujimori and Oguni Ref. [15], by permission)

temperature" (which was characterized by a relaxation time of 1,000 s in this case) were determined, and the results are shown in Fig. 2.3b for one case, bromothiophene [15]. Clearly the glass transition is a very weak phenomenon in this case (and others like it [16]), and accordingly the relaxation times are in good accord with the Arrhenius law.

Other cases of disordering processes with relaxation time behavior that follows the Arrhenius law, also have very weak thermal signatures of structural arrest, but an overall more interesting heat capacity behavior. One of the more relevant is the large quasi-spherical molecule, $C_{60}$, the packing of which has a dipolar excitation and so can be studied by dielectric relaxation. The relaxation process also follows the Arrhenius law, as might be expected from the tiny change of heat capacity at its "glass transition". The heat capacity in this case is shown in Fig. 2.4.

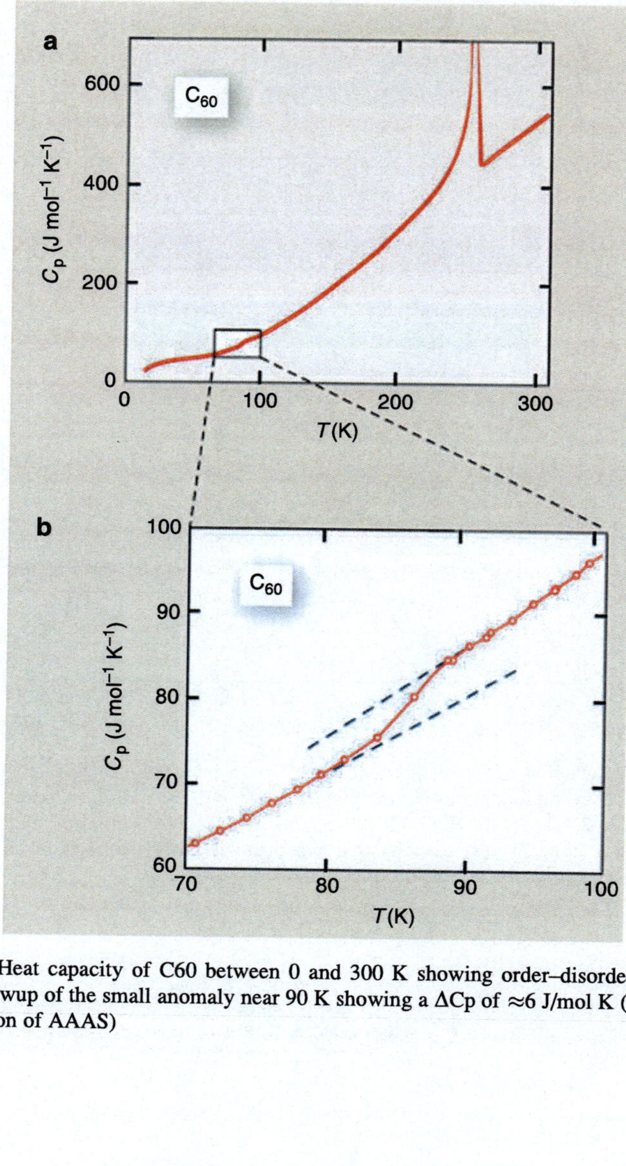


**Fig. 2.4** **(a)** Heat capacity of C60 between 0 and 300 K showing order–disorder transition at 260 K **(b)** Blowup of the small anomaly near 90 K showing a ΔCp of ≈6 J/mol K (From Ref. [4] with permission of AAAS)

The effect of rotational disorder freezing at $t = 1{,}000$ s, is again extremely weak, but nevertheless has been precisely determined by the adiabatic calorimetry of the Osaka laboratory [17], and its relaxation time could be determined, and shown in accord with the dielectric data [17]. The interesting point here is that the heat capacity function proceeds to a sharp peak at a temperature *almost three times* $T_g$.

There are many other sharply peaked disordering transitions that have almost no disorder left when they finally undergo a configurational arrest. A good example to compare with the rotational disordering of the substances in Fig. 2.2 is that of $NaNO_3$ which again has the classical lambda form but no low temperature arrest that can be observed with standard calorimetry. On the other hand, in the case of $TlNO_2$ for which relaxation time data are seen in Fig. 2.3a, the arrest can be detected and has been described by Moriya et al. [18]. In this case there is no melting phenomenon as in Fig. 2.3b until much higher temperature, and the heat capacity function can be seen to develop into a lambda peak like that of $C_{60}$.

These systems are clearly characterized by energy landscapes that are rather different in character from those discussed in the literature of glassforming sub stances [19]. While high states of configurational excitation can obviously be obtained, the states excited are apparently not characterized by energy minima that are separated from neighboring minima by significant energy barriers, by means of which the system can be trapped when the temperature is lowered. Such states are more closely related to anharmonic excitations than to true microstates in the sense that they are usually discussed. Later we will consider the case of a lambda transition in a metallic alloy superlattice that is rather different in character and will be quite useful to our broadened discussion.

Finally there are the intermediate cases where the disordering elements are arranged, center-of-mass-wise, on a crystal lattice like the above cases but can undergo a rotational disordering excitations that have larger volume requirements than the above and are accompanied by much larger heat capacity steps, when the disorder becomes frozen, than the ones discussed above. There seems to be a continuous series here in which the increasing heat capacity jump is accompanied by an increasing departure from Arrhenius behavior of the relaxation. A collection of data taken from the extensive study Brand et al. [20] – to which cases have been added at either extreme by us [21], is shown in Fig. 2.5. This pattern is very similar to that known for the glassforming liquids, but is richer in "strong" glassformer cases. Understanding the manner in which systems make the transition from the lambda type to the typical glassformer type is one of the major statistical thermodynamic problems in this field.

## 2.2 Glassforming Liquids, Strong and Fragile

With this background we now show the contrasting behavior of different members of the conventional type of glassformer, viz., liquids that slow down with decreasing temperature until they vitrify at $T_g$, as in Fig. 2.1. To make the immediate contrast with

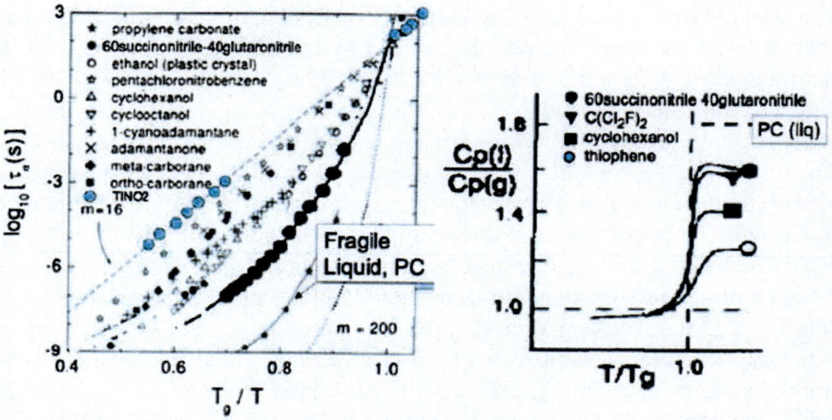


**Fig. 2.5** Plastic crystal properties. *Left*: $T_g$-scaled Arrhenius plot for the relaxation times (mostly dielectric); *right*: $T_g$-scaled temperature dependence of the heat capacity, relative to that of the glassy state at $T_g$. (From Ref. [21] by permission) note the correlation of $\Delta$Cp with the position in the scaled Arrhenius plot

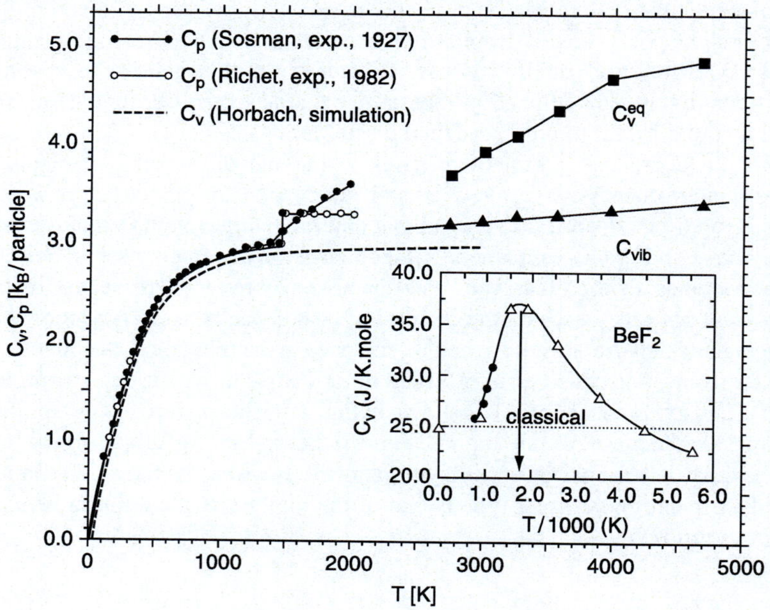


**Fig. 2.6** Heat capacity of $SiO_2$ from experiment (points below 2,000 K) and simulation (points above 2,500 K). The latter are conducted at constant volume which will depress the value below its constant pressure value. A point at 6,100 k that establishes a peak in Cp (Like that seen in the inset for BeF2) was omitted from this figure in ref. [23]. A peak value for $SiO_2$ was also evident in earlier studies, by Soules. *Inset*: heat capacity of $BeF_2$ by experiment (low T points) and simulation (From Ref. [4] by permission)

Fig. 2.1 we show in Fig. 2.6, the available data, experimental [22] and computer simulation [23], for the classical case, $SiO_2$. Because of its high glass temperature, the experimental data are limited in range, but they are well supplemented by the computer simulation data on the BKS model of silica [24] that reproduces many silica properties faithfully [25]. The striking difference from the case of the molecular liquid of Fig. 2.1 is reinforced by the data on the weak field analog of $SiO_2$, $BeF_2$, for which data far above the glass temperature, to $2T_g$, are available [26]. The small jump in $C_p$ for laboratory $SiO_2$ that was reported in the studies of Ref. [22] almost certainly owes its strength to the presence of some residual OH in the structure, as high temperature DSC studies of truly dry $SiO_2$ [27] detect essentially no "jump" signal at $T_g$, despite their superior sensitivity over the drop calorimetry method [22]. This is the implication of the simulation data of Fig. 2.6, when extrapolated back to 1473 K where the viscosity reaches $10^{12}$ Pa.s, commonly associated with $T_g$ in inorganic glasses. The striking aspect of Fig. 2.6 is not so much the small value of the excess $C_p$ of liquid over crystal at $T_g$, but the fact that the excess (often called "configurational") component of the heat capacity rises with increasing temperature, and then peaks at a temperature far above both $T_g$ and $T_m$. More extensive, but so far unpublished, simulations on $BeF_2$ (P. H. Poole, private communication) confirm the peak value seen in Fig. 2.6 (insert) [28], and also verify the link-up with the experimental values. Strong liquids apparently behave more like the crystalline materials that exhibit lambda transitions, than they behave like the normal fragile liquids.

We will come back to this similarity a little later but first, for the purposes of this chapter, we need to examine the other extreme in a little more detail. Note in Fig. 2.1 how the difference between liquid and crystal heat capacities is increasing as the temperature decreases. It has been noted by a number of workers [29, 30] that this excess $C_p$ can be well described by a hyperbolic function of temperature, $\Delta C_p = k/T$. Indeed, it is this form of temperature dependence that allows the transformation of the Adam-Gibbs equation for the relaxation time, into the familiar Vogel–Fulcher–Tammann equation [31]. Modeling this form of excess heat capacity has proven rather difficult. The fact that it shows no maximum eliminates the possibility of accounting for the excess with any simple two-state model, as adequately demonstrated on many occasions but particularly by Moynihan and the author [10]. One recent and provocative attempt that *does* account quantitatively for both the excess heat capacity and the excess entropy with the same parameter set, is that of Matyushov and the author [32] whose "Gaussian excitations" model fitting of the data is shown in Fig. 2.7. What is provocative is that the equations that fit the data so well insist that, at a temperature some 10–20% below $T_g$, the system in equilibrium would undergo a first order transition to a low entropy state, thereby resolving the Kauzmann paradox in an unconventional way.

Before looking further into this matter, we will look at the excess entropy functions seen in Fig. 2.7, in a different way, so as to compare them with the corresponding liquid relaxation time behavior. We plot them in such a way that they have a common value, unity, at the glass transition temperature. This requires a scaling by the value each liquid possesses at $T_g$.

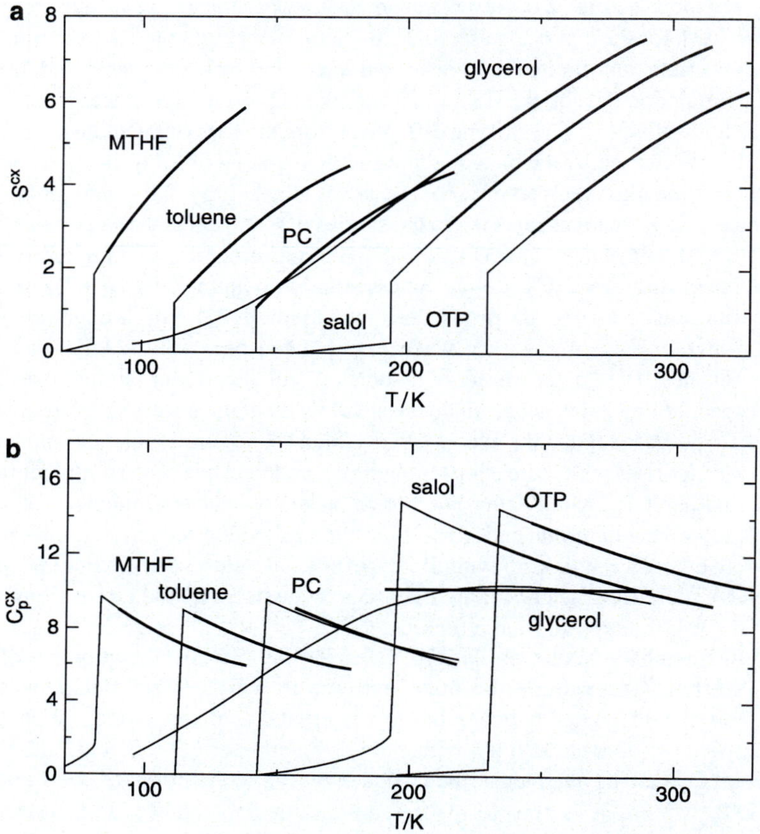


**Fig. 2.7** The excess entropy (part (**a**) and excess heat capacity part (**b**) of a series of fragile liquids ($m > 85$). The thick lines are the experimental data and the thin lines are the theoretical best fits continued to lower temperatures to show the predicted first order phase transitions. Only the intermediate liquid, glycerol, the excess heat capacity of which changes little with temperature, offers different behavior. The hyperbolic relation predicts extreme behavior will emerge as liquids of increasingly low cohesion are studied from Ref. [32] by permission

The value of the excess entropy relative to its value at $T_g$, is presented as a function of inverse absolute temperature scaled by $T_g$ itself, in order that the plot of excess entropies can have the same relation to temperature as does the viscosity in the so-called fragility plot (sometimes given the author's name). This plot, an example of which appears in Fig. 2.5 for plastic crystals, is now used to compare different rates of excitation of the disorder in liquids. Figure 2.8 shows that, as T rises above $T_g$, some liquids approach the tops of their energy landscapes more rapidly than others. At the top of the landscape, ToL, the entropy per rearrangeable unit $k_B \ln W$ reaches $k_B$, because all of the $e^N$ microstates are accessible and accordingly $\ln W = 1$. Thus this plot displays the "thermodynamic fragilities", of the liquids in a way that the jump in heat capacity does not. And the thermodynamic

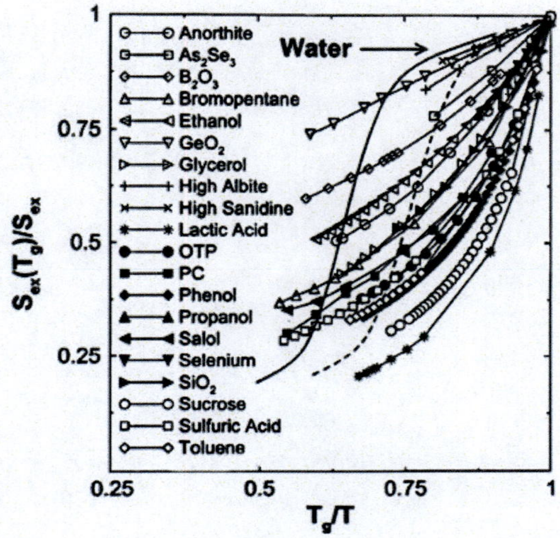


**Fig. 2.8** Thermodynamic fragility plot showing the relative rates of excitations of the excess entropy. The case of water is anomalous and implies a lambda like peak in the heat capacity to rationalize the high and low temperature branches of the relative entropy behavior. Reproduced from Angell, C. A. Chem. Rev. 102,2627 (2002) by permission

fragilities correlate with the kinetic fragilities [33] in a manner that the jump in heat capacity does not [34]. The scaling by the excess entropy at $T_g$ makes the display independent of the choice of number of "beads" [32, 35–37] per mole of liquid, but does not safeguard against anomalous entropies of fusion due to anomalously high entropies in the crystalline state, and the latter will invalidate the assumption that the crystalline state entropy represents the vibrational entropy of the liquid at its melting point. Thus the position of $SiO_2$ in this plot is anomalous, unless the excess entropy is assessed by thermodynamic integration from the ideal gas state, and subtraction of a harmonic entropy, as was done in Ref. [38].

There is one liquid in this plot that behaves in a highly anomalous manner. That liquid is water, about which much is known but much also remains mysterious. Water, which can be vitrified by a number of alternative routes, is known to have an extremely weak glass transition, but it also has, according to two independent assessments [39, 40] a very small entropy at its $T_g$. Thus at the extreme of high temperature its relative entropy is expected to be large. At the same time, its extremely weak calorimetric glass transition, comparable to that seen in Fig. 2.3 for dipole disordering in $C_{60}$, means that the rate of excitation of entropy immediately above $T_g$ is very small. Thus, it appears like the strongest liquid near $T_g$, but higher in temperature, in the moderately supercooled liquid region where its entropy relative to ice is well known, its excess entropy is very high. Accordingly, in the intermediate range where direct observations cannot be made because of the instability of the liquid against crystallization (hence the description "no-man's land"), the excess entropy must undergo a very rapid change. This entropy would be associated with a hidden lambda transition, or to a hidden first order transition like that seen in supercooled silicon in the Stillinger–Weber model [41], and more

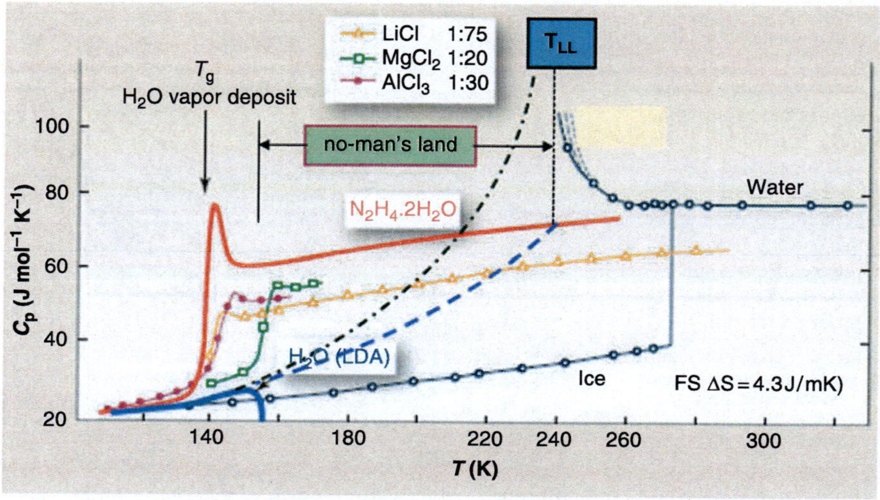


**Fig. 2.9** Heat capacity of water in normal, supercooled liquid and glassy states, showing two possible interpolations to preserve a continuity that must exist in view of the vitrifiability of liquids water under sufficiently high cooling rates (From Ref. [4] by permission of AAAS)

recently in ab initio simulations. These two options are shown in Fig. 2.9 and are seen to make a very plausible rationalization of the observations made in the two accessible regions (near the melting point on the one hand and near $T_g$ on the other).

Figure 2.8 contains only experimental data except for the case of water where there a data gap between high a low T (discussed below). The presence of peaks in the high temperature heat capacity of $BeF_2$, and to a lesser extent $SiO_2$ from simulation studies (Fig. 2.6), suggests that these (and perhaps other strong liquids as well), might show water-like behavior at higher temperatures when the extended measurements (particularly at higher pressures) become available.

Water and its anomalies are often discussed in terms of the presence of a second critical point that occurs just beyond the range of observability [42–45]. Opinions differ on whether it lies at positive or negative pressures [4], with some suggesting it might even be subsumed into the liquid–gas spinodal as suggested by the stability limit conjecture of Speedy [46] (in updated form to include a second spinodal for the low density liquid). Such a "critical point-free" scenario [4] is after all, the form that is suggested by all the empirical equations of state for water [47], though these seem to be generally discounted by workers in the field. On the other hand, if the second critical point not only exists but also were, serendipitously, to fall at ambient pressure, then the heat capacity curve for *ambient pressure* water would (in absence of crystallization) have *exactly* the lambda form, since a critical point transition is one example of this very general cooperative transition. This is indicated as the upper (dash-dotted) curve in Fig. 2.9. The lower dashed curve corresponds to the case of a weak first order transition that would occur at ambient pressure if the critical point lies at negative pressure or is subsumed into the liquid–spinodal.

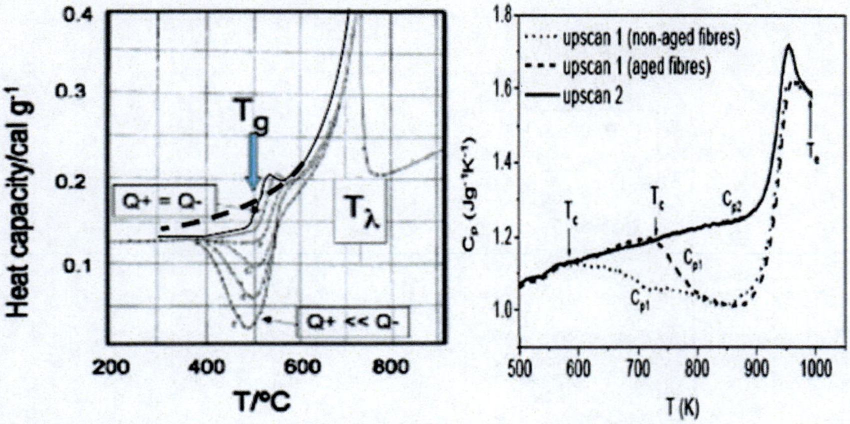


**Fig. 2.10** Comparison of the apparent heat capacity responses of CoFe alloy and a silicate glas after comparable thermal treatments. The darkest curves, in each case, are the standard glass transitions obtained when heating and cooling rates are the same. The lowermost curves are when the cooling rates were the most different (very fast in each case, corresponding to quenching in the most disorder. In each case the most disordered structure is the one trapped in a basin of smallest depth, but the difference in temperature between the standard $T_g$ and the temperature where the quenched system stats when the real glass starts to relax, is much greater in the case of real glass. Notably, glassy water exhibits behavior closer to that of the alloy, where the thermal relaxation peak for the quenched state, has a sharper peak starting at T closer to $T_g$ (From Ref. [48, 49] respectively, with the author's permission)

To complete our development of this increasingly broad picture of the glass transition under development in this chapter, it is important to add one further, and very informative, example of the lambda transition. This is the transition that occurs in the alloy superlattice, $Co_{50}Fe_{50}$. It is needed in order to make clear that, when the elementary step of the excitation process in a lambda type order–disorder transition is one which involves barrier crossing with substantial activation energy, this type of thermodynamic transition can support an ergodicity-breaking that has all the rate-dependent phenomena of the usual glass transition.

The heat capacity of this system was studied by Kaya and Sato [48], using different, increasingly severe, quenching procedures that arrested the ordering process in different states of excess entropy. The behavior of these were then evaluated during reheating at a fixed rate for comparison with the behavior observed when the cooling rate was the same as the standard heating rate (the darkened curve in Fig. 2.10). In Fig. 2.10 the observations are displayed alongside the more recent study of Yue and Jensen [49] for a silicate glass subject to a similar procedure (namely quenching very rapidly from well above $T_g$ followed by reheating at the standard rate of 20 K/min) that has qualitatively similar consequences (see also [50]). The darker curve is, in each case, for the condition, *cooling rate* = *subsequent heating rate*. In the case of Fig. 2.10a, the dashed line recalls what we know would happen if no ergodicity-breaking were to occur, since (unlike the glass transition equilibrium heat capacity) we know the lambda heat capacity form.

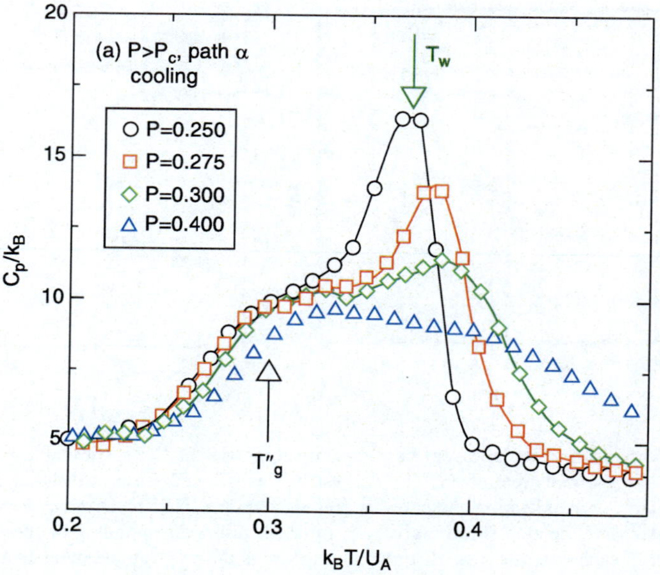


**Fig. 2.11** The constant pressure liquid heat capacity of the attractive Jagla model, for isobars of pressure close to, and increasingly above, the critical pressure of 0.235, showing the peak in heat capacity associated with super-critical fluctuations, which diminish with increasing distance from the critical point. The peak temperatures define the so-called Widom line. note that the cooling glass transition are always below the peak temperature (From Ref. [51], by permission of the American Physical Society)

With this background on ergodicity-breaking in a system with a lambda transition, let us now return to what happens when the conditions in a liquid system with a critical point are changed so that the critical point is narrowly missed and only the supercritical fluctuations are encountered. The behavior may be seen in the results of a recent study by Xu et al. [51] extended by Buldyrev et al. [52], of a model system that has been parameterized to have a liquid–liquid transition in the stable liquid domain [53]. This is the ramp model of Jagla [54] with attractive component of the potential included. The temperature dependence of the isobaric heat capacity for this system, at a series of pressures increasing in value above the critical pressure, is shown in Fig. 2.11. The point to be emphasized here is the increasingly rounded forms as the isobars cross the extension of the coexistence line beyond the critical point (now becoming known as the Widom line). The strength of the transition that "gathers-in" the peak of the cooperative transition at $T_c$ is seen to dissipate to higher temperatures as the isobar departs increasingly from the critical point. Sufficiently far from $T_c$ the effect of critical point vanishes and one is left with something reminiscent of the metallic glassformer heat capacity. The similarity of the lowest pressure isobar (the $p = 0.240$ "scan") to the heat capacity behavior of the Co–Fe alloy (Fig. 2.10 left) when heating and cooling rates are the same, can hardly be missed.

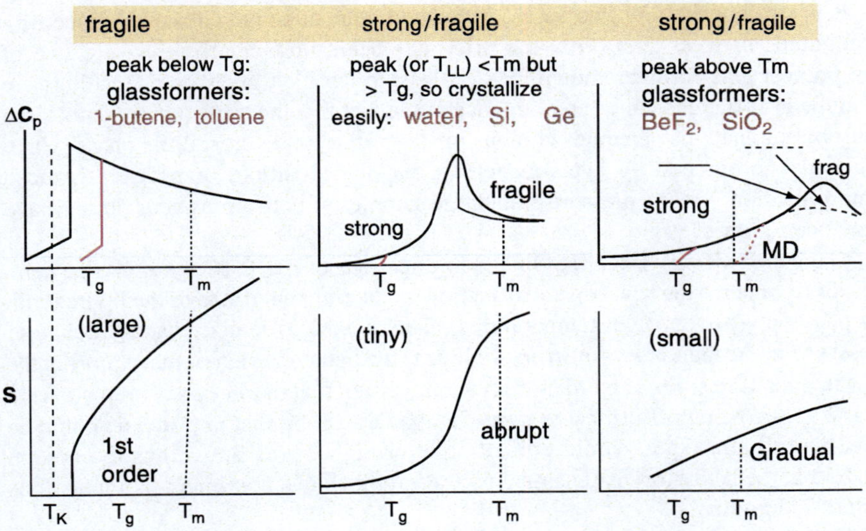


**Fig. 2.12** Changeover in the forms of excess heat capacity, and excess entropy, above the glass transition on passing from "strong" inorganic network glasses to "fragile" molecular glasses. Strong network liquids appear like expanded order–disorder transitions (see final section) and when pure may have tiny glass transitions, while fragile molecular liquids have large $\Delta C_p$ glass transitions and their ordering limits are depressed below $T_g$. Water, a tetrahedral network based on hydrogen bonding, lies at the crossover between the two classes of behavior. This can be interpreted in terms of the increasing Gaussian width in the distribution of excitation energies, and consequent increasing disorder stabilization of the excitations. This implies that the ordering in liquids during temperature decrease is an increasingly cooperative de-excitation process when changing from network to molecular liquid glassformers (From Ref. [2] by permission of Materials Research Society)

With these examples in mind we can then present the series of heat capacity and entropy functions that we have recently included in articles that attempt to describe a broader view of the glass transition phenomenon [2, 55]. The pattern of heat capacities has been viewed in terms of the effects of increasing cooperativity in the excitation process as we pass from strong network glasses, at one extreme, through the intermediate strength network represented by water, to non-network liquids like the preponderance of glassformers of common experience, ending in the very fragile liquids in which only the high temperature side of the transition is seen, because the low temperature side is hidden by the glass transition. Actually, most of the low temperature side of the transition in these cases, according to theory [32], is cut off by first order transitions to a de-excited low entropy state. Being, in most cases, a structure that is closer to that of the crystal form, than the viscous liquid from which it formed, this low enthalpy phase can often serve as a stepping stone to the crystalline state (the Ostwald rule of stages, in action). The rapid crystallization of second liquid phases, when they form above the glass transition, [56, 57] is one of the factors that has made the study of these phenomena confusing.

It is striking that the two cases, water and silicon, in which the presence of a continuous or weak first order transition has been most clearly seen, are two of the poorest glassformers known [55] (at least when high pressure quenchings are excluded [58]), this being a consequence of the fact that the transition, with its large entropy fluctuations generates copious nuclei of the stable crystalline phase which can only be avoided by hyperquenching, vapor deposition, or pressure-induced amorphization. Via the pressure path, large samples of both amorphous phases have now been grown [59].

There are two very important linked implications of Fig. 2.12 to emphasize here. The first concerns the heat capacity function on the right hand side of the figure (with its peak only at very high temperatures). This makes us realize that the behavior observed in the laboratory for *strong* liquids is the behavior of systems exploring the low temperature side of an order–disorder transition that has been smeared out as in Fig. 2.11, by being off-critical to a true lambda transition that in principle could be experienced at a liquid–liquid critical point at higher pressures. Thus the system studied in the laboratory is sensing only the approach to a maximum in fluctuations at the Widom line.

The second is the linked implication that, just as the correlation length for fluctuations increases as a critical point in a single phase system (e.g. $Co_{50}Fe_{50}$ of Fig. 2.10a, or the lambda transition in $C_{60}$) is approached from below, so must the correlation length for fluctuations increase with increasing temperature in the range explored by the laboratory strong liquids. This is of particular interest in the case of silica where the fluctuations, via their Fourier components, will affect the scattering of light because the scattering of light is of central concern in the optical fibers used for information transfer (not to mention the silica glass lasers used in laser fusion technology). Indeed, the evidence from light scattering studies conducted in relation to fiber fictive temperature, confirms our expectations [60].

The importance of this is that it is just the opposite temperature dependence of correlation length to that supposed by glass theorists who, it must be recognized, have always been concerned with fragile liquids. The theorists, however, have tended to adopt the increasing correlation length idea as a fundamental interpretation of viscous slow-down, hence of glass formation, and the fact that the relation between correlation length and relaxation time is inverted when it comes to the classical silicate glasses (and other strong liquids, presumably) is a warning that the universality aspects of the glass transition must be sought elsewhere. For strong liquids it can now be appreciated that any effect of correlation length on the dynamics will be in opposition to that of any natural barrier- crossing kinetics that exist, hence will oppose the slowdown, while for fragile liquids it will enhance the slowdown. Whether this can be considered as an explanation of the difference between strong and fragile liquid kinetics is a little difficult to say. We note, however, that in lambda transitions, the simple Arrhenius form of kinetics is followed over many orders of magnitude (e.g. $TlNO_2$ [18] in Fig. 2.3a, and $C_{60}$ [61]) while the magnitude of the fluctuations increases and the correlation length accordingly starts to diverge.

The presence of first order transitions below $T_g$, predicted by the analysis of Ref. [32] would be consistent with the current findings of ultrastable [62], and ultradense [63], glasses formed by vapor deposition processes on controlled temperature substrates, when the temperature is controlled at some 10–20% below the standard $T_g$. These are reported [64] to convert back to viscous liquids via nucleation and growth processes – the hallmark of a first order transition. The one micron length scale reported for the process [64] is reminiscent of the homogeneous nucleation of ice crystals from aqueous LiCl solutions near $T_g$ [65], now known to be consequent on the prior polyamorphous transformation of vitreous water [66]. Whether these new phases are truly amorphous or are some higher order disordered crystal form transitional to the ground state crystal (hence another example of the Ostwald rule of stages in action) has yet to be definitively decided. A true first order character would be expected if the liquid–liquid transition implicit in the behavior of $SiO_2$ at high pressure becomes modified to occur near ambient pressure for the weaker network $H_20$ and then passes to negative pressure domain for the more weakly interacting but more cooperative van der Waals liquids. It will require much more work to establish whether or not such a simple set of systematic changes across such a broad swath of liquids, can be supported.

Irrespective of the outcome on the latter question, it should be clear from the material of this chapter that a rich panoply of thermodynamic behavior accompanies the transition of non-ergodic to ergodic states of condensed matter, in particular the case of glassy solid to non-viscous liquid, and that much systematic work remains to be done before the complex patterns of behavior can be fully understood.

**Acknowledgments** Support of the NSF, DMR (Solid State Chemistry) and Chemistry divisions, Grant numbers 0454672 and 0404714, is gratefully acknowledged. We have profited from helpful discussions with Dmitry Matyushov and Ranko Richert. Yuanzheng Yue, Thomas Loerting and group, Masaharu Oguni Gene Stanley Sergey Buldyrev and group, Srikanth Sastry, Pablo Debenedetti and Mark Ediger.

## References

1. Debenedetti PG (1976) Metastable liquids: concepts and principles. Princeton University Press, Princeton, NJ
2. Angell CA (2008) Glassformers and viscous liquid slowdown since David Turnbull: enduring puzzles and new twists (text of Turnbull lecture). MRS Bull 33:545–555
3. Takeda K, Yamamuro O, Oguni M, Suga H (1995) Calorimetric study on structural relaxation of 1-pentene in vapor-deposited and liquid-quenched glassy states. J Phys Chem 99:1602–1607
4. Angell CA (2008) Insights into phases of liquid water from study of its unusual glass-forming properties. Science 319:582–587
5. Debenedetti PG, Stillinger FH (2001) Supercooled liquids and the glass transition. Nature 410:259–267
6. Angell CA (1995) Formation of glasses from liquids and biopolymers. Science 267:1924–1935

7. Cohen MH, Turnbull D (1959) Molecular transport in liquids and glasses. J Chem Phys 31:1164–1169
8. Adam G, Gibbs JH (1965) On temperature dependence of cooperative relaxation properties in glass-forming liquids. J Chem Phys 43:139–146
9. Dyre JC, Olsen NB, Christensen T (1996) Local expansion model for viscous-flow activation energies of glassforming molecular liquids. Phys Rev B 53:2171–2174
10. Moynihan CT, Angell CA (2000) Bond lattice or excitation model analysis of the configurational entropy of molecular liquids. J Non-Cryst Solids 274:131–138
11. Angell CA, Rao KJ (1972) Configurational excitations in condensed matter and the bond lattice model for the liquid-glass transition. J Chem Phys 57:470–481
12. Saika-Voivod I, Sciortino F, Poole PH (2004). Free energy and configurational entropy of liquid silica: fragile-to-strong crossover and polyamorphism. Phys Rev E 69:041503(13)
13. Faupel F, Frank W, Macht MP, Mehrer H, Naundorf V, Ratzke K, Schober HR, Sharma SK, Teichler H (2003) Diffusion in metallic glasses and supercooled melts. Rev Mod Phys 75:237–280
14. Fujimori H, Oguni M (1995) Correlation index (Tga-Tgb)/Tg and activation energy ratio as parameters characterising the structure of liquid and glass. Solid State Commun 94:157–162
15. Fujimori H, Oguni M (1993) Construction of an adiabatic calorimeter at low temperatures and the glass transition of crystalline 2-bromothiophene. J Phys Chem Solids 54:271–280
16. Fujita H, Fujimori HO, Oguni M (1995) Glass transitions in the stable crystalline state of dibenzofuran and fluorene. J Chem Thermodyn 27:927–938
17. Matsuo T, Suga H, David WIF, Ibberson RM, Bernier P, Zahab A, Fabre C, Rassat A, Dworkin A (1992) The heat-capacity of solid C-60. Solid State Commun 83:711–715
18. Moriya K, Matsuo T, Suga H (1983) Phase transitions and freezing of ion disorder in CsNO2 and TlNO2 crystals. J Phys Chem Solids 44:1103–1119
19. Wales DJ (2003) Energy landscapes, Cambridge molecular science series. Cambridge University Press, Cambridge
20. Brand R, Loidl A, Lunkenheimer P (2002) Relaxation dynamics in plastic crystals. J Chem Phys 116:10386–10401
21. Mizuno F, Belieres J-P, Kuwata N, Pradel A, Ribes M, Angell CA (2006) Highly decoupled ionic and protonic solid electrolyte systems, in relation to other relaxing systems and their energy landscapes. J Non-Cryst Solids 352:5147–5155
22. Richet P, Bottinga YD, Denielou L, Petitiet JP, Tegui C (1982) Thermodynamic properties of quartz, cristobalite and amorphous $SiO_2$: drop calorimetry measurements between 1000 and 1800 K and a review from 0 to 2000 K. Geochim Cosmochim Acta 46:2639
23. Scheidler P, Kob W, Latz A, Horbach J, Binder K (2005) Frequency-dependent specific heat of viscous silica. Phys Rev B 63:104204(14)
24. van Beest BWH, Kramer GJ, van Santen RA (1990) Force fields for silicas and aluminophosphates based on *ab initio* calculations. Phys Rev Lett 64:1955–1958
25. Vollmayr K, Kob W (1996) Investigating the cooling rate dependence of amorphous silica: a computer simulation study. Ber. bunsenges. Phys Chem 100:1399–1401
26. Tamura S, Yokokawa T, Niwa K (1975) The enthalpy of beryllium fluoride from 456 to 1083 K by transposed-temperature drop calorimetry. J Chem Thermodyn 7:633
27. Videa M, Angell CA (unpublished work)
28. Hemmati M, Moynihan CT, Angell CA (2001) Interpretation of the molten BeF2 viscosity anomaly in terms of a high temperature density maximum, and other waterlike features. J Chem Phys 115:6663–6671
29. Privalko Y (1980) Excess entropies and related quantities in glass-forming liquids. J Phys Chem 84:3307–3312
30. Alba C, Busse LE, List DJ, Angell CA (1990) Thermodynamic aspects of the vitrification of toluene, and xylene isomers, and the fragility of liquid hydrocarbons. J Chem Phys 92:617–624

31. Angell CA, Bressel RD (1972) Fluidity and conductance in aqueous electrolyte solutions. An approach from the high concentration limit. I. Ca(NO3)2 solutions. J Phys Chem 76:3244–3252
32. Matyushov D, Angell CA (2007) Gaussian excitations model for glassformer thermodynamics and dynamics. J Chem Phys 126:094501(19)
33. Martinez L-M, Angell CA (2001) A thermodynamic connection to the fragility of glass-forming liquids. Nature 410:663–667
34. Huang D-H, McKenna GB (2001) New insights into the fragility dilemma in liquids. J Chem Phys 114:5621–5630
35. Wunderlich B (1960) Study of the change in specific heat of monomeric and polymeric glasses during the glass transition. J Phys Chem 64:1052–1056
36. Moynihan CT, Angell CA (2000) Bond lattice or excitation model analysis of the configurational entropy of molecular liquids. J Non-Cryst Solids 274:131–138
37. Stevenson JD, Wolynes PG (2005) Thermodynamic-kinetic correlations in supercooled liquids: critical survey of experimental data and predictions of the random first order transition theory of glasses. J Phys Chem B 109:15093–15097
38. Saika-Voivod I, Poole PH, Sciortino F (2001) Fragile-to-strong transition and polyamorphism in the energy landscape of liquid silica. Nature 412:514–517
39. Whalley E, Klug DD, Handa YP (1989) Entropy of amorphous ice. Nature 342:782–783
40. Speedy RJ, Debenedetti PG, Smith RS, Huang C, Kay BD (1996) The evaporation rate, free energy, and entropy of amorphous water at 150 K. J Chem Phys 105:240–244
41. Sastry S, Angell CA (2003) Liquid–liquid phase transition in supercooled liquid silicon. Nat Mater 2:739–743
42. Poole PH, Sciortino F, Essmann U, Stanley HE (1992) Phase-behavior of metastable water. Nature 360:324–328
43. Tanaka H (2002) Simple view of water-like anomalies of atomic liquids with directional bonding. Phys Rev B 66:064202(8)
44. Brovchenko I, Geiger A, Oleinikova A (2005) Liquid–liquid phase transitions in supercooled water studied by computer simulations of various water models. J Chem Phys 123:044515(16)
45. Poole PH, Sciortino F, Grande T, Stanley HE, Angell CA (1994) Effect of hydrogen bonds on the thermodynamic behavior of liquid water. Phys Rev Lett 73:1632–1635
46. Speedy RJ (1982) Stability-limit conjecture. J Phys Chem 86:982–989
47. Wagner W, Pruss A (2002) The IWAPS formulation 1995 for the thermodynamic properties of ordinary water substance for general and scientific use. J Phys Chem Ref Data 31:387–535
48. Kaya S, Sato H (1943) Superstructuring in the iron–cobalt system and their magnetic properties. Proc Physico-Math Soc Japan 25:261–273
49. Yue YZ (2004) Influence of physical ageing on the excessive heat capacity of hyperquenched silicate glass fibres. J Non-Cryst Solids 348:72–77
50. Angell CA, Yuanzheng Y, Wang L, Copley JRD, Borick S, Mossa S (2003) J Phys cond Matt 15: S1051–S1068
51. Xu L-M, Buldyrev SV, Giovambattista N, Angell CA, Stanley HE (2009) A monatomic system with a liquid–liquid critical point and two glassy states. J Chem Phys 130:054505(12)
52. Buldyrev SV, Malesio G, Angell CA, Giovambattista N, Prestipino S, Saija F, Stanley HE, Xu L (2009) Unusual phase behavior of one-component systems with two-scale isotropic interactions. J Phys Condens Mat 21:504106(18)
53. Xu L, Buldyrev SV, Angell CA, Stanley HE (2006) Thermodynamics and dynamics of the two-scale spherically-symmetric Jagla model of anomalous liquids. Phys Rev E 74:031108(10)
54. Jagla EA (1998) Phase behavior of a system of particles with core collapse. Phys Rev E 58:1478–1486
55. Angell CA (2008) Glass formation and glass transition in supercooled liquids, with insights from study of related phenomena in crystals. J Non-Cryst Solids 354:4703–4712

56. Mizukami M, Kobashi K, Hanaya M, Oguni M (1999) Presence of two freezing-in processes concerning a-glass transition in the new liquid phase of tri-phenylphosphite. J Phys Chem B 103:4078–4088
57. Aasland S, Mcmillan PF (1994) Density-driven liquid–liquid phase-separation in the system Al2O3-Y2O3. Nature 369:633–636
58. Bhat H, Molinero V, Soignard E, Solomon VC, Sastry S, Yarger JL, Angell CA (2007) Vitrification of a monatomic metallic liquid. Nature 448:787–790
59. Loerting T, Winkel K, Kohll (submitted for publication: private communication) 2010
60. Sakaguchi S, Todoroki S-I (1998) Rayleigh scattering of silica core optical fiber after heat treatment. Appl Opt 37:7708–7711
61. Mondal P, Lunkenheimer P, Boehmer R, Loidl A, Gugenberger F, Adelmann P, Meingast C (1994) Dielectric relaxation dynamics in C-60 and C-70. J Non-Cryst Solids 172:468–471
62. Swallen SF, Kearns KL, Mapes MK, Kim YS, McMahon RJ, Wu T, Yu L, Ediger MD (2007) Organic glasses with exceptional thermodynamic stability and kinetic stability. Science 315:354–356
63. Ishii K, Nakayama H, Hirabayashi S, Moriyama R (2008) Anomalously high-density glass of ethylbenzene prepared by vapor deposition at temperatures close to the glass-transition temperature. Chem Phys Lett 459:109–112
64. Kearns KL, Ediger MD, Huth H, Schick C (2010) One micrometer length scale controls kinetic stability of low energy glasses. J Phys Chem Lett 1:388–392
65. Dupuy J, Chieux P, Calemzuk R, Jal JF, Ferradou C, Wright A, Angell CA (1982) Controlled nucleation and quasi-ordered growth of ice crystals from low temperature electrolyte solutions: a small angle neutron scattering study. Nature 296:138–140
66. Mishima O (2007) Phase separation in dilute LiCl–$H_2O$ solution related to the polyamorphism of liquid water. J Chem Phys 126:244507(5)

# Chapter 3
# Vibration Forms in the Vicinity of Glass Transition, Structural Changes and the Creation of Voids When Assuming the Role of Polarizability

**Jaroslav Šesták, Bořivoj Hlaváček, Pavel Hubík, and Jiří J. Mareš**

## 3.1 Movements Below the Critical Temperature

Under the certain so called *critical temperature* [1], the liquid phase becomes factually prearranged and separated into solid-like structures. Certain unoccupied vacancies existing within the space are called voids (in the obvious meaning of opening, hollowness or cavity) and are packed with gas-like molecules (so called "wanderers"). This realism has been known for a long time [2]. Some of the modern structural theories (such as the so called "mode coupling theory" – MCT, which is describing the structural phenomena of liquid state at lower temperatures) are also based on a similar scheme of the local density fluctuation [3]. Such a conjecture of heterogeneities in liquid phase goes back to the assumption of semi-crystalline phase published early by Kauzman [4], as well as to the assumptions of coexistence of gas–liquid semi-structures [5, 6] as related to numerous works of Cohen, Grest and Turnbull [7–10].

We can extend our vision toward heterogeneity of larger areas, which is often termed the medium range order (MRO) and adjacent to the existence of nano-solid-like domains. Below the critical temperature, the liquid structures can be empathized as a certain, mechanically sub-divided structure formed by blocks (in the conspicuous meaning of domains, icebergs or clusters) [6, 11] (see Fig. 3.1) and, on the other side,

J. Šesták (✉)
New Technologies Research Centre, University of West Bohemia, Univerzitní 8, CZ-30614 Plzeň, Czech Republic
e-mail: sestak@fzu.cz

B. Hlaváček
University of Pardubice, Studentská 573, CZ-53210 Pardubice, Czech Republic

P. Hubík and J.J. Mareš
Institute of Physics of the Academy of Sciences ČR, v.v.i., Division of Solid-State Physics, Cukrovarnická 10, CZ-16200 Praha, Czech Republic

J. Šesták et al. (eds.), *Glassy, Amorphous and Nano-Crystalline Materials*, Hot Topics in Thermal Analysis and Calorimetry 8, DOI 10.1007/978-90-481-2882-2_3,

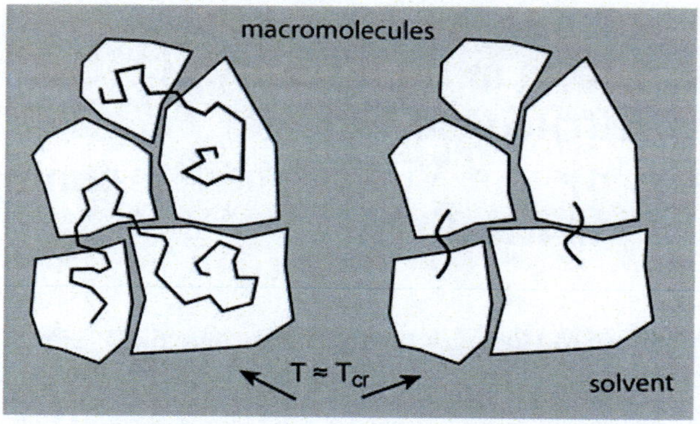


**Fig. 3.1** Illustration of a medium range structure. Schematic picture shows a liquid-like (and also glassy structures) composed of variously bonded blocks displaying the degree of possible coupling. When the separation of macro-molecularly interconnected blocks occurs above the cross-over temperature, $T_{cr}$, it produces rubber elasticity, which is often associated with entanglements. However, the low molecular compounds [14–16] (Fig. 3.2 at right) never achieve such a rubbery state. The blocks' partition can be found by adequate rheological measurements of a shear viscosity [17–23] at very small gradients. The separation of blocks for low molecular weights structures is connected with loss of shear module (cf. Fig. 3.3) and with the onset of fluidity (Note that the entropic elasticity of macromolecules is not incorporated herewith)

as an assortment of individual "semi-evaporated" units, which are subjected to non-linear anharmonic motions at high amplitudes [5, 6].

For the displacement of block as an entirety, the certain (often maximum) retardation time, $\tau_{max}$, reveals the state of interconnections inside of the block structure, characteristic of a given block size and its variability with temperature. It is assumed that the block size, together with maximum block retardation time decreases with increasing temperature. The blocks progressively disintegrate as the critical temperature is gradually approached [12, 13]. On contrary, the blocks bind themselves to a more rigid structure as the temperature decreases below the crossover temperature $T_{cr}$ and further on below the glass transition temperature $T_g$. Down from the critical temperature $T_c$ we can contemplate the four subsequent and most important temperatures, related to the characteristic points, which are gradually linking liquid and glassy states:

1. *Boiling/condensation/liquid temperature,* $T_b$, characteristic for the first-order transition. At this temperature, vibrating particles are capable to depart away from their typically nonlinear motion around its vibration focus (further loosing it at all to become a part of gaseous phase). Gaseous molecules are moving in both the translation and rotation modes within the whole volume, which are missing vibrational settings (characterized with energy minima) and vibration focusing (characteristic for a condensed state).
2. *Crossover temperature,* $T_{cr}$, at this temperature liquid starts to reveal an elastic response in the shear stress becoming thus viscously-elastic, which is characterized

by a nonzero, internal energy-related shear modulus build on energetic basis (an entropic elasticity of a rubbery-like network is not accounted for [17–23] but can be graphically illustrated upon the comparison of Figs. 3.2 and 3.3, in particular see the area "b" in Fig. 3.2).

3. *Glass transition temperature,* $T_g$. The shear modulus at this temperature overcome the value ~$G \geq 10^8$ N/m$^2$ having a tendency to level off with a declining temperature and further going ahead toward its maximum value of ~$10^9$– $10^{10}$ N/m$^2$
4. *Vogel's temperature* $T_V$. Below this temperature the Brownian movement restricts its progress only within isolated regions mostly along the domain's interfaces. The spatial adjustment also occurs at certain domains fixation (by, e. g., chemical bonding) making a three-dimensional web within the sample. In the macro-scale the sample undertakes the property of a solid phase incapable of the wholesale creep flow. Note that the time-honored Kauzmann temperature, $T_K$, may get coincidental with $T_v$ if derived from the entropy plot.

The subsequent succession $T_c \geq T_b \geq T_{cr} \geq T_g \geq T_v$ can be well recognized and the following model anticipated, which bears some distinctive characterization of vibration modes at different temperatures (or at associated temperature regions) – worth of

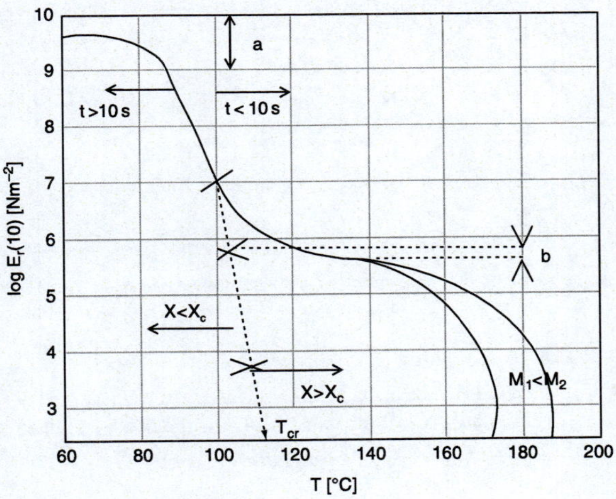


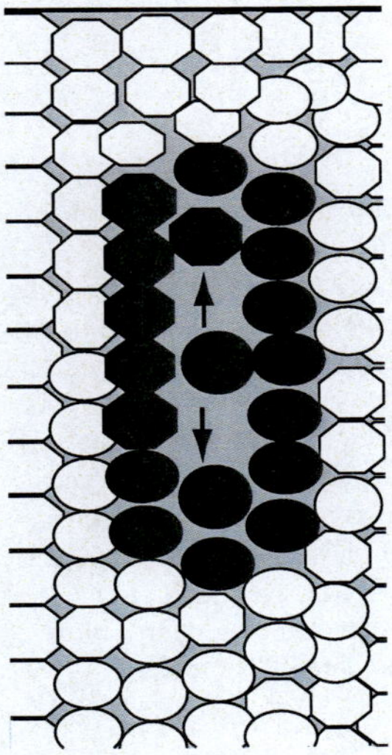

**Fig. 3.2** *Left*: the course of a shear modulus, $G \cong E_r/3$ for relaxation of an amorphous high-molecular compound around $T_g$ (according to Tobolsky-Alkonis [22, 23] for typical polystyrene). Note that in the glass region (area "a") the temperature $T_g$ is characterized by the shear modulus $G \cong 10^9$– $10^{10}$ N/m$^2$, which decreases with increasing temperature and disappears at (or closely above) the crossover temperature $T_{cr}$ where the proportionality, $T_g \cong 0.8\ T_{cr}$, is approximately applicable. The area "b" represents the rubberlike plateau, which is not taken into account in this study. The crossover temperature for the polystyrene is localized as $T_{cr} \cong 120$ °C while $T_g \cong 80$ °C At *right*: the scheme of non-linear oscillator, which existence is causing the relaxation module decline in main transition zone

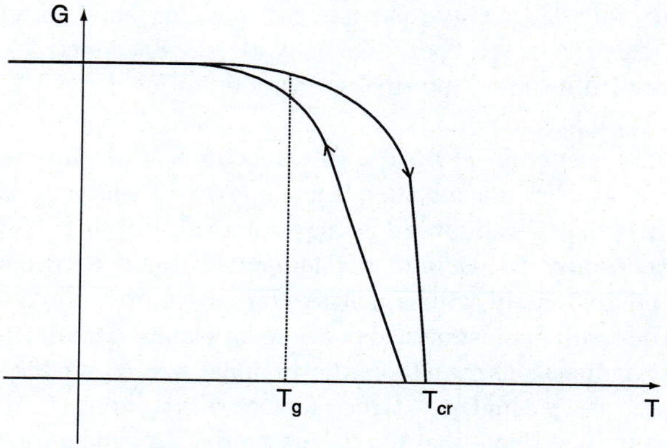


**Fig. 3.3** Course of a shear modulus G for amorphous low-molecular compound around $T_g$ [14–16]. Note that the glass transition region, $T_g$, is characterized by the shear modulus $G \cong 10^9 - 10^{10}$ [N/m$^2$], which decreases with increasing temperature and disappears at or closely above the crossover temperature $T_{cr}$ for which the proportionality $T_g \cong 0.8\ T_{cr}$ is approximately applicable (match up to Fig. 3.2, which is relevant for high-molecular compounds). Typical allied facts are the time-dependent hysteresis particularly evident during the sample consequent cooling

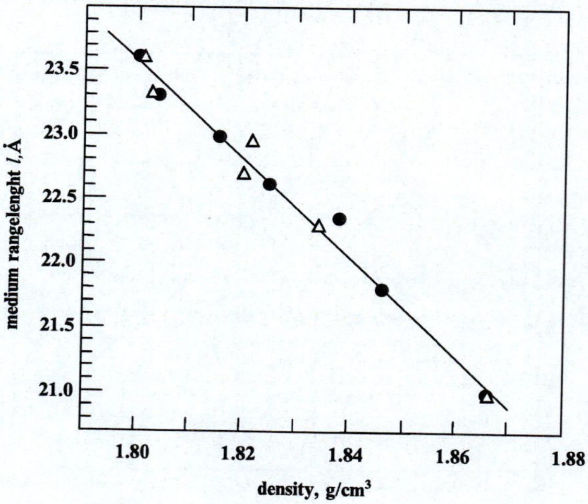


**Fig. 3.4** Mean length of the glass heterogeneity fluctuation as a function of the glass density (assumed in the terms of MRO) estimated on the basis of sound velocity for two types of glassy materials displayed by the dense and empty triangles (Adopted from Surovcev and Novikov [29–31], see [12, 13], the dependence compares with the data in Table 3.1)

mentioning beforehand of a needful identification. Searching for the size of the holes we refer to the simplest expression for P–V–T state equation. For example, the van der Waals equation is suitable to affirm the size of a minimum volume for the associated van der Waals isothermal curve (and its minimal spinodal points) while setting the

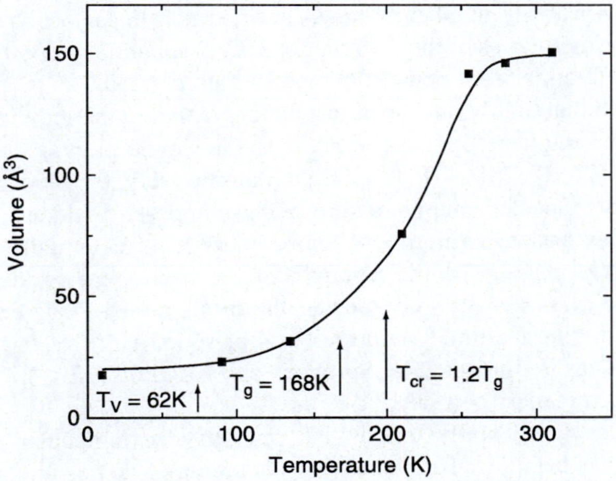


**Fig. 3.5** Vertically is given the size of growing voids (in cubic angstroms) for polybutadiene polymer measured along with increasing temperature (horizontal) starting from the Vogel's temperature, over $T_g$, up to the crossover temperature $T_{cr}$ (adjusted by means of a positron annihilation spectroscopy [60–63], see [5])

limits to the upper boundary of possible amplitudes of the oscillators' vibrations. In the moment when the limit is prevailed over such a crucial volume occupied by nonlinear oscillator, the system becomes unstable and can eventually expand toward the volume of gaseous phase. Below $T_b$, however, such a process (of maximum volume expansion) can undergo only those particles, which are existing on the liquid surface. The bottom component on the liquid side of saddle part of the van der Waals curve (approximately at 0.385 $V_{c,}$ i.e. just a portion of the critical volume $V_c$) plays an important role for the size evaluation of non-linear vibration amplitude. This estimate is not far away from that, which was given by B. Wunderlich and H. Baur [24, 25] at the turn of seventies.

Below the $T_{cr}$ the non-zero shear module has a drastic impact on the void size and on the extent of maximum diffusion jumps, which are depressed (or even eliminated) when temperature is dropping off. The presence of inter-block bonding will raise the shear module to a non-zero level forming thus an elastic network. As illustrated by Buchenau for the $T_g$ vicinity [26], the changes in average vibration amplitudes can be determined by experimental measurements of the Debye-Waller factor. Under $T_g$ the extrapolation of entropy towards negative values is somehow impossible (worth noting the Kauzmann paradox [4], which was misinterpreted in that sense documented in vast literature citations). In our view, the stepwise change of the entropy derivation occurs as a result of compositional and/or structural changes in liquid state when "semi-evaporated" particles turn out to be associates to individual non-linear oscillators. From mechanical point of view the matter above $T_{g,}$ became composed by the other type of oscillators than those existing under the glass formation temperature.The notable increase of entropy above $T_g$ is thus caused by growing number of newly formed non-linear oscillators endowed with the migrating focuses of their oscillations. Consequently above $T_g$ the liquid is structurally different as new degrees of freedom are created.

It is assumed that highly non-linear oscillators maintain the individual character of their motions and herewith are perceived as the simple individual units about a monomer size [11, 27, 28], cf. Fig. 3.2 right. Alternatively, the blocks are assumed to possess interconnected microstructure being composed of identical elastically bonded particles packed to the high compactness level eventually reaching the density of glassy phase. The mean length of the heterogeneity fluctuation as a function of the glass density (assumed in the terms of MRO) is experimentally determined and is best estimated on the basis of sound velocity for two types of glassy materials as is illustrated by Surovcev and Novikov [12, 13, 29–31] and is left for a detailed readers perusal.

The existence accidentally distributed, random size molecular clusters in a viscous liquid has been discussed since the beginning of sixties until present days [5–10, 29, 32–35]. Inter-clusters regions, which are containing molecules whose vibrations are strongly anharmonic, are considered herewith to provide a mount towards non-deterministic vibration movements. The characteristic of such movements of the amplitude switch of non-linear oscillators in liquid state gives rise to erratic characteristic of Brownian movements as well as to structural irregularities of glassy state formation.

The blocks, in other hand, are responsible for complex relaxation effects; because the interconnected linear oscillators are forming their structures interact. On the other hand, the semi-evaporated particles (acting as non-linear oscillators) are responsible for an erratic character of displacements due to the Brownian motion in liquids [36]. The semi-evaporated particles are correspondingly acting against the external pressure acting on the sample from outside, and exercise a so called 'push-aside' effect upon the individual blocks effective in their vicinity. Thus created gaps caused by straightforward amplitude jump (cf. Fig. 3.2 right), can be directed to any direction within amorphous phase. For the non-linear oscillators of semi-evaporated particles, the minimal change in the initial coordinates in phase space brings consequential and substantial changes in the motional trajectories of particles [37–41]. Therefore, particles can perform their oscillations on several different amplitudes [12, 13] and these motions bring, in the same time, apparent elements of uncertainty. In most cases it is describable by the non-linear and non-deterministic theories [42–50] of chaos. For non-liner systems, characterized with the different initial coordinates, their subsequent positions and momentums cannot be determined beforehand and the differential changes and variations in initial conditions bring completely different trajectories in the phase space expression. The general rule for the non-deterministic chaos theories [37–57] is consequently reflected into the structure of amorphous soul of glassy state, which durably depends on the experimental course of action deliberated for cooling. It will surface its irregular character, which is subject of discussion connected with the uplift of potential bottom well (and associated particle displacement) and its consecutive diffusion type displacement represented by the general length parameter $\lambda$ as the shift in bottom potential well displacement $\lambda \neq \sigma$ is much different than as the particle diameter $\sigma$ from Ref. [58, 59], giving thus more flexibility to correlations of Parkhurst and Zvanzig type [58, 59], dealing with viscosity and diffusivity.

## 3.2 Distinguishing Local Heterogeneity Using the Positron Annihilation Spectroscopy

The volumes of voids were exactly determined and reported in numerous papers on basis of positron annihilation spectroscopy – PASCA [60–64]. These PASCA experiments provide the unusually high coefficients of thermal expansion in the vacancies comprising region where the volumes of voids are exceedingly sensitive to the temperature changes above the glassy transition as well as to the external pressure changes in the boiling point area. The coefficient of thermal expansion in vacancies areas is about ten to a hundred times as high as that in the block-containing areas. Possible discontinuity in properties for liquid structure turns out to be apparent locally at expanding spots, which bear a responsibility for a high coefficient of their thermal expansion. These locally expanding structures appear particularly above $T_g$ and their number increases with rising temperature and the sample commences to consists of, at least, two kinds of particles. In the presence of such mechanically different units above $T_g$, the matter turn out to be rather mechanically heterogeneous. As can be seen, PASCA method cannot reach higher temperatures in structural description (exceeding the crossover temperature $T_{cr}$). Nevertheless, PASCA give good indication how the structure approaching the critical temperatures $T_c$ can look like. This temperatures areas has been visualized and described in models presented in previous papers [5, 6, 12, 13] in relation to the majority of typical equations of state [65–67].

## 3.3 Three Types of Entropy Contributions in Main Transition Zone

For the inherent temperatures regions the entropy main contributions can be portrayed in three gradual ways [68]. Below the Vogel's temperature $T_v$ the entropy $S \cong k_B \ln W_{th}$ is solely composed by the vibration energy of particles [52]. In the region above the Vogel's temperature, $T \geq T_v$, but below $T_g$ the entropy value enlarges by conformation part, i.e., $S \cong k_B [\ln W_{th} + \ln W_{conf}]$, which is associated with the release of particle's migration freeing thus their originally fixed positions in the sample space. Finally at $T \geq T_g$ where $S \cong k_B [\ln W_{th} + \ln W_{conf} + \ln W_{semievap}]$, the overall contribution extends as a result of semi-evaporated non-linear particles at a higher energy vibration level. The number of excited particles increases with rising temperature and acutely above $T_{cr}$ it turns out to form the diffusion process due to high-amplitude changes.

These types of entropy changes brought together by three distinct contributions have been experimentally confirmed in recent works of Johari [68]. Thanks to the reports made on the basis of PASCA experiments, the vacancies areas have a rather distinct dimension, which is larger than the van der Waals volume, but smaller than the 0.385-fraction of critical volume $V_c$ of the each particle's involved. Accordingly, the estimate of the enthalpy change can be connected with the semi-evaporated state. As shown by Hirschfelder [2], the total amount of evaporation enthalpy is required to produce an expansion of cavity to the level twice as high as is the particle diameter [69]. As the first estimate for the energy (of void creation producing a cavity of the size of about 0.385 of the fraction of critical volume $V_c$) a proportionally smaller part of evaporation enthalpy, $\Delta H_{evap}$, can be used. Therefore, the assumption of $\Delta H_{semievap} \cong \Delta H_{evap}/n$, where n stays from 2 to 4 (or even higher), is an agreement with the PASCA experiments as well as with the estimate of viscosity figured out by Eyring [70, 71] and his co-workers. Remarkable displacements, $\lambda$ – shifts of the potential well bottoms occur above the $T_{cr}$ with the onset of diffusion and progress of the liquid state connected to vanishing shear modulus.

For the heterogeneity size, the polarizability plays an important role because it is related to the critical volume of the matter which limits dimensions of possible voids. For example, the molecular polarizability, $\alpha$, of a monomer can be calculated by simple adding up the tabulated polarizabilities of individual atoms [72, 73] ultimately using empirical corrections in the line with Verköczy [6, 73] such as $\alpha_{molecul} = \alpha_{C_n} + \alpha_{H_m} + \alpha_{C=C} + \alpha_{C\equiv C}$ where $\alpha_C = 1$ Å$^3$ for bonded carbon in chained organic molecules and $\alpha_H = 0.4$ Å$^3$ for hydrogen. Special attention must be paid to double and triple carbon bonds, if isolated, the former increases polarizability by 0.2 Å$^3$ while the latter by 0.4 Å$^3$. Polarizability of nitrogen is about 0.9 Å$^3$ in molecules of organic compounds and of oxygen about 0.8 Å$^3$ whilst two coupled free electrons (~$NH_3$) bring a volume enlargement by 0.158 Å$^3$.

## 3.4 Linear Cooperative Phenomena in the Built-In Blocks

### *3.4.1 Limiting Values of Block Time Characteristic*

For limits of the time dependence, which is solicitous for a fastest time interval of retardation and/or relaxation, the time interval calculated from the sound velocity, $v_{sound}$, is most relevant. For solid blocks, $v_{sound} \approx 10^5 - 10^6$ cm s$^{-1}$ possess the minimal length of acoustic waves $\lambda_{min}$, which are determined by the mesh size, $r_o$. The relations $1/\nu_{max} \cong \tau_{min}$ and $v_{sound} \cong \lambda_{min}\ \nu_{max}$, where $\lambda_{min} \cong 2\Delta\ r_o$ become relevant. Substituting from the value estimate (as outlined above) the maximum frequency, $\nu_{max} = 10^{12}$–$10^{13}$ s$^{-1}$ in the sample body, the fastest time of material response in block, $\tau_{min} \approx \nu_{max}{}^{-1} = 10^{-12} - 10^{-13}$ s can be obtained. On behalf of

**Table 3.1** Some characteristic figures of various inhomogeneities domain size calculated base upon molecular polarizabilities for different levels of chain interconnection

| Brand name of polymer or type of the substance | Approximation of monomer polarizability (Å$^3$) | Linear length of a domain for ~600 segments (nm) | Linear length of a domain for ~300 segments (nm) | Linear length of a domain for ~100 segments (nm) | Linear length of a domain for ~20 segments (nm) | Temperature of glass transition, $T_g$ (K) |
|---|---|---|---|---|---|---|
| Polystyrene | 11.6 | 1.90 | 1.52 | 1.04 | 0.61 | 353 |
| Polyisobutylen | 7.2 | 1.62 | 1.29 | 0.89 | 0.52 | 198 |
| Polybutadiene | 6.4 | 1.55 | 1.24 | 0.85 | 0.5 | 168 |
| Polyvinylacetate | 8.0 | 1.68 | 1.34 | 0.92 | 0.54 | 298 |
| Polymethyl-metakrylate | 9.8 | 1.95 | 1.56 | 1.07 | 0.62 | 374 |
| $SiO_2$ | 6.7 | 1.59 | 1.27 | 0.87 | 0.51 | 1473 |
| $GeO_2$ | 7.6 | 1.65 | 1.32 | 0.91 | 0.53 | 957 |
| $B_2O_3$ | 4.4 | 2.05 | 1.64 | 1.13 | 0.66 | 523 |

the longest time characteristic of a given material we can take the typical time connected with the permanent displacement in series of elements inside of the block space to be bonded together through the elastic force $\tau_{max} \cong n^2 \tau_{min}$ [5]. Using from the subsequent text the most suitable number of interconnections $n \cong 600$, we can adapt Table 3.1. A limiting particularity of picking up "n" value is detailed below.

### 3.4.2 Estimate on the Number of N-Elements in the Built-In Blocks and, Accordingly, on Their Size as Related to the Shear Viscosity Data

Numerous predictions regarding the size of the blocks are reported in literature by various authors [32, 74–80] based upon different experimental techniques. Focusing our attention in a direction not yet fully familiar for the block size estimate we can directed our approximations towards rheology searching for yet unfamiliar parameters. The first one is characterized by the number of linear elements in a block at which blocks start to be interconnected in shear flow and become open for mutual interference when producing a non-zero contribution to the response of elastic shear module. A useful source of data, for the block's interactions, is the measurements of shear viscosity when the data are obtained at zero velocity gradients (already acknowledged for almost 40 years regarding different structures of polymer melts with various molecular weight values [17–19, 81]). These data show one common feature: the dependence of shear viscosity on the number of roughly monomer (acoustical [27]) units in a chain with X sequence. As the first instance, the X-number is small and remains below a certain limit ($X \leq X_c$. where

$X_c \approx 500$–$625$) where the polymer melt is characterized by constant and gradient-independent viscosity. In this particular case the module of shear elasticity is zero, as well as the first normal stress difference remains zero. The macromolecules are smaller than the critical block size ($X \leq X_c$) and the blocks, if subjected to shear flow, do not exhibit any mutual interference. A different level of viscosity, which rises upon a linear dependence on X, characterizes each polymer at this particular case. However, at the point when the critical value, $X_c$, is reached along the polymer chains, the changes in the rheological properties occurs. The block inter-connections start to interfere through the overgrown size of macromolecules (cf. Fig. 3.1). For polymers with $X \geq X_c$, the melt viscosity in reliance on shear gradient starts to possess a non-Newtonian character as the gradient grows and the first normal stress difference starts to become an important flow characteristic. This effect is usually explained in connection with the chain entanglements, the concept of which was originally introduced upon sketchy pictures [20, 21, 82] (such as an evident phenomenon). On the other hand, the nature of couplings [20] was left undeveloped resuming consequently an alternative in relation to account cooperative motions of molecular areas (also suited for the low molecular structures). The latter case corresponds more closely to a scheme presented herewith (cf. Fig. 3.1) being supported by results of Achibat [33, 34] when reporting a relationship between the domains and the entanglements. This model attributes the property of inter-blocks bonding either to the entanglements (as the individual blocks start to be interwoven by polymer network) or to the inflexible bonds of van der Waals type (or even to the bonds of a chemical nature). The inter-block bonding has a consequence in the development of non-zero shear module. Although different polymers have different levels of viscosity (or else chain flexibility), different molecular weight exhibits different levels of normal stresses. The numerical value of $X_c$ (for the blocks' mechanical interference in shear flow) holds a universal character. Just quoting Ref. [19], it subsists for polystyrene the value $X_c = 600$, for poly(vinyl acetate) $X_c = 600$, for polyisobutylene $X_c = 500$, for polydimethylsiloxane $X_c = 625$. If the level of critical interconnection is achieved in a linear sequence for polymers, the blocks become inter-bonded and get exhibiting the shear elasticity represented by shear modulus or by the first normal stress difference. Such elasticity created by flexible macromolecules stays entropy related. This type of interconnection has a "long distance character" relative to the block size and one single macromolecule can "over-bridge" several blocks, as well. In contrast to the blocks coupled by flexible macromolecules, the blocks can also be inter-connected by inflexible chemical bonds. This is the case for the low molecular weight substances under the crossover temperature. The blocks connected by inflexible bonds cannot be stretch to higher deformations and have to break up as the temperature is increased above $T_{cr}$. The inter-connection of inflexible blocks reflects such a self-character toward the internal energy as related to the contribution of shear module under the $T_{cr}$.

The other parameter not yet determined herewith is the crossover temperature itself. For the measurement purposes, the dependence of shear relaxation module on temperature in area of main phase transition can provide such a needed capture information [22, 23] (cf. Fig. 3.2). For clarity reasons of having a better explanation within the text, some data of Schultz [19] for the main phase transition can provide such a desirable capture of information; Alkonis [23] and Tobolsky [22] are exposed for an ideal illustration (cf. Fig. 3.2). The crossover temperature $T_{cr}$ is determined on basis of the shear elasticity of relaxation module related to the internal energy. The value $X_c$ can be determined by sourcing the zero gradient viscosity curves. At this level of internal connection the blocks start to interfere in their flow properties. At the temperature $T_{cr}$ the characteristic domain size become estimable. For lower temperatures, $T \leq T_{cr}$, the part of force responding to the external deformation has a component related to internal energy. Taking the approximate size of one oscillator unit in an interconnecting sequence from the expedient polarizability values [83, 84] and using the interconnection number of $X_c$ equal the typical 600, the characteristic size of a domain of about 6 $nm^3$ is realistically obtained (see Tables 3.1 and 3.2). For the linear dimension of a domain following number $\sqrt[3]{6} = 1.81$ nm can be situated as a most typical. This seems to be in good agreement with the published values [32, 74–80] obtained by different experimental techniques. As the polarizability of molecules is a simple additive property, the domain size can be easily estimated from the

**Table 3.2** Some characteristic accounts for selected compounds

| Substance | Critical volume ($cm^3$) $V_C \approx 55\ \alpha_{molecule}\ N_{AV}$ | Polarizability $\alpha_{molecule}$ ($Å^3$) ($[Å^3] = 10^{-24}\ cm^3$) | Critical volume divided by $\alpha_{molecule}\ N_{AV}$ as related to the polarizability | Notes: Remarkable deviations from the average value |
|---|---|---|---|---|
| Ethane | 148 | 4.47 | 55.1 | |
| Propane | 200 | 6.3 | 52.8 | |
| Butane | 255 | 8.2 | 52.1 | |
| Pentane | 311 | 10.0 | 51.9 | |
| Hexane | 368 | 11.9 | 51.9 | |
| Water | 45 | 1.45 | 51.5 | |
| Ethanol | 168 | 5.41 | 51.7 | |
| Propanol | 220 | 6.74 | 54.4 | |
| Acetone | 213 | 6.4 | 55.0 | |
| Methyl-propylketon | 301 | 9.93 | 50.5 | |
| Propion acid | 230 | 6.9 | 55.5 | |
| Butter acid | 290 | 8.38 | 57.6 | |
| Methylacetate | 230 | 6.9 | 55.5 | |
| HCN | 139 | | 89.3 | Upper anomalies |
| Methanol | 117 | | 59.0 | Upper anomalies |
| Toluene | 316 | | 44.4 | Foot anomalies |
| p-Xylene | 378 | | 45.8 | Foot anomalies |
| Benzene | 260 | | 41.6 | Foot anomalies |
| Naphthalene | 408 | | 38.7 | Foot anomalies |

polarizability values of individual monomer units and from the $X_c$ values, as hinted in Table 3.1.

## 3.5 Polarizability of Molecules Determining the Critical Volume

Polarizability represents the volume response of electron cloud of atom or molecule against the effect of a weak external electric field (where the notion "weak" means the electric field of a much smaller intensity in comparison with the electric interaction of the atom nucleus with orbiting electrons). For some organic molecules the Table 3.2 provides important information showing that for most compounds the volume allocated to a molecule at a critical temperature is about 50–55 times larger than the volume of its polarizability. Polarizability of a molecule can be calculated by a simple adding of the contributions of individual atoms [72, 73] whilst it is important to follow deviations both upwards as (HCN) → $(HCN)_3$ or downwards (naphthalene) occurring due to the orientation of aromatic nuclei at the molecule contacts.

It follows that at the critical volume is the average volume related to a single molecule described by the value equal to about 50 times of the relevant volume of electron cloud of each molecule [73]. The table provides beneficial estimate on critical volumes and basis of polarizability, which has correlation not only to critical volume but also dielectric constant, density of cohesion energy and so mutual solubility of liquids and its refraction index [72].Upper deviation shows possible interconnection of molecules lasting up to the critical temperature (e.g. $(HCN)_3$). Bottom deviation for aromatic compounds reveals a certain orientation or better a mutual approach of the "face-to-face" arrangement. It is worth noting that polarizability is a decisive parameter adjacent to the notion of liquid state because liquid cannot bears a greater volume than is its critical volume. The least volume of liquid follows from approximations used in the state equation, which consequently provides the adjacent volume of resulting glass.

## 3.6 Alpha Process and Beta Slow Processes

For both the explanatory reasons and the basis of simplicity, we have provided our illustrations using only examples of translation forms of diffusion displacements. However, the rotational forms and the orientation forms of diffusion displacements in the angular coordinates $\Phi$ and $\Psi$ (which signify rotation around the chemical bonds) possess the equivalent importance. It was Bueche [82] and consequently Eisenberg [85] as well as Boyer [86, 87], Helflend and Meier [88–91] and Ilavsky [92], who correctly explained the origin of so-called $\beta$ slow processes connected with the orientation and side chains movements. These authors explanations were similar to those given much later on by Rossler [93–95]; Vogel, Rossler and coworkers [96–100] related β slow process on Arrhenius a non-Arrhenius thermal relaxations. According to other authors [82, 101–103], the reorientation motions of molecules are associated with a secondary beta relaxation taking place in super-cooled liquids. We should note that besides the slow β process some literature unveil fast β process, which, however, is not accounted for herewith though having some particularities such as time $\tau_{\beta fast} <<< \tau_{\beta slow}$ is rather shorter in turn of several orders and involves processes $\tau_{\beta slow} \approx \tau_{min}$. Relaxation time $\tau_{\beta slow}$ is much longer due to the limited mechanical succession of bonding elements.

Another mechanism, which cannot be excluded, is associated with an explanation given by Johari and Goldstein [104–106] who suggested that the β slow processes are connected with the "islands-of-mobility" persisting even under $T_g$. Both approaches seems be acceptable. In our approach, however, the α process is connected with an infinite number of structural network "inter-connections", which directly contribute to the elastic shear modulus under $T_{cr}$. As the structural skeletally networks of the macro-sample are formed under cooling, two different polymer motion are freeze-in at $T_g$. Namely we speak about either the rotational diffusion around chemical bonds at polymer chains and translation motion within certain macro-scale. It follows that one type of diffusion can become structurally deficient or better is bypassed and disappears at lower temperatures or can proceed simultaneously with the α-process. Thus some elements, which participate in β slow processes, are structurally bypassed and below $T_g$. are considered as still mobile, Therefore they cannot contribute in a full extent to the shear modulus increment within $T_g$ area and adhere them self fully to the elastic network, which forms the whole sample.

The n-number of elements connected by α process is rather unspecified or better is only restricted by the sample dimensions. Chaining and interconnecting of the elements, which compose the α process go imaginary toward infinity under forming rigid viscous-elastic web of a solid sample. For the β slow process, the limited interconnection of joined elements, "n" is restricted inside of blocks (cf. Fig. 3.1) by limited number of particles in linear block's sequentionality. Thus the basic reciprocal temperature dependence of β slow processes is often expressed by prime line of Arrhenius relation [74, 93–95, 98–100] as the temperature declines under $T_g$ temperature. In this region the ability to have controlled motion of blocks through elastic network is for β-process different than that for α -process. Considering the variable inter-connection, the number "n" brings an additional point to the perception of differences of alpha and beta relaxation processes.The relaxation times, τ, of components with decreasing temperature (for the α-process) increases up to infinity as a result of enormous growth of elements in the relaxation series ($\tau_{max} \cong n^2\, \tau_{min}$) [5, 6].

The accrual of relaxation time τ due to the increment of number of elements "n" is the main reason for the disposition of shear modulus during cooling below $T_{cr}$. For the α process the number of n-elements can increase up to infinity when an elastic web inside the sample is accomplished. Below $T_{cr}$ the cause of relaxation time increases strengthening the mechanical interconnection of network connections. On the other hand, the β slow process is different because the interconnection characterized by "n" is adjusted at a relatively minor value $n_1 =$ *const.*, just composed by several units. Only for some exemptions "n" can reach a higher values and such a fragment have no other choice than to behave Arrhenian. With an increase of the environs density, which is surrounding mechanically interconnected friction centres, the whole aggregate behaves adequately frictionally. For the β slow process the elements growth terminates with the interference of blocks' boundary below $T_{cr}$ as a consequence of dawdling of n-numeral, which is exhibited by a standard Arrhenius relation upon decreasing temperature. For this reason $\tau_{\beta slow}$ is not increasing along with the escalating collection but follows mere Arrhenius growing trend at decreasing temperature.

**Acknowledgments** The study was supported by the Czech Science Foundation, Project No. FR-TI 1/335.

## References

1. Stanley HE (1971) Introduction to phase transitions and critical phenomena. Oxford University Press, New York
2. Hirschfelder JO, Curtiss CF, Bird RB (1954) Molecular theory of gases and liquid Wiley, New York, pp 122, 291
3. Gotze W, Sjogren L (1992) Relaxation processes in supercooled liquid. Rep Prog Phys 55:241–376
4. Kauzmann W (1948) The nature of the glassy state and the behavior of liquids at low temperatures. Chem Rev 43:219–256
5. Hlaváček B, Šesták J, Koudelka L, Mošner P, Mareš JJ (2005) Forms of vibrations and structural changes in liquid state. J Therm Anal Calorim 80:271–283
6. Hlaváček B, Mareš JJ (2008) In: Šesták J (ed) Fyzika struktur amorfních a krystalických materiálů (Physics of Strutures of Amorphous and Crystalline Materials). Publ. House Pardubice University, Pardubice
7. Cohen MH, Turnbull D (1959) Molecular transport in liquids and glasses. J Chem Phys 31:1164–1169; Turnbull D, Cohen MH (1961) Free-volume model of the amorphous phase: glass transition. J Chem Phys 34:120–125
8. Turnbull D (1965) Free volume model of the liquid state. In: Hughel TJ (ed) Liquids: structure, properties and solid interaction. Elsevier, Amsterdam, pp 6–24
9. Cohen MH, Grest GS (1979) Liquid–glass transition, a free-volume approach. Phys Rev B 20:1077–1098, 21, 4113 (1980)
10. Grest GS, Cohen MH (1981) In: Prigogine I, Rice ST (eds) Advances in chemical physics, vol XLVII. Wiley, New York, pp 455–525
11. Hlaváček B, Šesták J (2000) Vibration perception of Tg transition. In: Proceedings of the first slovak glass conference. Slovak Glass Society, Trenčín, pp 168–172
12. Hlaváček B, Křesálek V, Souček J (1997) The anharmonicity of motion in the liquid state and its consequences. J Chem Phys 107:4658–4667
13. Hlaváček B, Šesták J, Mareš JJ (2002) Mutual interdependence of partitions functions in vicinity of $T_g$ transition. J Therm Anal Calorim 67:239
14. Jackle J (1986) Theory of glass transitions. New thoughts about old facts. Phil Mag 56:113–127

15. Jackle J (1986) Models of glass transition. Rep Prog Phys 49:171–232
16. Piccireli R, Litowitz TA (1957) Ultrasonic shear and compressional relaxation in liquid glycerol. J Acoust Soc Am 29:1009
17. Allen VR, Fox TG (1964) Viscosity–molecular weight dependence for short chain polystyrene. J Chem Phys 41:337
18. Allen VR, Fox TG (1964) Dependence of the zero shear melt viscosity and related friction coefficient and critical chain length on measurable characteristics of chain polymers. J Chem Phys 41:344
19. Schulz J (1974) Polymer material science. Prentice-Hall, Englewood Cliffs, p 337
20. Ferry JD (1961) Viscoelastic properties of polymers. Wiley, New York, p 225
21. Bueche F (1956) Viscosity of polymers in concentrated solution. J Chem Phys 20:599–600
22. Tobolsky AV (1960) Properties and structure of polymers. Wiley, New York, p 75
23. Aklonis JJ, MacKnight WJ, Shen M (1972) Introduction to polymer viscoelasticity. Wiley-Interscience, New York, p 42
24. Wunderlich B, Baur H (1970) Heat capacities of linear high polymers. Springer, Berlin
25. Wunderlich B (1960) Study of the change in specific heat of monomeric and polymeric glasses during the glass transition. J Chem Phys 64:1052–1056
26. Buchenau U, Zorn M (1993) A relation between fast and slow motions in glassy and liquid selenium. Europhys Lett 18:523–528
27. Heuer A, Spiess HW (1994) Universality of glass transition temperature. J Non-Cryst Solids 176:294
28. Hlaváček B, Souček J, Prokůpek L, Večeřa M (1997/1998) The thermal entropy concept and $T_g$ transition. J Polym Eng 17:111–137
29. Surovtsev NV, Wiederich J, Batalov AE, Novikov VN, Ramos MA (2004) Inelastic light scattering in $B_2O_3$ glasses with different thermal histories. J Phys Condens Matter 16:223
30. Novikov VN, Ding Y, Sokolov AP (2005) Correlation of fragility of supercooled liquids with elastic properties of glasses. Phys Rev E 71:061501
31. Malinovsky VK, Surovtsev NV (2000) Inhomogeneity on the nanometer scale as a universal property of glasses. Glass Phys Chem 26:217–222
32. Donth E (2001) The glass transition, relaxation dynamics in liquids and disordered materials. Springer, Berlin
33. Achibat T, Boukenter A, Duval E, Lorentz G, Etienne S (1991) Low-frequency Raman scattering and structure of amorphous polymers: stretching effect. J Chem Phys 95:2949
34. Duval E, Boukenter A, Achibat T (1990) Vibrational dynamics and the structure of glass. J Phys Condens Matter 2:10227–10234
35. Porai-Koshitz EA (1964) Structure of glass. In: Proceedings of the 4th All-Union Conference on the Glassy State, Leningrad
36. Einstein A (1956) In: Furth R (ed) Investigation on the theory of the Brownian movement. Dover, New York
37. Peitegen HO, Jurgens H, Saupe D (1992) Chaos and fractals, new frontiers of science. Springer, New York
38. Feigenbaum MJ (1983) Universal behavior in nonlinear systems. Physica D 7:16
39. Walgraef D (1997) Spatio-temporal pattern formation. Springer, New York, pp 242–301
40. Chvoj Z, Šesták J, Tříska A (1991) Kinetic phase diagrams; non-equilibrium phase transitions. Elsevier, Amsterdam
41. Swiney HL, Gollub JP (1978) Hydrodynamic instabilities and the transition to turbulence. Physics Today 41
42. Rabinowitz MI, Ezersky AB, Weidman PD (2000) The dynamics of patterns. World Scientific, Singapore, pp 239–279
43. Paine HJ (1997) Physics of vibrations and waves. Wiley, New York
44. Ueda U (1980) New approaches to non-linear dynamics. SIAM, Philadelphia
45. Tabor M (1989) Chaos and integrability in non-linear dynamics. Wiley, New York
46. Prigogine I, Stengers I (1984) Order out of chaos. Bantam Books, New York

47. Kondepudi D, Prigogine I (1998) Modern thermodynamics. Wiley, New York, pp 409–452
48. Rössler OE (1976) An equation for continuous chaos. Phys Lett A 57:397–398
49. Lorenz EN (1963) Deterministic non-periodic flow. J Atmos Sci 20:130
50. Šesták J (2004) Heat, thermal analysis and society, Chap. 14 Oscillations modes and modern theories of glassy state. Nucleus, Hradec Králové
51. Hlaváček B, Shánělová J, Málek J (1999) Discontinuities in amplitudes of particles micromotions characterizing phase transitions in liquid state. Mech Time-Depend Mater 3:351–369
52. Gurney RW (1949) Introduction to statistical mechanics. Dover, New York
53. Martienssen O (1910) Über neue Resonanzerscheinungen in Wechselstromkreisen. Phys Z 11:448–460
54. Appleton EV (1924) On the anomalous behavior of vibration galvanometer. Philos Mag Ser 6(41):609–619
55. Minorsky N (1983) Nonlinear oscillations, vol 47. D. Van Nostrand, Toronto, p 15
56. Haken H (1983) Synergetics. Springer, Berlin
57. Bender CM, Orszag SA (1978) Advanced mathematical methods for scientists and engineers. McGraw-Hill, New York
58. Parkhurst HJ Jr, Jonas J (1975) Dense liquids. I. The effect of density and temperature self-diffusion of tetramethylsilane and benzene-$d_6$. J Chem Phys 63:2699–2709
59. Zwanzig R (1983) On the relation self–diffusion and viscosity of liquids. J Chem Phys 79:4507–4508
60. Bartoš J (1996) Free volume microstructure of amorphous polymers at glass transition temperatures from positron annihilation spectroscopy data. Colloid Polym Sci 274:14
61. Bartoš J, Banduch P, Šauša O, Krištiaková K, Krištiak J, Kanaya T, Jenniger W (1997) Free volume microstructure and its relationship to the chain dynamics in cis-1,4- polybutadiene as seen by positron annihilation lifetime spectroscopy. Macromolecules 30:6906
62. Račko D, Chelini R, Cardini G, Bartoš J, Califano S (2005) Insights into positron annihilation lifetime spectroscopy by molecular dynamics simulation. Eur Phys 32:289–297
63. Jean YC (1990) Positron-annihilation spectroscopy for chemical-analysis – a novel probe for microstructural analysis of polymers. Microchem J 42:72; Jean YC, Deng Q, Nguyen TT (1995) Free-volume hole properties in thermosetting plastics probed by positron annihilation spectroscopy: chain extension chemistry. Macromolecules 28:8840–8844
64. Jean YC (1996) Comments on the paper "Can positron annihilation lifetime spectroscopy measure the free-volume hole size distribution in amorphous polymers?". Macromolecules 29:5756–5757
65. Dieterici C (1899) Ueber den kritischen Zustand. Ann Phys 305:685–705
66. van der Waals JD (1873) Over de continuiteit van den gas/en Vloeisfoestand. PhD thesis, Leiden
67. Peng DY, Robinson DB (1976) A new two-constant equation of state. Ind Eng Chem Fundam 15:59–64
68. Johari GP (2000) Contributions to the entropy of a glass and liquid, and the dielectric relaxation time. J Chem Phys 112:7518–7523
69. Frenkel J (1955) Kinetic theory of liquids. Dover, New Jersey, p 12
70. Eyring H, Shik Jhon M (1969) Significant liquid structures. Wiley, New York
71. Glasstone S, Laidler KJ, Eyring H (1941) The theory of rate processes. McGraw-Hill, New York
72. Le Fevre RJW (1965) Refractivity of organic molecules. Adv Phys Org Chem 3:1–35
73. Verkoczy B (1986) Saskoil Rep No.113. Saskoil, Regina, Saskatchewan
74. Donth E (1982) The size of cooperatively rearranging regions at the glass transition. J Non-cryst Solids 53:325–330
75. Dollase T, Graf R, Heuer A, Spiess HW (2001) Local order and chain dynamics in molten polymer blocks revealed by proton double-quantum NMR. Macromolecules 34: 298–309

76. Bartoš J, Šauša O, Krištak J, Blochowitz T, Rossler E (2001) Free-volume microstructure of glycerol and its supercooled liquid-state dynamics. J Phys Condens Matter 13:11473–11484
77. Beiner M, Huth H, Schroter K (2001) Crossover region of dynamics glass transition: general trends and individual aspects. J Non-cryst Solids 279:126–135
78. Hempel E, Beiner M, Renner T, Donth E (1996) Linearity of heat capacity step near the onset of $\alpha$ glass transition in poly-(n-alkylmethacrylates). Acta Polym 47:525
79. Mohanty U (1994) Inhomogeneities and relaxation in supercooled liquids. J Chem Phys 100:5905–5909
80. Reinsberg SA, Qiu XH, Wilhelm M, Spiess HW, Ediger MD (2001) Length scale of dynamic heterogeneity in supercooled glycerol near $T_g$. J Chem Phys 114:7299–7302
81. Doi M, Edwards SF (1986) Theory of polymer dynamics. Clarendon, Oxford
82. Bueche F (1962) Physical properties of polymers. Interscience, New York, p 106
83. Hlaváček B, Černošková E, Prokůpek L, Večeřa M (1996) The transition points in liquid state and their molecular modeling. Thermochim Acta 280/281:417
84. Lide DR (2002) Handbook of chemistry and physics, vol 10, 82nd edn., pp 160–174. CRC Press LLC, Washington DC; DiMarzio EA, Dowell F (1979) Theoretical prediction of the specific heat of polymer glasses. J Appl Phys 50:6061–6066
85. Eisenberg A (1993) The glassy state and the glass transition. In: Mark JE (ed) Physical properties of polymers. United Book, Baltimore
86. Boyer RF (1963) The relation of transition temperatures to chemical structure in high polymers. Rubber Chem Technol 36:1303–1421
87. Boyer RF (1973) $\Delta C_P$ $T_g$ and related quantities for high polymers. J Macromol Sci Phys B 7:487–501
88. Helfand E (1984) Dynamics of conformational transitions in polymers. Science 226: 647–650
89. Meier DJ (1978) Molecular basis of transitions and relaxations. Gordon and Breach Science, London
90. van Krevelen DV, Hoftyzer PJ (1979) Properties of polymers. Wiley, New York
91. Curtis AJ (1960) Dielectric properties of polymeric systems. In: Birks J, Schulman JH (eds) Progress in dielectrics, vol 2. Wiley, New York, p 29
92. Ilavský M, Hasa J, Janáček J (1968) Secondary and tertiary $\beta$ and $\gamma$ maxima in amorphous polymers. Collect Czech Chem C 33:3197
93. Rossler E (1990) Indication for change of diffusion mechanism in supercooled liquids. Phys Rev Lett 65:1595–1598
94. Blochowitz T, Brodin A, Rossler EA (2006) Evolution of the dynamic susceptibility in supercooled liquids and glasses. In: Coffey T, Kalmykov YP (eds) Advances in chemical physics, vol 133, part A, Fractals, diffusion, and relaxation in disordered complex systems. Wiley, Hoboken, pp 127–256
95. Brodin A, Rossler EA (2005) Depolarized light scattering study of glycerol. Eur Phys J B 44:3–14
96. Rossler E (1990) Indications for a change of diffusion mechanism in supercooled liquids. Phys Rev Lett 65:1595–1598; Rossler E (1992) Comment on "decoupling of time scales of motion in polybutediene close to the glass transition. Phys Rev Lett 69:1620
97. Sokolov AP, Rossler E, Kisliuk A, Quitman D (1963) Dynamics of strong and fragile glass formers: differences and correlation with low-temperature properties. Phys Rev Lett 71:2062–2065
98. Vogel M, Rossler E (2000) On the nature of slow $\beta$-process in simple glass formers: A $^2$H NMR study. J Phys Chem B 104:4285–4287
99. Wiederich J, Surovtsev NV, Rossler E (2000) A comprehensive light study of the glass former toluene. J Chem Phys 113:1143–1153
100. Vogel M, Rossler E (2001) Slow $\beta$ process in simple organic glass formers studied by one- and two-dimensional $^2$H nuclear magnetic resonance. J Chem Phys 114:5802–5815, and 115: 10883–10891

101. Gibbs JH (1956) Nature of glass transition in polymers. J Chem Phys 25:185–186; DiMarzio EA, Gibbs JH (1963) Molecular interpretation of glass temperature depression by plasticizers. J Polym Sci A 1:1417–1428
102. Gibbs JH, DiMarzio EA (1958) Nature of glass transition and the glassy state. J Chem Phys 28:373–383
103. Adam G, Gibbs JH (1965) On the temperature dependence of cooperative relaxation properties in glass-forming liquids. J Appl Phys 43:139–146
104. Goldstein M (1976) Viscous liquids and glass transition. Sources of the excess specific heat of the liquid. J Chem Phys 64:4767–4767; Goldstein M (1977) Viscous liquids and glass transition. Molecular mechanism for a thermodynamic second order transition. J Chem Phys 67:2246–2253
105. Johari GP (1985) Low frequency molecular relaxations in disordered solids. J Chim Phys 82:283–291
106. Prevosto D, Capaccioli S, Lucchesi M, Rolla PA, Ngai KL (2009) Does the entropy and volume dependence of the structural α-relaxation originate from the Johari–Goldstein β-relaxation? In: Wondraczek L, Conradt R (eds) Glass and entropy, special issue of J Noncryst Solids 355:705–711

# Chapter 4
# Some Aspects of Vitrification, Amorphisation and Disordering and the Generated Extent of Nano-Crystallinity

Jaroslav Šesták, Carlos A. Queiroz, Jiří J. Mareš, and Miroslav Holeček

## 4.1 Introduction: Some Particularities of Natural and Artificial Disordered Materials

Historically, glass is often viewed as a remarkable translucent substance though usually made from the simplest raw materials upon the effect of firing. Nature, itself, is the best instance to learn how different temperature changes can provide various glassy states, altering from very slow rates occurring within geological time scales (e.g. obsidians – glassy volcanic rocks consisting of natural acidic silicate glasses) to extremely fast, occurring as a result of fast energetically driven collapse (e.g., by impact of meteorites, yielding melted droplets then cooled to various tektites). Mimicking evolution, however, man became responsible for the creation of further families of a wide variety of glassy and amorphous materials (geopolymers), which have gradually appeared through human creativity, particularly during last 100 years. Properly chosen procedure of rapid extraction of heat (often called *quenching*) turned up to be a efficient route for successful glass-formation (*vitrification*, as a repeatable process) of almost all substances (thus allowing for the preparation of glasses from different sorts of various inorganic materials, including metals) in contrast to the traditional chemical approach, seeking just for an appropriate composition to vitrify under a customary self-cooling procedure [1, 2].

J. Šesták (✉) and M. Holeček
New Technologies Research Centre, University of West Bohemia, Univerzitní 8, CZ-30614 Plzeň, Czech Republic
e-mail: sestak@fzu.cz

C.A. Queiroz
VICARTE, Faculty of Sciences and Technology, New University of Lisbon, P-2829-516 Caparica, Portugal

J.J. Mareš
Institute of Physics of the Academy of Sciences ČR, v.v.i., Division of Solid-State Physics, Cukrovarnická 10, CZ-16200 Praha, Czech Republic

J. Šesták et al. (eds.), *Glassy, Amorphous and Nano-Crystalline Materials*, Hot Topics in Thermal Analysis and Calorimetry 8, DOI 10.1007/978-90-481-2882-2_4,

The first natural glasses were formed as the Earth cooled and, therefore, predate creation of living organisms by about 1.5 billion years. Such primordial glasses were limited in composition and versatility just as were the first primitive unicellular organisms. However, some compositions (in an unstable state of glass) survived unchanged for a enormously long period of time, similarly to certain strains of bacteria. A still open intellectual terrain ahead is apparent in each of the material classes associated with real bodies of living organisms, being either promoted naturally or created synthetically – particularly interesting are *glasses* when assuming its rather curious but apparent degree of similarity with certain *biopolymers* (carrying out genetic information) [3]. Both are in a reinforced (thermodynamically unstable) state formed due to kinetic reasons, which remain influential until glasses transform into a stable structure (a melt of the lowest free energy) while biopolymers (occurring in living organisms) stop their role in life by death. Both glasses and biopolymers adapt to their environment and previous history. There is an almost limitless range of compositions of glasses and countless numbers of different living species, each family having particular molecular building blocks and severe structural rules, although some compositional restrictions of glasses were gradually overcome. Although the origin of glass is nevertheless lost in prehistorically obscurity, all imaginable glasses have been increasingly artificially synthesized, making thus the natural and man-made glasses indistinguishable. Another intriguing parallel can be drawn in their spontaneously apparent emergence of new *phyla* or groups of species involving both the disordered state of glasses and highly ordered situation of biopolymers as well as they are often highly non-equilibrated status. It may involve our better understanding of possible convergence of glass and genes assuming known bioactivity of certain inorganic glasses, particularly their silanol-rich (Si–OH) active surface capable of binding to living tissues owing to their specific composition, porosity, self-similarity, fractality and thermal pretreatments.

Various other silicate glasses were created naturally [4] by accidental melting due to, e.g., the charge of lightening (‘*lechatelierite*’). Worth of noting are ‘*obsidians*’ (glassy volcanic rocks consisting of natural acidic silicate glasses exhibiting a high melting point) and ‘*pumice*’ (fumed volcanic glass), which were formed by terrestrial upshot of volcanism and have attracted the men’s attention since prehistoric times (as they were used as cutting tools, amulets or cult objects). Interesting enough is that the extent of natural glass on earth is in the range of a tenth of a percent with the ratio of about 4,000 minerals versus few types of natural glasses while on the moon (and possibly also other planets) it is possible to identify only ~60 minerals where half of them are glasses. The frequency of glass deposits is at least one order of magnitude higher than that found on the Earth.

The melt-cooling processes, traditionally seen as the solely responsible for the formation of vitreous states, are not the only methods which generate glassiness. Other non-crystalline materials are various alumina-silicates (based on the mixture of meta-kaolin with an alkaline activator, producing thus hardening pastes), which are frequently termed either as alkaline inorganic polymers, geo-polymers or as hydro-ceramics; exhibiting, among other interesting properties, amalgamated qualities, intermediate between those peculiar to cements with that characteristic of glass-ceramics and zeolites.

Herewith we can specifically recognize various manners of disordering a regular network, which thus ensues *disordered states* of various cause and portrayal, often in low-dimensional systems (as thin layers), and including the particularity of quantum dots or wells, which, however, falls beyond the framework of this brief introduction. These processes are principally of macroscopic origin, bearing dynamic perspectives, although they can be related with the microscopic structure of matter owing to adequate theoretical models. Quite generally, one may state that non-crystalline states can be achieved by certain (seemingly contradictory) methods:

1. By maintaining already existing disorder in the liquid phase when applying fast heat withdrawal (such as a drastic freeze-in by melt radical undercooling), this way preventing phase (ordering) transition
2. By vapour (chemical/physical) deposition of substance from their gaseous state in order to form a disordered thin-film over a solid (cooled) substrate, such as hydrogenated amorphous silicon, a-Si:H, deposited by plasma-enhanced chemical vapour deposition, PE-CVD, etc. [1, 5–7], thus avoiding structural ordering of the deposited phase
3. By implementing disorder on a previously ordered network upon energetically driven structural disintegrative collapse, e.g., by intensive milling, particle bombardment, or by pressure induced amorphization
4. By creation of a disordered structure through variously accomplished chemical reactions including low temperature diagenetic and biotic processes, such as by sol–gel or trough aqueous amorphization response creating, e.g., natural hyalites and opals.

Hereafter, we would like to pay attention to the first category of non-crystalline materials only, which involves the particularity of traditional technology of vitrification of silicate melts often upon self-cooling as well as the unconventional technique of metallic glasses by particularly designed equipment for their quenching [1, 2, 5].

## 4.2 Some Structural Features of Non-crystallinity and Nano-crystallinity

The basic reason for the failure of a liquid to crystallize can be attributed to the problem in the molecular arrangement, not so much in the liquid state (clustering) but predominantly in the formed crystalline state during the process of ordering, i.e.; the higher the crystalline symmetry of the phase to appear, the better the glass-forming ability of its parent melt [5–7]. Furthermore, the probability of glass-formation increases with the formation of liquid mixtures in which the Gibbs energy of the non-crystalline arrangement decreases, while that for corresponding crystalline phase to precipitate remains unchanged. The stronger the interaction among the mixture components, the deeper the depression of the freezing point of the solvent; while the viscosity increases, thus diminishing any possible nucleation. The component interaction, however, should not become too strong as to generate a new competing crystalline phase which would strolls toward decreased vitrification.

A noteworthy alternative element crucial for non-crystalline randomness is the variation of bond angles sometimes assumed to become central in further distinguishing special (constrained) states of glassy and amorphous materials. The flexibility of covalent bond is largest for the two-fold coordination of group VI-elements and is lowest for the tetrahedrally coordinated group IV-elements. For instance, in the $SiO_2$ glasses the oxygen atoms are bridging the Si-tetrahedral providing the essential flexibility, which is considered necessary to form a random covalent network (without exhibiting excess of strain). However, if such a network is formed without the flexing bridges contribution of the group VI-elements, the structure becomes amorphous (as in the deposited strain-confined films of, e.g., $As_2S_3$), which can exist in many various forms of non-crystalline configurations (often experimentally irreproducible).

The highly constrained nature of the variously obtained amorphous films suggests that defects might not be randomly distributed but, instead, can be predominantly located at internal blocks, voids and as strain-relief interfaces between domains. In contrast to vitrified glasses, the CVD-produced amorphous films can thus exist in many non-crystalline configurational states, the thermal annealing of which can lower its strain energy, but, however, cannot transform the over-constrained amorphous configuration from one atomic layer to another. A drastic atomic rearrangement would need to be enforced to accomplish such an 'unstructured' reconstruction, which would, instead, materialize overlapping by more pertinent as well as unprompted crystallization.

However, a possible interference of the so called 'medium-range order' or 'modulated structures' (often bordering upon nano-crystallinity) becomes particularly common in resolving various non-crystalline materials (pertinent for semiconductors), where the concept of homogeneous random network and its associated defects was most extensively studied. It is closely connected with the fashionable use of the prefix 'nano' (*nano-technology*, *nano-materials*) touching the limits where the ordered and disordered states transpire factually their guaranteed delimitability. The standard observations, based on measurable crystallographic characteristics (XRD diffractograms) can provide data on the amount of crystalline phases (capable to detect the crystalline phase down to about 2% within the glassy matrix, certainly under certain crystal-size detectability). If not assuming here the capability to distinguish a minimum crystallite size (by traditional XRD peak broadening), nor we account for a specialized diffraction measurement at low diffraction angles (aiming to calculate the radial distribution function), we can direct our attention toward the critical amount of crystalline phase in the glassy sample. This issue is yet befitting the crucial question how to relevantly define the boundary between yet 'true glassiness' and already 'nano-crystallinity'. There became known a few proposals, but the generally accepted figure is, for long, the value of $10^{-6}$ vol.% (less commonly $10^{-3}$%), of crystallites to physically exist distributed within the glass matrix, as not yet disturbing its experimental non-crystalline characterization and consequent definition of glassines. The appropriateness of this value is, however, difficult to authorize, persisting its maintenance on basis of acute convenience and reiterated use [2].

By the same argument, glasses exhibiting the so-called 'short-range order' can be viewed to also encompass intrinsic defects (interconnected islands, voids, domains interfaces, etc.) defined as a deviation from the lowest energy bonding arrangement. The extent and size of such structural inhomogeneities are specified within the so called medium-range-order, which is neighbouring the nano-crystalline approach. Throughout freezing of such defected inhomogeneities certain complex processes can subsist, which stages we use to associate with a certain assortment of temperatures. Experimentally, they are determinable all the way down from the states of liquid equilibrium through the non-equilibrium undercooled liquids, down to the constrained state of glasses (below $T_g$).

A needed definition of short-range order imposes that each constituent species (e.g. atoms) fulfils its chemical valence requirement [7] (according to the so-called 8-N rule); however, the stronger heteropolar bonds (e.g. As – S) are favoured over the homopolar bonds (e.g. S – S or As – As). It is worth noting that the constrained bond length survives within ~1% equivalency of the bonds existing in the long-range order on the corresponding crystal. We can distinguish a topological short-range order, in which the structural defects are subject to relaxation and annihilation at annealing temperatures close to $T_g$, which is usually associated with collective movements of constitutional species related to both the atomic or the whole cluster hauling. On the other hand, there is commonly a compositional short-range ordering, in which chemically similar species (such as Ni, Fe or Co in metallic glasses) can mutually exchange their atomic positions, often resulting in re-arrangement of the nearest neighbours, thus ensuing magnetic anisotropy, etc. Finally, there can happen a diffusional-driven ordering process, connected with mutual survival and disappearance of various structural defects, often shaped when the samples are inflicted to great temperature gradients (e.g. quenching), which subsequently yields various structural adjustments due to relaxation of mechanical stresses and deformations.

## 4.3 Specificity of Non-crystalline Chalcogenide Materials

Structure modelling has been proven particularly useful in the case of chalcogenide glasses. However, their representation is difficult (or better too-all-embracing) due to the great flexibility in the range of the structural units forming their random network. The general concept of a homogeneously accidental and covalent network has often been explored and re-examined [6]. Nevertheless, it is difficult to encompass all differences flanked by various characters of non-and-nano crystalline chalcogenides, which incongruity is due to the bonding constrains and to the number of degrees of freedom allowed by their three-dimensional network, arranged towards the flexibility required accommodating the given random mismatch [7, 8]. For a binary $A_xB_{x-1}$ exhibiting only a short range order, the average coordination number is $m = x\,N_{cn}(A) + (1 - x)\,N_{cn}(B)$ where the number of constrains ($N_{co}$) per atom (A and B) is given by $N_{co}(m) = m/2 + m(m - 1)/2 \cong m^2/2$. When equating the number of constrains to the ideal three spatial degrees of freedom, Phillips [8] showed that a most favourable average coordination is $m_c = \sqrt{6} = 2.45$ corresponding to an optimal connectivity. As a consequence, the glass-forming tendency occurs greatest when the short-range order imposed by bond stretching and bending forces is just sufficient to exhaust the local degrees of freedom. The internal strain increases with $m$ while the entropy follows the opposite trend because the non-crystalline state becomes *insufficiently* interconnected ('cross-linked'). Therefore, the conventional state of chalcogenide glasses is typically restricted to lie in the region ~$3.3 > m > 2$; while with $m > 3.3$ becomes *over-constrained* amorphous; yet higher, with $m > 4.3$, associates with unusual non-crystalline metals. On the other hand, those having the lower connectivity ($m < 2$) are assumed to be under-cross-linked amorphous materials, such as thin films of inert gases or halides. As any other possible criteria, this one suffers from uncertainty around the borderlines ($m \cong 2$ or $m \cong 3.3$), since the long-range interactions, ionicity, size effects, etc., have been neglected, as well as its applicability to the class of quenched tetrahedral glasses (e.g., $A^{II}B_2{}^{V}$ like $CdAs_2$ and their derivatives with admixtures of Ge, P, etc.).

Even glasses with the average coordination $m = m_c$ contain a considerable strain, which magnitude can be accessed from the difference in heats of solution, such as, that for vitreous silica ($m = 2.67$) 2.16 kcal/mol, while for crystalline α-christobalite one obtains 1.52 kcal/mol. The flexibility of covalent bond is largest for the two-fold coordination groups of VI-elements and is lowest for the tetrahedrally coordinated groups of IV-elements. For instance, in the $SiO_2$ glasses the oxygen atoms are bridging the Si-tetrahedral providing the essential flexibility, which is considered necessary to form a random covalent network. However, if such a covalent random network is formed without the flexing bridges the structure tends to become amorphous (as the deposited strain-confined films of, e.g., $As_2S_3$), which can exist in many various forms of non-crystalline configurations (often experimentally irreproducible). Structural ordering associated with a medium-range order (i.e., sizable nano-crystalline heterogeneity) is known to exist in many deposited-like structures, such as hydrogen-rich interconnecting tissue of a thin film deposited by glow-charge over a-Si:H alloys [9], or domains of ~600 Å found at evaporated thick films of $As_2Se_3$, or smaller domains existing at most amorphous structures based on $As_2S_3$ or $GaSe_2$. It is known that such chalcogenide melts variously thermally annealed exhibit already various stages of melt structural ordering (clusters, nano-crystalline regions, etc.).

## 4.4 Pressure and Temperature-Induced Amorphization

Worth special noting is the transformation of a crystalline-to-amorphous phase, which can be achieved without going through the conventional steps of bulk-liquid melting-and-quenching. Such methods are envisaging the micro-processes involved, where, e.g., powdered particles could melt in their neighbourhood thanks to high temperatures locally achieved due to the plastic deformation. Such processes are hauling a lot of energy and, moreover, may include many other processes of interstitial incorporation and segregation into non-stoichiometric sites, defective links, etc. Such a process of amorphization can result from chemical, irradiative, thermal or pressure-induced disruption of the crystalline order, when the free energy of the crystal obviously exceeds that of an allied amorphous phase [10–12]. Compressive amorphization was first discovered already 25 years ago when ice-crystals were observed to amorphize at 77 K and 1 GPa [10, 13]. There are numerous examples of amorphization of originally genuine well-crystallized materials, which is inducible by a variety of processes, ranging from bombardment with high energy particles, spray- and freeze-drying, dehydration of hydrates (common in pharmaceuticals [14]), to continuous grinding and milling (grains miniaturization below perceptible nano-crystallinity), cold-rolling (common with metals' tryout [12, 15–17]), and experimentally recognized and thus applicable to an inquisitive assortment of materials [e.g. 18].

Rather different routes to non-crystallinity can be best monitored in the ring statistics of silica, which gets recovered after amorphization. These cases involve the breaking and reforming of bonds and its reconstruction. Usually, pressure induced amorphization signifies the existence of a metastable amorphous phase with a lower free energy compared to the parent crystal and that transformation into a more stable high-pressure crystalline phase is kinetically hindered [18, 19]. This metastable amorphous phase may correspond to high-pressure thermodynamic melting of the parent crystal phase at a significantly low temperature, where the Clapeyron slope for the melting transition ($dP/dT$) is negative. For instance, the metastable extension of the melting curve of $SiO_2$ to higher pressures [19] yields reasonable estimates of the amorphization pressure of quartz, which occurs around 30 GPa at ambient temperature [6]. Some correlations indicate that the denser the starting ambient for the crystalline phase, the higher the mechanical work needed to complete the route to obtain a denser glass.

Worth of attention is the functional model by Ponyatovsky and Barkolov [20], where crystalline instability under thermobaric stress (shown by perturbation of $P$ and/or $T$) is attributed to the coexistence of a low-density amorphous phase (= LDA) and a high-density amorphous phase (= HDA). The LDA and the HAD phases represent polymorphs with the same composition as the crystal but different density. Thermobaric amorphization was experimentally studied in zeolites [11, 21, 22], which readily convert to amorphous alumina-silicates under heating and/or at modest pressures. In the case of a zeolite collapse, order–disorder transitions are found dependent on the rate at which the thermobaric stress is applied [11]. Whether amorphization is thermally induced isobarically or pressure-induced isothermally, the low-density crystalline structure should convert to a final high-density HDA phase via a low density LDA phase. The zeolite–LDA transition is expected to be of the displacive type, while the LDA-HDA transition is reconstructive.

Another fascinating finding is the noticeable fact that the temperature of isobaric amorphization, $T_A$, shows a significant depression as the heating rate is reduced. This type of kinetic behaviour seems reminiscent of the comparable trend followed by the glass transition temperature, $T_g$, on varying the cooling rate. The relatively large scale of the changes in $T_A$, is though to suggest that this effect might be controlled by the viscosity of a very strong liquid. The idea of a perfect glass, with zero configurational entropy, $S_c$, was first discussed by Kauzmann [24], in the context of melt-quenching, and coincides in Adam–Gibbs theory [25] and with the discontinuity at the Vogel temperature ($T_v$) in the Vogel–Fulcher–Tammann–Hesse (VFTH) three-parameter viscosity relation:

$$\eta(T) = \eta_\circ \exp\left(\frac{B}{T - T_v}\right) \tag{4.1}$$

where $T_v > 0$ K, $B$ and $\eta_o$ (pre-exponential factor) are fitting parameters, being $T_v$ usually determined from adjusted experimental viscosity data reported for $T > T_g > T_v$.

An interesting outcome of the comparisons between the dynamics of zeolite collapse, and that of structural relaxation in conventional melts, is that below $T_g$, i.e., for very slow heating or for low pressure-increase rates, respectively for the isobaric and for the isothermal amorphization processes, zeolites are expected to collapse more rapidly than the final HDA phase can respond. In either case, highly ordered glasses with very low configurational entropy, $S_c$, might be obtained if the LDA phase could be isolated [11]. Indeed, Angell and co-workers [23] have speculated that the ideal glass state might be better approached via such a phase transition involving a kind of 'superstrong' liquid. If the LDA phase is chemically ordered, then the network will comprise only even membered rings, and this compares with the chemically disordered HDA phase where odd membered rings are required topologically to accommodate bonding contacts (as Al – Al or Si – Si). Zeolites and glasses amorphized from them show a clear relationship between the size of secondary building units and features in the Boson peak, i.e., the Boson peak, accessed in the low-frequency range of vibrational spectroscopy, is hugely enhanced in low-density networks [22]. If amorphization can ever bestow a practical route to the state of an "ideal" glass accomplished directly from the crystalline state, rather than via freeze-in from the liquid state, then the idea of Kauzmann glass [24, 25] might be better realized, though the specificity of such an ideal glassy state as recently have been argued [26].

## 4.5 Vitrification Versus Crystallization: Reduced $T_g$ and the Hrubý Parameter

For a quenched glass, the energy radial distribution function (often investigated by X-ray diffraction) depends on a general configurational coordination, showing a minimum, which bottom lies higher than that of the minimum of a corresponding crystalline state because of the inherent strain energies. The barrier between the two main minima represents the kinetic hindrance preventing crystallization of the glass below $T_g$, which surmounting is only possible by a substantial atomic rearrangement. The amorphous deposited solids, in contrast, can exist in many near-crystalline configurationally coordinations exhibiting thus multiple (but smaller) energy distribution of minima. Over-constrained state can be lowered by suitable thermal treatment (e.g. common annealing, which supports compacting the structure thus lowering and even overlapping the minima). In some cases, the mere temperature increase may yield a catalyst-like pseudo-crystallization to occur yet below $T_g$ [1, 27, 28, 31].

Šesták [28] revealed that quenched glasses disclose upon heating a regular glass transformation, $T_g$, which is a noticeable step-wise deviation from the base line measured by, e.g., differential thermal analysis (DTA). Upon further temperature increase $T_g$ is followed by crystallization as a base-line reversal deviation – peak (Fig. 4.1). On contrary, in amorphous materials the $T_g$ effect is often overlapped by a too early crystallization, the last being shown by non-matching onset and outset of the DTA peak base lines. Worth repeating is that rapidly quenched metallic glasses does not habitually expose distinguishable thermal changes, which can be noticeably associable with the entire glass transition.

Well before the development of any generalized nucleation theory for condensed systems, Tammann [29, 30] already called earlier attention to a tendency revealing that the higher the melt viscosity at the melting temperature ($T_m$), the lower its crystallizability. Qualitatively, this tendency can be explained by an increased inhibition of motion or molecular rearrangement of the basic units of any melt with increasing viscosity. The quality of the glassy state is possible to quantify by examining the reduced quantities [5] such as typically glass transition temperature ($T_{gr}$) given by the ratio $T_g/T_m$; a lower ratio reveals a higher difference of $T_g$ and $T_m$ thus showing a greater stability of the glassy state. Referring to a large set of available experimental data obtained for nucleation of several silicate glasses, Zanotto [32–34] concluded that glasses having $T_{gr}$ higher than ~0.58–0.60 display only surface (mostly heterogeneous) crystallization, while glasses showing volume (homogeneous) nucleation have $T_{gr} < 0.58 - 0.60$. The reduced glass transition temperature was examined in more details by Sakka and Mackenzie [36], and for metallic glasses by Davis [37]. Determination of theoretical values of reduced temperatures were approached by Angell [25, 35] based on extension and extrapolation by means of application of the VFTH viscosity equation.

Glass forming ability is noticeably related to the ease of the reverse process of devitrification possibly evaluable on basis of the difference between crystallization

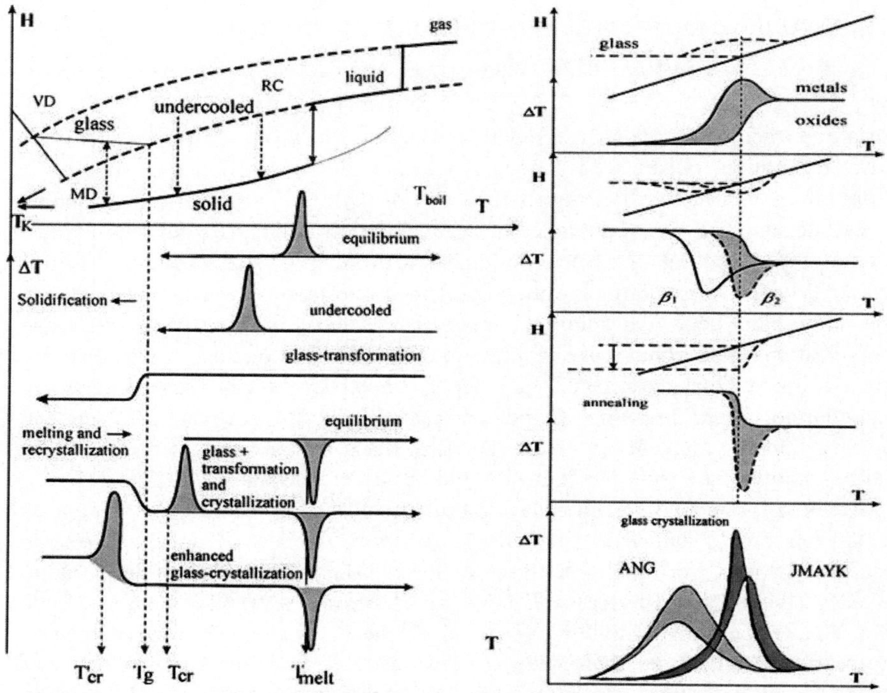


**Fig. 4.1** *Left*: Plot of enthalpy, $H$, versus temperature, $T$ (*upper*), which is accessible and thus characteristic as an experimentally obtainable DTA/DSC curve (factually representing $dH/dT \approx \Delta T$). Solid line indicates regions of stable gaseous, liquid and solid phases; dashed lines designate regions of metastability (*left*: undercooling; *upper*: superheating). Thin line displays the unstable territory of constrained state of glass. Equilibration and rapid liquid cooling (RC) can result to either equilibrium or delayed crystallization (exothermic peak at $T'_{cr}$), or to non-equilibrium vitrification through the step-wise glass transition ($T_g$). Under consequent reheating DTA/DSC shows exothermic crystallization ($T_{cr}$) below the melting temperature ($T_m$). These points are serviceable in the determination of some material's characteristics, such as reduced temperatures ($T_g/T_m$), or Hrubý coefficient: $K_H = (T_{cr} - T_g)/(T_m - T_g)$. A special meaning has the point $T_K$ (often called Kauzman temperature), estimated by intersection of extrapolated line for the equilibrium liquid with that of equilibrium solid (obtainable at the extrapolation limit of a fictitious null cooling rate). The distinctive amorphous phases (obtained, e.g., by vapour deposition, VD; or by mechanical disintegration of the crystalline state, MD) can behave differently because their enhanced reactive state is capable to promote (or catalyze) too early crystallization, thus overlapping with glass transition (an often neglected or experimentally misinterpreted effect). *Right*: typical traces depicted to illustrate a possible span of the glass transformation region distinctive for various dissimilar materials (oxides/non-oxides) and/or for different heating rates ($\beta_2 > \beta_1$). If additional temperature treatment occurs, such glass gets pseudo-equilibrated (vertical shift of its glassy state: *horizontal dashed line*) and such a process is called thermal annealing. The DTA/DSC peaks responsible for glass crystallization (*exotherms bottom left*) can be of two-fold character (*bottom right*) given by the simultaneous and/or consequent amalgamation of elementary processes involved in either the nucleation-and-growth (JMAYK) or the normal-grain-growth (ANG) kinetics

temperature $T_{cr}$ (exothermal maximum) and the glass transition temperature $T_g$. Various examples reveal that this difference $T_{cr} - T_g$ varies repeatedly with composition reaching its maximum value in a composition range which appears to provide best glasses. In order to make possible comparisons with glasses showing different $T_g$ some authors weighted this difference $T_{cr} - T_g$ by the reciprocal value of $T_g$ leading thus the dimensionless factor $(T_{cr} - T_g)/T_g$. Even a more sensitive interrelation to the glass formation peculiarities can be found on basis of Hrubý coefficient, beforehand developed mostly for chalcogenide systems [38], which, however, is typically available only upon physical preparation of a given type of glass: $K_H = (T_{cr} - T_g)/(T_m - T_{cr})$, where $T_{cr}$ is the crystallization temperature and $T_m$ is the melting temperature. This criterion was thoroughly analysed [39–41] having almost the same meaning as the difference ($T_{cr}$ – $T_g$) alone varying, however, more rapidly when crystallization peak is shifted and taking into account melting temperature, which may not be too significant advantage as $T_g$ and $T_m$ are usually correlated. In order to make the criterion more sensitive, Poulain [42] took into consideration the width of a DTA/DSC peak, i.e., the difference between the onset of crystallization, $T_x$ and its maximal value, $T_{cr}$, accounting, among other features, that more stabile glasses exhibit broader isotherms and thus a larger crystallization time and therefore a smaller crystallization rate. The new criterion $(T_{cr} - T_x)\cdot(T_{cr} - T_g)/T_g$ owns units of [K] but able to be encoded dimensionless if weighted by ${T_g}^2$. It is clear that the evaluation move a step forward to further sophistication providing more refined criteria for the use of DTA/DSC traces toward the resolution of glass-formation region, which is the subject of recent analysis [31, 35, 45–48].

It is worth noting, however, that the kind of crystallization kinetics may essentially change the curve shape and thus its testifying character [43, 44]. For example, see the differential scanning calorimetry (DSC) curves obtained for two Fe-based metallic alloys of comparable composition and which are depicted for portrayal in Fig. 4.1 (right column, bottom). The quenched alloy (Fe75Si15B10, with relative atomic composition at.%) was produce as the fully amorphous ribbons and examined twofold: as-quenched and isothermally annealed, indicating a sharper (so called '*Johnson–Mehl–Avrami–Yerofeev–Kolmogorov*' JMAYK-type of nucleation-growth) crystallization curve [1, 43], which upon annealing shifts the peak position (as depicted in Fig. 4.1: right column, bottom). For the analogous (but now already nanocrystalline) ribbon of the comparable composition (Fe74Cu1Nb3Si13B9 at.% again – notice a difference in a small contribution of Cu and Nb), a broader DSC peak is obtained (indicating thus a different, so-called '*Atkinson-normal-growth*' ANG-type crystallization kinetics), as depicted in Fig. 4.1 (bottom right), associated with a nanocrystalline grain-structure formation [49] (often called FINEMET-type alloys), which thermal annealing restrains that changes in the peak size, only.

Various connotations and alternative forms of various reduced quantities were analysed in detail by Šesták [5, 43]. Several models and limitations towards clarification of various features of vitrification were extensively described [25, 26, 45–49], just mentioning Parthasarathy et al. [50], who presented glass transition as a transition from ergodic to nonergodic behaviour and discussed the meaning of residual entropy.

## 4.6 Practical Aspects of Freeze-In Process Mostly Applied to Metallic Alloys

For a general material outlook there subsists a widespread acquaintance with the measurable reality that near the melting temperature, $T_m$, the melts of most ready-to-crystallize solids possess viscosities around 0.1 Pa·s (1 P), while, for comparison, the anomalous water has that of mere $10^{-3}$ Pa·s (1 cP) [7]. It is worth to remark that metallic glasses do not show this characteristic reduction on viscosity; their melts remain very fluid, which requests the application of extremely high quenching rates ($> 10^6$ $Ks^{-1}$) in order to ever produce a noncrystalline solid. However, for the alloy melt compositions in the neighbourhood of deep eutectics, such as 80% of metals (Fe, Ni, Co, Mo, Cu, Cr) combined with 20% of glass-forming metalloids (B, Si, P, C, Ge) the viscosities stand for a moderate melt quenching in order to produce technologically beneficial glasses (often in quest for specific magnetic properties), but a clear glass transition is seldom observed.

Increased quenching rates can diminish the clustering size for nano-crystals, while increasing the chemical inhomogeneity, super-saturation and ease the formation of glassy state, which is usually achievable in such systems. The components should possess reciprocal low solubility, the highest difference between their melting points (of alloy and the pure components), a suitable ratio between their atomic radii and also strong mutual interactions. An empirical relation in the form of a power law holds for the dependence of a suitable characteristic distance ($d$) of clustered dendrites on the local solidification time, $\tau$, or on the average cooling rate, $\boldsymbol{\phi}$, *i.e.*, $d = a\boldsymbol{\phi}^{-n} = a'\tau^{\ n}$, where $a$, $a'$ and $n$ are characteristic parameters and $d$ might well be taken as the initial side branch spacing of the dendrites [1, 43, 51–53]. The local conditions of solidification to form glasses by melt vitrification gives up only its specific heat, whereas coupled crystallization is accompanied by additional release of the heat of crystallization.

In order to reach a sufficiently high quenching rate, it is mandatory that the heat transfer coefficient (at the boundary between the melt and the cooling substrate) ought to be high enough, and the cooled melt layer sufficiently thin, so that the heat can be adequately removed from the whole volume of the sample away to the coolant substrate in a sufficiently short time. Therefore, the most suitable substrates are those metals possessing higher values for the heat capacity, thermal conductivity, and surface heat transfer coefficient (often reduced in practice due to the oxidation layer over most metal surfaces). The cooling rate is essentially influenced by the heat transfer coefficient ($\lambda$), by the thickness of cooled sample ($w$), and relatively less, by its actual temperature. At the condition of ideal cooling, where we assume infinitely high coefficient of heat transfer, the cooling rate is proportional to $1/w^2$ while for Newtonian cooling controlled by real phase boundary, $\phi$ correlates with $1/w$, only. In practice, we may adopt the power relation $\boldsymbol{\phi} \sim 1/w^n$ (where $1 \leq n \leq 2$ is a experimentally determined exponent). For illustration we may present the example of cooling, which is close to ideal, with the proportionality coefficient, $\lambda = 10^4$ J $m^{-2}$ $K^{-1}$ $s^{-1}$, frequently exhibited for several real materials.

We can guess the minimum glass-forming quenching rates of $10^2$, $10^6$ and $10^{10}$ $Ks^{-1}$ for a critical sample thickness of 10, $5 \times 10^{-2}$ and $10^{-3}$ mm, respectively. The last value is very small indeed, hardly attainable even by intensive *surface laser glazing*, which works with very thin layers and assures intimate contact of the in situ melt with the entire surface of the solid coolant support [53]. It, however, lies within the possible quenching rates assumed for vapour/gas deposition on a solid surface, which can attain values as high as about $10^{10}$ $Ks^{-1}$.

Successful mastering the technology of forced cooling is long time known in metallurgy where the meaning 'rapid' can be taken to imply a short-time interval between initiation and completion of solidification with respect of propagation of the advancing solidification front [53, 54]. It occurs either directly, as a result of coupling between external heat extraction and the transport of latent and specific heat required to advance the solidification front; or indirectly, during the recalescence that follows nucleation of solidification at large undercoolings. Experimental evidence can help to compare various cooling rates approximately, estimated for quenching of the melt of silicates down from the temperature of about 1,200 °C, which can achieve as much as $10^4$ $Ks^{-1}$ when applying either the melt pressing between two moving metallic surfaces (Fig. 4.2), which forms a ribbon with the thickness of about 0.2 mm, or the melt crumbling to drops of about 0.5 mm in diameter and their centrifugation and splashing against metallic walls. Pouring a fibber into a water bath (with a diameter less than 0.5 mm) decreases the cooling rate by more than one order of magnitude (to about $10^3$ $Ks^{-1}$) and that would be similar to the helium fast-flow cooling of thin layers or wires. When dropping droplets of melt (~3 mm in diameter) into oil, a further decrease of cooling rate is experienced, to about $10^2$–$10^3$ $Ks^{-1}$, while melt-bulk self-cooling slows down the

**Fig. 4.2** Some fast-quenching devices formerly used at the Institute of Physics. *Left*: a modified twin-roller quenching device, comprising a thin metallic ribbon belt; the melt is poured and squeezed between the belt and the opposing drum. This adjustment helps to assure prolonged contact of the quenched melt with the belt's coolant surface, thus avoiding supplementary crystallization of the glass once it leaves the contact line between the two opposing revolving drums. *Middle*: assembly for splat quenching; the melt is poured and spreads over the bottom surface just before being quickly squeezed under the falling slanted copper plate. *Right*: device for the 'single roll technique' used for the preparation of metallic ribbons [49]; the melt is continuously poured and casted over the coolant rotating drum

cooling rate to the order of a few $Ks^{-1}$. Any cooling in liquid media is often ineffective, due to the immediate creation of gaseous envelop upon immersing, which obstructs the heat exchange (despite the coolant low temperature – often liquid nitrogen).

Most common is the so called '*twin roll technique*', forming the basis for the continuous production of metallic glass ribbons with homogeneous thickness and compositions, comprising as well smooth surfaces on either side of the ribbon. A relatively short time of effective cooling at the contact line of the rolling wells often causes the temperature of the departing (cooled) ribbons to remain high enough to enable re-crystallization processes to evolve, or otherwise promotes undesirable relaxation. That, together with other disadvantages, such as the difficult synchronization of the outlet velocity of the melt with the tangential speed of both wheels (to about 15–30 $ms^{-1}$), steered preference to the practical exploitation of the single wheel quenching. These difficulties were, however, successfully counteracted by a specially designed set-up of the so-called *continuous belt-running method*, which was developed in the laboratory of the Institute of Physics, as shown in Fig. 4.2.

Of an upmost technological interest is the still lacking proper understanding of the balance between thermal stability against micro-crystallization and nano-crystallization for metallic glasses [5, 43]. For instance, soft-magnetic properties of Fe-based ferromagnetic amorphous alloys might be significantly improved, or even fine-tuned to special requirements, by a properly heat treatment schedule (encompassing a tuned crystallization process), since such properties are known to depend strongly on the nano-crystalline structure. Therefore, the investigation of the complexity of crystallization processes occurring in metallic glasses is important to better understand the mechanisms involved when trying to produce designed micro- and nano-structures aimed to enhance the required magnetic properties. Among metallic glasses, the Fe–Si–B-based alloys produced by melt-spun technique attracted substantial recent investigation owing to their excellent soft magnetic properties (e.g. high magnetic permeability). These are capable to be modified by properly chosen annealing schedules, so that they can be optimized aiming a number of technological applications, as their inclusion in components for magnetic sensors or in information handling devices.

For example, the above mentioned Fe75Si15B10 alloy prepared by single roll quenching exhibits a rather understandable process of crystallization when two exothermic interconnected contributions are attributed to the crystallisation of micro-crystalline α-Fe(Si), $Fe_3Si$ and self-possessed composite (the inner $Fe_3B$ core with a α-Fe(Si) envelope). Further temperature increase is followed by the transformation of metastable boride ($Fe_3B$) to more stable one ($Fe_2B$) and finally to a α-Fe [55, 56]. When such a primary composition is adjusted under comparable quenching by Cu and Nd melt doping, a novel internal order appears what makes difficult to distinguish between *non-* and *nano-*crystalline states. The previous thermally activated crystallization process is thus converted into a more multifaceted make-up mainly controlled by the particles' nano-grain-growth and their interface properties enabling thus to produce more specific magnetic behaviour.

**Acknowledgements** The study was supported by the following scientific foundations: GAČR: P204/11/0964, MŠMT: 1M 06031 and MPO: FR-T11/335.

## References

1. Šesták J (2004) Heat, thermal analysis and society. Nucleus, Hradec Králové; (2005) Science of heat and thermophysical studies: a generalized approach to thermal analysis. Elsevier, Amsterdam
2. Šesták J (1997) Art and horizon of non-equilibrated and disordered states by quenching and the new phase formation. Glass Sci Technol 70C:153; (2000) Miracle of reinforced states of matter: glasses – old and new structures for the third millennium. J Thermal Anal Calor 61:305 (2000)
3. Hench L (1996) Life and death: the ultimate phase transformation. Thermochim Acta 280/281:1; (1997) Glasses and genes: a forecast for the future. Glass Sci Technol 70C:439
4. Bouška V (with contributions by Borovec Z, Cimbálníková A, Kraus I, Lajčáková A, Pačesová M) (1993) Natural glasses. Academia/Prague/Ellis Horwood, London
5. Šesták J (1985) Some thermodynamic aspects of glassy state. Thermochim Acta 95:459–471; (1988) Role of thermal annealing during processing of metallic glasses and resulting magnetic properties. Thermochim Acta 116:427
6. Greaves GN, Davis EA (1974) A continuous random network model with three-fold coordination. Philos Mag 29:1201–1206
7. Fritzche H (1981) What are non-crystalline semiconductors. In: Yonezawa F (ed) Fundamental physics of amorphous semiconductors, Springer series in solid-state sciences, vol 25. Springer, Heidelberg
8. Phillips JC (1979) Topology of covalent non-crystalline solids I: short-range order in chalcogenide alloys. J Non-Cryst Solids 34:153–181
9. Reimert JA, Vaughan RW, Knights J (1980) Proton magnetic resonance spectra of plasma-deposited amorphous Si:H films. Phys Rev Lett 44:193–196
10. Greaves GN, Sen S (2007) Inorganic glasses, glass-forming liquids and amorphizing solids. Adv Phys 56:1–166
11. Greaves GN, Meneau F, Sapelkin A, Colyer LM, Gwynn I, Wade S, Sankar G (2003) The rheology of collapsing zeolites amorphized by temperature and pressure. Nat Mater 2:622–629
12. Sieber H, Wilde G, Perepezko JH (2000) Solid state amorphization by cold-rolling. In: Wood JV, Schultz L, Herlach DM (eds) Materials development and processing – bulk amorphous materials, undercooling and powder metallurgy. Series Euromat 99; vol 8, pp 3–9. Wiley/VCH
13. Zheng W, Jewitt D, Kaiser RI (2008) Amorphization of crystalline water ice. http://arxiv.org/ (2008); submitted to Astrophys J
14. Willart JF, Descamps M (2008) Solid state amorphization of pharmaceuticals. Mol Pharm 5:905–920
15. Perepezko JH, Hamann J, Hebert RJ, Rösner H, Wilde G (2007) Amorphization and devitrification reactions in metallic glassy alloys. Mat Sci Eng A 449/451:84–89
16. Suryanarayana C (2004) Solid-state amorphization. In: Suryanarayana C (ed) Mechanical alloying and milling. Marcel Dekker, New York, pp 269–362
17. Nagase T, Hosokawa T, Umakoshi Y (2006) Solid state amorphization and crystallization in Zr66.7Pd33.3 metallic glass. Intermetallics 14:1027–1032
18. Rao CNR, Govindaraj A, Aiyer HN, Seshadri R (1995) Polymerization and pressure-induced amorphization of C60 and C70. J Phys Chem 99:16814–16816

19. Richet P, Gillet P (1997) Pressure-induced amorphization of minerals; a review. Eur J Mineral 9:907–933
20. Ponyatovsky EG, Barkolov OI (1992) Pressure induced amorphous phases. Mater Sci Rep 8:147–191
21. Ovsyuk NN, Goryainov SV (2006) Amorphous-to-amorphous phase transition in zeolites. JETP Lett 83:109–112; translated from Russian: Pisma v Zh Eksp Teor Fiz 83:138–142
22. Greaves GN, Meneau F, Majérus O, Jones DG, Taylor J (2005) Identifying vibrations that destabilize crystals and characterize the glassy state. Science 308:1299–1302
23. Angell CA, Moynihan CT, Hemmati M (2000) 'Strong' and 'superstrong' liquids, and an approach to the perfect glass state via phase transition. J Non-Cryst Solids 274:319–331
24. Kauzmann W (1948) The nature of the glassy state and the behaviour of liquids at low temperatures. Chem Rev 43:219–256
25. Angell CA (1968) Oxide glasses in light of the "ideal glass" Concept: I, ideal and nonideal transitions, and departures from ideality. J Am Ceram Soc 51:117–124
26. Simon SL, McKenna GB (2009) Experimental evidence against the existence of an ideal glass transition. J Non-cryst Solids 355:672–675
27. Adam G, Gibbs JH (1965) On the temperature dependence of cooperative relaxation properties in glass-forming liquids. J Chem Phys 43:139–146
28. Šesták J (1996) Use of phenomenological enthalpy versus temperature diagram (and its derivative – DTA) for a better understanding of transition processes in glasses. In: Šesták J (ed) Vitrification, transformation and crystallization of glasses, special issue of Thermochim. Acta, vol 280/281. Elsevier, Amsterdam, pp 175–191
29. Tammann G (1897) Über die Grenzen des festen Zustandes. Annalen der Physik und Chemie (Wiedemann), Bd. 298/N.F.Bd. 62:280–299; (1906) Über die Natur der 'flüssigen Kristalle' III. Annalen der Physik 324:421–425
30. Tammann G, Hesse W (1926) Die abhängigkeit der viskosität von der temperatur bei unterkühlten flüssigkeiten. Zeitschr Anorg Allgem Chemie 156:245–257
31. McKenna GB (2008) Glass dynamics: diverging views on glass transition. Nat Phys 4:673–674
32. Zanotto ED (1987) Isothermal and adiabatic nucleation in glass. J Non-Cryst Solids 89:361–370
33. Zanotto ED, Weinberg MC (1989) Trends in homogeneous crystal nucleation in oxide glasses. Phys Chem Glasses 30:186–192
34. Fokin VM, Zanotto ED, Schmelzer JWP (2003) Homogeneous nucleation versus glass transition temperature of silicate glasses. J Noncryst Solids 321:52–65; (2006) Homogeneous crystal nucleation in silicate glasses: a 40 years perspective. J Noncryst Solids 352:2681–2714
35. Angell CA (2008) Glass formation and glass transition in supercooled liquids, with insights from study of related phenomena in crystals. J Non-Cryst Solids 354:4703–4712
36. Sakka S, Mackenzie JD (1971) Relation between apparent glass transition temperature and liquids temperature for inorganic glasses. J Non-Cryst Solids 6:145–162
37. Davies HA (1975) The kinetics of formation of a Au – Ge – Si metallic glass. J Non-Cryst Solids 17:266–272
38. Hrubý A (1972) Evaluation of glass forming tendency by means of DTA. Czech J Phys B 22:1187–1193; (1973) Glass-forming tendency in the $GeS_x$ system. Czech J Phys B 23:1263–1272
39. Thoronburg DD (1974) Evaluation of glass formation tendency from rate dependent thermograms. Mat Res Bull 9:1481–1487; Lu ZP, Liu TC (2003) Glass formation criterion for various glass-forming systems. Phys Rev Lett 91:115505
40. Cabral Jr, AA, Fredericci C, Zanotto ED (1997) A test of the Hrubý parameter to estimate glass-forming ability. J Non-Cryst Solids 219:182–186; Avramov I, Zanotto ED, Prado MO (2003) Glass-forming ability versus stability of silicate glasses: theory. J Non-Cryst Solids 320:9–20

41. Kozmidis-Petrović AF (2010) Theoretical analysis of relative changes of the Hruby, Weinberg, and Lu–Liu glass stability parameters with application on some oxide and chalcogenide glasses. Thermochim Acta 499:54–60
42. Saad M, Poulain M (1987) Glass forming ability criterion. Mater Sci 19/20:11–13; Poulain M (1996) Crystallization in fluoride glasses. Thermochim Acta 280/281:343–251
43. Šesták J (1991) Glasses: phenomenology of non-isothermal glass formation and crystallization. In: Chvoj Z, Šesták J, Tříska A (eds) Kinetic phase diagrams: non-equilibrium phase transitions. Elsevier, Amsterdam, pp 164–277
44. Šesták J (1976) Study of glass-formation and crystallization by DTA. Sklář a keramik 26:307 (in Czech); (1986) Applicability of DTA to the study glass-formation and non-isothermal crystallization of glasses. Thermochim Acta 98:339–358
45. Badrinarayanan P, Zheng W, Li Q, Simon SL (2007) The glass transition temperature versus the fictive temperature. J Non-Cryst Solids 353:2603–2612
46. Hutchinson JM (2009) Determination of the glass transition temperature: Methods correlation and structural heterogeneity. J Therm Anal Calorim 98:579–589
47. Louzguine-Luzgin DV, Inoue A (2009) The outline of glass transition phenomenon derived from the viewpoint of devitrification process. Phys Chem Glasses – Eur J Glass Sci Technol Part B 50:27–30
48. Schröter K (2009) Glass transition of heterogeneous polymeric systems studied by calorimetry. J Therm Anal Calorim 98:591–599
49. Illeková E (2005) Kinetic characterization of nanocrystal formation in metallic glasses. In: Idzikowski B, Švec P, Miglierini M (eds) Properties and applications of nanocrystalline alloys from amorphous precursors, NATO Science Series II: mathematics, physics and chemistry. Kluwer Academic, Dordrecht, vol 184, p 79
50. Parthasarathy R, Rao KJ, Rao CNR (1983) The glass transition: salient facts and models. Chem Soc Rev 12:361–385
51. Šesták J (1984) Thermophysical properties of solids: their measurements and theoretical thermal analysis. In: Svehla G (ed) Thermal analysis, vol. XIII of comprehensive analytical chemistry – Part D. Elsevier, Amsterdam
52. Šesták J (1978) Magnetic properties and glass-formation of doped oxide glasses prepared by various methods of rapid quenching: Part I. Sklář a Keramik 28:321 and Part II. Sklář a Keramik, 28:353, both in Czech
53. Šesták J, Strnad Z (1993) Preparation of fine-crystalline and glassy materials by vitrification and amorphization. In: Šesták J, Strnad Z (eds) Special technologies and modern materials. Academia, Prague, p 176, in Czech
54. Glicksman ME, Voorhees PW (1984) Ostwald ripening and relaxation in dendritic structures. Metall Trans A 15:995–1001
55. Illeková E, Czomorová A, Kuhnast FA, Fiorani JM (1996) Transformation kinetics of the Fe73.5 Cu1Nb3Si13.5B9 ribbons to the nanocrystalline state. Mater Sci Eng A 205:166–179
56. Illeková E, Matko I, Duhaj P, Kuhnast F (1997) The complex characteristics of crystallization of the Fe75Si15B10 glassy ribbon. J Mater Sci 32:4645–4654

# Chapter 5
# Basic Role of Thermal Analysis in Polymer Physics

**Adam L. Danch**

## 5.1 Introduction: Misunderstanding on Thermal Analysis

'*Works in the field of calorimetry were very appreciated in the physics of the nineteenth century but for decades are not a part of physics and belong to the engineering science...*' [1], this sentence is the most nonsensical one which has been ever written about calorimetry. One can easy find that although calorimetry is a somewhat 'primitive' experimental technique, it, however, is the one which does not disturb a sample physical state during the preparation process that is necessary in polymer physics. The more complicated measurement apparatus, the more the system under investigation is disturbed. Moreover, one can easy show that there is a set of experimental techniques of thermal analysis (TA), which, if applied correctly, give us a comprehensive description of the studied system under study. There are only two questions: which methods of TA and how they should be used. Certainly, it is not a problem for an experienced experimentalist who understands the basis of thermodynamics and who is able to apply the basic rules of physics in practice. It is true that some knowledge about the technical aspects of the instrument construction is required. It means that we should hardly work in our laboratories in order to improve our knowledge about the techniques used. Some incidental experiment, performed by technicians, is not sufficient for so called 'theoreticians' who try to involve in experimental comprehension. We will not improve any theory if we do not understand experiments and, likewise, we are unable to interpret measured parameters. There is no sense to 'produce' theories if they are not applicable, if they do not reflect reality. We also know that it is not easy to find an adequate theory which is confirmed totally by an experiment, especially, if we take into account 'many-body systems'. Physics is not able to describe completely (and without approximations) systems which include more

A.L. Danch (✉)
Research and Development Center, STOMIX, 790 65 Žulová, Czech Republic
e-mail: danch@stomix.cz

J. Šesták et al. (eds.), *Glassy, Amorphous and Nano-Crystalline Materials*,
Hot Topics in Thermal Analysis and Calorimetry 8, DOI 10.1007/978-90-481-2882-2_5,

than three bodies. Therefore any theory or formula describing a polymeric system will be only a more or less good approximation reflecting a real situation occurred during an experiment.

In this chapter, it will be emphasised that only those methods of TA are useful in the case of polymer physics, which are applied properly. It is high time we understood thermal analysis as any experimental technique which gives information about physical parameter with temperature changing. The key aspect of the chapter is the relation between supermolecular structure and physical parameters of the polymeric system. Also, structural heterogeneity of the amorphous phase in semi-crystalline and nano-crystalline systems (the three-phase model) will be discussed and new interpretation of the pretty old experimental data obtained for poly (4-methyl-1-pentene) (PMP), as an example of TA proper using, will be presented. Moreover, two very important (if not the most important) physical phenomena: glass transition and α relaxation (structural relaxation) will be discussed on the molecular level using new definition and new formula. Precise carrying out of the calorimetric measurements and the proper interpretation of the calorimetric results are the clue to the riddle of the both physical phenomena. The chapter shows how the methods of TA can give a successful answer toward the fundamental questions of polymer physics. Unfortunately, here, the potential reader attention will be restricted to the chosen experimental methods only, however, the methods are complementary.

## 5.2 Applicability of the Chosen Methods

### *5.2.1 Calorimetry*

'*...Calorimetry means the measurements of heat...*' [2]. We may possibly classify the calorimeters according to various criteria. Owing to the fact that there are a large number of available calorimeters for the scientific, engineering or technological applications, one can find many different classifications in literature [2–4]. The most important feature of the instrument from the physics point of view seems to be the most precise determination of heat effects or a phase situation of studied object in a quantitative manner in order to compare the experimental results with theoretical predictions. Basically, two types of DSCs are widely used for polymers study: heat-flux DSC and power-compensation DSC. The principle of the first measurement system is similar to that of DTA; instead ΔT differences the apparatus is calibrated in terms of heat fluxes and the mathematical description remains equal to that used for DTA [4]. In order not to mislead both types, the heat-flux DSC is named DTA in the following text. Both DTA and DSC are the most generally applicable of all thermal analysis methods and it is not a purpose of this chapter to discuss which one is better and why. It is worth noting, however, that only a proper application of both instruments done by a well educated and experienced researcher exhibits the desired differences between them [4]. In other case the 'myths and legends', advocated by some groups, are repeated credulously. There are numerous problems in polymer physics which still require further investigation. Most of them are strictly related to the proper using of calorimetry which is the only one technique giving evidences on thermodynamics of the polymeric systems.

The DSC/DTA curve may show an exothermic or endothermic peak which are characterised by: position (i.e., start, end, extrapolated onset and peak temperatures); size (related to the amount of material and energy of the reaction or the phase transition); shape (which can be related to the kinetics of the process) [4]. The other feature of the curves is their run, especially, the jump reflecting the step transition occurred in the studied system. The proper result analysis requires the knowledge of the base line for the sample that is ignored very often in many publications. Frequently, physicists and chemists, who are not strictly related to TA business, treat the curves as a supplement and as a 'nice picture'. They apply mathematical tricks for the curve smoothing losing a lot of information about the investigated system often employing the luxury of sophisticated TA instruments capable to execute it alone. Moreover the second heating run is also taken into consideration. The reason is the only one: the curve of the second run looks 'nicer' as compared with the curve of the first run. Such a sample treatment makes that we obtain information about two various systems. The first run informs us about a thermal history of the polymeric sample before the heating run. The second run, which is always done after the cooling run is performed with the some cooling rate, gives merely information about the sample phase situation occurred during the specific cooling run. Nothing is wrong with this methodology when investigators have knowledge about the difference between the first and the second runs. Unfortunately, we can easy find in vast literature a lot of examples when the investigators compare the results obtained by various techniques of TA studying samples of different thermal histories. The results obtained for polyethylene (PE) are presented as an example of various sample treatment, see Fig. 5.1 [5] (samples were thermally modified). We can see that although all of them were prepared from the same piece of polyethylene of low density, each of them exhibited different thermal behaviour, it means different phase situation. It is especially seen in the temperature range of the sample melting. This example shows that there is no sense to compare results obtained for any polymeric samples if their thermal histories differ. Therefore we should not say about thermodynamic parameters of polymer but we should say about thermodynamic parameters of the polymeric system with some thermal history. Any comparison between our results and the results presented in the databases makes no sense if the thermal history of the sample is not precisely describe. Although DSC (DTA) seems to be a suitable method for the quick and cheap structure analysis, we face one serious problem. It is a question of material homogeneity and whether the results obtained for the small part of the material can be extended on the whole final product. However,

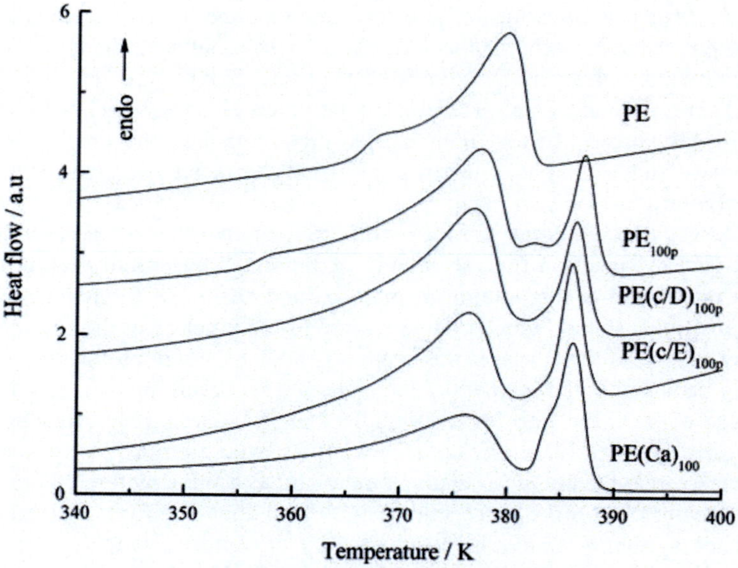


**Fig. 5.1** DSC traces of samples cut out from as produced ($PE_{100p}$), drawn ($PE(c/D)_{100p}$) and electron irradiated ($PE(c/E)_{100p}$) materials annealed at 373 K over 1 h: PE- commercial product; $PE(Ca)_{100}$− composite material annealed at 373 K over 1 h

this problem can be resolved by testing several pieces cut out from different part of the same product. It is possible if we take into account that the weight of one piece can be less than 1 mg for a high quality apparatus. It must be emphasised that we do not describe the supermolecular structure of the polymeric system in this case, at all. For this, DSC is not proper, as everyone has been able to easy find out. We describe the relative changes of the structure of the PE products based on the changes of the thermodynamic parameter and the general knowledge on the possible PE morphology. The morphology description by means of DSC is speculative as much as every description done for other experimental technique. However, for the property description of plastic commercial products, we propose to use very simple and cheap method (it is not time consuming), DSC or DTA, against very expensive one. In conclusion, we should consider the influence of some techniques on the obtained results at last [5].

Owing to the fact mentioned above, we usually obtained a set of glass transition temperatures ($T_g$), specific heat changes at $T_g$ ($\Delta C_p(T_g)$) or melting enthalpies ($\Delta H_m$) for one polymer which formed various phase systems from the physical point of view, where one $T_g$, one $\Delta C_p(T_g)$ and one $\Delta H_m$ reveal only the feature of one thermodynamic system. The situation is more complicated if we take into account the composites. A small amount of additive may perturb the polymeric system drastically. Someone can give the question: 'What is the sense to apply DSC/DTA if the results are not univocal?'. DSC is one of the most sensitive methods which is able to detect even small changes occurred in the investigated system and does not disturb it. When we use DSC/DTA, thermal history of the sample should be kept in mind. There are kinetic and thermodynamic contributions to the heat capacity value in the transition range due to the fact that below $T_g$, the glass is trapped kinetically in one thermodynamic metastable state and above $T_g$, the supercooled liquid may explore all the configurations at equilibrium. Consequently, the heat capacity, entropy (its temperature change is proportional to the measured $\Delta C_p(T_g)$ value) and any other thermodynamic quantity are dependent on the thermal history of the system. Hence, the heat capacity posses a special importance because it is the thermodynamic measure of temperature-induced structural changes in liquid [6].

### 5.2.2 *Thermogravimetry*

Thermogravimetry (TG) is a well-known thermoanalytical technique [4] in polymer research. The mass of a material is continuously registered as a function of time or temperature. Owing to the fact that polymers do not evaporate, the change of mass provides information about low-mass molecular substances removal that may be related to either the thermal degradation of the sample or some substance (solvent, catalyst, initiator, monomer, etc.) evaporation. Not only the component variety but also the structure heterogeneity can be detected by means of TG [7, 8]. Although the material is homogenous from the chemical point of view, 'different' kind of molecules can be found [9–11]. It is a result of physical heterogeneity, it means, the variety of supermolecular structure of the studied system. The structure of the polymeric matrix, which some solvent molecules are occluded in, influenced the diffusion process of the molecules in such a manner that the boiling temperature of solvent might be shifted towards higher temperature. TG is another one method of TA which is able to study phase situation of the polymeric sample. This simple technique may indirectly reveal the supermolecular structure complexity as well as the more advanced one, which very often requires special sample treatment. The ranges of the mass change are precisely determined from the first derivative of TG signals (DTG). A simple fitting procedure (the Gaussian peaks) was proposed for the DTG curve interpretation in order to find the mass loss occurred in each of the range for studied PMP samples. The typical TG and DTG curves are presented in Fig. 5.2. Although the TG signal reflects one step mass loss between 50°C and 250°C, the DTG curve exhibits three processes distinctly [11].

Owing to the fact that two fractions of the amorphous phase of different morphologies and one crystalline phase coexist in one polymeric system, we can attribute the processes to the removal of the solvent occluded in each of the morphologically different areas. This untraditional using of TG led to the description of the supermolecular structures of the polymeric systems. The experimental results were used in a computer simulation that gave quite good agreement with an analytical equation which was obtained from theory [12].

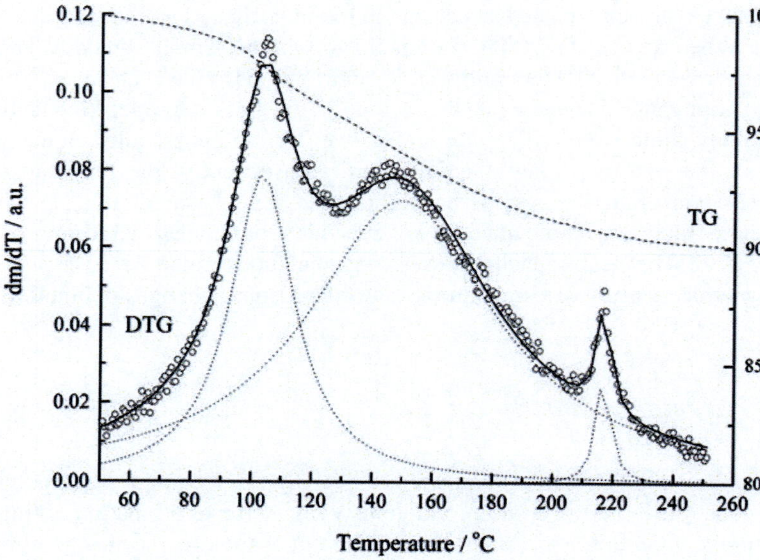


**Fig. 5.2** A chosen example of the - - - - TG and ○ – DTG curves obtained for the studied membranes. The *dotted lines* (Gaussian curves obtained from the fitting procedure of the DTG signal) represent the amounts of the solvent occluded in different areas of the membrane; the *solid line* is a fitting curve

### 5.2.3 Dilatometry and Positron Annihilation Lifetime Spectroscopy

It is well known that every structural change is reflected in measurements of the adequate physical parameters. One can easy find that if we measure specific (dilatometry, DIL) or free (positron annihilation lifetime spectroscopy, PALS) volume, we are able to estimate the glass transition temperature, $T_g$. The model relationship between free ($V_F$), occupied ($V_O$) and specific ($V_S$) volumes has been presented/discussed in handbooks [13, 14] and papers [15, 16]. It is commonly accepted that all of the volumes should increase with rising temperature. The model relationship is presented in Fig. 5.3.

It is very difficult, due to strong anisotropy of polymeric systems, to perform volumetric investigations. Moreover every technique requires some simplicity for the evaluation of the parameters. PALS assumes the spherical shape of the cavities, unoccupied places by the matter, and that the intensity of the measured relative intensity of the longest lived component of positron lifetimes is proportional to the number of the cavities [17]. In this way, indirectly, we receive information about a free volume of the system. The volume increases with temperature rising, changing a slope of the line representing the temperature function, $V_F \rightarrow f(T)$. A similar situation is in the case of the specific volume evaluated from dilatometric measurements. However in this case we conclude indirectly about the thermal properties of

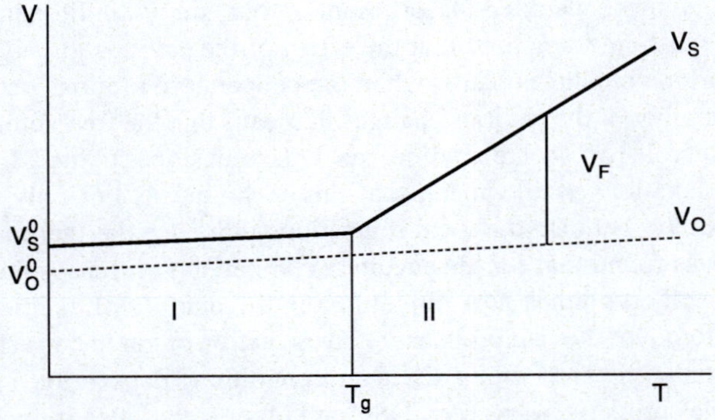


**Fig. 5.3** A general scheme of relation between $V_S$, $V_O$ and $V_F$ in two temperature regions, below (I) and above (II) Tg

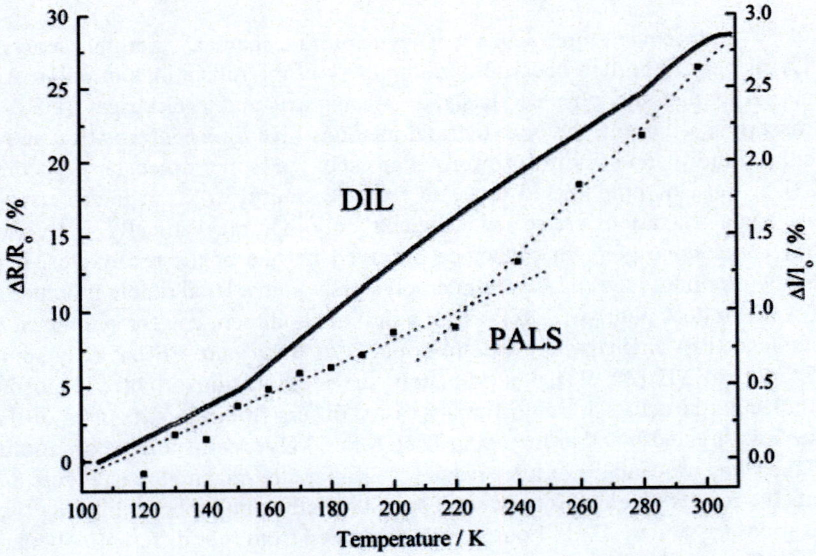


**Fig. 5.4** Relative changes of the parameters obtained from PALS (*left scale*) and DIL (*right scale*). The relative units, in both cases, are applied in order to better show the parameter changes for PALS (R – diameter of the average cavity) and DIL (l – sample elongation). The obtained data were scaled at 100 K for the better results comparison. It means that the relative values are zero for both data at 100 K

the studied system only base on one dimensional test. The simplicities taken in the both methods make the problem in data interpretation. It is very difficult to secure the same sample orientation for the both techniques. Lack of the same orientation may lead to the false conclusions. The selected results obtained from PALS and DIL are presented in Fig. 5.4.

One can find more detailed description of our study in literature [17–19]. However it is worth noticing here that the effect of the cavities joining not only in the glass transition zone takes place, when the cooperative motions occur, but also below *Tg* the number of the cavities changes. It means that the free volume seems to play a significant role for local relaxation, too. Different slopes of the DIL curve after the β and γ relaxations would mean that this technique is not only sensitive to structural relaxation but also the local relaxations influence the thermal expansion coefficient. It was found that for the adequate comparison of the absolute values of the free and specific volumes new model giving the number of cavities should be applied [18]. Moreover the comprehensive investigation ought to be performed with using the samples prepared under the same condition. Especially, one must be careful when the comparison between DSC and other techniques is done [5].

### 5.2.4 Dynamic Mechanical Thermal Analysis

Frequency and temperature dependent dynamic mechanical thermal analysis (DMTA) is very important because of complexity of the relaxations in polymeric systems. Although the physical basis of mechanical and dielectrical (DETA) spectroscopy is different, the both thermal methods give direct information about molecular fluidity and indirect information about a supermolecular structure [20–23]. It must be emphasised that one must be careful in result interpretation because some relaxations are stronger dielectrically than mechanically. Moreover in many cases some of them cannot be observed by one of the techniques. It is especially important for semi-crystalline polymers where a local dipole moment is rather weak or does not exist. DETA is a 'blind' method then, e.g. for symmetrical polyethylene (PE) chains where we are not able to study pure PE but only some modification of PE [20, 21]. Unfortunately the modification might change the supermolecular structure which, in many cases, differs from the structures of the samples investigated by other methods. Therefore DMTA seems to be more useful technique when we want to compare results of different thermal methods. Figure 5.5 presents the results obtained for different semi-crystalline polymers exhibiting two α relaxations ($\alpha_g$ and $\alpha_{c,}$ [23]). Four samples prepared from four different polymers PMP-poly(4-methyl-1-pentene), PA6- polyamide 6, LDPE- polyethylene of low density, PVDF- poly(vinyl diflouride) exhibit various dependence of the $\alpha_c$ relaxation on the degree of sample crystallinity.

## 5.3 Supermolecular Structure of Semi-Crystalline Polymers

The methods presented in Sect. 5.2 should be completed by another one very important method, i.e. X-ray scattering, small (SAXS) or wide (WAXS) angle X-ray scattering. It would not be possible to perform any structural study without this method. Unfortunately, very often, it is impossible to prepare the sample exactly

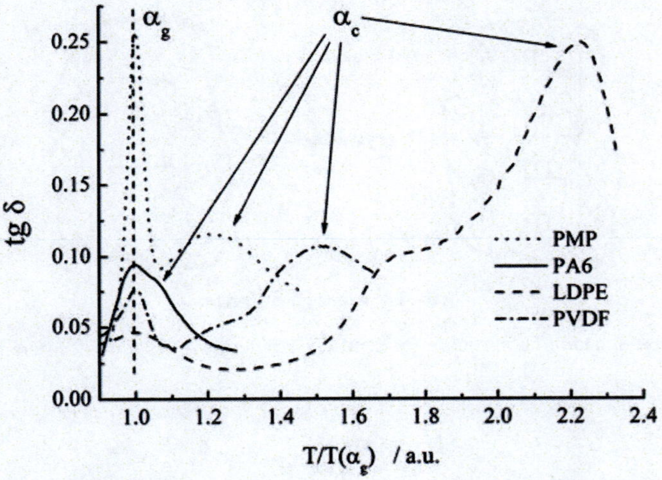


**Fig. 5.5** Tangent delta (tg = E″/E′) as a function of scaled temperature. The scaling was done to the temperature of the $\alpha_g$ relaxation (T[$\alpha_g$]) for the better results comparison

in the same way for WAXS/SAXS and other methods of thermal analysis, especially like for DMTA/DETA or DIL. Therefore this method is only used in a restricted way in a complementary study for polymer. X-ray scattering was applied by us in the cases when the sample could be prepared at least in the same way as a one for DMTA/DETA and PALS/DIL. As was mentioned above, DSC/DTA and TG can be used always that makes them very attractive for experimentator. As we know, only a small piece of sample prepared for other methods must be used. The complementary methods of thermal analysis and comprehensive results gave a general picture of supermolecular structure of semi-crystalline polymers. The structure is presented in a schematic way in Fig. 5.6. The supermolecular structure of semi-crystalline polymers can be described as a system of various phases (amorphous and crystalline) interlocked by chain molecules which traverse the phases. It means that some part of the chain is located in the amorphous region while the other part builds the crystalline domains [24]. It was shown; the domains of long-range order distinctly influence the arrangement of the chains within the amorphous regions [25, 26]. It resulted in the appearance of two amorphous fractions in one polymeric specimen. The fractions were named: 'real' and 'ordered' amorphous phases, RAP and OAP, respectively [27]. The co-existence of the amorphous and crystalline phases resulted in different properties of the 'real' and 'ordered' amorphous phases. It was shown that the 'ordered' amorphous phase could be treated as a 'fingerprint' of the possible crystallisation, when the crystallisation process is not completed due to fast cooling regime. The experimental evidences showed that two fractions of the amorphous phase were created in one polymeric material that is schematically presented in Fig. 5.6. There are two possible scenarios of the morphology depended on the degree

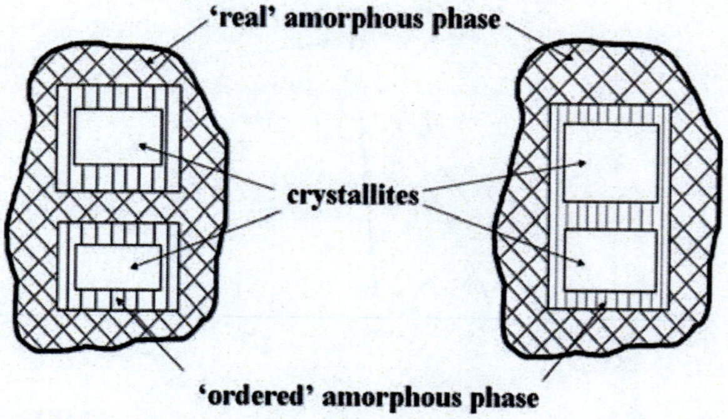


**Fig. 5.6** Phase situation taking place for semi-crystalline polymers with a different degree of crystallisation

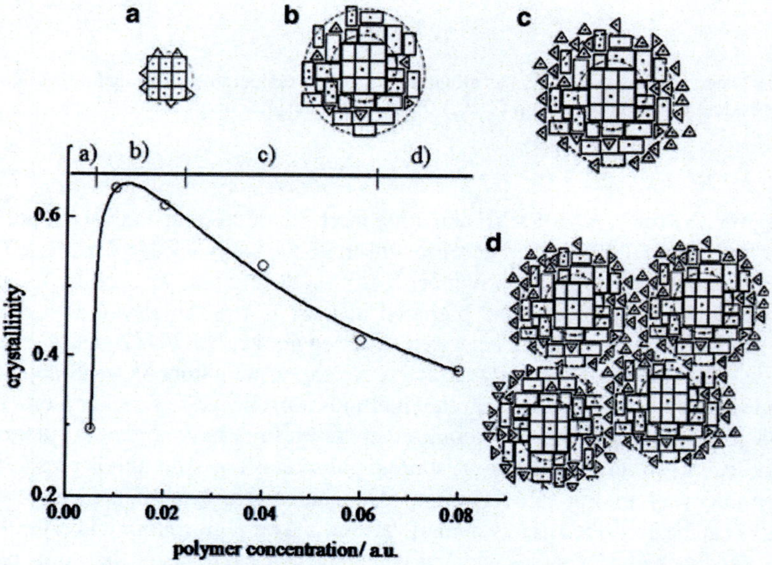


**Fig. 5.7** The stages of the crystallites formation in terms of the polymer concentration in the solution. *Squares*, *rectangles* and *triangles* represent the crystal unit-cells, OAP and RAP, respectively. *Dots* mean that this area is supposed to contain the molecules of the solvent

of crystallinity of the polymeric sample. The left morphology is characteristic for low crystallinity whereas the right one can be found in highly crystallised polymers. The role of the crystal-amorphous regions was already discussed in 1962 [25]. The interphase is necessary for the dissipation of the order existing on the crystal surface because the long chains must traverse the ordered and disordered regions many times creating some kind of network composed of crystalline, amorphous and intermediate regions. Spectacular presentation of the interphase existence in real systems was done for dielectric spectroscopy applied to liquids [28] and polar oligomeric systems (polymers with a low molecular mass) [29]. The studied systems were prepared as a glass (there was no sign of crystallites) in nanoporous media, hence, biphasic nature of the supermolecular structures of the oligomers (only 'real' and 'ordered' fractions) could not be a result of the influence of the arrangement of the chains built the crystal phase on the amorphous regions. However the similar situation was found by us and described in the paper concerning formation of PMP membranes [12]. We showed experimentally (using the methods of TA) and theoretically (crystallisation and lamellae aggregation processes) how the membrane morphology varies with the degree of crystallinity (see Fig. 5.7). Two main topics were addressed: a three-phase model ('real' and 'ordered' amorphous phases, and a crystal phase), and factors affecting the supermolecular structure of the membranes. Special attention was paid to lamellae perfection in the solution of different polymer concentrations and its relation to the membrane morphology. The stages of the crystallisation were proposed mainly in terms of the role played by the solvent molecules.

## 5.4 Structural Relaxations and Glass Transition Temperatures

Semi-crystalline polymers exhibit the morphology of lamellar crystals located in an amorphous matrix. The matrix is not homogenous in a morphological sense. Although some fraction of the chains attains the mobility level of the liquid-like state as temperature is increased above $T_g$, it has been shown, that a portion of the amorphous phase (the interphase) remains still rigid above $T_g$ [24, 27, 30, 31]. The crystalline phase reduces the segmental mobility of the chains located in the interphase. The main consequence of this is that the glass transition of the interphase is extended towards the high temperature and the second α relaxation appears. The correlation between different results obtained by various techniques of TA is presented in Fig. 5.8. Now, $T_g$ is defined as a temperature at which 50% of the mers (segments) take part in the collective motion that is show schematically in the right side of the picture. Such a definition correlates with the results obtained by mechanical and/or dielectrical spectroscopy. The relation between the local (β) and the structural (α) relaxations is shown in the diagram of relaxation frequencies as a function of temperature (log f→f(T)). Also, the temperatures in this diagram correlate with the temperatures at which the adequate number of the mers is activated to the collective motions that further is revealed as a step transition in the DSC curve. It is worth noticing that the shape of the DSC curve is well reproduced by the new formula discussed in our earlier paper [16, 30]. The formula is an alternative for the formulae which readers can find in literature. It is the most important that the parameters included in the formula possess physical interpretation and can be evaluated by independent measurements. This empirical relationship combines the Arrhenius and Vogel–Fulcher equations for the temperature dependence of the relaxation time. It must be emphasised that the formula describes the dynamics of one system, it means, the thermodynamic parameters, taken from the calorimetric

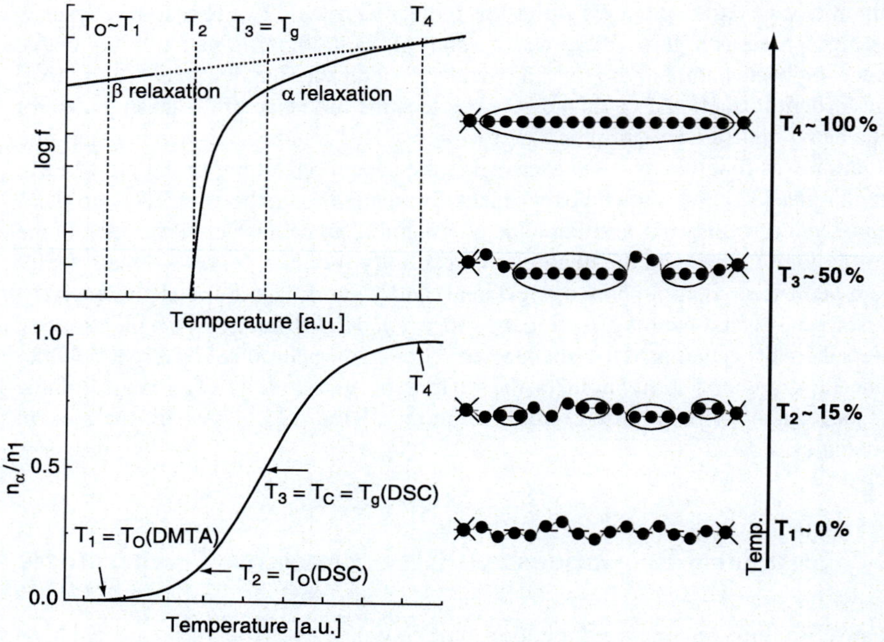


**Fig. 5.8** Comparison of various thermal methods: DMTA/DETA, DSC. For more detailed description see [16, 30, 31]

study, must be exactly pointed out for the system which is the subject of the mechanical or dielectric spectroscopy. The same must be true of the free volume and the X-ray measurements. The new formula of the relaxation time exhibits the existence of some correlation between thermodynamic and structural parameters. Moreover, it distinguishes two fractions of the amorphous phase that gives better description of polymer properties.

## 5.5 Concluding Remarks: 'Myths and Legends'

It is very often called by some authors that PMP is unique due to its room-temperature density of the amorphous phase exceeding that of the crystalline one. Among many papers, these authors have cited the volume-temperature studies [32, 33]. It should be mentioned that the densities of the PMP samples, prepared in highly specific and different ways for various experimental methods, were determined either by the capillary dilatometer or by the X-ray measurements of the lattice parameters. Reading the original papers [32, 33], one can easy find that the conclusion concerning the PMP unusual property was drown indirectly from the experimental data. Griffith et al.

determined the density of the pure amorphous sample, which was prepared under some condition. It is commonly accepted now that the amorphous state, thus its density and the thermodynamic parameters (e.g. $C_p$), strongly depend on thermal sample treatment. It means that the densities of the pure amorphous and, especially, the crystallised samples were not the properties of the amorphous and crystalline phases of PMP, at all. Only the densities of some PMP samples were determined and not the density of the polymer. One can suppose that various sample preparations, however producing only the amorphous samples always, would have resulted in their different densities. Very often in discussion, the density of the sample of 76% crystallinity has been taken by some authors as a density of the crystalline phase. It should be emphasised that the crystallinity in the original paper was obtained from the density measurements taking the two-phase model for determination of the volume of the crystalline phase. Griffith et al. showed only that the density of the PMP samples was lower for the samples with higher crystallinity. However, this tendency was not regular (see Table in original paper). Luck of the regularity in the density results might be explained by the presence of the third phase: 'ordered' amorphous phase, OAP, and the different ratio of both amorphous phases in the sample. It should be mentioned that the three-phase model for polymer had been suggested, without experimental confirmation, in 1962 by Flory [25]. It was experimentally confirmed almost 20 years later although the confirmation was not strictly observed by the structural study, performed next 20 years later. Now, the usefulness of three-phase model is not disputed, especially, that there are a lot of structural studies (NMR, X-ray) giving evidences of the three-phase structure.

The second disputed problem concerning PMP is the high temperature relaxation signed as the $\alpha_c$ relaxation. Some authors have attributed this relaxation to the motion of the chains within the crystalline phase. As a proof, the temperature change of the lattice parameters, measured in X-ray study, is called. However, other interpretation of this relaxation is given [23]. It was proposed that the $\alpha_c$ relaxation occurs due to the motion of the chains formed OAP and the second glass transition at high temperature, related to this relaxation, was defined. Thus X-ray study, presented by Ranby et al. [33], would give the proofs confirming the latter idea. Let us assume that the three-phase model is correct for PMP supermolecular structure. It was proposed [27], and recently showed [34], that the specific volume of OAP exceeds the specific volume of RAP. Firstly, it would explain why the density of the amorphous PMP samples exceeding that of the crystalline one, as was discussed above. Secondly, the origin of the $\alpha_c$ relaxation concerning the motion within OAP would explain the thermal expansion of the lattice parameters. Taking into account the difference in the activation of the large scale motion within RAP and OAP, the change in the slope of the lattice expansion-temperature relationship is understandable. The large scale motion of the chains (cooperative motion) is activated above some characteristic temperature. Owing to the fact that the slope change does not occur in the temperature range of the first glass transition (about 40°C), we can suppose that the crystallites are enclosed by OAP and not by RAP. This would be a proof confirming the three-phase model importance for semi-crystalline polymer.

## References

1. Hołyst R (2007) The review report sent to Inst. Phys. of Silesian University (in Polish)
2. Mathot VBF (ed) (1994) Calorimetry and thermal analysis of polymers. Hanser, Munich
3. Zielenkiewicz W (2000) Pomiary efektów cieplnych (in Polish). Warszawska drukarnia naukowa PAN, Warszawa
4. Šesták J (1982) Měření termofyzikálních vlastností pevných látek (in Czech). Academia, Praha; (1984) Thermophysical Properties of Solids: their measurements and theoretical thermal analysis. Elsevier, Amsterdam; (1988) Teoreticheskij termicheskij analyz (in Russian). Mir, Moskva
5. Danch A, Osoba W (2006) DSC monitoring of supermolecular structure damage of polyethylene products: academia and industry challenges. J Therm Anal Calorim 84:331–337
6. Danch A (2006) The glass transition-finite size effect. J Therm Anal Calorim 84:663–668
7. Saaoudi M, Chassaing E, Cherkaoui M, Ebntouhami M (2002) Hydrogen incorporation in Ni–P films prepared by electroless deposition. J Appl Electrochem 32:1331–1336
8. Markovic N, Ginic-Markovic M, Dutta NK (2003) Mechanism of solvent entrapment within the network scaffolding in organogels: thermodynamic and kinetic investigations. Polym Int 52:1095–1107
9. Ngui MO, Mallapragada SK (1999) Mechanistic investigation of drying regimes during solvent removal from poly(vinyl alcohol) film. J Appl Polym Sci 72:1913–1920
10. Janowska G, Rybiński P (2004) Thermal properties of swollen butadiene-acrylonitrile rubber vulcanizates. J Therm Anal Calorim 78:839–847
11. Wolnik A, Borek J, Sułkowski WW, Żarska M, Zielińska-Danch W, Danch A (2007) Thermogravimetric evidences of supermolecular structure variety of PMP membranes. J Therm Anal Calorim 90:237–242
12. Kruszewska N, Danch A, Zielińska-Danch W, Wieczorek E, Sułkowski W, Gadomski A (2009) Supermolecular structure formation of PMP membranes: theoretical argumentation in terms of the experimental evidences. Mater Sci Eng B 163:105–113
13. van Krevlen DW (1990) Properties of polymers. Elsevier, Amsterdam
14. Ward IM (1971) Mechanical properties of solid polymers. Wiley, London
15. Richet P (2002) Enthalpy, volume and structural relaxation in glass-forming silicate melts. J Therm Anal Calorim 69:739–750
16. Danch A (2003) On the influence of the supermolecular structure on structural relaxation in the glass transition zone: free volume approach. Fibres Text East Eur 11:128–131
17. Danch A, Osoba W (2003) The temperature dependence of free volume in polymethylpentene by positron annihilation. Radiat Phys Chem 68:445–447
18. Danch A, Osoba W (2006) Stability of supermolecular structure below Tg – a role of free and specific volumes in local relaxations. J Therm Anal Calorim 84:79–83
19. Danch A, Osoba W, Wawryszczuk J (2007) Comparison of the influence of low temperature and high pressure on the free volume in polymethylpentene. Radiat Phys Chem 76:150–152
20. McGrum NG, Read BE, Williams A (1967) Anelastic and dielectric effects in polymeric solids. Wiley, London
21. Graff MS, Boyd RH (1994) A dielectric study of molecular relaxation in linear polyetylene. Polymer 35:1797–1801
22. Danch A, Osoba W, Stelzer F (2003) On the alpha relaxation of the constrained amorphous phase in poly(ethylene). Eur Polym J 39:2051–2058
23. Danch A (1998) Dynamic mechanical relaxation in the opaque and transparent PMP films. J Therm Anal 54:151–159
24. Struik LC (1987) The mechanical and physical ageing of semicrystalline polymers: parts 1, 2, 3, 4. Polymer 28:1521–1533; (1987) 28:1534–1542; (1989) 30:799–814; (1989) 30: 815–830
25. Flory JP (1962) On the morphology of the crystalline state in polymers. J Am Chem Soc 84:2857–2867

26. Flory JP, Yoon DY (1978) Molecular morphology in semicrystalline polymers. Nature 272:226–229
27. Danch A (2001) Effect of supermolecular structure changes on the glass transition of polymer. J Therm Anal Calorim 65:525–535
28. Kremer F (2002) Dielectric spectroscopy – yesterday, today and tomorrow. J Non-Cryst Solids 305:1–9
29. Schönhals A, Goering H, Schick C, Frick B, Zorn R (2005) Polymers in nanoconfinement. J Non-Cryst Solids 351:2668–2677
30. Danch A (2005) Thermodynamics and structure of the ordered amorphous phase in polymer. J Therm Anal Calorim 79:205–221
31. Danch A (2008) Some comments on nature of the structural relaxation and glass transition. J Therm Anal Calorim 91:733–736
32. Griffith JH, Ranby BG (1960) Dilatometric measurements on poly(4-methyl-1-pentene) glass and melt temperatures, crystallization rates, an unusual density behavior. J Polym Sci 44:369–381
33. Ranby B, Chan KS, Brumberger H (1962) Higher-order transition in poly(4-methyl-1-pentene). J 3Polym Sci 58:545–552
34. Dlubek G, Gupta AS, Pionteck J, Haessler R, Krause-Rehberg R, Kaspar H, Lochhaas KH (2005) Glass transition and free volume in the mobile and rigid amorphous fractions of semicrystalline PTFE: a positron lifetime and PVT study. Polymer 46:6075–6089

# Chapter 6
# Phases of Amorphous, Crystalline, and Intermediate Order in Microphase and Nanophase Systems

**Bernhard Wunderlich**

## 6.1 Solids, Liquids and Their Transitions

To describe matter, one can use two levels, the microscopic one, which requires identification on a molecular scale, and the macroscopic one, which can make use of the identification of the phase of the sample. The International Union of Pure and Applied Chemistry, IUPAC, has provided a binding scientific definition of the phase [1]. It is to be "an entity of a material system which is uniform in chemical composition and physical state."

Next, one should describe the state of the just defined phase [2]. For solids and liquids, however, there seems to be no operational definition of the type suggested by Bridgman [3]. Such would require a quantitative experiment, an operation, which can answer the question whether a given phase is solid or liquid. The chosen experiment should provide a simple yes or no result. Turning to the common meaning of the words solid and liquid, one may turn to Meriam-Webster's Collegiate Dictionary [4]. It suggests a solid is "a substance that does not flow perceptibly under moderate stress, has a definite capacity for resisting forces (as compression or tension) which tend to deform it, and under ordinary conditions retains a definite size and shape." A liquid, in turn, is "a fluid (as water) that has no independent shape but has a definite volume and does not expand indefinitely and that is only slightly compressible." As scientific definitions, these two statements are insufficient. Expressions like "... does not flow perceptibly, ... moderate stress, ... definite capacity, ... ordinary conditions, ...



B. Wunderlich (✉)
The University of Tennessee, Knoxville, TN, USA
and
Rensselaer Polytechnic Institute, Troy, NY, USA
e-mail: Wunderlich@CharterTN.net

J. Šesták et al. (eds.), *Glassy, Amorphous and Nano-Crystalline Materials*,
Hot Topics in Thermal Analysis and Calorimetry 8, DOI 10.1007/978-90-481-2882-2_6,

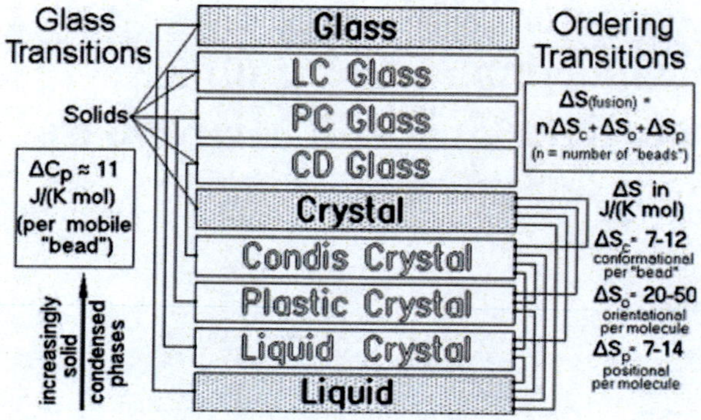


**Fig. 6.1** Today's nine condensed phases of matter, their states and transitions. First suggested in [7] and detailed in [5] and [6]

indefinitely, ... slightly" are imprecise. To improve on these definitions, specific limits need to be set.

When the molecular structure of matter was better understood, it became common practice to link the solid state to the fully ordered crystal structure. It is obvious, that the amorphous glass is also a solid, although its structure is that of a liquid, i.e., *structure alone cannot predict the solid or liquid nature of a phase*. In addition, over the last 150 years, a number of intermediate phases were discovered [5, 6], the so-called mesophases. They are of intermediate order and also solidity or fluidity. Figure 6.1 is an attempt to list the types of condensed phases [7]. In addition, on the left side of the schematic, the solidity trends are indicated. The further details in the figure are discussed in the bulk of the chapter.

The connections left and right of the phase boxes in Fig. 6.1 indicate the possible transitions. The dictionary meaning of the word transition is a "passage from one state, stage, subject, or place to another" [4]. In the present chapter, the two types of transitions listed will be discussed, namely, the glass and ordering transitions. Going from the liquid in Fig. 6.1 upward toward the crystal, the phases are listed on the left as becoming "increasingly solid," and the transitions describing the changes to the upper solids are the glass transitions. *This makes it reasonable to choose the measurement of the glass transition temperature, $T_g$, as the operation to define a solid.* The glass transition, the crystal phase, and the ordering transitions are discussed with their thermal behavior in the next six sections.

## 6.2 Phase Types, Distinguished by Their Sizes and Molecules Lead to 57 Different Entities

Phases, as just defined, must be sufficiently large so that the inhomogeneity due to the size of its constituent atoms is insignificant. Since atoms are smaller in diameter than one nanometer, and the smallest phase that can be seen with the unaided eye is bigger than 0.01 mm (= 10 μm = 10,000 nm), the homogeneity is not endangered by the molecular structure when working with 'visible phases.' To distinguish phases of this size from smaller phases, to be discussed next, they will be called 'macrophases.' The thermal properties of the macrophases are represented by the well-established thermodynamics, developed more than 100 years ago [8].

It already was recognized by early thermodynamicists that small phases may have different properties than larger ones. It was proposed in Gibbs' "theory of capillarity [8]" that the "influence of surfaces of discontinuity upon the equilibrium of heterogeneous masses" must be considered. Two observations can be clarified by considering the effect of surfaces. One is the existence of 'colloidal particles' with typical sizes of less than 1 μm. Today, phases of this size are called 'microphases.' Colloidal particles or droplets are kept in metastable dispersion by their surface charge, which prohibits coagulation to a more stable, larger phase. The other observation is a change in the overall thermodynamic quantities due to the surface free energy. One can use the Gibbs–Thomson equation [9–11] to calculate the lowering of the melting temperature, $\Delta T_{\mathrm{m}}$, for known, small volumes, $V$, or in turn, to calculate the surface free energy from the measured $\Delta T_{\mathrm{m}}$:

$$\Delta T_{\mathrm{m}} = \left(\sigma T_m^{\circ} A\right) / \left(\Delta h_{\mathrm{f}} \rho V\right), \tag{6.1}$$

where σ is the specific surface free energy; $T_{\mathrm{m}}$, the bulk equilibrium melting temperature; $A$, the surface area; $\Delta h_{\mathrm{f}}$, the specific heat of fusion (per gram); and ρ, the mass density of the crystal. Equation 6.1 can easily be modified if more than one specific surface free energy is found on a crystal, for different phase geometries, and even for the presence of internal, non-equilibrium defects.

The lowering of the melting temperature due to microphase formation is of particular importance for the understanding of semicrystalline macromolecules [2]. For a typical, 10 nm thick, laterally large, chain-folded lamella of a polyethylene crystal, for example, $\Delta T_{\mathrm{m}}$ is 26 K [7]. In this case, only the two fold-surfaces of the lamellae contribute significantly to the surface free energy. Clearly, when increasing the lamellar thickness (= $V/A$) beyond that of the microphase dimension, let us say to 10 μm (thicker than the micrometer limit of the microphase), the lowering of the melting temperature, $\Delta T_{\mathrm{m}}$, decreases to a negligible 0.026 K. Such phase is then a 'macrophase,' practically unaffected in its thermal properties by the presence of the surface.

In the last 20 years, it became customary to use the term 'nanophase' [12, 13] for phases which approach a nanometer-size scale. For a long time, the development of this terminology was not given an operational definition [14, 15]. Feynman when he speculated about 'very small phases,' suggested in 1959 already that "on a small scale we will get an enormously greater range of possible properties that substances can have" [16]. This suggestion can be the basis for an operational definition.

In the field of flexible macromolecules (polymers), it is obvious that rather small, disordered entities existed which are larger than typical crystal defects and smaller than microphases. These defects were initially called 'amorphous defects' [7]. Ultimately, it became clear, that these amorphous defects behave like very small phases with different properties than the bulk-amorphous microphases. On hand of many literature examples on phase separation of small and large molecules, an operational definition was suggested which can distinguish nanophases from microphases on the basis of its properties [14]: "In nanophases, the opposing surfaces of a phase area are sufficiently close to interact." While the microphase is also described as a phase with a sufficiently large surface area to cause changes in the properties, it can still be considered a bulk phase (with macrophase properties) enveloped by a distinct surface layer.

The nanophase, in contrast to a microphase, has *no* remaining bulk phase. Its properties change gradually from the outside to the center, without reaching the plateau of the bulk phase observed in the microphase. The operation to distinguish a nanophase from a microphase, thus, must involve the detection of the presence or absence of bulk phase in their centers. An experiment most suited for this, is the measurement of the glass transition temperature, $T_g$. The glass transition of a phase is set by the properties around a 'hole' of perhaps 0.1–1.0 nm in diameter [17–19], close to atomic dimensions.

Nanometer-sized structures were suggested in 1963 to account for certain defects in semicrystalline fibers [20]. This structure was similar to the amorphous defect proposed at the same time [7, 21]. Later it became clear these were nanophases.

In macromolecular nanophases the molecules are sufficiently long to cross the phase boundary multiple times. Semicrystalline polymers are usually a globally metastable aggregate of two or more phase types. Similarly, amphiphilic crystals, of small, flexible molecules, synthesized out of two or more incompatible, chemically different segments, can be considered as being molecularly coupled nanophases [6, 14]. A crystal structure of alternating aliphatic and aromatic layers, for example, was found for 4-*n*-octyloxybenzoic acid [22]. A number of such amphiphilic crystals undergo separate solid–liquid phase transitions at different temperatures in their different phase layers [5, 6].

To characterize a material one must know which of the phase or phases listed in Fig. 6.1 are present and then specify their sizes and interconnections. In addition, the 'class of the molecules' must be given. After the last class of molecules, the flexible, linear macromolecules, had been recognized by Staudinger [23], it became obvious, that *there are three distinct classes of molecules*, 'small molecules,' 'flexible macromolecules,' and 'rigid macromolecules' [7, Vol. 3, pp. 4–5]. This classification scheme is closely connected to the phase properties [24]. Rigid macromolecules can only exist as solids. On liquefaction or evaporation, the strong bonds defining the molecules are broken, destroying the integrity of the molecule. Flexible macromolecules can be disordered without losing their integrity, so that for many, liquid as well as solid states are possible (but no gaseous states). Small molecules under proper conditions may exist in all three states, solid, liquid, and gaseous. Examples of the three classes of molecules are found in all of the historically defined inorganic, organic, and biological molecules. It is now known that none of these historically developed classes of molecules show unique characteristics and, thus, do not help in the understanding of chemistry.

With the newly selected definitions, materials can now be identified as being made up of three classes of molecules, three phase sizes, and ten phase types. Gases, however, must be of small molecules, and because of the lack of self-sustaining surfaces cannot be microphases or nanophases. Rigid macromolecules may not become liquids or mesophases. Finally, mesophases may be coupled to specific molecule types. This leaves *57 different possible condensed states* in addition to the gas [14].

Thermal analysis is the macroscopic tool to analyze these 57 condensed states to be discussed next.

## 6.3 Measurement of Thermal Properties

With the description and classification of the molecule and phases, one can return to Fig. 6.1 and consider the measurements that identify the properties listed on the sides of the schematic. On the left, the changes when going from liquid to solid are indicated. The thermal property of interest is the change in heat capacity, $\Delta C_p$ (at constant pressure, $p$, and composition, $n$), when becoming a solid at $T_g$. The heat capacity, $C_p$, is equal to $(\partial H/\partial T)_{p,n}$, where $H$ is the enthalpy ($= U + pV$, the internal energy, $U$, plus the product of pressure, $p$, and volume, $V$). The change in $H$ is equal to the measured 'heat' gained or lost by the system [1].

On the right of Fig. 6.1, the ordering transitions are indicated which have a more abrupt change in enthalpy, $H$. This change is related to the change in entropy, $\Delta S$ ($= L/T$), to be calculated for equilibrium transitions from the heat of transition, $\Delta H$, the latent heat, $L$ [1].

To understand the experimental basis of Fig. 6.1, calorimetry, as the tool to measure heat will be considered in this section and an interpretation in terms of molecular structure and motion will be given thereafter. Today it is well known, that 'heat' is a form of energy exchanged between systems [1] and has its microscopic origin in molecular motion [2].

The measurement of unknown heat effects always involves its comparison to known effects. The SI unit for energy, work and heat, is the joule (kg $m^2$ $s^{-2}$) [25]. The most common calorimeters are the adiabatic calorimeter, differential scanning calorimeter (DSC), temperature-modulated differential scanning calorimeter (TMDSC) and quasi-isothermal, non-scanning version, TMDC. Classical, precision adiabatic calorimetry starting with measurements at temperatures close to 0 K was accomplished in the early twentieth century [26]. The total $H$ is accessible by measuring heat exchanges from 0 K to the temperature of interest. Increasingly more automated calorimeters have been developed over the years [27–29].

Quantitative DSC and modulated calorimetry developed only in the second half of the twentieth century. Since then, these more convenient techniques also have reached considerable precision. An example of a calorimeter capable to perform standard DSC as well as TMDSC and TMDC is shown in Fig. 6.2. The roots of DSC are the qualitative cooling and heating curves of the eighteenth century [2].

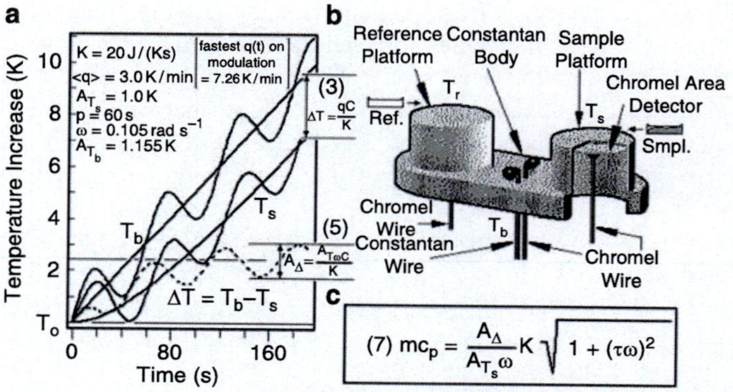


**Fig. 6.2** Differential calorimeter of type TA Instruments Q1000. (**a**) Graph of the change of sample and body temperatures, $T_s$ and $T_b$, during a heating scan as given by Eqs. 6.3 for DSC and 6.5 for TMDSC. (**b**) The measuring principle. The calorimeter assembly is placed in a temperature-controlled enclosure, filled with slow-flowing $N_2$ gas free of turbulence, also kept at $T_b$. (**c**) The TMDSC Eq. 6.7 for $C_p$ (expressed by sample mass, $m$, in g, specific heat capacity, $c_p$, in J $K^{-1}$ $g^{-1,}$ and the measured amplitudes $A$, in K at $T_s$, the temperature difference $\Delta T$, $= T_r - T_s$, the angular modulation frequency $\omega$, in rad $s^{-1}$, and the two calibration constants, $K$ and $\tau$)

When inserting a sample into a bath of constant temperature, $T_o$, Newton's law allows to describe its measured temperature, $T(t)$, as a function of time, $t$. The value of $T(t)$ exponentially approaches $T_o$:

$$dT/dt = K(T_o - T). \tag{6.2}$$

Over small temperature ranges, $K$ is constant and accounts for the nature, geometry, heat capacities, and thermal conductivities of sample and container.

The calorimeter in Fig. 6.2b can perform the modern version of such 'Newton's Law' measurement and then permits the extraction of the changes in $H$ as a function of temperature. The key to quantitative DSC is proper calibration and comparison to a standard of known $C_p$, often sapphire (single crystals of $Al_2O_3$). The function of the constant-temperature bath is taken over by the constantan body of temperature $T_b$, changing linearly with temperature at the rate $q = dT_b/dt$. Figure 6.2a illustrates that about 60 s into the measurement, steady state is reached, i.e., thereafter $\Delta T$ changes parallel to the changes of $C_p$ with temperature. The thin cylindrical walls supporting the sample and reference platforms cause the major temperature lags of the sample temperature, $T_s$, and reference temperature, $T_r$, relative to the constantan body temperature, $T_b$.

As indicated in the figure, the sample (of typically 1–20 mg) is enclosed in a sample pan (of high thermal conductivity, usually Al or Au). This configuration is to keep the temperature gradient within the sample pan small (perhaps <0.5 K, defining the error limit in temperature determination). The asymmetry of the platforms must be established by calibration. Experience has taught, that extra time spent on calibrations repays with increased precision. The heat capacity of sample *and* pan, $C$ (at $T_s$ or time $t$) can then be extracted from Eq. 6.3 in Fig. 6.2a. By performing DSC, i.e., also measuring the reference temperature, $T_r$, the $C_p$ ($= mc_p$) of the sample can be obtained based on Eq. 6.3:

$$\Delta T = T_r - T_s = q\,mc_p/K, \tag{6.4}$$

where $m$ is the sample mass and $c_p$ the specific heat capacity of the sample in J $K^{-1}$ $g^{-1}$.

The Newton's law constant $K$ needs a second calibration, namely the conversion of the measured quantities ($\Delta T/q$, in s) into $C_p$, in units J $K^{-1}$. This is done by performing a separate run on a reference material of known $C_p$. This should be done under identical conditions and with a similar magnitude of $\Delta T$ as the sample run. Careful handling of the calorimeter and samples, carrying out the measurement and calibrations, equalizing the pan weights, etc., are the main issues for a quantitative DSC [2]. Unfortunately, qualitative runs, of value only for preliminary information, still find their way into the literature. Under optimum conditions, accuracies in $C_p$ could be $\pm 1.0\%$ or better. Data can easily be obtained from 100 K, using liquid $N_2$ as a coolant, to temperatures above 1,000 K.

Standard DSC has been expanded in the last 15 years to TMDSC. Figure 6.2b indicates that no new hardware is needed. The linear increase of $T_b$ is now modulated with a periodic change (commonly with an amplitude between 0.1 and 5.0 K) and a fixed period (usually between 10 and 500 s). The heating rate of interest is the 'underlying heating rate,' $< q>$. In case the response of the calorimeter to the modulation is strictly linear, a sliding average over the time of one modulation period yields the underlying quantities, indicated by the angular brackets, $< >$. The TMDSC values of $< q>$, $<T_r>$, and $< T_s >$ correspond to the standard DSC values of $q$, $T_r$, and $T_s$. The dashed curve in Fig. 6.2a indicates that the steady state of TMDSC is reached after 2 min, later than in the standard DSC mode.

By subtraction of $< T >$ from the instantaneous, modulated value of $T$, one can extract the effect of modulation as a function of temperature (or time), usually being called the 'reversing temperature.' Its analysis is done using a pseudo-isothermal method since the underlying changes have been removed [30, 31]. The result is the reversing heat capacity, $C$, indicated by the Eq. 6.5 of Fig. 6.2a. Changing, as before, to a differential measurement with $\Delta T = T_r - T_s$, the TMDSC equation results:

$$mc_p = \{A_\Delta/(A_{Ts}\omega)\}K \tag{6.6}$$

where $\omega$ is the angular modulation frequency (in rad $s^{-1}$). If steady state and linearity are preserved and the heat capacities computed from Eqs. 6.4 and 6.6 are identical, $C_p$ is reversible. Complications arise in the transition regions, to be discussed below. In the latter cases, the response is often non-linear. As long as reversibility is not proven, the heat capacity by TMDSC must be called reversing (and usually is time-dependent). The basic calibrations of TMDSC remain similar

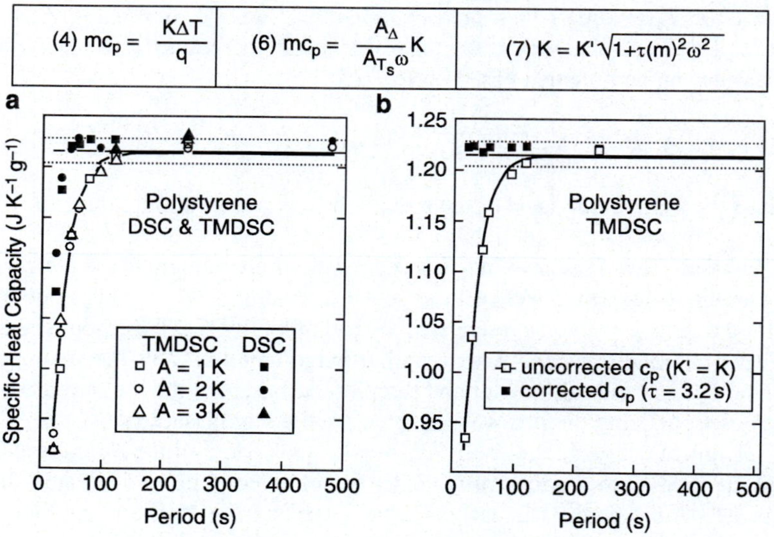


**Fig. 6.3** Comparison of DSC and TMDSC data [33]. The equations at the top correspond to the standard DSC analysis, to the TMDSC analysis with K = K', and fitted to a constant τ, as in Eq. 6.7, suggested in Fig. 6.2c. For periods less than 10 s, τ changes with frequency and mass, can, however, still be calibrated by evaluating its change with ω [34]

to that of the standard DSC, but gets more involved since the amplitude responses are identical for positive and negative deviations, an effect which can be assessed by considering the phase shift of the response [2, 32].

In order to correct for the frequency-dependence of *K*, an additional calibration constant τ is introduced in Fig. 6.2c by Eq. 6.7 [31, 33]. Figure 6.3 illustrates its evaluation. In Fig. 6.3a, a comparison of heat capacity by DSC and TMDSC is shown. In DSC, the measurement was made at constant *q* at the end of the indicated time period and $c_p$ was calculated using Eq. 6.4. The reversing $c_p$ by TMDSC at the indicated amplitudes of modulation was calculated with Eq. 6.6. As expected from Fig. 6.2a, the standard DSC reaches steady state faster than the TMDSC. In Fig. 6.3b, the TMDSC data for a 2.0 K modulation are extracted from Fig. 6.3a. In addition, the corrections, as given in Fig. 6.3 by Eq. 6.7, are marked by the filled squares. Periods as short as 10 s can generate acceptable data when properly corrected. But even at much shorter periods, quantitative information can be gained by calibration of τ as a function of not only sample mass, but also frequency [34]. It is important, that all calibration and measurement runs must be independently corrected for frequency.

The quasi-isothermal mode, TMDC, is carried out at an underlying constant temperature, i.e., the measurement is performed by modulation without scanning ($<q> = 0$,). The TMDC results are derived by similar procedures, as derived in Figs. 6.2 and 6.3 [31] and can be extended to very long times to analyze the kinetics of slow changes.

## 6.4 Interpretation of the Heat Capacity of Solids

In this section, an attempt will be made to *link the macroscopically measured heat capacity with its microscopic, molecular origin.* The first success in this endeavor was Einstein's discussion of the possible vibrations in crystalline metals and salts [35]. It was shown that the vibrations of each atom or ion are determined by the force field of its 6–12 symmetrically placed neighbors. It was proposed then to approximate the force field with a spherical symmetry, giving each vibration in the solid the same frequency, the Einstein frequency. Calorimetry revealed that this approximation was valid only at intermediate temperatures, and even then only for crystals of the highest symmetry and coordination number for the atoms or ions. The problem was resolved by replacing the single Einstein function [36] by a three-dimensional Debye distribution [37, 38]. This distribution of frequencies was derived from a macroscopic description of acoustic vibrations, extended to higher frequencies until the maximum number of degrees of freedom of an atomic assembly [1] was accounted for. This treatment described the heat capacities of many metals and salts over wide temperature ranges by specifying only the end-frequency of the spectrum, $\upsilon_{(Debye)}$, the Debye temperature $\Theta_3$ is represented by $h\upsilon_{(Debye)}/k$, in kelvin (h = Planck's constant, k = Boltzmann's constant, 1 Hz corresponds to $4.8 \times 10^{-11}$ K). An extensive discussion with data comparisons is available in [39].

Solid linear macromolecules, however, do not fit such an analysis. Strong deviations occur, starting at rather low temperatures. For polyethylene, for example, only the crystalline solids yield the expected increase of heat capacity at low temperature with a $T^3$ temperature dependence, and even this, only up to about 10 K! Figure 6.4 illustrates a frequency spectrum for polyethylene, suitable to understand IR and Raman spectra [40]. This spectrum fits $C_p$ at higher temperatures, but not at low temperatures.

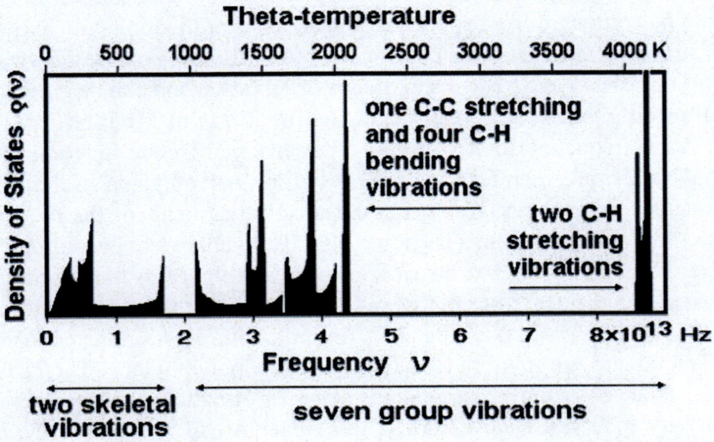


**Fig. 6.4** Vibrational spectrum of crystalline polyethylene, derived from normal-mode calculations based on a fit to the measured infrared and Raman frequencies [40]

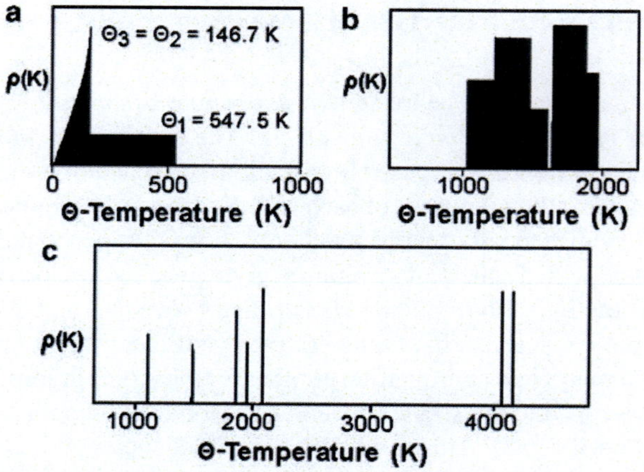


**Fig. 6.5** Various approximations of the vibrations of crystalline polyethylene $(CH_2\text{–})_x$ [46]. The repeating unit has nine normal modes of vibration. (**a**) The two skeletal vibrations (overall chain torsion and bending in addition to the inter-chain acoustic vibrations) approximated by the three Θ-parameters of the Tarasov treatment [41]. (**b**) Five partial, coupled modes approximated by box-distributions (mainly consisting of $CH_2$ wagging, twisting, and rocking modes, total modes 2.4) [46]. (**c**) The remaining eight partial and full modes of the group vibrations (The three highest frequencies are the complete $CH_2$ stretching and bending modes, total modes 4.6) [46]

Empirical equations for the approximation of low-temperature heat capacities for linear and two-dimensional macromolecules were suggested by Tarasov [41] and are shown in Fig. 6.5a. They were based on a three-dimensional Debye function for the lowest-frequencies [38], starting with the acoustical vibrations. This is followed by a two-dimensional and/or a one-dimensional Debye function to average the rest of the so-called skeletal vibrations marked in Fig. 6.4 [42, 43]. The remaining vibrations are group vibrations, known to change only insignificantly for the same chemical grouping in different molecules. Their contribution to the heat capacity can be computed from spectroscopic analysis of the molecule in question, or even of model compounds. Because of the rather limited coupling between the group vibrations, they are narrow local modes of vibration and can be either treated as single Einstein modes [36] or approximated by a box distribution fitted at the upper and lower frequency limit with a one-dimensional Debye function [44–46]. Figure 6.5a–c illustrate such a fitting for crystalline polyethylene in the different frequency ranges. In Fig. 6.5a, the general Tarasov treatment for the two skeletal modes with three Θ-temperatures is shown [46]. The remaining seven group vibrationsare approximated as five box distributions (b), and eight Einstein vibrations (c). A comparison with the spectrum in Fig. 6.4 allows to judge the simplifications.

Figure 6.6a illustrates $C_p$ (solid) and $C_p$ (liquid), and Fig. 6.6b the contributions from the skeletal and group vibrations for crystalline polyethylene. The difference between $C_p$ and $C_v$ can be computed from information on compressibility and expansivity [47, 48]. Below 200 K, this difference is negligible. Up to 150 K the $C_v$ ($\approx C_p$) is almost fully accounted for by the skeletal vibrations and calorimetry permits an easy fit to the approximate frequency spectrum in Fig. 6.5a.

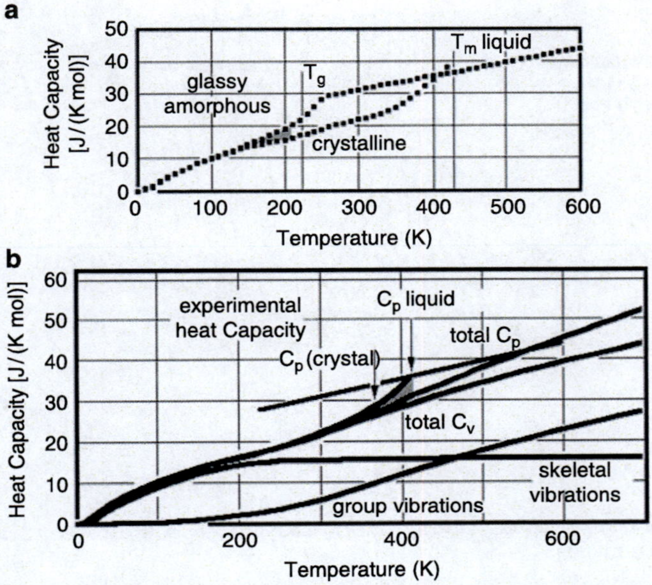


**Fig. 6.6** Measured and calculated heat capacities of glassy, liquid, and crystalline polyethylene. (**a**) Measured data, extrapolated to 100% amorphous and 100% crystalline content, based on about 100 publications reviewed for the ATHAS Data Bank [49]. (**b**) Comparison $C_p$ with the vibrational heat capacity (total $C_p$) calculated from an approximate frequency spectrum

The agreement between measured and calculated data from the approximate frequency spectrum of Fig. 6.5 is ±3%. At the melting temperature (414.6 K), the measured heat capacity of the crystal and liquid intersect.

When sufficient data on heat capacities of linear macromolecules were measured [49] and their link to the vibrational motion was established, it was possible to generate a reliable Advanced Thermal Analysis Scheme (ATHAS) to evaluate the approximations of the skeletal vibrations [2, 50]. After conversion of $C_p$ to $C_v$ [47, 48], the group vibration contributions to $C_v$ are subtracted, and the remaining skeletal contributions are fitted to the proper Tarasov equation [45, 46]. Figure 6.7 illustrates the quality of one of the most complicated Tarasov fits yet attempted, that for bovine α-chymotrypsinogen type II protein [51]. This molecule consists of 245 amino acid repeating units with a total molar mass of 25,646 Da and 3,005 skeletal vibrations. The minimization of the error in the figure shows a unique solution and allows a reproduction of the experimental data. Such data are now available for more than 100 linear macromolecules in their solid states. A number of small molecules, as well as rigid macromolecules have also been analyzed.

Overall, these skeletal frequency spectra reveal that the vibrations below $10^9$ Hz (Θ-temperature ≈ 0.05 K), with a time scale larger than one ns ($10^{-9}$ s), which ultimately lead below ≈ $2 \times 10^4$ Hz to the acoustic vibrations, are negligible with respect to their contributions to the integrated thermodynamic functions $H$, $S$,

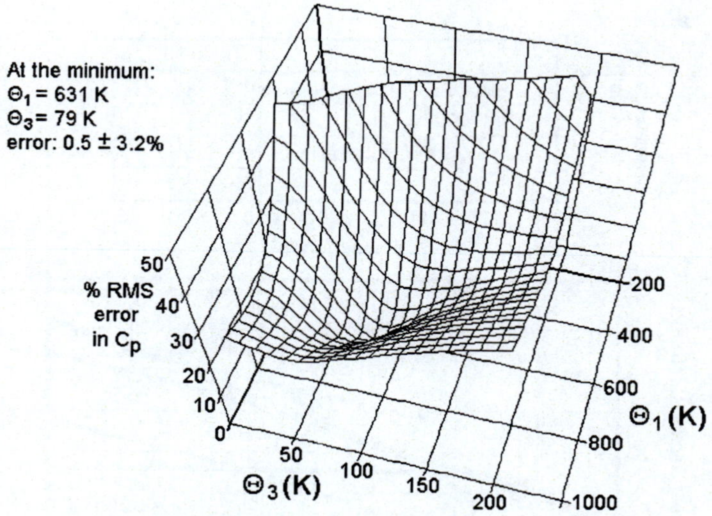


**Fig. 6.7** Fit of the skeletal heat capacity of α-chymotrypsinogen with the ATHAS data using a minimization algorithm [51]

and $G$. This means that in calorimetry, the heat capacity of solids is describable by vibrations which react instantaneously to the changes in temperature. Any lags are due to heat conduction delays and slow transitions.

Because of the great similarity of the weak intermolecular forces in polymers and their strong C, N, O intramolecular backbone-bonding, the ATHAS Data Base can also be used to estimate the $C_p$ of samples which have not been measured yet. The overall error is usually less than ±5%. This scheme is valuable to assess the unlimited numbers of proteins and synthetic copolymers [2]. All solid phases of the same polymer have a closely similar $C_p$ down to about 50 K. Below 50 K, $C_p$ for glasses yields a lower $\Theta_3$-temperature. For glassy polyethylene, $\Theta_3$ is 80 K, compared to 147 K for the crystals (see Fig. 6.5a, $\Theta_1$ is identical for both, crystal and glass).

Liquid, flexible macromolecules have a long temperature-range of linearly changing $C_p$ [49]. In addition, they are additive with respect to their structure units. To develop a more precise theoretical description, however, has proven difficult because of problems to assess the large-amplitude motion in liquids with a wide variety of intermolecular barriers to translation and rotation [52, 53].

## 6.5 Large Amplitude Motion

Besides the small-amplitude vibrational molecular motion about an equilibrium position, there are a number of large-amplitude motions. Easiest is the description of the three *translational* degrees of freedom known from the ideal gas theory via:

$$\tfrac{1}{2}M\overline{v^2} = 3/2RT = U, \qquad (6.8)$$

where $M$ is the molar mass of the particle in question, $\overline{v^2}$ the mean square translational velocity. The left third of the equation represents the kinetic energy, linked to the internal energy of the gas, $U$, on the right. The connection between energy and temperature is made by the gas constant, $R$, in the center (= 8.314334 J K$^{-1}$ mol$^{-1}$). The translational heat capacity at constant volume is then simply: $C_v = (\partial U/\partial T)_v = 3/2\ R$. A similar expression can be derived for the *rotational* degrees of motion. Both, translational and rotational energies refer to the molecule as a whole and one does not expect either of these motions in the solid state without additional potential energy contributions.

The intramolecular *conformational* rotation is the basic internal large-amplitude motion of flexible molecules. It represents a hindered rotation of parts of a molecule about covalent bonds. The different conformational isomers reached by this internal rotation have usually several well-defined potential-energy minima and maxima which define the symmetry of the motion.

Within crystals, the process of conformational motion can be simulated by large-scale molecular-dynamics calculations [54]. In polyethylene-like solids, this main motion involves a twisting of the backbone chain, ultimately producing defects consisting of various combinations of gauche- and trans-conformations which leave the chains largely parallel [55]. At room temperature, such defects have a lifetime of the order of magnitude of $10^{-12}$ s and a concentration of ≈ 0.5% [54]. The calculations substantiated that the deviations of $C_p$ from the vibration-only value, seen in Fig. 6.6b to start for the crystals at about 300 K, agrees with the defect formation [56]. In the amorphous glass, the first deviations are seen in Fig. 6.6a below 150 K. Reasonable agreement of the computed concentrations of gauche conformations exists also with measurements by IR spectroscopy on paraffins [57] and is discussed in [58].

The contribution to $C_p$ of an isolated, internal rotation at low temperature is similar to a torsional vibration. At higher temperature, when the potential energy barrier to rotation into the next minimum can be overcome, it reaches a maximum, and finally it drops to that of a free rotator with half the vibrational $C_p$ [59]. The internal rotations involving cooperative motion of neighboring molecules are sufficiently slow to be measurable by DSC and TMDSC. Model calculations made use of the hole model [18, 19]. It describes the configuration involved in the cooperative, large-amplitude conformational as motion of a ‘hole’ with a 1-nm or smaller radius and can also be used to describe the glass transition [17].

## 6.6 Ordering Phase Transitions

The integral calorimetric functions, $H$, $S$, and $G$, are summarized in Fig. 6.8 and expressed there by Eqs. 9–11. They will be the basis for the description of the ordering transitions. The data for polyethylene were taken from the ATHAS Data

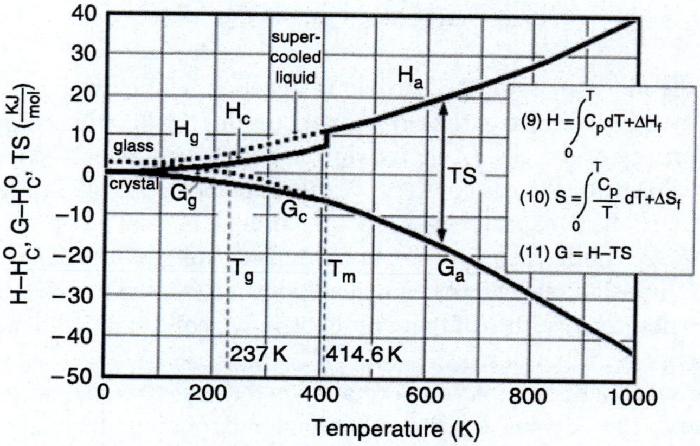


**Fig. 6.8** The integral thermodynamic functions of amorphous (a), crystalline (c), and glassy (g) polyethylene, based on calorimetric measurements

Bank [49]. Information on both, the fully crystalline and amorphous sample is given in the figure, normalized to zero for the enthalpy of the crystalline state at 0 K, $H_c^{\circ}$. The contribution to the enthalpy change, d$H$, during transition is represented by two terms. The first is due to the heat capacity $C_p$ $\{= (\partial H/\partial T)_{p,n}\}$, the second to the latent heat, $L$ $\{= (\partial H/\partial n)_{p,T}\}$:

$$\mathrm{d}H = (\partial H/\partial T)_{p,n}\mathrm{d}T + (\partial H/\partial n)_{p,T}\mathrm{d}n \qquad (6.12)$$

In standard DSC experiments, one has to separate the two contributions from the measured, apparent heat capacity, $C_p^{\#}$ (= d$H$/d$T$). The second contribution to $C_p^{\#}$ depends according to Eq. 6.12 on d$n$/d$T$, the amount of phase transformations during the change of temperature. This can be assessed by TMDSC with a proper choice of frequency and underlying heating or cooling rate. It can be written as (d$n$/d$t$)/(d$T$/d$t$) and introduces the time, $t$, in form of the ratio of rate of transformation and the rate of temperature change $q$ = d$T$/d$t$ (see Figs. 6.2 and 6.3). Only in case of continuous equilibrium is d$n$/d$T$ time and frequency independent.

A schematic of the free enthalpy as a function of temperature is drawn in Fig. 6.9, allowing the discussion of equilibrium states (dotted lines), as well as metastable or unstable states of higher free enthalpy (continuous lines). The equilibrium melting temperature, $T_{\mathrm{m}}$, is easily recognized in Figs. 6.8 and 6.9 at the temperature were $G_{\mathrm{a}} = G_{\mathrm{c}}$. At this temperature $H_{\mathrm{a}} - H_{\mathrm{c}}$ represents the equilibrium latent heat of fusion ($L = \Delta H_{\mathrm{f}}$) and the corresponding entropy of fusion ($\Delta S_{\mathrm{f}} = \Delta H_{\mathrm{f}}/T_{\mathrm{m}}$) can be connected to the increase in disorder. The entropy contribution introduced during the transition has predictable limits, as shown on the right-hand side of Fig. 6.1. The various values have been established over the years [7, Vol. 3, pgs. 5–23], $\Delta S_{\mathrm{p}}$ is Richards' rule, $\Delta S_{\mathrm{o}}$ is Walden's rule, and $\Delta S_{\mathrm{c}}$ was established based on the ATHAS Data Bank. Also shown on the right-hand side of Fig. 6.1, are the possible connections between the various phases via order/disorder transitions. Besides the melting and crystallization transitions, one can note partial ordering and disordering involving mesophases (at $T_o$ and $T_d$) and isotropization of mesophases (at $T_i$). The boiling and sublimation, involving the gas phase at a fixed pressure are not further discussed, they are also connected with a largely fixed entropy contribution, $\Delta S_e \approx 100$ J K$^{-1}$ mol$^{-1}$ (Trouton's rule).

The transitions characterized by an entropy of transition have a discontinuity in the slope of $G$, $\partial G/\partial T = -\Delta S$, when progressing along the curve of Figs. 6.8 and 6.9, but not in $G$ itself. *Such transitions where were called by Ehrenfest 'first order transitions'* [60]. A first order transition was to be distinguished from a 'second order transition' which has a discontinuity in curvature, $\partial^2 G/\partial T^2$ $(= -\Delta C_p/T)$, but not in slope. These definitions apply for systems which stay in equilibrium throughout the transitions, a condition which can be realized for systems of simple structure. For flexible, linear macromolecules this formalism is, at best an approximation. All transitions marked on the right side of Fig. 6.1 have been analyzed by assuming such a first-order formalism for the order/disorder transitions.

Only one metastable crystal is marked in Fig. 6.9, naturally many might exist. A series of lamellar crystals, for example, distinguished by different lamellar thicknesses would lead to parallel states with increasing metastability, fixed in metastability by the decreasing lamellar thickness and calculated with Eq. 6.1. In case the degree of order of the metastable crystal is different from the equilibrium crystals, as in a mesophases, the slope of $G$ $\{(\partial G/\partial T)_{p,T} = -S\}$, would vary in addition to the level of $G$. Under proper conditions, the metastable state may then cross $G$ of the crystal, as well as $G$ of the melt and reach equilibrium at a limited range of intermediate temperatures [2].

It is possible, to follow $G$ of the metastable crystal in Fig. 6.9 during a transformation with the non-equilibrium Eq. 6.13 (given in the figure), which is linked to the Gibbs–Thomson equation 6.1. Inspecting the point of non-equilibrium melting marked zero-entropy-production melting, one notes that formally, this point is similar to equilibrium melting. In case of folded-chain crystals, the degree of metastability is set by the fold length and must at this point be identical to the metastability of the supercooled melt.

The main issue in using nonequilibrium thermodynamics is to avoid the ever-present possibility that the metastable states become unstable and change during measurement [61]. For the analysis, unstable systems must be followed as a function of time. Useful calorimetric techniques are then to follow the process with TMDC until a new metastable state is reached for analysis based on the observed changes [32]. The second technique is to speed up the analysis such, that the change during the analysis is negligible, a technique which by now has reached thermal analyses of up to $10^6$ K s$^{-1}$ with superfast chip calorimetry [62]. While the lamellar crystals of linear, flexible macromolecules are frequently metastable and melt quickly at the zero-entropy-production $T_m$, the superheated crystals are usually unstable and their kinetics must be followed [63].

Above the glass transition of the surrounding amorphous phases, semicrystalline macromolecules, being metastable, become increasingly unstable with increasing temperature. On approach of the melting temperature, for example, a multitude of

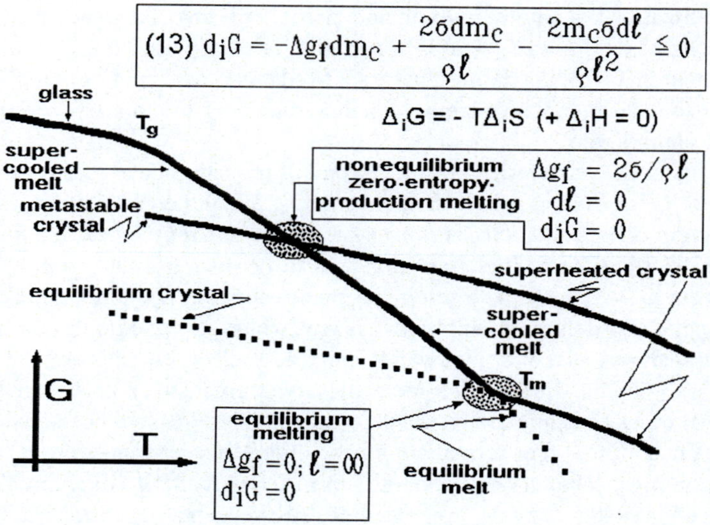


**Fig. 6.9** Schematic of the free enthalpy of a system capable to display equilibrium and nonequilibrium states. The equations are derived from the Gibbs–Thomson equation 6.1 expressed for lamellar crystals with negligible side-surface effects and thickness $\ell$. The latent heat effect on fusion $\Delta h_f = \Delta g_f + T\Delta s_f$. The lower-case letters signify specific quantities, subscript c refers to crystals, the prefix 'i' indicates the "production" quantities (deviation from equilibrium). The glass transition temperature is marked as $T_g$. The first law of thermodynamics forbids enthalpy production ($\Delta_iH = 0$), the second law upward motion in the diagram to reach a different phase line ($\Delta_iG \leq 0$, or $\Delta_iS \geq 0$)

reorganization, irreversible melting, and recrystallization may occur. These effects cause changes in $C_p^{\#}$ of Eq. 6.12 with a much longer time scale than the fast vibrations with frequencies in the THz region ($10^{12}$ Hz), shown in Fig. 6.4 for polyethylene. For the interpretation of the measured data, they must be compared to the thermodynamic functions caused by vibrations only, which represent a hypothetical, solid equilibrium crystal. A larger number of experimental data have been collected and discussed in [32]. With the modern modulated calorimetry and the ultrafast calorimetry for small sample mass, much progress is expected not only by understanding the thermal behavior, but also by the link of mechanical properties and large-amplitude molecular motion to thermal properties.

## 6.7 Glass Transitions

The glass transitions are marked on the left side of Fig. 6.1, producing a jump in heat capacity, $\Delta C_p$, as can also be seen in Fig. 6.6a for glassy polyethylene. At $T_g$, the function of $G$ shows a change in curvature, but no change in its slope, as can be seen in Figs. 6.8 and 6.9, i.e., there is a change in $C_p$, but no change in entropy, $S$ at the given temperature. Both of these observations are the requirements of a second order transition [60], *but the glass transition is not an equilibrium* transition, rather a kinetic transition. The glass transition temperature, $T_g$, is located best at the mid-point of the change in $C_p$, at half-completion of the transition at the given heating or cooling rate, $q$. It also depends on the thermal history of the sample [2]. An empirical analysis of many glasses of flexible molecules suggests, that $\Delta C_p$ depends on the number of 'beads' that gain mobility at $T_g$, being linked to a number of internal, conformational rotators.

Besides with calorimetry, the glass transition can also be identified by its jump in thermal expansivity at $T_g$. Furthermore, the glass transition can be recognized by the change it causes in response to simple mechanical tests. Based on these, the glass transition has also been called the 'brittle point,' the 'softening point,' the 'thread-pull temperature,' the 'maximum in the loss tangent,' etc. All these point to the glass transition as being an easy operation to distinguish solids from liquids, as suggested in the Introduction. Of special interest is the observation, that the viscosity of a liquid (which increases on cooling), rapidly approaches a value of about $10^{12}$ Pa s at the glass transition temperature. Viscosity of such magnitude is also observed, for example, in ice crystals close to their melting temperature. Based on these experiments, *a glass can be identified as a solid that changes on heating at its transition temperature, $T_g$, to a more mobile phase*, such as a liquid or a mesophase.

The solidity of many crystals must next be questioned. Often, orienting or ordering of the molecules increases the glass transition. Ultimately, this may move the glass transition to the melting temperature. In such cases, melting and devitrification or crystallization and vitrification may occur simultaneously. If this is the case, the crystal is a solid. Not because it is ordered, but because its glass transition is increased to the melting point.

The glass transition of amorphous polystyrene, PS, was one of the earliest analyzed in detail by calorimetry [64]. Its kinetics could be linked to the hole model of Hirai and Eyring, mentioned above [17]. An exponential decrease was observed for $T_g$ with decreasing cooling rate. This might suggest that at infinitely slow cooling one may retain the $C_p$ of the liquid to absolute zero. Such slow experiments, however, are impossible to extend far below $T_g$ and it is erroneous to extrapolate the experimental, linear $C_p$ of the liquid to temperatures below $T_g$ to assess the thermodynamic functions of a hypothetical, supercooled liquid far below $T_g$. If such erroneous extrapolations are done, they result at sufficiently low temperature in a lower entropy for the glass than for the crystal and yield the so-called Kauzman paradox [65]. With better estimates of $C_p$ of the liquid, it could be shown, at least for polyethylene, that this paradox does not seem to exist [66]. Even when avoiding the glass transition, the $C_p$ of the liquid decreases sufficiently quickly on cooling so that the amorphous solid and supercooled liquid have similar $C_p$s and the liquid retains a positive entropy at 0 K, in addition to the substantially positive $G_a$, signaling its metastability.

Looking to the onset of the glass transition, one notes a similarity of glass and crystal. For polyethylene, one can see in Fig. 6.6 that up to 150 K below $T_g$ and $T_m$ there are only vibrational contributions. As the transitions are approached, defect-based, large-amplitude motion is noted for both the glass (at ≈ 150 K) and the crystal (at ≈ 300 K). For the glass transition, close to $T_g$, it gradually turns into the characteristic cooperative motion. For the crystals, melting intervenes before a glass transition is reached. It was found, however, that for crystals of some nylons [67] and poly-(oxymethylene) [68] the full glass transition can be reached before melting of the crystals occurs, i.e. these polymers have a separate $T_g$ and $T_m$ [69].

Analyzing the glass transition of different polymers with TMDC using a simple, first-order kinetics, based on the hole theory yields different relaxation times for the different samples, accounting for the broadening of the transition when going from quenched amorphous, to slowly cooled or annealed glasses, and finally to semi-crystalline samples [70, 71]. In addition to the broadening of the glass transition, the $\Delta C_p$ at the glass transition decreases more than linearly with crystallinity. This suggests that a sizable amorphous fraction exists that does not participate in the measured glass transition. This fraction remains rigid on heating and shows a separate glass transition at higher temperature. It was identified as a *'rigid-amorphous fraction,' RAF* of nanophase dimension [32, 72]. Both crystallinity and RAF, and their transition behavior must be known to judge the mechanical properties of semicrystalline polymers.

## 6.8 Conclusions

The characterization of the phases between solid and liquid using thermal analysis was begun with a review of the definitions and classifications. Of particular importance were changes suggested for the *definition of the solid state, types of molecules, and of small phases*. A sample of condensed phases to be analyzed by thermal analysis is suggested to be identified in terms of 1 of 57 types, based on three molecule classes (small molecules, flexible macromolecules, and rigid macromolecules), the nine phase types of Fig. 6.1, and three phase sizes (macrophase, microphase, and nanophase). In case the sample is heterogeneous, the global fitting of the different phases must be identified by their shapes and possibly molecular coupling across the interfaces, which calls for a rather extensive analysis program [2, 32].

Next, the new experimental tools of calorimetry which permit the measurements of heat capacity and latent heat, have been summarized with Figs. 6.2 and 6.3. Modern developments were detailed, and methods available to not only measure equilibrium properties, but also to handle non-equilibrium and kinetic processes were displayed. The conclusion is that *the accuracy of the data in differential calorimetry lies in the quality of the calibration.*

Molecular motion was linked to $C_p$. The vibrational motion in the solid state with a time scale shorter than 1 ps accounts for most of the enthalpy of the solid state, as illustrated with Figs. 6.4–6.6, 6.8. The large-amplitude conformational motion evolves at higher temperature out of torsional oscillations. Depending on the molecular structure, it may begin with the creation of isolated, intramolecular,

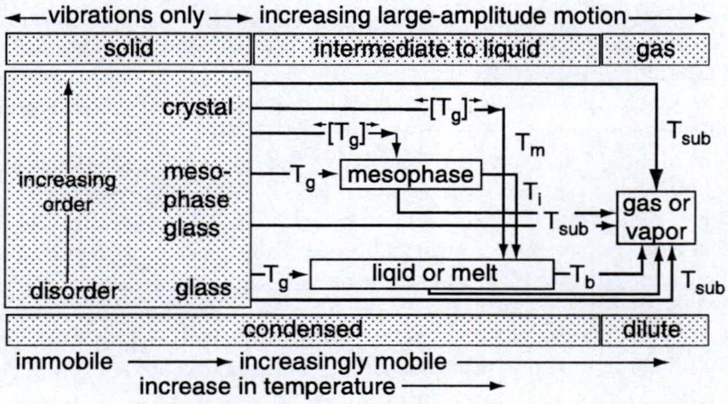


**Fig. 6.10** Schematic of the changes in molecular motion and structural order on going through the various phase transitions indicated in Fig. 6.1

conformational defects. At this stage the large-amplitude motion may also have a timescale in the picosecond range. The larger potential energy needed for the defect creation is detectable by a gradual deviation of $C_p$ beyond the vibrational level, as shown in Fig. 6.6. At higher temperatures, the large-amplitude motion expands into intermolecular, cooperative, liquid-like motion, starting when approaching $T_g$ with a high activation energy. In case the molecules are sufficiently small to undergo rotation and translation, these additional large-amplitude motions also begin when approaching the glass transition.

These conclusions are combined into the scheme of Fig. 6.10. Under the heading *vibrations only* one finds the solid (molecularly) immobile, condensed phases. The transition behavior changes when going from a disordered glass (bottom, left) to increasingly ordered mesophase glasses and ultimately to the crystal (top, left). *Increasingly large-amplitude motion* makes the condensed states (molecularly) increasingly mobile, ending with the liquid or melt. The transformation of the solid to the liquid (melt) occurs at $T_g$, where $T_g$ may occur at lower temperature or simultaneous with $T_d$ or $T_m$.

To complete the phase picture, the dilute gas (vapor) phase is also included in Fig. 6.10. It is linked to all condensed phases, either at well-defined boiling temperatures, $T_b$, or via the sublimation at the temperature ranges $T_{sub}$.

## References

1. McNaught AD, Wilkinson A (1997) IUPAC. Compendium of Chemical Terminology (the "Gold Book"). Blackwell Scientific, Oxford (1997); XML on-line corrected version: http://goldbook.iupac.org, created by Nic M, Jirat J, Kosata B (2006-) updates compiled by Jenkins A, doi: 10.1351/goldbook
2. See, for example, Wunderlich B (2005) Thermal analysis of polymeric materials. Springer, Berlin. See also the Computer course: thermal analysis of materials. http://athas.prz.rzeszow.pl, or www.scite.eu.

3. Bridgeman PW (1927) The logic of modern physics. MacMillan, New York
4. Merriam Webster's Collegiate Dictionary, 11th edn. Merriam-Webster, Inc., Springfield, MA, (2003); see also: http://www.m-w.com/.
5. Wunderlich B, Grebowicz J (1984) Thermotropic mesophases and mesophase transitions of linear, flexible macromolecules. Adv Polym Sci 60/61:1–59
6. Wunderlich B, Möller M, Grebowicz J, Baur H (1988) Conformational motion and disorder in low and high molecular mass crystals. Springer, Berlin (Adv Polym Sci 87)
7. Wunderlich B (1973–1980) Macromolecular physics, vols 1–3. Academic, New York; A pdf reprint with a new Preface and electronically searchable index was republished, available from http://athas.prz.rzeszow.pl, or www.scite.eu.
8. Gibbs JW (1875–1876, 1877–1878) On the equilibrium of heterogeneous substances. Trans. Conn. Acad. III, 108–248 and 343–524. An extended abstract was published in Am J Sci, Ser 3 16:441–458 (1878); reprinted in Bumstead HA, Gibbs van Name R (1961) The Scientific Papers of J Willard Gibbs 1, Thermodynamics. Dover, New York
9. Hill TL (1962) Thermodynamics of small systems. J Chem Phys 36:3182–3197; see also: Thermodynamics of small systems, Parts I and II. Benjamin, New York (1963, 1964). Reprinted by Dover, New York (1994)
10. For an early discussion see Tammann G (1920) A method for determining the relationship between the melting point of a crystal lamella and its thickness. Z Anorg Allg Chemie 110:166–168; and also Tolman RC (1948) Consideration of the Gibbs theory of surface tension. J Chem Phys 16:758–774
11. For early experiments see Meissner F (1920) The influence of state of division on the melting point. Z Anorg Allg Chemie 110:169–186
12. Eicke H-F (1987) Aqueous nanophases in liquid hydrocarbons stabilized by ionic surfactants. Surf Sci Ser 21:41–92
13. Siegel RW, Ramasamy S, Hahn H, Li Z, Lu T, Gronsky R (1988) Synthesis, characterization, and properties of nanophase titanium dioxide. J Mater Res 3:1367–1372
14. Chen W, Wunderlich B (1999) Nanophase separation of small and large molecules. Macromol Chem Phys 200:283–311
15. For a recent description of nanophases see: Wunderlich B (2008) Thermodynamics and properties of nanophases. Thermochim Acta 492:2–15
16. Quoted from the lecture as recorded at: http://www.its.caltech.edu/~feynman/plenty.html
17. Hirai N, Eyring H (1959) Bulk viscosity in polymeric systems. J Polymer Sci 37:51–70; Bulk viscosity of liquids. J Appl Phys 29:810–816
18. Eyring H (1936) Viscosity, plasticity, and diffusion as examples of absolute reaction rates. J Chem Phys 4:283–291
19. Frenkel J (1946) Kinetic theory of liquids. Clarendon, Oxford
20. Hosemann R (1963) Crystalline and paracrystalline order in high polymers. J Appl Phys 34:25–41
21. Wunderlich B, Poland D (1963) Thermodynamics of crystalline linear high polymers. II. The influence of copolymer units on the thermodynamic properties of polyethylene. J Polym Sci Part A 1:357–372
22. Bryan RF, Hartley P, Miller RW, Shen M-S (1980) An X-ray study of the p-n-alkoxybenzoic acids. Part VI. Isotypic crystal structures of four smectogenic acids having seven, eight, nine, and ten alkyl chain carbon atoms. Mol Cryst Liq Cryst 62:281–309
23. The term "Macromolecule" was first used by Staudinger H, Fritschi J (1922) Isoprene and Rubber. V. Reduction of rubber and its constitution. Helv Chim Acta 5:785–806; In his Nobel Lecture of 1953 Staudinger sets the limit of small molecules at 1,000 atoms: Staudinger H (1961) Arbeitserinnerungen, p. 317. Hüthig, Heidelberg
24. For a summary of these classifications see Wunderlich B (1999) A classification of molecules and transitions as recognized by thermal analysis. Thermochim Acta 340/41:37–52
25. Mills I, Cvitaš T, Homan K, Kallay N, Kuchitsu K (1993) Quantities, units and symbols in physical chemistry (Green Book, IUPAC), 2nd edn. Blackwell, Oxford

26. Nernst WH (1911) The energy content of solids. Ann Physik 36:395–439
27. Worthington AE, Marx PC, Dole M (1955) Calorimetry of high polymers. III. A new type of adiabatic jacket and calorimeter. Rev Sci Instrum 26:698–702
28. Gmelin E, Rödhammer P (1981) Automatic low temperature calorimetry for the range 0.3–320 K. J Phys E Sci Instrum 14:223–228
29. Tan Z-C, Shi Q, Liu B-P, Zhang H-T (2008) A fully automated adiabatic calorimeter for heat capacity measurement between 80 and 400 K. J Therm Anal Calorim 92:367–374
30. Boller A, Jin Y, Wunderlich B (1994) Heat capacity measurement by modulated DSC at constant temperature. J Therm Anal 42:307–330
31. Wunderlich B, Jin Y, Boller A (1994) Mathematical description of differential scanning calorimetry based on periodic temperature modulation. Thermochim Acta 238:277–293
32. Wunderlich B (2003) Reversible crystallization and the rigid amorphous phase in semicrystalline macromolecules. Prog Polym Sci 28/3:383–450
33. Androsch R, Moon I, Kreitmeier S, Wunderlich B (2000) Determination of heat capacity with a sawtooth-type, power-compensated temperature-modulated DSC. Thermochim Acta 357/358:267–278
34. Androsch R, Wunderlich B (1999) Temperature-modulated DSC using higher harmonics of the Fourier transform. Thermochim Acta 333:27–32
35. Einstein A (1907) Planck's theory of radiation and the theory of the specific heat. Ann Physik 22:180–190, 800
36. Sherman J, Ewell RB (1942) A six-place table of the Einstein functions. J Phys Chem 46:641–662
37. Debye P (1912) To the theory of the specific heat. Ann Physik 39:789–839
38. Beattie JA (1926) Tables of three dimensional debye functions. J Math Phys (MIT) 6:1–32
39. Schrödinger E (1926) Thermische eigenschaften der stoffe. In: Geiger H, Scheel K, Henning F (eds) Handbuch der physik, vol 10. Springer, Berlin
40. Barnes J, Fanconi B (1978) Critical review of vibrational data and force field constants for polyethylene. J Phys Chem Ref Data 7:1309–1321
41. Tarasov VV (1950) Theory of the heat capacity of chain and layer structures. Zh Fiz Khim 24:111–128; Heat capacity of chain and layer structures 27:1430–1435 (1953); Tarasov VV, Yunitskii GA (1965) Theory of heat capacity of chain-layer structures. Zh Fiz Khim 39:2077–2080
42. Crystals with planar molecules and tables for the two dimensional Debye functions are given by Gaur U, Pultz G, Wiedemeier H, Wunderlich B (1981) Analysis of the heat capacities of group IV chalcogenides using debye temperatures. J Thermal Anal 21:309–326
43. A first analysis of the heat capacity of crystalline polyethylene and tables of one dimensional Debye functions are available in: Wunderlich B (1962) Motion in polyethylene. II. Vibrations in crystalline polyethylene. J Chem Phys 37:1207–1216
44. For initial computer programs and discussions of the fitting of $C_p$ of linear macromolecules, see: Cheban YuY, Lau SF, Wunderlich B (1982) Analysis of the contribution of skeletal vibrations to the heat capacity of linear macromolecules. Colloid Polymer Sci 260:9–19
45. Zhang G, Wunderlich B (1996) A new method to fit approximate vibrational spectra to the heat capacity of solids with Tarasov functions. J Therm Anal 47:899–911
46. The use of 3-, 2-, and 1-dimensional Debye functions is described in: Pyda M, Bartkowiak M, Wunderlich B (1998) Computation of heat capacities of solids using a general Tarasov equation. J Thermal Anal Calorim 52:631–656
47. Grebowicz J, Wunderlich B (1985) On the $C_p$ - $C_v$ conversion of solid linear macromolecules. J Therm Anal 30:229–236
48. Pan R, Varma M, Wunderlich B (1989) On the $C_p$ to $C_v$ conversion for solid linear macromolecules II. J Therm Anal 35:955–966
49. Gaur U, Shu H-C, Mehta A, Lau S-F, Wunderlich BB, Varma-Nair M, Wunderlich B (1981, 1982, 1983, 1991) Heat capacity and other thermodynamic properties of linear macromolecules. Parts I–X. J Phys Chem, Ref. Data 10:89–117, 119–152, 1001–1049, 1051–1064 ; 11:313–325,

1065–1089; 12:29–63, 65–89, 91–108; 20:349–404; an internet version is available through: http://athas.prz.rzeszow.pl
50. Wunderlich B (1995) The athas data base on heat capacities of polymers. Pure Appl Chem 67:1019–1026; see also The advanced thermal analysis system, ATHAS. Shin Netsu Sokuteino Shinpo 1:71–100 (1990)
51. Zhang G, Gerdes S, Wunderlich B (1996) Heat capacities of solid globular proteins. Macromol Chem Phys 197:3791–3806
52. Pyda M, Wunderlich B (1999) Computation of heat capacities of liquid polymers. Macromolecules 32:2044–2050
53. Loufakis K, Wunderlich B (1988) Computation of heat capacity of liquid macromolecules based on a statistical mechanical approximation. J Phys Chem 92:4205–4209
54. Sumpter BG, Noid DW, Liang GL, Wunderlich B (1994) Atomistic dynamics of macromolecular crystals. Adv Polym Sci 116:27–72
55. Blasenbrey S, Pechhold W (1970) Theory of phase transitions in polymers. Ber Bunsenges 74:784–796
56. Wunderlich B (1962) Motion in polyethylene. III. The amorphous polymer. J Chem Phys 37:2429–2432; Motion in the solid state of high polymers. J Polymer Sci Part C 1:41–64 (1963)
57. Kim Y, Strauss HL, Snyder RG (1989) Conformational disorder in crystalline n-alkanes prior to melting. J Phys Chem 93:7520–7526
58. Wunderlich B, Pyda M, Pak J, Androsch R (2001) Measurement of heat capacity to gain information about time scales of molecular motion from pico to megaseconds. Thermochim Acta 377:9–33
59. Herzberg G (1945) Infrared and Raman spectra of polyatomic molecules. Van Nostrand, New York
60. Ehrenfest P (1933) Phase changes in the ordinary and extended sense classified according to the corresponding singularities of the thermodynamic potential. Proceedings of the Academic Science, Amsterdam vol 36, pp 153–157. Suppl 75b, Mitt Kammerlingh Onnes Inst, Leiden
61. Wunderlich B (1964) The melting of defect polymer crystals. Polymer 5:125–134, 611–624
62. Pyda M, Nowak-Pyda E, Heeg J, Huth H, Minakov AA, Di Lorenzo ML, Schick C, Wunderlich B (2006) Melting and crystallization of poly(butylene terephthalate) by temperature-modulated and superfast calorimetry. J Polym Sci B 44:1364–1377
63. Hellmuth E, Wunderlich B (1965) Superheating of linear high-polymer polyethylene crystals. J Appl Phys 36:3039–3044
64. Wunderlich B, Bodily DM, Kaplan MH (1964) Theory and measurements of the glass-transformation interval of polystyrene. J Appl Phys 35:95–102
65. Kauzmann W (1948) The nature of the glassy state and the behavior of liquids at low temperatures. Chem Rev 43:219–256
66. Pyda M, Wunderlich B (2002) Analysis of the residual entropy of amorphous polyethylene at zero Kelvin. J Polymer Sci B 40:1245–1253
67. Wunderlich B (2008) Thermal properties of aliphatic nylons and their link to crystal structure and molecular motion. J Therm Anal Calorim 93:7–17
68. Qiu W, Pyda M, Nowak-Pyda E, Habenschuss A, Wunderlich B (2005) Reversibility between glass and melting transitions of poly(oxyethylene). Macromolecules 38:8454–8467
69. Wunderlich B (2006) The glass transition of polymer crystals. Thermochim Acta 446:128–134
70. Wunderlich B, Boller A, Okazaki I, Kreitmeier S (1996) Modulated differential scanning calorimetry in the glass transition region II. The mathematical treatment of the kinetics of the glass transition. J Therm Anal 47:1013–1026
71. Wunderlich B, Okazaki I (1997) Modulated differential scanning calorimetry in the glass transition region, VI. Model calculations based on poly(ethylene terephthalate). J Therm Anal 49:57–70
72. Suzuki H, Grebowicz J, Wunderlich B (1985) The glass transition of polyoxymethylene. Br Polym J 17:1–3

# Chapter 7
# Thermal Portrayal of Phase Separation in Polymers Producing Nanophase Separated Materials

**Ivan Krakovský and Yuko Ikeda**

## 7.1 Polymers

Differential scanning calorimetry (DSC) and other methods of thermal analysis provide a lot of information about phase behaviour and physical properties of heterogeneous systems. This information is usually supplied by information provided by other methods, e.g., optical microscopy, X-ray and neutron diffraction, etc. Advantage of methods of thermal analysis consists in small amount of material necessary for the measurement, simple sample preparation and short measuring time.

Polymers represent a class of materials in which methods of thermal analysis are very popular. Today, a large variety of polymeric materials is available and used in modern technologies. Synthetic polymers like polyethylene, polypropylene or polystyrene are exploited intensively in everyday life. Polymers are also abundant in nature, e.g., polysaccharides such as cellulose represent main constituents of wood and paper. Other natural polymeric materials, e.g., caoutchouc have been used by mankind for centuries. Proteins, nucleic acids are biopolymers which play a crucial role in life processes.

Despite a great difference in their chemical composition, polymers are distinguished from other materials by the spatial structure of their molecules, namely, a long linear sequence of atoms or groups of atoms which is referred to as polymer chain or macromolecule [1]. In rubbers or resins, polymer chains are linked together to give rise to one giant molecule of macroscopic dimensions – polymer network.

I. Krakovský (✉)
Department of Macromolecular Physics, Charles University, V Holešovičkách 2, 180 00 Prague 8, Czech Republic
e-mail: ivank@kmf.troja.mff.cuni.cz

Y. Ikeda
Graduate School of Science and Technology, Kyoto Institute of Technology, Matsugasaki, Sakyo, Kyoto 606-8585, Japan

J. Šesták et al. (eds.), *Glassy, Amorphous and Nano-Crystalline Materials*,
Hot Topics in Thermal Analysis and Calorimetry 8, DOI 10.1007/978-90-481-2882-2_7,

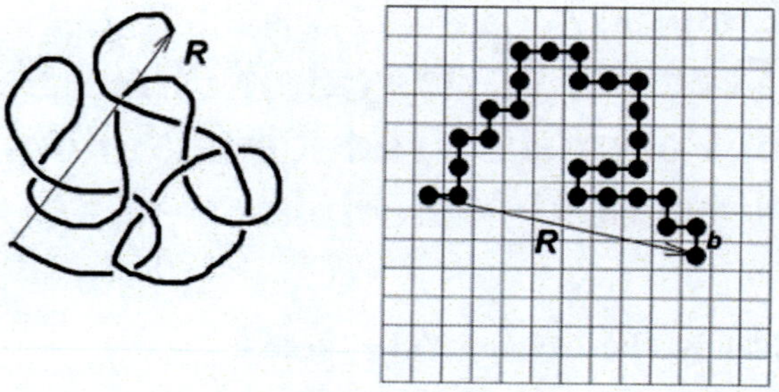


**Fig 7.1** Illustration of the analogy between the space form of polymer chain (polymer coil in 3D) and random walk (in 2D)

There is something universal for all polymer chains: despite the details of their structure which is reflected in flexibility, any *sufficiently long* polymer chain obtains a form of an interwoven coil resembling a trail of a randomly flying object. As a consequence, many physical properties of polymers can be explained exploiting the analogy between polymer chain and random walk [2], see Fig. 7.1.

For example, the average size of a polymer coil can be estimated from the mean-square root of the end-to-end distance of the random walk, $\sqrt{\langle \mathbf{R}^2 \rangle}$:

$$\sqrt{\langle \mathbf{R}^2 \rangle} = b\sqrt{N} \tag{7.1}$$

where the number of steps, $N$ ($N>>1$), and their length, $b$, correspond to the number of (statistical) segments and persistent length of the polymer chain, respectively [2].

Obviously, polymer coil, typically formed by a linear chain composed of hundreds of statistical segments of the length of few nanometres has a characteristic size of a few nanometres which explains interest in polymers in nanotechnology[1].

Many typical properties of polymeric materials originate from the chain structure of their molecules, e.g.:

- Polymers are poor in configurational entropy.
- Main part of elasticity of polymers at temperatures above their glass transition temperature is of entropic origin.
- In polymers there is a wide spectrum of processes – from very slow translational and rotational dynamics of whole chains realized by means of torsional movements of their segments of large amplitude to very fast vibrations of bond length and angles of small amplitude.

Thermal properties of an amorphous polymer (i.e., polymer unable of crystallization) resemble properties of any glass-forming substance. A big increase in viscosity and response time to external perturbations is observed when the temperature is decreased

[1]At the same time, the length of the same chain in fully stretched state, $L$, is: $L = bN>>b\sqrt{N}$ for $N>>1$.

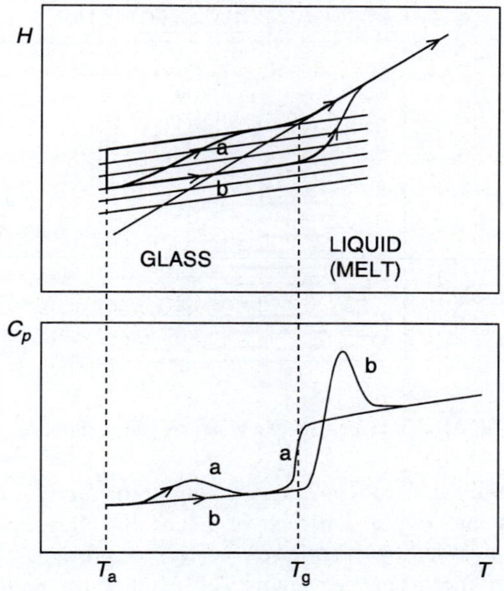


**Fig 7.2** Two possible ways of enthalpy relaxation during heating of a polymer annealed for a time in the glassy state (Adapted from [5])

below the glass transition temperature, $T_g$. The material is not able to attain its equilibrium state in experimentally observable time (see Fig. 7.2). This state is referred to as glassy state.

Despite many efforts, the exact nature of the glass transition has not yet been satisfactorily clarified. In polymers, the reduction in configurational entropy of the system seems to play an important role [3, 4]. If a polymer is cooled down at a constant rate to a temperature below $T_g$ and the system is annealed for a time, a slow relaxation to an "equilibrium" state occurs which is reflected in a decrease of the enthalpy, $H$. When reheated again at a rate used typically in DSC, enthalpy of the system can evolve in two ways as shown in Fig. 7.2. A recovery peak is found in heating DSC curves which represent temperature dependences of specific heat at constant pressure, $c_p = (\partial H / \partial T)_p$. The peak can occur either before glass transition (a) or (more often) it is superimposed on it (b).

## 7.2 Polymer Solutions and Blends

Polymer solutions and blends represent mixtures of two or more components and exhibit a rich variety of phase behaviour. One special example of the phase diagram for a binary mixture exhibiting upper critical solution temperature (UCST) is shown in Fig. 7.3.

At temperatures higher than the critical temperature, $T_c$, the one-phase state of the binary mixture is stable for all compositions, described by, e.g., molar fraction

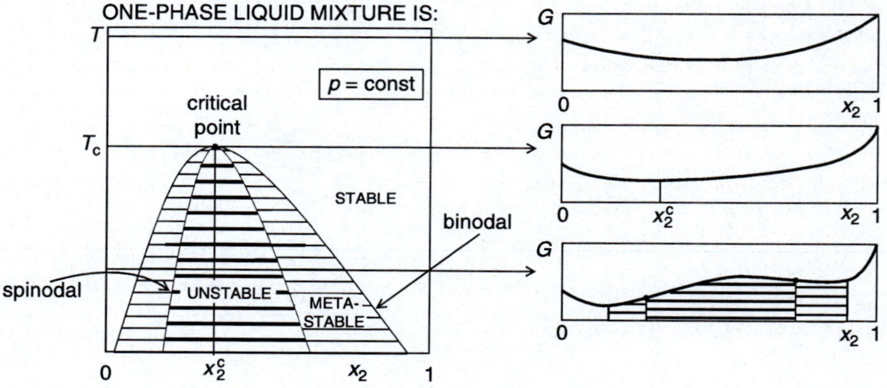


**Fig. 7.3** Phase diagram of a binary mixture exhibiting upper critical solution temperature (UCST)

of the component 2, $x_2$. Cooling of the mixture with a certain composition can bring it into region where it is metastable. Separation of the system into two phases separated by interphase is started and droplets of a new phase appear. In this case, an energy barrier against diffusion of the components into new phases has to be overcome. The curve allocating the onset of phase separation in the phase diagram is referred to as cloud point curve or binodal line (Fig. 7.3). If viscosity of the system is low as it is the case in polymer solutions, the droplets coalesce and the two phases can be isolated. However, in polymer blends the viscosity of phases is usually very high and full demixing would require unrealistic time. Eventually, heterogeneous material with a particulate morphology which is in non-equilibrium state is obtained.

A different situation occurs when the system is cooled rapidly into unstable state which is marked out by spinodal line (Fig. 7.3). In unstable state, the energy barrier against diffusion vanishes and the system spontaneously separates into two phases by a mechanism known as spinodal decomposition. Similarly to the previous case, the process can be again hindered by high viscosity of the system. Fixation of the system in this state by freezing or a chemical reaction provides a way of preparation of the heterogeneous materials with bicontinuous morphology.

Spinodals and binodals touch each other in the critical point where they have also common tangent (Fig. 7.3).

For a binary mixture the binodal, spinodal lines and critical point in the phase diagram can be determined from the Gibbs free energy of the mixture, $G$, as

$$\begin{aligned} \mu_1^I &= \mu_1^{II} \\ \mu_2^I &= \mu_2^{II} \end{aligned} \quad \text{binodal (equality of chemical potentials in individual phases)} \tag{7.2}$$

$$\frac{\partial^2 G}{\partial x_2^2} = 0 \quad \text{spinodal} \tag{7.3}$$

$$\frac{\partial^2 G}{\partial x_2^2} = 0 \quad \text{and} \quad \frac{\partial^3 G}{\partial x_2^3} = 0 \qquad \text{critical points} \tag{7.4}$$

Therefore, if an expression for the Gibbs energy of a binary mixture as a function of temperature and composition is available from a microphysical model the phase diagram as well as its thermodynamic properties can be calculated. First model of this kind for polymer systems was developed by Flory et al. almost 60 years ago [6–8], see also [1]. The model is a lattice model: polymer chains and solvent molecules are "inweaved" into a simple cubic lattice such as in Fig. 7.1 and the sum over states of the system is calculated in the mean-field approximation. Final expression for the Gibbs energy of the polymer solution derived by Flory reads

$$G = N_1\mu_1^0 + N_2\mu_2^0 + k_\mathbf{B}T[(N_1 + rN_2)\chi v_1 v_2 + N_1 \ln v_1 + N_2 \ln v_2] \tag{7.5}$$

where $N_1$, $N_2$ are numbers of solvent and polymer molecules, $\mu_1^0$,$\mu_2^0$ their chemical potentials in pure form, and $v_1$, $v_2$ their volume fractions. In discussion of the phase behaviour of mixtures involving polymers volume fractions are preferred to molar fractions due to high molecular weight of polymers.

In the derivation of Eq. 7.5 it is assumed that polymer chains consist of equal number of segments, $r$, which are linked into a flexible array. The segment is supposed to occupy the same volume as a solvent molecule. Interaction parameter, $\chi$, represents a measure of readiness of polymer segments to mutual mixing with solvent molecules.

Similar expression can be also derived for binary mixture of two polymers with the numbers of segments $r_1$ and $r_2$, respectively:

$$G = N_1\mu_1^0 + N_2\mu_2^0 + k_\mathbf{B}T[(r_1N_1 + r_2N_2)\chi v_1 v_2 + N_1 \ln v_1 + N_2 \ln v_2] \tag{7.6}$$

In the original treatment by Flory and Huggins, $\chi$ is assumed to be a function of temperature, only:

$$\chi(T) = \chi_S + \frac{\chi_H}{T} \tag{7.7}$$

where $\chi_S$ and $\chi_H$ are entropic and enthalpic part of the interaction parameter.

Generally, a large variety of phase diagrams found for polymer solutions and blends can be explained by assuming following temperature dependence of the interaction parameter [9]:

$$\chi(T) = A + \frac{B}{T} + \frac{C}{T^2} \tag{7.8}$$

with three constants $A$, $B$ and $C$.

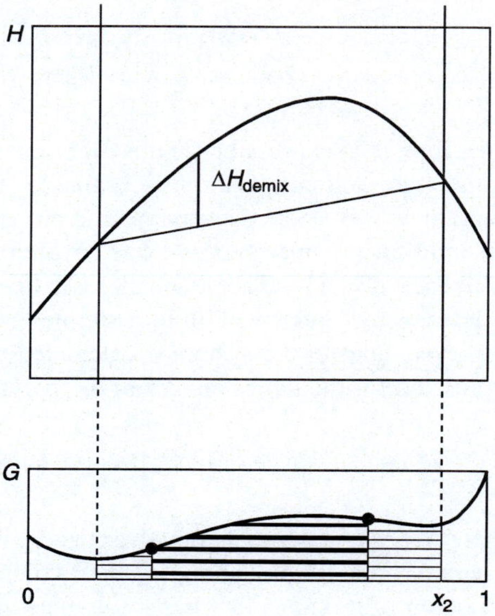


**Fig. 7.4** Compositional dependence of the enthalpy and Gibbs free energy of a binary mixture at a temperature where the system is partially miscible

However, it was found experimentally that $\chi$ may be also composition-dependent. For example, Šolc et al. [10] expressed this dependence by

$$\chi(T, v_2) = \left(\chi_S + \frac{\chi_H}{T}\right) + av_2 + bv_2^2 \tag{7.9}$$

where $a$, $b$ are constants. This enabled an explanation of peculiarities of phase diagrams found in some systems, e.g., existence of double critical points in aqueous solutions of poly(vinylmethylether) (PVME).

In polymer solutions, binodals can be determined by observation of cloud points. For experimental location of spinodals methods like pulse-induced critical scattering has to be used [11]. Calorimetry can be also exploited for determination of binodal lines. The method is based on the measurement of the change of enthalpy that occurs during demixing process induced by heating or cooling as illustrated in Fig. 7.4.

In metastable or unstable one-phase state the enthalpy of the mixture is larger (thick curve in Fig. 7.4) than the enthalpy in two-phase state (thin line in Fig. 7.4). Therefore, demixing is accompanied by an enthalpy jump, $\Delta H_{\text{demix}}$, which can be measured by a calorimetric method. This is illustrated in Fig. 7.5 where DSC traces obtained in heating and cooling of an aqueous solution of PVME are shown. Demixing in heating and remixing in cooling of the system is clearly visible. Note that this system has lower critical solution temperature (LCST) unlike the phase diagram with UCST shown in Fig. 7.3. Enthalpy of demixing for a polymer solution

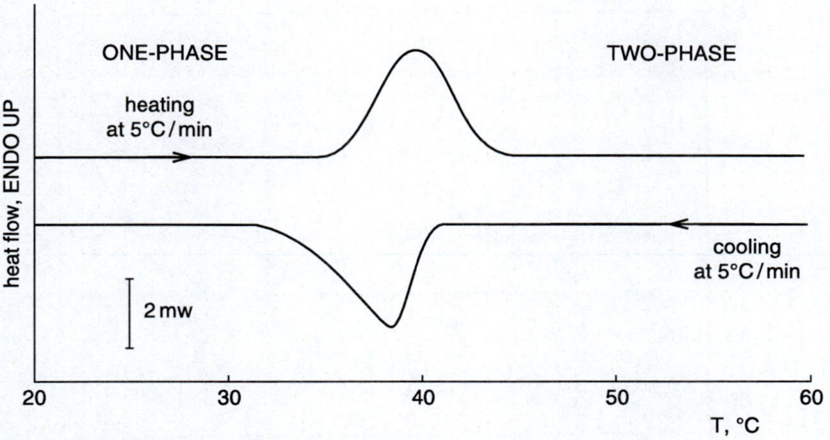


**Fig. 7.5** DSC scans of poly(vinylmethylether) aqueos solution (volume fraction of polymer, $v_2 \approx 0.07$) (Adapted from [12])

or blend can be also calculated from the Flory–Huggins formulas (Eqs. 7.5, 7.6 or 7.7) by virtue of the Gibbs-Helmholtz relation.

So far we dealt with phase behaviour of liquid binary systems – polymer solutions or blends in liquid state. What will happen if these systems are cooled to lower temperatures? In the case of the system with UCST, the components in individual phases pass into glassy state eventually. The glass transition can be preceded by crystallization if some components have a suitable regular molecular structure.

Systems with LCST which are one-phase liquids below the critical temperature exhibit phase diagrams similar to those found for low-molecular-weight mixtures, as illustrated in Fig. 7.6 by the phase diagram of the aqueous solution of Jeffamine ED2003. This polymer is basically poly(oxyethylene) and crystallization (melting) of water and polymer, glass transition of polymer and formation of the eutectic mixture are found on DSC scans as shown in Fig. 7.7.

Actually, reheating of a system from its glassy state and determination of the glass temperatures of phases is the simplest and most often way used for investigation of phase behaviour and miscibility of polymer blends.

An accepted unambiguous indication of one-phase state, i.e., that the components are miscible, is a single $T_g$ which is close to a value calculated from $T_g$ 's of the components by means of additivity rules [14, 15]. Detection of multiple transitions, coincident with or shifted from those determined for the neat components, proves that the system is in the multi-phase state.

Implementation of this procedure requires that glass transitions of phases are separated by a sufficiently large gap, at least 10–20°C. The domains of individual phases should also have a size bigger that a critical size to manifest a unique glass transition. The critical size was estimated to be about 10–15 nm [16].

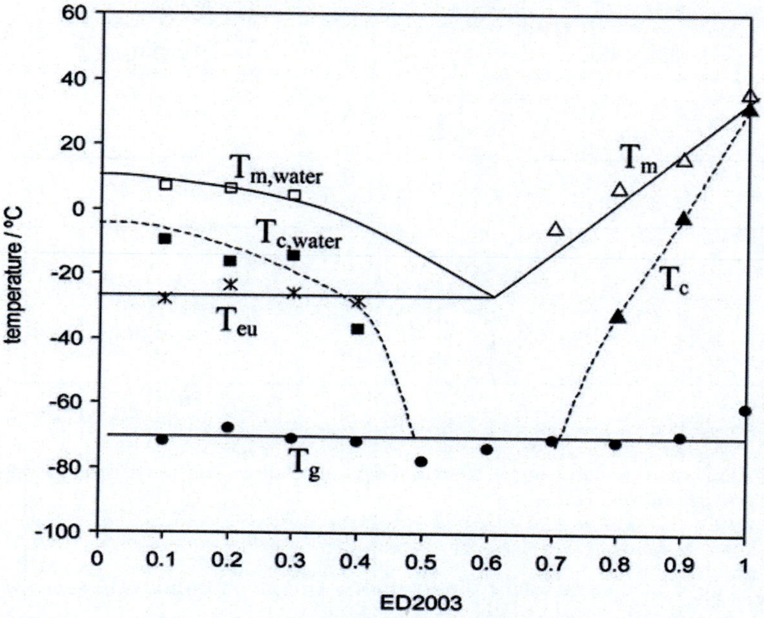


**Fig. 7.6** Experimental phase diagram for the Jeffamine ED2003 aqueous solutions; (●) glass transition, (□) water melting, (■) water crystallization, (Δ) polymer melting, (▲) polymer crystallization, (*) eutectic. *Lines* are a guide to eye (Reproduced from [13] with the permission of Elsevier)

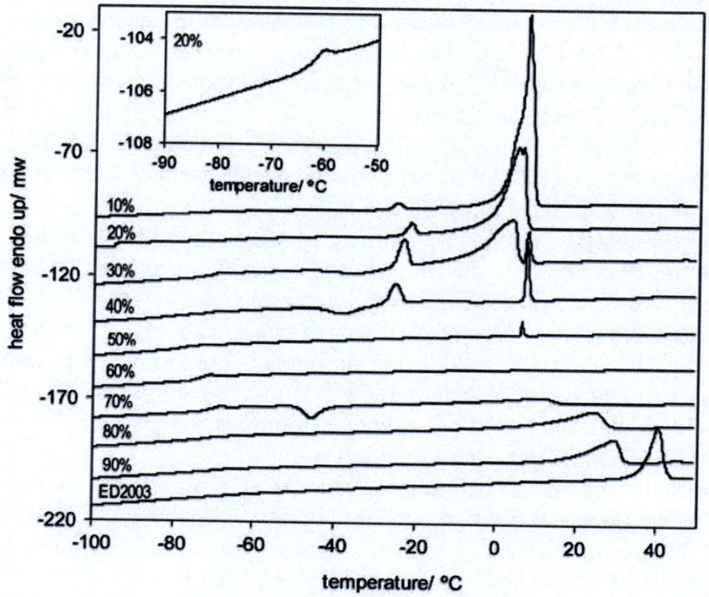


**Fig. 7.7** DSC thermograms at a heating rate of 10°C/min for the Jeffamine ED2003 aqueous solutions. The number on each curve represents the weight percentage of polymer in the solution. The insert shows a detail of the glass transition region for one of the solutions with high water content (Reproduced from [13] with the permission of Elsevier)

Enthalpy relaxation illustrated in Fig. 7.2 can be used in identification of multiphase structure in cases when the difference in glass transitions of components is small. The method is attractive due to its simplicity as illustrated in Fig. 7.8 where thermal behaviour of the polystyrene (PS), poly(methylmetacrylate) (PMMA) and their blend (67/33 % by weight) is shown [17]. The samples were first heated to temperature 150°C which is higher than glass transition temperature of both polymers (106°C and 126°C for PS and PMMA, respectively). Before the heating scans shown in Fig. 7.9, the samples were subject to three different thermal histories: annealing at 92°C for 24 h (Fig. 7.8a), annealing at 98°C for 320 h (Fig. 7.8b) and annealing at 112°C for 24 h (Fig. 7.8c).

In the case of short annealing time (Fig. 7.8a), distinct enthalpy relaxation peak is observed for PS during heating scan. The peak for PMMA is much smaller because of higher $T_g$ of PMMA and correspondingly much slower relaxation time at the annealing temperature. This is also reflected in the heating scan of PS/PMMA blend which has to be phase separated at 150°C. Longer annealing at somewhat higher temperature makes the enthalpy relaxation peak of PMMA more distinct

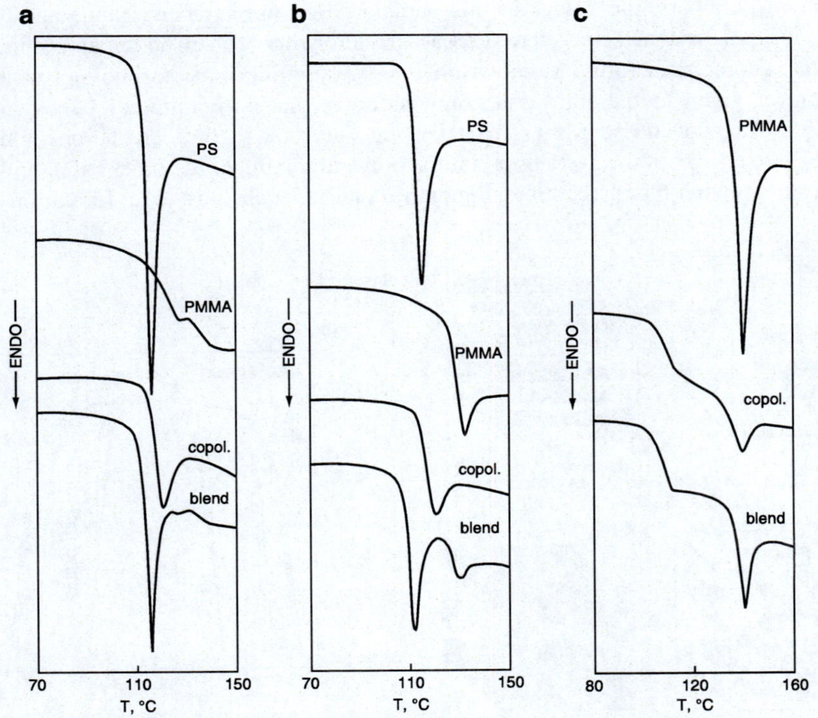


**Fig. 7.8** DSC traces from polystyrene, polymethylmetacrylate, their blend (67/33 by weight) and block copolymer: (**a**) annealed at 92°C for 24 h, (**b**) annealed at 98°C for 320 h, (**c**) annealed at 112°C for 24 h. Note that in this figure endotherms are oriented down (Reproduced from [17] with the permission of The American Chemical Society)

(Fig. 7.8b). Annealing at a temperature between the glass transitions of PS and PMMA for 24 h suppresses (Fig. 7.8c) the height of enthalpy relaxation peak of PS and further sharpens the peak of PMMA.

## 7.3 Block Copolymers

A mixing of polymer chains of two different homopolymers results usually in a two-phase blend which is heterogeneous on micrometer scale. If the pairs of different homopolymer chains are linked covalently (giving rise to block copolymer chains) the phase separation on micrometer scale is not allowed since it would require splitting of the copolymer chains. Instead of that, the material separates into domains of the size commeasurable with the characteristic length of the chain blocks. The magnitude of this length can be estimated using the mean root square of the end-to-end distance of chain blocks, $\sqrt{\langle \mathbf{R}^2 \rangle}$. As it was mentioned above, for common polymers, the values of this parameter are in nanometre range and the domains can be therefore referred to as nanodomains. The nanodomains obtain various geometrical forms as shown in Fig. 7.9. Equilibrium morphology of the system at given temperature and composition is that with minimal Gibbs free energy. Three phenomena must be taken into account in the calculation of the Gibbs free energy of the system: (a) interaction and mixing of the blocks analogical to that in polymer blend, (b) stretching of the copolymer chains and (c) formation of

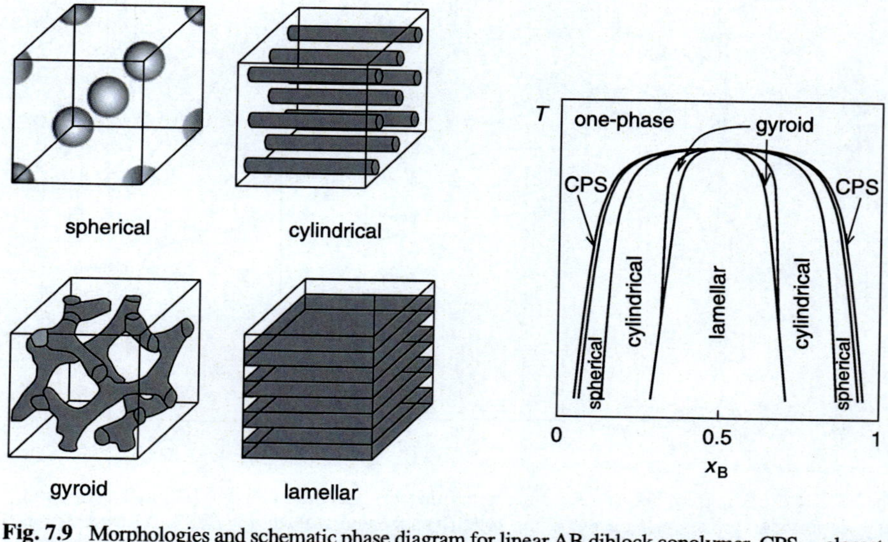


**Fig. 7.9** Morphologies and schematic phase diagram for linear AB diblock copolymer. CPS = closest packed spheres. $x_B$ denotes the molar fraction of B-block (Adapted from [18])

the new surface between (nano)phases, see, e.g. [18]. Of course, morphology of the copolymer has to be identified by other method, such as X-ray scattering or transmission electron microscopy.

## 7.4 Nanophase Separation in Polymer Networks

Polymer networks represent a system in which polymer chains are linked into one molecule of macroscopic dimensions (therefore, "infinite" on the molecular length scale). They can be prepared by cross- or end-linking of existing polymer chains. Cross- or endlinking of one of the components of the (multicomponent) reaction mixture is accompanied by change of the phase diagram of the system as illustrated in Fig. 7.10.

With increasing molecular weight the binodals are shifted to higher temperature and initially homogeneous mixture can become unstable in the course of the reaction. This is illustrated in Fig. 7.11 where changes of structure during end-linking of the mixture of poly(butadiene) diol (PBD) and poly(oxypropylene) triol (POPT) by diphenylmethane diisocyanate (MDI) [19] were monitored by small-angle X-ray scattering (SAXS). Similarly as in block copolymers, the size of phase separation is restricted to nanometer level because of the formation of covalent polymer network.

The phases can be identified using DSC by glass transitions of the components as shown in Fig. 7.12 for a series of nanophase separated polyurethane networks.

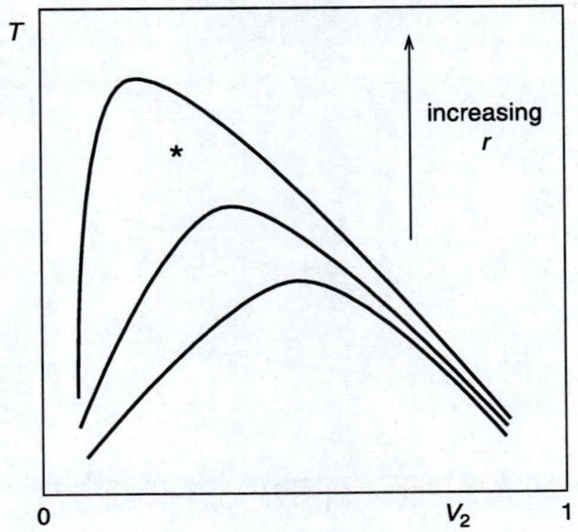


**Fig. 7.10** Change of the phase diagram (shift of bimodal line) with increasing molecular weight of polymer formed by cross- or endlinking (*r* denotes number of chain segments in the Flory-Huggins model). The *asterisk* denotes the state of the reaction mixture at the beginning of the reaction

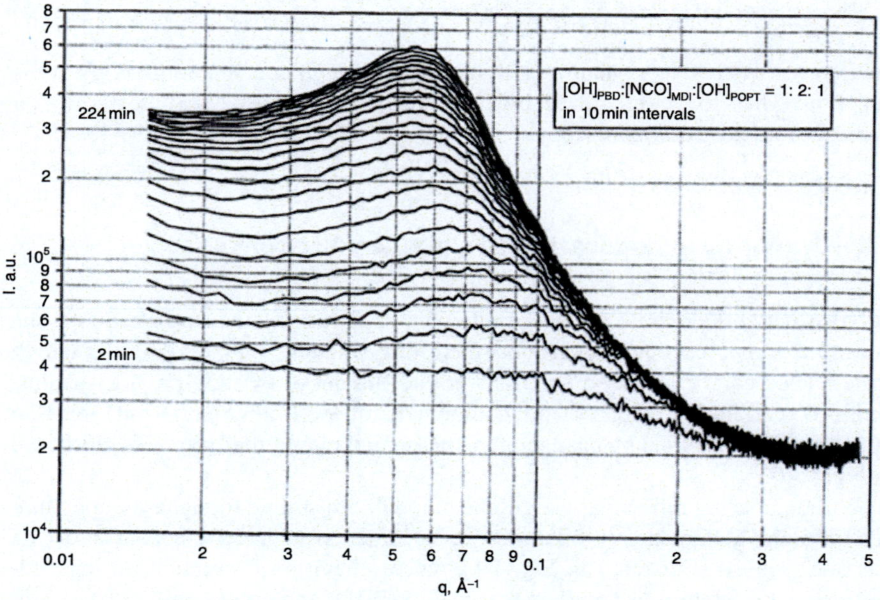


**Fig. 7.11** Time-resolved SAXS patterns from the formation of nanophase separated polyurethane network by endlinking reaction of PBD and POPT with MDI (Reproduced from [19] with the permission of Elsevier)

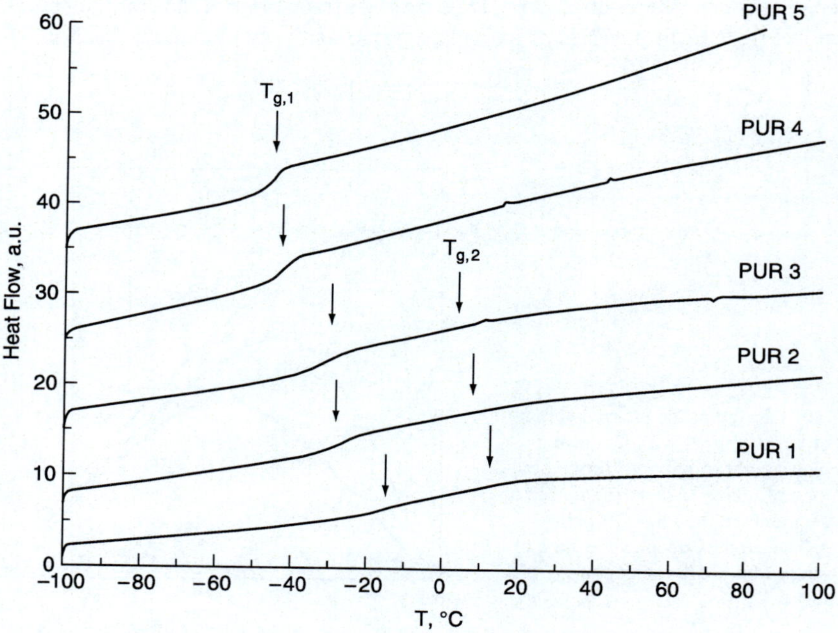


**Fig. 7.12** DSC traces (second heating scans) from the series of nanophase separated polyurethane networks formed by endlinking reaction of PBD and POPT with MDI. The networks differ in molecular weight of PBD which increases from PUR1 to PUR5. The *arrows* indicate phase transitions of soft and hard segment nanophases (Reproduced from [20] with the permission of Elsevier)

**Acknowledgements** Financial support from the Ministry of Education of the Czech Republic (project MSM 0021620835) is gratefully acknowledged.

## References


1. Flory PJ (1993) Spatial configuration of macromolecular chains. In: Forsen S (ed) Nobel lecture. Chemistry 1971–1980. World Scientific, Singapore, pp 147–178
2. Kawakatsu T (2004) Statistical physics of polymers. Springer, Berlin
3. Gibbs JH, Di Marzio EA (1958) Nature of the glass transition and the glassy state. J Chem Phys 28:373–383
4. Di Marzio EA, Gibbs JH (1958) Chain stiffness and the lattice theory of polymer phases. J Chem Phys 28:807–813
5. ten Brinke G, Oudhuis L, Ellis TS (1994) The thermal characterization of multicomponent systems by enthalpy relaxation. Thermochim Acta 238:75–98
6. Flory PJ (1941, 1942) Thermodynamics of high polymer solutions. J Chem Phys 9:660; 10:51–61
7. Huggins ML (1941) Solutions of long chain compounds. J Chem Phys 9:440, Some properties of solutions of long-chain compounds. J. Phys. Chem. 46, 151–158 (1942)
8. Miller AR (1948) The theory of solutions of high polymers. Clarendon, Oxford
9. Eitouni HB, Balsara NP (2007) Thermodynamics of polymer blends. In: Mark JE (ed) Physical properties of polymers handbook. Springer, New York, pp 339–356
10. Šolc K, Dušek K, Koningsveld R, Berghmans H (1995) "Zero" and "off-zero" critical concentrations in solutions of polydisperse polymers with very high molar masses. Collect Czech Chem Commun 60:1661–1688
11. Kiepen F, Borchard W (1988) Light scattering as tool for the determination of adiabatically performed temperature changes. Makromol Chem 189:1543–1550
12. Arnauts J, De Cooman R, Vandeweerdt P, Koningsveld R, Berghmans H (1994) Calorimetric analysis of liquid–liquid phase-separation. Thermochim Acta 238:1–16
13. Gómez Ribelles JL, Salmerón-Sánchez M, de la Torres Osa L, Krakovský I (2005) Thermal transitions in a,w-diamino terminated poly(oxypropylene)-block-poly(oxyethylene)-block-poly(oxypropylene) aqueous solutions and their epoxy networks. J Non-Cryst Solids 351:1254–1260
14. Utracki L (2002) Polymer blends handbook. Springer, New York
15. Olabisi O, Robeson LM, Shaw MT (1979) Polymer–polymer miscibility. Academic, New York
16. Utracki L (1990) Polymer blends and alloys. Hanser Gardner, New York
17. Tsitsilianis C, Staikos G (1992) Phase behavior in PS-b-PMMA block copolymer by enthalpy relaxation. Macromolecules 25:910–916
18. Bates FS, Fredrickson GH (1999) Block copolymers – designer soft materials. Phys Today 52:32–38
19. Krakovský I, Urakawa H, Kajiwara K (1997) Inhomogeneous structure of polyurethane networks based on poly(butadiene)diol. 2. Time-resolved SAXS study of the microphase separation. Polymer 38:3645–3653
20. Krakovský I, Pleštil J, Baldrian J, Wübbenhorst M (2002) Structure of inhomogeneous polymer networks prepared from telechelic polybutadiene. Polymer 43:4989–4996

# Chapter 8
# Solid Forms of Pharmaceutical Molecules

**Bohumil Kratochvíl**

## 8.1 Introduction

A drug discovery is characterized by two stages. The first in terms of time is called "lead structure", followed by a so called "drug candidate" stage. The lead structure stage involves selecting the optimum molecule of the pharmaceutical, while drug candidate stage means selecting the optimum solid form. Usually, five to ten candidates pass to the drug candidate stage and the result is the selection of the final solid API (Active Pharmaceutical Ingredience) for the ensuing formulation of the solid dosage form. The lead structure stage concerns only the discovery of the original drug, the drug candidate stage may concern also generics (a drug which is bioequivalent with original and is produced and distributed after the patent protection of the original).

The choice of the optimal API for a specific solid drug formulation means the optimization of its properties. The most important properties of API include its solubility, dissolution rate and permeability, which are closely related to the oral bioavailability of the drug. Apart from these, there are other properties influencing functional and technological parameters of the API and its patent non-collision status (Table 8.1).

For the selection of the optimal API, several dozens of solid forms may be available from one molecule. An example is atorvastatin calcium, a drug used for the treatment of high cholesterol, for which more than 60 solid forms are patented [1]. Piroxicam, a non-steroidal anti-inflammatory drug was synthesized in more than 50 forms [2] and more than 100 forms have been described for sulphathiazol [3], a local antimicrobial agent. A review of possible chemical and physical types of pharmaceutical solid forms is given in Table 8.2. In the case of multi-component compounds the reduction in number of solid forms is given by the condition of pharmaceutical acceptability of the fellow component (e.g. counterion in the case of salts), see GRAS (Generally Recognized as Safe [4]).

B. Kratochvíl (✉)
Prague Institute of Chemical Technology, Technická 5, CZ-166 28 Praha 6, Czech Republic
e-mail: kratochb@vscht.cz

J. Šesták et al. (eds.), *Glassy, Amorphous and Nano-Crystalline Materials*,
Hot Topics in Thermal Analysis and Calorimetry 8, DOI 10.1007/978-90-481-2882-2_8,

**Table 8.1** The most important parameters for the selection of optimum API

| |
|---|
| Solubility, dissolution rate |
| Hygroscopicity |
| Crystal design or amorphous state |
| Chemical purity (including chiral purity) and physical purity (polymorphism) |
| Physical a chemical stability |
| Powder flow |
| Static charge |
| Porosity |
| Robust manufacture reproducibility |
| Mechanical stress resistance |
| Taste acceptability |
| Must not corrode the tablet machine |
| Non-collision patent status (generic companies) |

**Table 8.2** Chemical and physical types of pharmaceutical solids (API)

| Crystalline forms (all are potentially polymorphic) | Semicrystalline forms | Non-crystalline forms |
|---|---|---|
| Anhydrates | Semicrystalline phases | Amorphous phases |
| Multi-component phases: | | Amorphous hydrates |
| Hydrates (exceptionally ethanol solvates) | | |
| Salts | | |
| Cocrystals | | |
| Glycosylated derivatives | | |

## 8.2 Polymorphs

Most pharmaceutical molecules are polymorphic. Polymorphism (from Greek: *polys – multiple, morfé – shape*) is an ability of a chemical compound to crystallize – depending on crystallization conditions – in different crystal structures alias polymorphs. Molecules in the crystal structure of a polymorph are bonded by weak interactions (H-bridges, Van der Waals forces, π-π interactions). Two general categories of polymorphism are distinguished: packing polymorphism and conformational polymorphism. Packing polymorphs which differ by molecules packing in the crystal structure, are formed by a rigid molecule (e.g. sulphapyridine) while a flexible molecule existing in various conformers forms conformational polymorphs (e.g. L-glutamic acid).

In practice mixed types of polymorphism are often encountered. Labelling of polymorphs is not unified (e.g. I, II, III . . .; A, B, C. . .; α, β, γ) and it occasionally happens that identical polymorphs are named differently by different discoverers. The polymorphism of anhydrates (ansolvates) means that water (solvent) molecule is not involved in the crystal structure of the polymorph. The polymorphism of hydrates (solvates) is called pseudopolymorphism or solvatomorphism. Polymorphs may or may not differ by their crystal shape (habitus). An ability of a compound to form different crystal shapes, while its crystal structure remains the same, is not polymorphism, but crystal morphology (crystal design). Fundamental causes of polymorphism are not known. But a statement by W. McCrone from 1963 is ever confirmed that if a molecule becomes a focus of attention, further polymorphs are discovered. An example can be olanzapine intermediate, the so-called ROY (red-orange-yellow), Fig. 8.1, which is already described in ten polymorphs [5]. On the other hand a very well-known and many times crystallized molecule of sucrose is monomorphous.

Among pharmaceutical molecules, the most frequent case is dimorphism. A well-known example is the patent litigation between pharmaceutical companies Glaxo and Novopharm over two polymorphs of ranitidine hydrochloride [6], which decreases the production of stomach acid, or the problems of the company Abbot Laboratories concerning two polymorphs of ritonavir [7] – an inhibitor of HIV-protease. Since polymorphs differ by their crystal structures, they differ by their properties, of which solubility and dissolution rate are the most important. A typical ratio of solubility (beware various definitions of solubility) of two polymorphs is less than two, but there are exceptions, e.g. polymorphs of premafloxacin I/III or polymorphs of chloramphenicol A/B have this ratio larger than 10 [8]. Thus it can happen that a less soluble polymorph does not even reach the minimum medicinal concentration in blood.

An unwanted polymorph in the mixture is called a polymorph impurity. In a polymorph system only one polymorph is thermodynamically stable, the other are unstable. The stable polymorph is characterized by the lowest Gibbs energy, the lowest solubility in any solvent, the lowest dissolution rate and the lowest reactivity. For a drug formulation the original companies usually choose a stable polymorph, the generic companies then have to use even an unstable and less-lasting polymorph.

Uncontrolled phase transitions of unstable polymorphs into more stable ones are a big problem of pharmaceutical industry. Two types of polymorphous transitions are distinguished, the enantiotropic and the monotropic (Fig. 8.2). The enantiotropic transition is characterized by the transformation temperature $T_{A \to B}$ at which an originally more stable polymorph A transforms into a finally stable polymorph B. The enantiotropic transition is often reversible and well-defined. The monotropic

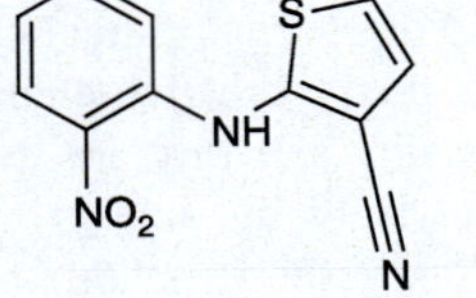


**Fig. 8.1** Molecule of "ROY"

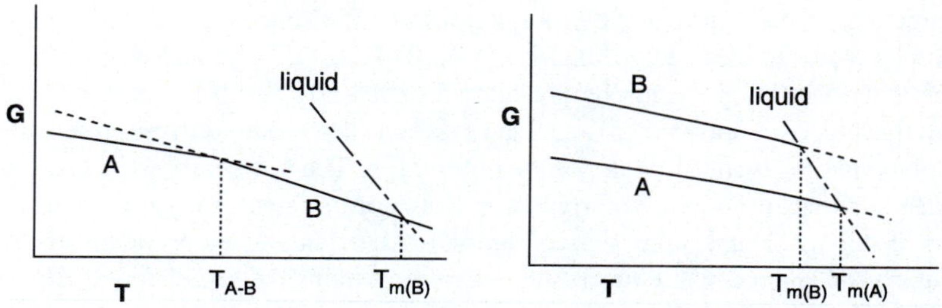


**Fig. 8.2** Enantiotropic (*left*) and monotropic (*right*) polymorphic transformations

**Table 8.3** The critical parameters affecting the controlled crystallization of a wanted polymorph

| |
|---|
| Temperature and pressure |
| Solution cooling rate |
| Solution saturation grade |
| Solvent (precipitant) or a mixture of solvents |
| Water content in solvent |
| Impurities |
| Crystallization additives |
| Saturation rate |
| Standing of product in mother solution |
| Stirring intensity |
| Concentration and temperature gradients |
| Ultrasound, microwave, laser and other shocks |
| Solution pH |

transition in solid state has no transformation temperature, so that the polymorph transition passes over the liquid phase. In practice this means the crystallization from a different solvent. Unfortunately, the polymorphous transitions of pharmaceutical substances are more often monotropic than enantiotropic and moreover hysteretic.

Uncontrolled polymorph transitions in pharmaceutical manufacture may happen during the final crystallization of API, during long-lasting standing of the product in the parent solution, during drying, micronization, tablet pressing, during wet granulation, or even in the tablet during storing. The most important for the production of the wanted polymorph is the final crystallization and the monitoring of all its parameters (Table 8.3) to prevent a potential creation of an unwanted polymorph. Since there are many variable parameters and it is difficult to monitor them all in cases of sensitive polymorph systems, a method of seeded crystallization is often used. In this case of seeding the requested product certain nuclei are added to the oversaturated solution. On them then the product grows.

For pharmaceutical companies, the problem of polymorphism is rather a blocking than a creative element. Sometimes the differences between two polymorphs are tiny and tiny are the differences in properties (e.g. polymorphs of aspirin – acetylsalycilic acid [9]). Nevertheless, polymorphism is closely watched by regulatory authorities and no pharmaceutical manufacturer can afford to ignore it.

## 8.3 Anhydrates and Hydrates

The first choice of API for a solid drug formulation is the anhydrate of active substance (free acid, free base or neutral compound). Anhydrates together with salts form the majority of all drug formulations. If the anhydrate for some reason is not suitable (e.g. it is little soluble, unstable, has complicated polymorphism etc.), then possible hydrates are monitored. The hydrate is most frequent a solvate containing water molecules in its crystal structure. Water molecules can be incorporated in the structure in a stoichiometric manner (stoichiometric hydrates) or non-stoichiometrically (non-stoichiometric hydrates), Fig. 8.3. For the formulation stable stoichiometric hydrates in a lower stage of hydration are chosen in which water molecules are bound to molecules of the active substance by H-bonds. The dehydration of a stoichiometric hydrate often results in the collapse of the crystal structure and the origin of an amorphous phase. Non-stoichiometric hydrates are not suitable for the formulation because the water content in them changes with the partial pressure of water vapour in the ambience and with temperature a thus they difficult to define. In non-stoichiometric hydrates, water is not bound very firmly, it rather fills present cavities in the structure, often without forming H-bridges. The dehydration of non-stoichiometric hydrates does not result in the origin of an amorphous phase but a crystalline anhydrate originates. An example of a non-stoichiometric hydrate is the interstitial water molecules in the cavity of β-cyklodextrin [10]. Other solvates (with the exception of ethanol solvates) are not used for the formulation but can be used as important precursors. For instance polymorhs which are otherwise difficult to attain can be obtained by their desolvation.

The stability of the system anhydrate/hydrate depends on the ambient relative humidity. Many active substances form hydrates, often in a various degree of hydration and stability. If the hydrate is the more stable in the system anhydrate/hydrate then the hydrate has all available reliable proton donors and acceptors better satiated compared to anhydrate (Etter's rule [11]). For instance ergot alkaloid tergurid exists as an anhydrate, a twothird hydrate and a monohydrate, and the stable phase is the monohydrate [12], Fig. 8.4. Tergurid tends to form hydrates eagerly which results in taking up residual water molecules from acetone during

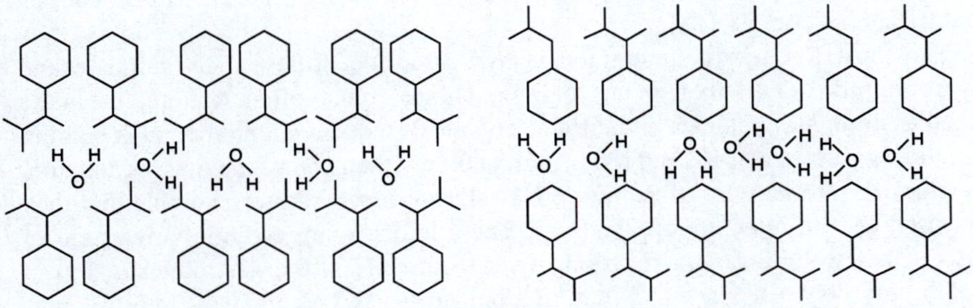

**Fig. 8.3** Stoichiometric hydrate with a regular H-bond network (*left*) and non-stoichiometric hydrate with water molecules in cavities (*right*)

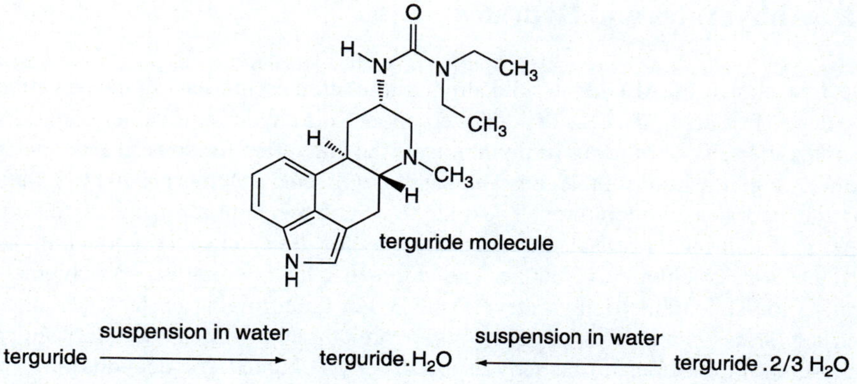


**Fig. 8.4** Transformation pathways among solid forms of terguride

crystallization. Formulations from hydrates are not very frequent and represent only several percent of the total number of APIs (e.g. chloral hydrate, levofloxacin hemihydrate, terpin hydrate and others). The reason is their thermal instability and possibility of the potential dehydration during drying. Of excipients, much used is the lactose monohydrate.

## 8.4 Salts

About a half of all APIs used today are salts. Salts represent a considerable enlargement of the portfolio of solid forms of pharmaceutical molecules. Salts are stable and well soluble in polar solvents (first of all in water), because they contain ionic bond. A necessary prerequisite for the formation of salts is the presence ionizable groups in the molecule (Fig. 8.5). A pharmaceutical substance then can be in the API either in the form of cation (about 75% of pharmaceutical salts) or in the form of anion (about 25% of pharmaceutical salts). The counterpartners must comply with the pharmaceutical acceptability (see GRAS). At present 69 cations and 21 anions comply [13].

The most frequent counteranion is hydrochloride, followed by sulphate and hydrobromide. Only then occur organic anions, most often tartrate, mesylate (methansulfonate), maleate and citrate. The most frequent countercation is sodium ion, followed by $Ca^{2+}$, $K^{+}$ and $Mg^{2+}$ ions, and only then comes the first organic ion, meglumine (N-methyl-D-glucamine). Na-salts are mostly so well soluble that they are used also in injection applications. There is one more essential advantage of salts – their solubility is a function of pH. Since pH in the gastrointestinal tract (GIT) vary between 1 and 7.5 (e.g. in stomach pH is 1–3, in small intestine it is 5–7), it is possible to optimize in GIT the location with the highest solubility by selection of suitable salt. Each salt has a $pH_{max}$ value with the maximum solubility.

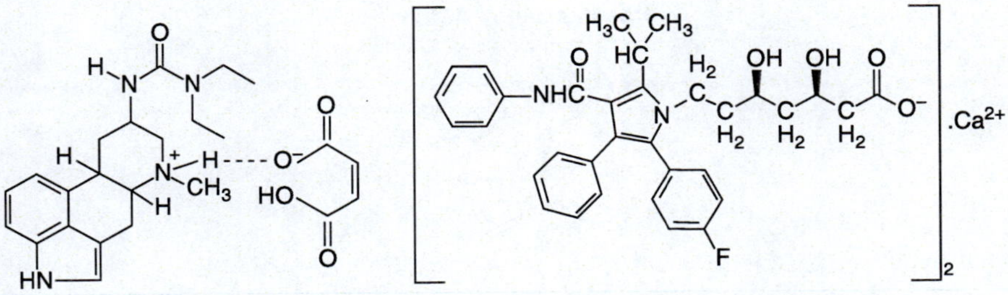


**Fig. 8.5** Salts – terguride hydrogenmaleate (*left*) and atorvastatin calcium (*right*)

The choice of an optimum salt for the solid drug formulation does not mean only finding a substance with the maximum solubility, but also with maximum stability. With growing solubility, diffusibility rises and stability decreases. The substance is easily diffusely dispersed in the organism and penetrates biological membranes. As a result it is less specific as to the site of action and eliminates more readily.

Salts show polymorphism as well but not so effuse as in the case of free acid, free base or neutral compound. The problem of polymorphism can be circumvented by choosing a suitable salt. For instance the ergot alkaloid terguride crystallizes in seven forms as a base, while converted to salt we obtain only one monomorphous terguride hydrogenmaleate monohydrate. The crystallization of API in the form of a salt can be used for the separation of the active substance from the mixture or for its purification. For instance a liquid valproic acid forms solid Na- and Mg-salts.

In a mixture of two or more API it is necessary to consider their mutual interaction. For instance the analgetic proxyfen was originally formulated as a hydrochloride and used together with aspirin in one formulation. But aspirin decomposed easily in the presence of propoxyfen hydrochloride, it was unstable. Only after re-formulation of propoxyfen into napsylate aspirin is stabilized (brand name Darvocet, marketed by Elli Lilly) [14]. Moreover, propoxyfen napsylate is more stable and less toxic compared to hydrochloride. Salts may also form hydrates which can be also used for the formulation. The best known example is atorvastatin calcium trihydrate (Sortis, Pfizer).

## 8.5 Cocrystals

Cocrystals are at present the most dynamically developing group of solid pharmaceutical substances. The definition of the term "pharmaceutical cocrystal" is still under discussion, but essentially it is a multi-component compound that is formed between a molecular or ionic API and a cocrystal former that is a solid under ambient conditions [15], Fig. 8.6. Pharmacodynamically, cocrystal former is a ballast molecule (the same applies to salts), and the GRAS rules apply. Nevertheless even a cocrystal former can be an active molecule.

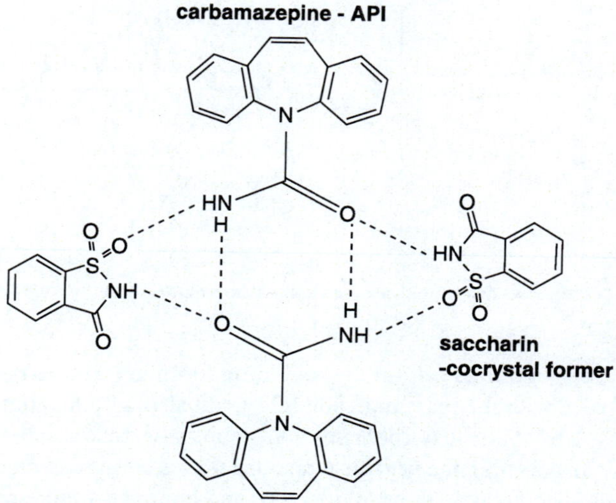


**Fig. 8.6** Cocrystal carbamazepin/saccharin (1:1), *dottes lines* are H-bonds

The stoichiometric ratio of API and cocrystal former in a pharmaceutical cocrystal is mostly simple (1:1, 1:2, 1:3 or vice versa). Cocrystals are not necessarily binary compounds, ternary and quarternary cocrystals are known. Cocrystals can be divided into: cocrystal anhydrates, cocrystal hydrates (solvates), anhydrates of cocrystals of salts and hydrates (solvates) of cocrystals of salts. The borderline between salts and cocrystals is blurred and can be distinguished by the location of the proton between an acid and a base. In salts, carboxyl proton is moved to the hydrogen of the base while in cocrystals the proton remains on the carboxyl of the acid. In cases when $\Delta pKa = pKa\ (base) - pKa\ (acid) = 0–3$, the transfer of proton is ambiguous and we talk about the salt-cocrystal continuum [16].

The cocrystallization potential of some active molecules is studied in detail, e.g. carbamazepine, itraconazole, piroxicam, norfloxacin, fluoxetin, caffein and others [17]. The reason is to achieve a wide variation in solid-state properties of APIs. These efforts stem from principles of supramolecular chemistry and crystal engineering to affect the properties of API through the "bottom up" approach. This is illustrated in the following examples. By the cocrystallization of antifungal drug itraconazole with 1,4-dicarboxylic acids (succinic acid, L-tartaric acid or L-malic acid) a modification of the dissolution profile is achieved compared to the amorphous form of itraconazole (Sporanox, Janssen-Cilag) [18]. A 1:1 carbamazepine/saccharin cocrystal compared to polymorph III of carbamazepine (anticonvulsant Tegretol, Novartis) shows no polymorphous behaviour and is not prone to hydration [19]. The cocrystallization of pregabalin with S-mandelic acid separates from the mixture of R and S isomers only the (1:1) cocrystal (S)-pregabalin/(S)-mandelic acid. This technology is used by Pfizer in manufacturing dosage form Lyrica [20]. The cocrystals of paracetamol show an improved tablet formation ability than free paracetamol, polymorf I (Panadol, GlaxoSmithKline) [21]. Caffein tends to form hydrates at high RH (relative humidity) while its cocrystals with oxalic acid or malonic acid do not have this unwanted property (never form hydrates) [21]. However, general trends of variation of properties during the transition from APIs to their cocrystals are not so far evident because fundamental causes of cocrystallization are not known so far.

The preparation of cocrystals involves a number of techniques, in gas, liquid or solid phase. The most important is the joint cocrystal growth from solution or joint solid state grinding, often with the addition of a small amount of a "molecular lubricant" (methanol, cyclohexan, chlorophorm etc.). Furthermore, cocrystals can be synthesized by evaporation, sublimation, melting, sonication etc. It often holds that identical starting components may not yield the same product under different cocrystallization techniques.

Although cocrystals are intensively studied and patented by both academic institutions and R&D departments of pharmaceutical companies, there is no medicament on the market formulated from a cocrystal. Nevertheless it turns out that some pharmaceutical salts should be re-classified as cocrystals. This is also important for patent litigation.

## 8.6 Glycosylated Derivatives

Glycosylated derivatives (acetals of saccharides) are not usually ranked among solid forms of pharmaceutical molecules in literature [22]. Certainly unjustly because in natural materials the molecules of active substances are often bonded to saccharides, e.g. digitoxin in the plant *Digitalis lanata*.

Glycosylated derivative can be obtained by adding saccharide (sugar) component to the molecules of active substances through a glycosidic bond, Fig. 8.7. This bond can be formed if a hydroxyl group is present in the molecule of the active substance which is bonded to the hemiacetal group of a saccharide. The presence of a saccharide component containing several OH-groups often increases solubility of API in polar solvents.

Moreover, the glycosylation often improves also pharmaco-dynamic properties of the active substance. A well-known example is the antibiotics vankomycin some of whose glycosylated derivatives are 500 times more efficient compared to vankomycin itself [24]. Apart from saccharides, it is possible to bond for instance peptide, or protein to the molecule of active substance and thus to change dissolution profiles and pharmaco-dynamics of these derivatives.

## 8.7 Amorphates

Amorphous forms are thermodynamically metastable which results from the disordering of their inner structure on molecular level. Compared to ordered crystalline phases, amorphates have better molecular mobility which results in a better

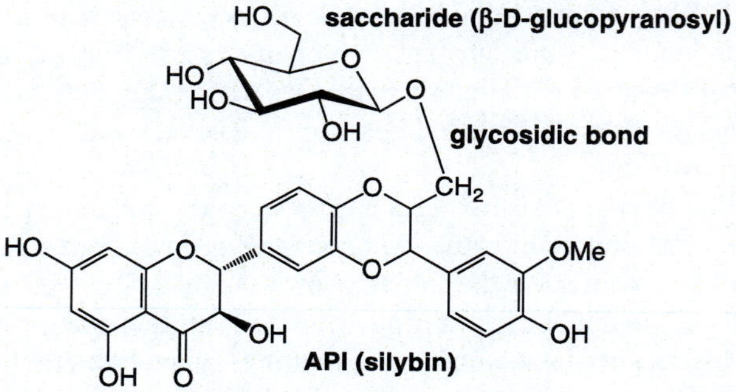


**Fig. 8.7** Glycosylated derivative of hepatoprotectivum silybin. Solubility of silybin is very low (430 mg/l). Silybin glycosides are 4–30 times more water-soluble [23]

dissolution profile and thus a better oral bioavailability. On the other hand this is compensated by lower chemical and physical stability (shorter expiration) and by greater demands on production and storing (e.g. protecting atmosphere).

An empirical rule applies to amorphates: the temperature of storing must be 50°C below the temperature of their glass transition $T_g$ [25]. The amorphous state has a higher energy than the crystalline state and therefore amorphous phases tend to turn into crystalline ones. The crystallization of amorphates is facilitated by their high hygroscopicity and absorbed water acts as a plasticizer increasing molecular mobility. The transition between amorphous and crystalline phases is not sharp and so called semicrystalline phases appear, e.g. atorvastatin calcium, V (Teva) [26].

Tiny differences between amorphous phases of one API (e.g. different methods of synthesis) initiate discussion about polyamorphism (the ability of a substance to exist in several different amorphous forms). Polyamorphism is well defined in inorganic phases (e.g. six- and four-coordinated amorphous silicon) but no polyamorphates have been so far proved in pharmaceutical substances. Current formulations from amorphous phases include asthma medicine, e.g. zafirlukast (Accolate, Astra-Zeneca [27]), quinapril hydrochloride (Accupro, Accupril, Pfizer [28]), antifungal drug itraconazole (Sporanox, Janssen-Cilag [29]) or non-steroidal anti-inflammatory drug indomethacin. (Indocin, Merck [30]). In solid drug formulations, amorphates are stabilized by suitable excipients (e.g. PVP, trehalose, sorbitol, etc.).

Depending on temperature and ambient relative humidity the water content in amorphous phases varies. However, pharmaceutical phases denoted as amorphous hydrates have been patented lately, e.g. amorphous esomeprazole hydrate [31], amorphous cephalosporine hydrate [32] or amorphous imatinibe mesylate hydrate [33]. Although a physical and chemical substance of the term amorphous hydrate is debatable, we can admit that in certain cases a relatively stable amorphous phase containing a defined amount of water may exist.

## 8.8 Conclusion

The portfolio of solid forms of pharmaceutical molecules is nowadays very wide and somehow difficult to overlook. A further increase in number of new co-crystals, or multi-component compounds generally, and their application in solid drug formulations are expected in future. Progress in the theory of chemical bond, prediction of crystal structures and the development of supramolecular chemistry enable better understanding of the fundamentals of polymorphism and control of crystallization processes. This will lead to a better orientation and targeted selection of the optimum solid form of a certain pharmaceutical molecule with requested technological and functional properties.

**Acknowledgments** This chapter was written in the framework of the project MSM 2B08021 of the Ministry of Education of the Czech Republic.

## References

1. Hájková M, Kratochvíl B, Rádl S (2008) Atorvastatin – the world's best selling drug. Chem Listy 102:3–14
2. Childs SL, Hardcastle KI (2007) Cocrystals of piroxicam with carboxylic acid. Cryst Growth Des 7:1291–1304
3. Bingham AL, Hughes DS, Hursthouse MB, Lancaster RW, Tavener S, Threlfall TL (2001) Over one hundred solvates of sulfathiazole. Chem Commun 7:603–604
4. http://www.fda.gov/Food/FoodIngredientsPackaging/GenerallyRecognizedasSafeGRAS/GRASListings/default.htm
5. http://www.pharmacy.wisc.edu/SOPDir/PersonDetails.cfm?ID=32
6. Bernstein J (2002) Polymorphism in molecular crystals. Oxford University Press, New York, p 298
7. Bauer J, Sponton S, Henry R, Quick J, Dziki W, Porter W, Morfia J (2001) Ritonavir: an extraordinary example of conformational polymorphism. Pharm Res 18:859–866
8. Pudipeddi M, Serajuddin ATM (2005) Trends in solubility of polymorphs. J Pharm Sci 94:929–939
9. Bond AD, Boese R, Desiraju GR (2007) On the polymorphism of aspirin. Angew Chem Int Ed 46:615–617
10. Steiner T, Kellner G (1994) Crystalline beta-cyclodextrin hydrate at various humidites - fast, continuous, and reversible dehydration studies by X-ray diffraction. J Am Chem Soc 116:5122–5128
11. Etter MC, Urbanczyk-Lipkowska Z, Zia-Ebrahimi M, Panunto TW (1990) Hydrogen bond directed cocrystallization and molecular recognition properties of diarylureas. J Am Chem Soc 112:8415–8426
12. Hušák M, Kratochvíl B, Císařová I, Cvak L, Jegorov A, Böhm S (2002) Crystal forms of semisynthetic ergot alkaloid terguride. Collect Czech Chem Commun 67:479–489
13. Stahl PH, Wermuth CG (eds) (2002) Handbook of pharmaceutical salts: properties, selection, and use. Wiley-VCH, Weinheim
14. http://en.wikipedia.org/wiki/Dextropropoxyphene

15. Vishweshwar P, McMahon JA, Bis JA, Zaworotko MJ (2006) Pharmaceutical co-crystals. J Pharm Sci 95:499–516
16. Childs SL, Stahly PG, Park A (2007) The salt-cocrystal continuum: the influence of crystal structure on ionization state. Mol Pharm 4:323–338
17. Schultheiss N, Newman A (2009) Pharmaceutical cocrystals and their physicochemical properties. Cryst Growth Des 9:2950–2967
18. Morissete SL, Almarsson Ő, Peterson ML, Remenar J, Read M, Lemmo A, Ellis S, Cima MJ, Gardner CR (2004) High-throughput crystallization: polymorphs, salts, co-crystals and solvates of pharmaceutical solids. Adv Drug Deliv Rev 56:275–300
19. Hickey MB, Peterson ML, Scoppettuolo LA, Morisette SL, Vetter A, Guzman H, Remenar JF, Zhang Z, Tawa MD, Haley S, Zaworotko MJ, Almarsson O (2007) Performance comparison of a co-crystal of carbamazepine with marketed product. Eur J Pharm Biopharm 67:112–119
20. Zaworotko MJ (2008) Crystal engineering of cocrystals and their relevance to pharmaceuticals and solid-state chemistry. In: XXI congress of the international union of crystallography, Book of Abstracts C11. Osaka
21. Jones W (2009) Multicomponent crystals in the development of new solid forms of pharmaceuticals. In: 25. European Crystallographic Meeting (ECM 25), Abstracts p. 102. Istanbul
22. Hilfiker R (ed) (2006) Polymorphism in the pharmaceutical industry. Wiley-VCH Verlag, Weinheim
23. Křen V et al (1997) Glycosylation of silybin. J Chem Soc, Perkin Trans 17:2467–2974
24. Nagarajan R (1993) Structure-activity relationship of vancomycin-type glycopeptide antibiotics. J Antibiot 46:1181–1195
25. Hancoek BC, Zografi G (1997) Characteristics and significance of amorphous state in pharmaceutical systems. J Pharm Sci 86:1–12
26. Teva Pharmaceutical Industries Ltd Patent WO 01/36384 A1
27. Accolate (2008) http://www.astrazeneca-us.com/pi/accolate.pdf
28. Accupro (Accupril) (2008) http://www.pfizer.com/files/products/uspi_accupril.pdf
29. http://www.janssen-cilag.com/product/filtered_list.jhtml?product=none
30. http://www.merck.com/product/usa/pi_circulars/i/indocin/indocin_cap.pdf
31. http://www.faqs.org/patents/app/20080293773
32. http://www.freepatentsonline.com/7244842.html
33. Parthasaradhi et al Novel polymorphs of imatinib mesylate. Patent US2005/0234069A1

# Chapter 9
# Chalcogenide Glasses Selected as a Model System for Studying Thermal Properties

**Zdeněk Černošek, Eva Černošková, and Jana Holubová**

## 9.1 Introduction

Chalcogenide glasses have been intensively studied from the seventieth of twentieth century as the important new class of promising high-tech materials for semiconducting devices and infrared optics. Chalcogenide glasses are formed by chalcogens, stoichiometric chalcogenides, e.g. germanium and/or arsenic sulfides or selenides or by non-stoichiometrics alloys whose composition (and physicochemical properties) can be modified in broad ranges. They have unique optical properties – low phonon energies as compared with oxide glasses, high refractive index, infrared luminescence and so on. The advantage of many chalcogenide glasses is that they can be obtained using very simple technologies.

Besides applicability in many areas outlined above chalcogenide glasses represent excellent model materials for thermoanalytical studies. Extremely stable glasses and undercooled melts can be prepared, as well as very temperature-sensitive unstable ones. Glass transition temperature can be changed easily from subambient temperatures (glassy sulphur, for example) up to temperatures over 500°C (germanium disulphide or diselenide). This make chalcogenide glasses ideal candidates for differential scanning calorimetry (DSC) studies, because these apparatus operate in mentioned temperature region.

Z. Černošek (✉) and J. Holubová
Faculty of Chemical Technology
Department of General and Inorganic Chemistry, University of Pardubice,
nám. Legií 565, 53210 Pardubice, Czech Republic
e-mail: Zdenek.Cernosek@upce.cz

E. Černošková
Joint Laboratory of Solid State Chemistry of Institute of Macromolecular Chemistry,
Academy of Sciences, Czech Republic and University of Pardubice,
Studentská 84, 53210 Pardubice, Czech Republic

J. Šesták et al. (eds.), *Glassy, Amorphous and Nano-Crystalline Materials*,
Hot Topics in Thermal Analysis and Calorimetry 8, DOI 10.1007/978-90-481-2882-2_9,

All chalcogenide glasses used in following studies were prepared by direct synthesis from high purity elements (semiconductor purity – 5N elements) in evacuated quartz ampoules in a rocking furnace and quenched in air or in cold water. Other glasses used (organic polymers, oxide glasses) were commercial products.

## 9.2 Differential Scanning Calorimetry

The development of differential scanning calorimetry (DSC) technique in last 2 decades was done by introduction of commercially available calorimeters with temperature modulation, MDSC, by TA Instruments. The temperature of conventional heat-flux DSC heating block is sinusoidally modulated and so the sample temperature is modulated in the same manner about a constant ramp. The resulting instantaneous heating rate varies sinusoidally about the underlying heating rate (average heating rate). The average heat flow is called total heat flow. This one is the only quantity that is available and hence it is the only quantity that is always measured in conventional DSC experiments. The sample temperature and the amplitude of instantaneous heating flow are measured and finally, using Fourier transformation of the experimentally obtained data, the quantity termed reversing heat flow is obtained. The nonreversing heat flow is the difference between the total heat flow and the reversing heat flow and represents heat flow due to kinetically hindered process. Process is called reversing if the system responds in a reversible way on the timescale of the experiment (or faster) and nonreversing if the system is either too slow to respond reversibly on the timescale of the experiment or if it is irreversible altogether (on any timescale). In case of MDSC it means that reversing process is in-phase with temperature modulation and nonreversing process with some phase lag is out-of-phase.

Reversing isobaric heat capacity can be determined by MDSC using the magnitude of heat flow and heating rate obtained by averaging over one modulation period. As a complex heat capacity has been defined, see [1], the nonreversing heat flow recalculation to the nonreversing heat capacity has also been used. Detailed information about MDSC one can found in [2–5] and references therein.

More recent and from MDSC essentially different technique, StepScan DSC by Perkin–Elmer, is based on enthalpic method of isobaric heat capacity, $C_p$, determination adapted to the high sensitive power-compensated apparatus. This method allows equilibration of the system after each step in a series of small step increases (decreases) in temperature. The area under the resulting curve (the total enthalpy change in the step) is evaluated and divided by the temperature step to give the heat capacity at the midpoint of the temperature step. Enthalpic changes connected with possible kinetic effects are recorded in the timescale during equilibration after each one temperature step. StepScan DSC method allows obtaining not only $C_p$ at the midpoint of the temperature step but also enthalpic changes connected with slow processes (compare to the time of temperature change) after temperature step. Two curves are obtained as a result. The first of them is the temperature dependence of

$C_p$ (reversible part) and the second one is the temperature dependence of slow processes' enthalpy changes (irreversible part); these parts have been termed by Perkin–Elmer as thermodynamic and kinetic ones, respectively.

The result obtained by StepScan DSC seems to be close to this one obtained by MDSC, but two crucial differences should be stressed. Firstly, when TMDS is used, the sample temperature is continually periodically changed, aside from the extent of possible kinetic effects, whereas in the case of StepScan DSC the software-controlled variable isotherm duration allows the sample to achieve the state close to the thermal equilibrium at each temperature step. Secondly, no special mathematical operation, like Fourier transformation, is needed to obtain results by StepScan DSC.

## 9.3 Glass Transition

### *9.3.1 Capability of Conventional DSC, Temperature Modulated DSC (MDSC) and Stepscan DSC for the Glass Transition Phenomenon Study*

Bulk glass of $As_2S_3$ was used as the model glass. The power-compensated differential scanning calorimeter Pyris 1 operated with Pyris software (both Perkin–Elmer) capable of working in all three modes under study was used. Conventional DSC mode was used with heating rates successively 1, 10, 20, 50 and 100 K/min. For experimental curves, see Fig. 9.1.

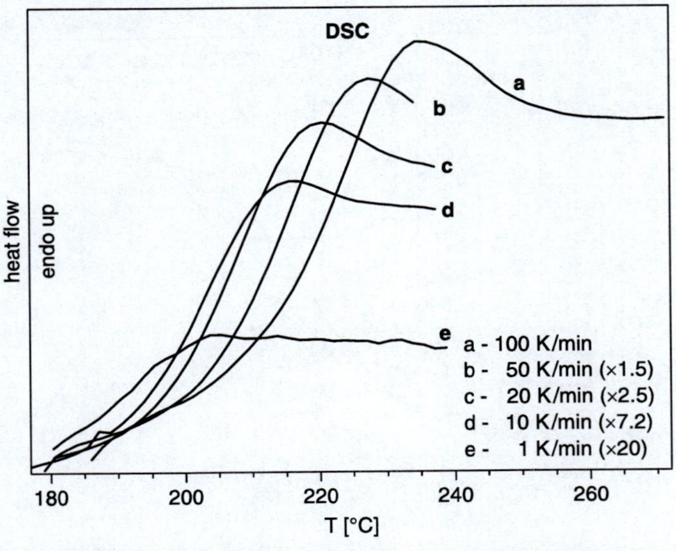


**Fig. 9.1** Conventional DSC results. Number in *brackets* is magnification factor

Dynamic DSC (DDSC) is the Perkin–Elmer version of MDSC. To avoid possible confusion the MDSC will be used as a common mark for temperature modulated DSC, despite of calorimeter manufacturer. Dynamic mode operated with saw-tooth modulation was used with period 40 s, temperature amplitude 0.75 K and heating/cooling rates 2/−1, 4/−2 and 10/−5 K/min. These experimental conditions correspond to underlying heating rate 0.5, 1.0 and 2.5 K/min, respectively. For temperature dependence of reversing $C_P$, see Fig. 9.2.

StepScan DSC experiments were carried out with temperature step 1 K, heating rates in the temperature step successively 1, 10 and 100 K/min and isotherm duration either 60 s. or with maximal allowed heat flow difference ±0.1 μW per approx. 2 s. before next step. The experimental setup corresponds to average underlying heating rate 0.50, 0.91 and 0.99 K/min, respectively. For results, see Fig. 9.3.

In the glass transition temperature range the influence of three available DSC methods on the determination of glass transition temperature, $T_g$, was studied. $T_g$ was determined as a temperature of half-change of heat flow (DSC) or isobaric heat capacity, $\Delta C_p$, (MDSC, StepScan DSC). Results are collected in Fig. 9.4. The $T_g$ dependence on the heating rate obtained by conventional DSC was significant as it is well-known and is discussed elsewhere [6–14].

Results obtained by MDSC show still distinguishable dependence of $T_g$ on heating rate (more correctly on underlying heating rate) even though these rates are slow at all. The $T_g$ values are from 10°C up to 22°C higher comparing with conventional DSC results at the comparable heating rates. Furthermore, it must be stressed that with increasing underlying heating rate besides the glass transition temperature also temperature dependence of isobaric heat capacity, $C_p$, has been

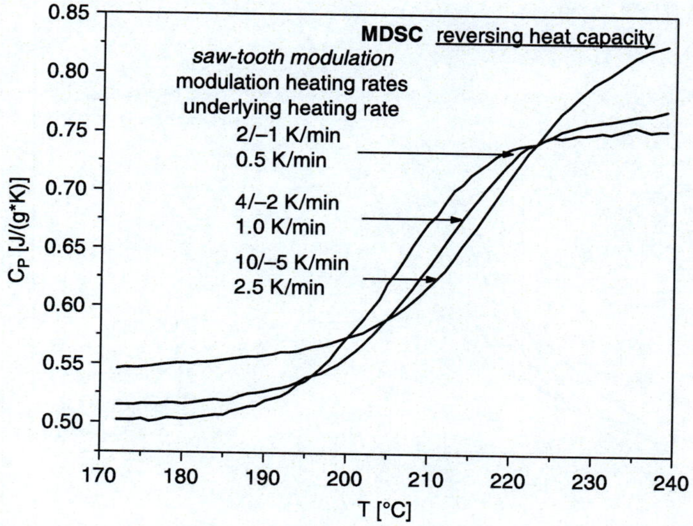


**Fig. 9.2** Reversing $C_P$ dependence on T obtained from MDSC measurements. For details, see text

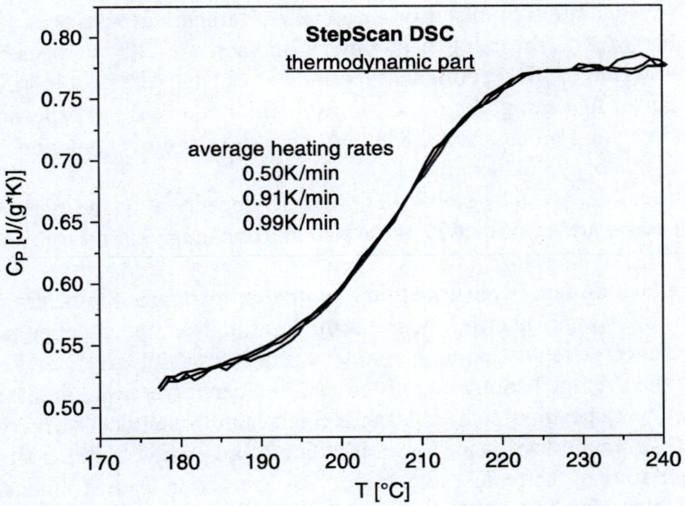


**Fig. 9.3** StepScan DSC thermodynamic (reversing) part at different average heating rates, see text

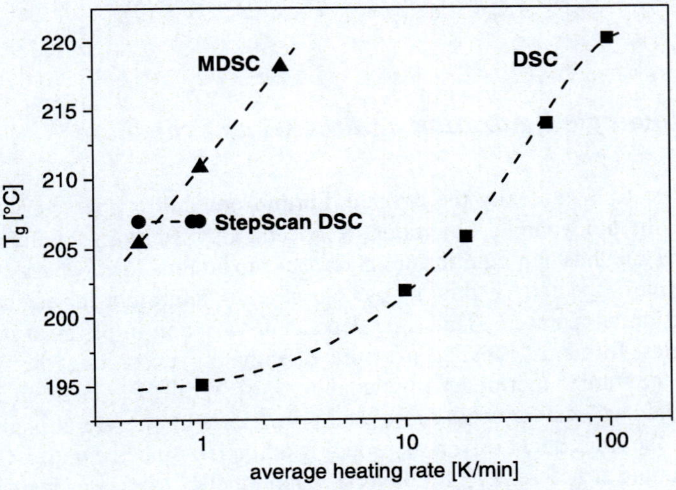


**Fig. 9.4** Heating rate dependencies of $T_g$, obtained by indicated methods

shifted up but without changing of isobaric heat capacity change at glass transition, $\Delta C_p$, see Fig. 9.2.

From above mentioned follows that experimental MDSC dependencies have been moved up in both axis when underlying heating rate increases. It means that MDSC requires both careful calibration and choice of experimental set up. It is well known that both $T_g$ and $C_p$ depend not only on frequency but also on amplitude of temperature modulation (it means on mean rate of temperature change). It seems to be clear that MDSC reversing heat flow is in fact not fully in phase with the temperature change. The permanent periodical change of sample temperature probably causes that sample is not close to thermal equilibrium, especially in the case when reversing process is attended by some slow nonreversing one. The glass transition is one of typical examples.

The values of $C_p$ given by StepScan DSC are independent on the average heating rate and in consequence of that $T_g$ value remains unchanged, see Figs. 9.3 and 9.4, and also [15].

Average heating rate is governed both by the temperature step heating rate and duration of following isotherm. The isotherm duration is variable, depending on the amount and average relaxation time of kinetic processes at the given temperature. It means that the average heating rate of the one step can differ more or less from the other one at every heating-isotherm step. It can be simply said that in every step the StepScan DSC method will wait for termination of all processes being slower then experimental time of the temperature step, of course within the instrument sensitivity. As follows, the heat flow of nonreversing (kinetic) process is effectively separated from reversing (thermodynamic) one and so the temperature dependence of isobaric heat capacity, and from it $T_g$, can be determined without influence of both thermal history of glass and experimental conditions.

### 9.3.2 *Enthalpic Relaxation and the Glass Transition*

The glassy state is generally the non-equilibrium one and is characterized by an excess of thermodynamic quantities (e.g. enthalpy, entropy, volume). The as-quenched and thus non-equilibrium glass seeks to attain a lower energy metastable equilibrium especially if this one is held at temperature not too far below the glass transition temperature. This time dependent variation in physical properties following glass formation is called structural or enthalpic relaxation if the change of enthalpy is the studied thermodynamic quantity [16]. While the glass transition has been described as “fast” process associated primarily with the vibrational degrees of freedom, the subsequent slow structural relaxation is connected with a change in the frozen liquid structure [17]. So the relaxation kinetics of glasses is determined not only by the thermodynamic temperature, T, but also by the instantaneous structure of the glass, which is characterized by fictive temperature, $T_f$, firstly introduced by Tool [18]. The fictive temperature is defined as the temperature at which the observed value of an intensive quantity would be the equilibrium one. During relaxation $T_f$ approaches relaxation temperature $T_r$. In the metastable equilibrium $T_f = T_r$ and the departure from equilibrium is measured by $|T_f - T|$. The initial non-equilibrium state of glass is not unique but depends on the conditions of glass formation and the relaxation process is strongly influenced by the complete thermal history of glass.

Besides the great interest of structural relaxation the considerable attention has been paid to the glass transition. Historically, the glass transition has been observed in undercooled liquids and, therefore, is regarded as a characteristic property of liquids. Later experiments, however, showed that this phenomenon is quite common for non-crystalline materials prepared by various methods other than traditional liquid cooling, see [19] and references cited in.

The glass transition temperature is "defined" by various inconsistent manners, for example:

1. The glass transition temperature denotes a temperature for which relaxation time in the supercooled liquid become longer than typical observation time. The ratio between these two times, the Deborah number, is approximately unity for $T = T_g$. (Prophetess Deborah is claimed to have stated, "...the mountains flowed before the Lord...", Old Testament, Judges 5, verse 5.)
2. The glass transition temperature is the one at which the viscosity of the supercooled liquid reaches $10^{13}$ Poise. It is important to note that the viscosity is continuous through $T_g$, exhibiting none of the discontinuities observed in heat capacity [20] and thus this definition has no physical meaning.
3. The glass transition temperature is the one at which configurational entropy vanishes during melt cooling.

Notwithstanding these "definitions" the glass transition is characterized by a gradual break in slope of extensive thermodynamic quantities (enthalpy, entropy and volume). The region over which the changes of slope occur is termed glass transition region. This region is usually characterized by midpoint temperature called glass transition temperature, $T_g$. Continuous change of extensive thermodynamic quantities through the glass transition implies that there must be a discontinuity in derivative variables at $T_g$, such as a heat capacity or a coefficient of thermal expansion. Such differences are used to distinguish two classes of glass forming liquids – strong and fragile [21]. For review on supercooled liquids and the glass transition refer to [22] and references cited in.

It is well-known that $T_g$ is not regarded as a material constant because when measured for instance by DSC it depends on many parameters as for example heating rate, $q^+$, [6–8], the cooling rate, $q^-$, [8, 9] and the physical aging [7, 10, 11]. If the $T_g$ is determined by heating the temperature obtained often differs from the one from cooling measurement. These values of $T_g$ may vary in the range of 10–20% depending on difference of cooling rates and heating rates.

The nature of the glass transition is very complex and poorly understood so far. It is clear that regarding long structural relaxation time relatively to laboratory time scale during measurement the material is out of thermodynamic equilibrium. Elimination of this influence of scanning rate on determining of $T_g$ was the main aim of some models. Expression relating dependence of $T_g$ on the cooling rate was derived by Kovacs [12]. More often the linear dependence of $T_g$ on $\ln(q^+)$ proposed by Lasocka [13] has been used. Extrapolation of the experimental results to $q^{\pm} = 0$ K/min in order to obtain the "correct" glass transition temperature has been suggested in [23]. It must be noted that logarithmic dependence lacks of physical meaning when $q^{\pm}$ is less then 1 K/min or limited to infinity and thus logarithmic model only hardly can be correct.

### 9.3.3 *Enthalpic Relaxation – The Model*

Enthalpic relaxation (physical aging, structural relaxation) of glassy materials has been studied by a number of techniques, but in particular the differential scanning calorimetry (DSC) has been used extensively to measure the kinetics of enthalpic relaxation of glasses.

According to the Tool's concept of fictive temperature, the specific enthalpy of a glassy sample can be expressed as a function of fictive temperature, $T_f$, and thermodynamic temperature, T:

$$H(T,T_f) = H(T_o,T_o) + \int_{T_r}^{T_f} C_{pm}(T)dT + \int_{T_f}^{T} C_{pg}(T)dT \tag{9.1}$$

where $C_{pm}$, $C_{pg}$ are specific isobaric heat capacities of metastable melt and glass, respectively, and $T_o$ is an arbitrary sufficiently high reference temperature at which the sample is in a metastable thermodynamic equilibrium. Narayanaswamy generalized Tool's model [24] by allowing for distribution of relaxation time and obtained the following expression for the fictive temperature that can be calculated for any thermal history:

$$T_f(t) = T(t) - \int_0^t dt' \left(\frac{dT}{dt}\right)_{t'} M_H[\xi(t) - \xi(t')] \tag{9.2}$$

T is time, $M_H$ is a Kohlrausch–William–Watts (KWW) relaxation function:

$$M_H(\xi) = \exp(-\xi^{\beta}) \tag{9.3}$$

$\beta$ is the non-exponentiality parameter ($0 < \beta \leq 1$), which is inversely proportional to the width of a distribution of relaxation times of independent relaxation processes. $\xi$ is the dimensionless reduced relaxation time:

$$\xi(t) = \int_0^t \frac{dt'}{\tau(t')} \tag{9.4}$$

The contribution to the relaxation time $\tau(T,T_f)$, simply $\tau$, from both the temperature and fictive temperature is controlled by a non-linearity parameter x ($0 \leq x \leq 1$) according to the Tool–Narayanaswamy–Moynihan (TNM) equation [9]:

$$\tau = \tau_0 \exp\left[\frac{x\Delta h^*}{RT} + \frac{(1-x)\Delta h^*}{RT_f}\right] \tag{9.5}$$

where $\tau_0$ is a constant, $\Delta h^*$ is an apparent activation energy, R is the universal gas constant.

### 9.3.4 *Enthalpic Relaxation*

Perkin–Elmer Pyris 1 DSC calorimeter was used for enthalpic relaxation measurements. All experiments were carried out without removing the sample from the instrument. To ensure good thermal contact of glass and aluminium pan and to minimize thermal gradient inside sample thin disks (thickness less than 1 mm) of glassy sample for relaxation study was prepared directly in calorimeter. Encapsulated powder of bulk glass (approx. 10 mg) was melted and equilibrated at 420°C ($T_m(As_2Se_3) = 375$°C) and subsequently cooled to 50°C with rate $q^- = -100$°C/min. Glass prepared by such a way was immediately heated by heating rate $q^+ = +100$°C/min onto relaxation temperature $T_r$. After isothermal relaxation the sample was cooled down to the temperature 50°C by cooling rate q = -100°C/min. After this the DSC curve was recorded up to 420°C by heating rate $q^+ = +20$°C/min. This scan was used for computer simulation. Glasses were isothermally relaxed at temperatures 145, 150, 155, 160, 165 and 170°C with duration between 15 min and 35 h. All in-instrument steps were computer controlled using Pyris 1 software.

Generally the relaxation enthalpy, ΔH, corresponds to area of so called overshoot on the DSC heating scan. The values of ΔH were obtained as a difference between overshoot areas of relaxed glass scan and non-relaxed glass one. The relaxation enthalpy, ΔH, increases with increasing time (or duration) of relaxation, $t_r$, at every relaxation temperature, $T_r$, used for isothermal aging. Relaxation enthalpy, ΔH, reaches its limit value, $\Delta H_{eq}(T_f = T_r)$ after the sufficiently long time, $t_r$, at each isothermal relaxation. This means that glass achieves a metastable equilibrium at given temperature. The dependence of obtained values of $\Delta H_{eq}$ on relaxation temperature is shown in Fig. 9.5. When relaxation temperature decreases the value of $\Delta H_{eq}$ increases, but not linearly. At relaxation temperatures sufficiently below $T_g$ ($T_r \sim T_g - 30$°C) the enthalpy changes from that one of non-relaxed glass to the enthalpy of metastable equilibrium of glass and achieves its final value $\Delta H_{eq}{}^{max}$. This value does not change with further decreasing of $T_r$, see Fig. 9.5. For $As_2Se_3$ glass the maximal enthalpic change is $\Delta H_{eq}{}^{max}(As_2Se_3) \sim 6.4$ kJ/mol.

DSC curves were normalized to pass from zero to unity as the sample goes from glassy state to the equilibrium undercooled liquid state. When above mentioned model of enthalpic relaxation was used and all normalised DSC scans were computer simulated the complete set of parameters of TNM model was obtained for each of relaxation. This set contains the parameter of non-exponentiality, β, non-linearity, x, fictive temperature, $T_f$, and apparent activation energy, $\Delta h^*$.

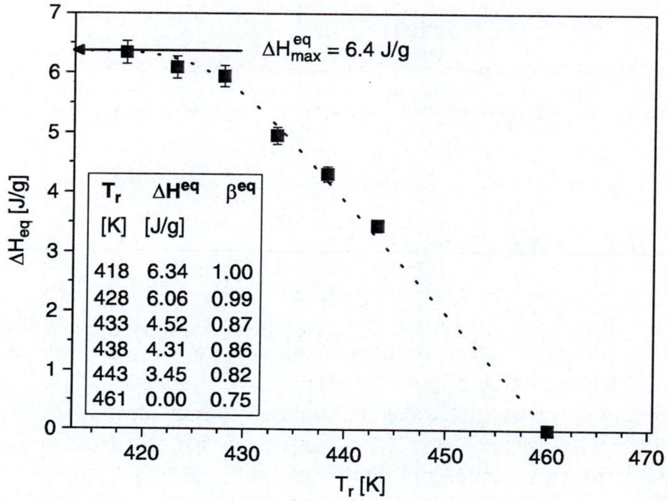


**Fig. 9.5** Dependence of total relaxation enthalpy, $\Delta H_{eq}$, on relaxation temperature $T_r$

Non-exponentiality parameter β was found not to be constant in both the time scale of isothermal relaxation and the temperature scale of the set of $T_r$. That parameter increases with increasing time of relaxation at every relaxation temperature used. The time dependence of β at two different relaxation temperatures is shown in Fig. 9.6. As one can see, after sufficiently long duration of relaxation (when $T_f = T_r$) the parameter β achieves its final value $\beta_{eq}(T_r)$ corresponding to the metastable equilibrium structure of glass at $T_r$. Value of $\beta_{eq}$ increases with decreasing $T_r$, see inset in Fig. 9.5. Based on that one can conclude that the non-exponentiality parameter is both time and temperature dependent, $\beta = f(t_r, T_r)$.

It was found that for given relaxation temperature the value of $\beta_{eq}(T_r)$ is independent on the way in which the metastable equilibrium was reached, Fig. 9.7. The as-quenched glass was subsequently completely relaxed at $T_f = T_r = 165°C$ and the value of $\beta_{eq} = 0.86$. Another sample of glass was completely relaxed at $T_f = T_r = 145°C$ ($\beta_{eq} = 1.00$) and after temperature jump it was completely relaxed again at $T_r = 165°C$. The $\beta_{eq}$ was then found 0.87, within error the same as it was found in previous experiment. It is necessary to emphasize that the first relaxation at 165°C is the exothermic process and the second relaxation is the endothermic one, see Fig. 9.7.

In the case when the change of the relaxation enthalpy reaches its maximal value, $\Delta H_{eq}{}^{max}$ (as it was mentioned above) the value of $\beta_{eq}$ reached its maximal value $\beta_{eq}{}^{max} = 1$.

Non-linearity parameter, x, increases as $T_f$ is getting near $T_r$ during isothermal relaxation. This parameter was found in range 0.60–0.75 and these values reflect relatively small influence of structure to relaxation time. During annealing this influence still decreases.

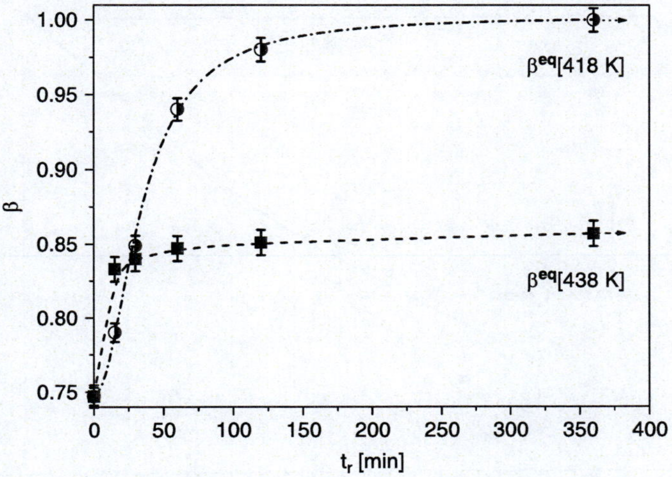


**Fig. 9.6** The time dependence of non-exponentiality parameter β

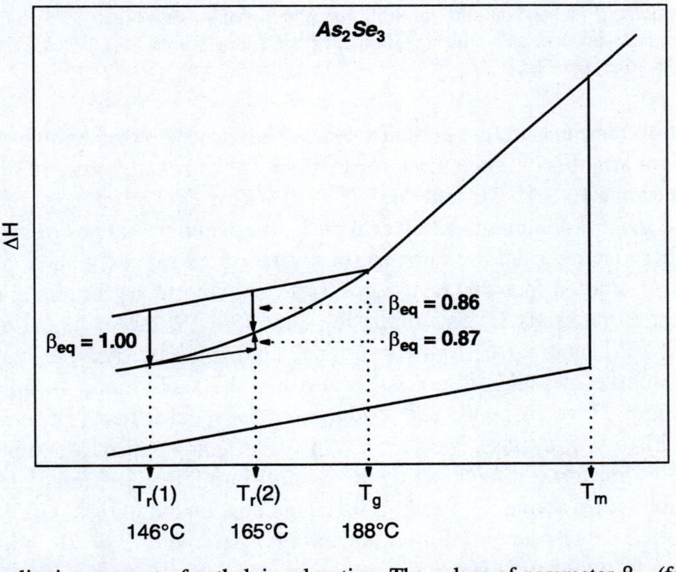


**Fig. 9.7** Qualitative concept of enthalpic relaxation. The values of parameter $\beta_{eq}$ (for $T_f = T_r$) were obtained experimentally

The apparent activation energy is nearly constant and independent on both temperature and time of relaxation, $\Delta h^*(As_2Se_3) = 263 \pm 15$ kJ/mol.

It was found that glass reached the metastable equilibrium at each of relaxation temperature used. Corresponding limiting value of the enthalpy change, $\Delta H_{eq}(T_r)$, are indispensable lower than the expected values from linear extrapolation of melt equilibrium enthalpy, see Fig. 9.8. At certain temperature, $T_o$, sufficiently below the

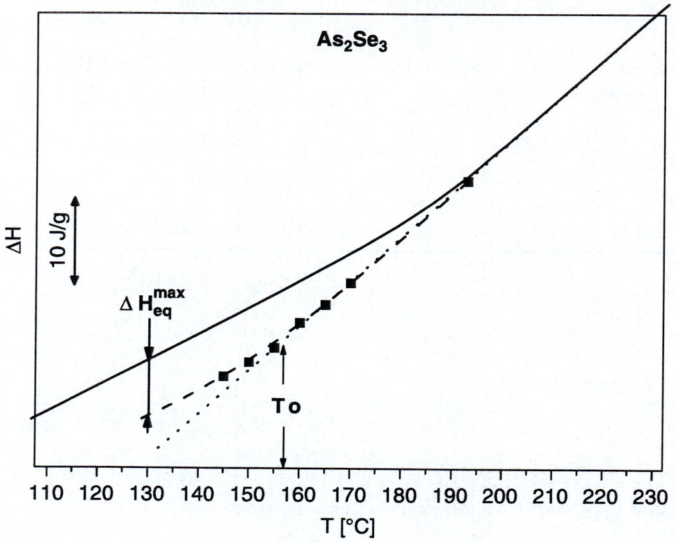


**Fig. 9.8** Temperature dependence of equilibrium enthalpy of $As_2Se_3$ bulk glass. *Full lines* are calculated from $C_p$ measurement during glass formation and points ($\Delta H_{eq}$) reflect the set of isothermal relaxation, see Fig. 9.5

glass transition temperature the enthalpy loss achieves its maximal value $\Delta H_{eq}{}^{max}$. This one is invariant at temperatures lower than $T_o$ for relaxed glass. For $As_2Se_3$ glass it was found $T_o \sim 157°C$ and $\Delta H_{eq}{}^{max} \sim 6.4$ kJ/mol. The changes of H(T) of fully relaxed glass at temperature lower than $T_o$ bear on the changes of vibrational enthalpy. These changes are the same as for crystal of the same chemical composition because of known fact that heat capacities for both the crystalline and glassy states of most of materials are essentially the same [20, 25, 26], except at ultra-low temperatures [25], and arise from vibrational contributions. As one can see in Fig. 9.8, the curve of metastable equilibrium has the same slope as this one of non-relaxed glass. It corresponds with finding that the specific heat is insensitive on the thermal history of a glass, see below. All these facts confirm assumption that metastable equilibrium is not identical with equilibrium linearly extrapolated of the equilibrium enthalpy above $T_g$, see Fig. 9.8. This conclusion agrees with [26–28].

The obtained dependence of non-exponentiality parameter $\beta = f(t_r, T_r)$ need to be interpreted in two steps. Firstly the attention was focused on $\beta = f(t_r)$, thus on results obtained from isothermal relaxation ($T_r = \text{const.}$).

The relaxation function, which is frequently simplified by KWW stretched exponential Eq. 9.3, may be expressed by a sum of exponential terms of N individual simultaneous relaxation processes [29]:

$$M_H = \sum_{i=1}^{N} w_i \exp\left[-\tau \frac{t}{\tau_{oi}}\right] \approx \exp\left[-\left(\frac{t'}{\tau_o}\right)^{\beta}\right] \tag{9.6}$$

where $w_i$ are weighting factors, $\tau_{0i}$ are independent relaxation times and t is duration of isothermal relaxation. In the right part of Eq. 9.6 the KWW parameter $\tau_o$ means relaxation time of all relaxation processes at actual time t' and β reflects the variation of weighting factors $w_i$ (the distribution of relaxation times) in this time. From the theories describing structural relaxation it follows that the non-exponentiality parameter is inversely proportional to the distribution of structural relaxation times. $\beta = 1$ corresponds to a single relaxation time and as β decreases the distribution broadens. It is reasonable assumption that in the course of relaxation to the metastable equilibrium, $T_f \rightarrow T_r$, the number of independent relaxation processes decreases in consequence of decreasing disorder. Therefore the experimentally found growth of parameter β during the isothermal relaxation, Fig. 9.6, may be interpreted as a consequence of restriction of relaxation times distribution when structure becomes relaxed.

Dependence of non-exponentiality parameter on relaxation temperature, $\beta = f(T_r)$, is interesting especially in case of its limit value $\beta_{eq}$, thus for $T_f = T_r$, see inset in Fig. 9.5. These values, inversely proportional to the distribution of relaxation times of glass in the metastable equilibrium, have shown namely that with decreasing temperature the metastable equilibrium structure approaches the state with only one relaxation time (${\beta_{eq}}^{max} = 1$ at the temperature $T_r \leqslant T_o$). This structure is characterized also by a final relaxation enthalpy, ${\Delta H_{eq}}^{max}$, Fig. 9.8, see above. In contrast to our results some researchers found, especially on organic polymers, that β decreases when temperature decreases [30, 31]. These results are only hardly compatible with the idea of structural relaxation. The metastable equilibrium structure becomes denser when temperature decreases and the number of independent relaxation processes decreases, as well. Consequently distribution of relaxation times becomes narrower and thus β rises up. Also it has found that $\beta_{eq}$ does not depend on fact whether metastable equilibrium was reached by exothermic or by endothermic relaxation, Fig. 9.7. Therefore non-exponentiality parameter of metastable equilibrium is path independent.

It can be concluded that it would be better to express the parameter β dependent on the structure of glass than on the time of relaxation $t_r$. While the time increases constantly from the beginning of process irrespective of relaxation extent, the change of the structure of glass is finite. The change of the structure is described by the change of fictive temperature, $T_f$. Then one may summarize that β depends on thermodynamic temperature and simultaneously on fictive temperature, $\beta(T_r, T_f)$.

### *9.3.5 The Description of Glass Transition*

Conventional DSC measurements were carried out using of DSC 7 calorimeter and stepwise technique StepScan DSC was carried on the Pyris 1 DSC (both PERKIN–Elmer) with special software. Powdered glassy samples (weigh around 4 mg for conventional DSC and 10 mg for StepScan DSC measurements, see below) were encapsulated into sealed aluminium pans. Conventional DSC scans were done using heating and cooling rates $q^{\pm} = (1 - 20)°C/min$.

To distinguish results of conventional DSC and StepScan DSC, the glass transition temperature, $T_g$, and $\Delta C_p$ measured by StepScan DSC are labelled by prime, i.e., $T_g'$, $\Delta C_p'$.

Typical StepScan DSC traces of $As_2S_3$ melt cooling through the glass transition region, as well as of glass heating are shown in Fig. 9.9. Reversible (thermodynamic) component, $C_p$ vs. T, and enthalpic change corresponding to irreversible (kinetic) one are separated. Both exothermic peak, $\Delta H = -3.2$ J/g, on the kinetic part of cooling scan and endothermic overshoot, $\Delta H = +3.5$ J/g, on heating scan are seen. Reversible parts are identical for cooling and heating.

Important result shown in Fig. 9.9 clearly demonstrates that the $C_p$ measurement close to isothermal equilibrium removes completely both well-known hysteresis of thermal capacity and shape difference during heating and cooling always obtained by conventional DSC. Temperature dependence of $C_p$ (StepScan DSC) is independent on both heating/cooling rate and thermal history. Similar result was found also for organic polymers [32]. In other words, the application of StepScan DSC eliminates influence of thermal history of glass and also influence of heating or cooling rate on the glass transition temperature, $T_g'$. It can be easily shown that the sum of thermodynamic and kinetic components is equivalent to the conventional DSC scan at the same heating rate.

It is well known that conventional DSC traces obtained during cooling and heating differ significantly in the shape and thus the value of glass transition temperature, $T_g$, depends on heating and cooling rate as well as the specific heat capacity change, $\Delta C_p$, e.g. [33–36]. With increasing rates (heating and/or cooling)

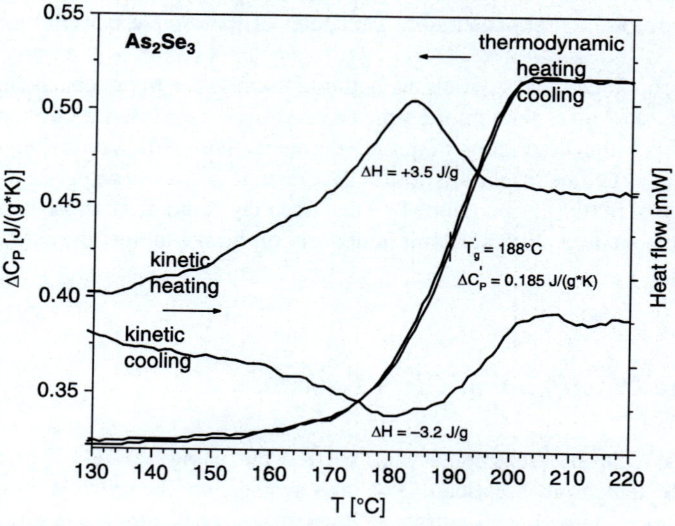


**Fig. 9.9** Typical results of StepScan DSC. For details, see text

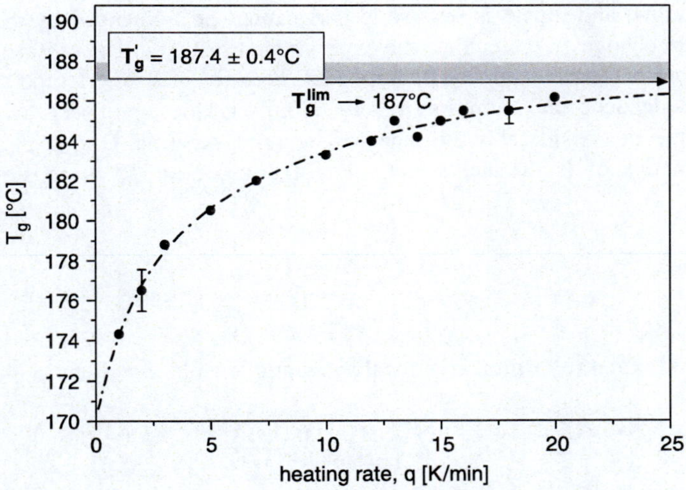


**Fig. 9.10** Heating rate dependence of the glass transition temperature. Conventional DSC, $T_g$, (*full circles*), StepScan DSC, $T_g'$, (shaded stripe of $2_*(T_g'$ std. deviation) width)

the glass transition temperature increases, see e.g. [13, 34]. Results of conventional DSC experiments carried out at different heating rates compared with StepScan DSC result, $T_g'$, are in Fig. 9.10. It should be stressed that dependence of $T_g$ on heating rate doesn't need to be logarithmic contrary to Lasocka's proposal [13]. The application of two-phase exponential association equation (also known as pseudo-first order association kinetics eq.) allows us to extrapolate the $T_g$ values from zero heating rate even to the infinity. The limit values of $T_g$ obtained in this manner are more realistic compared to application of logarithmic dependence.

According to the results there is directly proposed conception that in the glass transition region the glass may be viewed as an equilibrium mixture – supercooled liquid ↔ glass. Starting from the upper temperature end of a glass transition region when temperature decreases this equilibrium moves towards the glass and at the temperature $T_o$, see Fig. 9.8, the supercooled liquid completely disappears and vice versa. Temperature dependence of isobaric specific heat in the glass transition interval also supports this concept. In the case of studied glass this change of the total quantity of $\Delta C_p$ is also finished practically at the temperature $T_o$. The fact that the overwhelming majority of studies of the relaxation have been done in the glass transition region, e.g. [37], probably due to strongly increasing time-consumption at lower temperatures is worthy of remark.

Conventional DSC measurements of the glass transition temperature showed that the glass transition temperature $T_g$ approaches StepScan DSC $T_g'$ in the case of sufficiently high heating rates, Fig. 9.10. It is clear that the explanation of known heating/cooling rate dependence of the glass transition temperature should be searched in relaxation times of processes in the glass transition region. The temperature dependence of reversible (thermodynamic) $C_p$, essentially the change of vibrational amplitudes, is rapid enough in comparison with experimental time of

DSC technique and thus this reversible part cannot be influenced by the rate of temperature changes. To elucidate the experimental results it is necessary to focus attention on irreversible (kinetic) component, because of strong temperature and structure dependent relaxation time $\tau(T,T_f)$, simply $\tau$. One can expect that this fact should influence considerably the shape of the non-isothermal DSC scans.

The kinetics of isothermal structural relaxation, $T = T_r = \text{const.}$, can be expressed:

$$\frac{dT_f}{dt} \approx \frac{|T_f - T|}{\tau(T,T_f)} \tag{9.7}$$

For non-isothermal relaxation kinetics this equation can be rewritten in the form:

$$q = \frac{dT}{dt} \quad \frac{dT_f}{dT} \approx \frac{1}{q}\frac{|T_f - T|}{\tau(T,T_f)} \tag{9.8}$$

where q is a heating rate,$q = \frac{DT}{dt}$
For the forthcoming discussion on influence of non-isothermal structural relaxation on the shape of conventional DSC curve, and thus on the $T_g$ and $\Delta C_p$ values, refer to Fig. 9.11.

Heating scan wil be discussed at first. Provided that $(T_f > T)_{T<Tg}$, then the glass relaxes exothermally toward the equilibrium structure at T. It manifests itself as so-called undershoot on the non-isothermal DSC scan, e.g. [14]. On the other hand when $(T_f < T)_{T<Tg}$ the relaxation is endothermic and its typical overshoot is found at the temperature above conventional DSC glass transition temperature. In the limit case of $(T_f = T)_{T<Tg}$ the glass is in metastable equilibrium. It is important to point out here that this is the case only when $T_f$ crosses line $T = T_f$. At this moment exothermic relaxation changes to endothermic one, Fig. 9.11.

Because the relaxation time, $\tau$, increases exponentially with decreasing temperature so at glass formation structural relaxation time is much longer than experimental time, $\tau >> t$, for the temperatures not too lower than $T_g$ ($T \sim T_g - 10°C$). As a matter of fact the fictive temperature of non-relaxed glasses is practically always substantially higher than thermodynamic temperature ($T_f >> T$). It results in well-known fact that during glass heating the exothermic relaxation (undershoot) can be observed in many cases. Its magnitude depends on the difference between the rate of glass formation (the rate of cooling) and the rate of following heating. In case of a low heating rates the equilibrium structure can be reached, $(T_f = T)_{T<Tg}$, and crossed. In such case further increasing of temperature leads to departure from equilibrium and subsequent endothermic process proceeds toward a new equilibrium. Thus an overshoot begins to appear at DSC scan. Very slow heating of quickly cooled glass, represented by curve 2 in Fig. 9.11, leads to deep undershoot and very small overshoot when conventional DSC is used. Resulting shape of DSC scan is known as so-called pre-peak, or sub-$T_g$ peak. Origin of this is unclear up to now, see e.g. [16, 38]. When non-relaxed or only slightly relaxed glass is heated

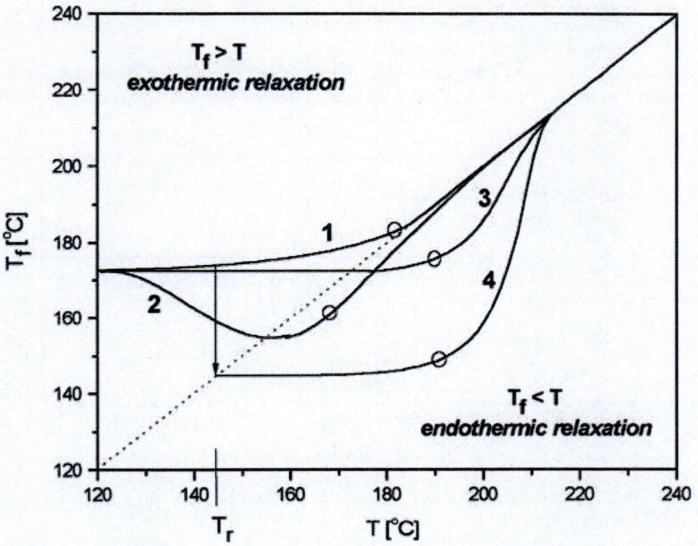


**Fig. 9.11** A schematic drawing of relation between fictive (structural) temperature, $T_f$, and thermodynamic temperature, T. *Circles*: conventional DSC glass transition. *Curves*: 1 – cooling, 2 – slowly heated non-relaxed glass, 3 – quickly heated non-relaxed glass, 4 – quickly heated glass fully relaxed at $T_r$, see text for details

with sufficiently high rate the exothermic undershoot decreases and at the same time the endothermic overshoot shifts up to higher temperature because the equality $(T_f = T)_{T<Tg}$ is reached at higher temperature, Fig. 9.11.

With regard to conventional DSC cooling scan the explanation is not prima facie so clear. During cooling of the melt the exothermic process (relaxation) takes place in and below the glass transition region, see Fig. 9.9. Now it should be pointed out that the conventional DSC doesn't measure directly the specific heat but only the heat flow and the later is higher in the glass than in the undercooled melt. Therefore when melt is cooled and glass is formed the heat flow of exothermal changes is subtracted from the growth of heat flow during glass transition, Fig. 9.12. It leads consequently to erroneous lower values of both $T_g$ and $\Delta C_p$ obtained from conventional DSC cooling scan, as it is demonstrated in Fig. 9.12.

The opposite sign of heat flow of the glass transition and of the exothermic structural relaxation during cooling and, on the other hand, the same sign of changes of both heat flow components, when the heated glass undergoes the glass transformation, causes the completely different shape of conventional DSC traces. These differences led to erroneous conclusion that when the melt is cooled through the glass transition region there is not exothermal counterpart to endothermic overshoot.

From the aforesaid it is obvious that the value of glass transition temperature obtained by conventional DSC strongly depends on the facts that (1) relaxation kinetic affects significantly shape of DSC scan, and that (2) glass transition

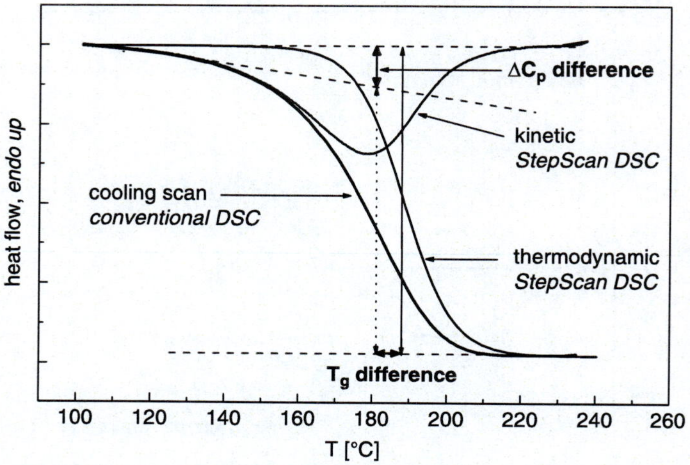


**Fig. 9.12** Glass transformation region-the influence of exothermic process on the shape of conventional DSC cooling scan [14]

temperature is actually determined on an ascending part of irreversible endothermic overshoot added on the reversible $C_p(T)$ dependence.

The heating rate dependence of the overshoot peak shift is widely used for determination the apparent activation energy of relaxation. Heating rate dependence of the glass transition temperature has been used to find the activation energy of either the glass transition [39] or the activation energy of relaxation [40]. From aforesaid results it is evident that both relaxation peak and $T_g$ shift bear on relaxation and thus it doesn't surprise that the obtained activation energies of relaxation and of glass transition are very close each other. Their difference is caused only by the fact that in the first case the peak shift and in the second one the shift of the ascending part of endothermic overshoot is used, but in all cases it is only more less good estimation of activation energy of relaxation.

It's clear that using conventional DSC it is possible only to approach more or less correctly the values of $T_g^{eq}$ and $\Delta C_p^{eq}$ (characteristic values of reversible changes).

Crucial question follows from obtained results – whether or not the glass transition is the kinetic effect at all. To solve this question the StepScan DSC heating scan was stopped in the glass transition region and after 30 min isotherm (sufficient long time to reach thermal equilibrium) heating was continued. In the case that experimental heat flow separation into reversing and nonreversing parts represents only misleading software separation both reversing and non-reversing parts should change during the isotherm. However, results show clearly that only kinetic overshoot disappeared after isothermal dwell whereas the sigmoidally shaped reversing part remains unchanged, Fig. 9.13.

This result, along with previous finding t hat reversible part (and consequently $T_g$) depends only on the chemical composition of glass [14, 15], shows that the commonly

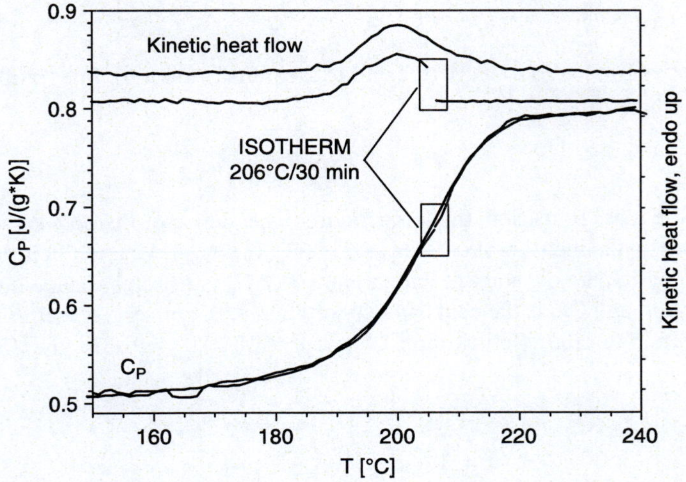


**Fig. 9.13** StepScan DSC of the glass transition of bulk $As_2S_3$ glass. Reversing parts are labelled $C_p$, the nonreversing ones are labelled Kinetic heat flow, see text

used concept that glass transition is entirely kinetic process is probably not fully realistic. It should be stressed at this moment that well known $T_g$ dependence on the thermal history of glass and on the experimental conditions, when conventional DSC is used, can be caused by superimposition of the slow kinetic effects on the sigmoidally shaped glass transition $C_p$ change, for more details, see [14].

### *9.3.6 Glass Transition Modelling*

Using StepScan DSC kinetic effects were separated and sigmoidal-shaped curves corresponding to the changes of isobaric heat capacity in the glass transformation region (reversible part) were obtained for wide class of non-crystalline materials, e.g. chalcogenide glasses ($As_2Se_3$, $As_2S_3$, $GeS_2$, $Ge_4As_4S_{92}$, S and Se bulk glasses), oxide glass LiPbPBO ($40Li_2O$:$10PbO$:$10B_2O_3$:$40P_2O_5$), lead-silica glass NBS 711, poly(styrene-*co*-acrylonitrile), 75/25, PSA, from BASF and poly(ethylen terephtalate), PET.

On the basis of above-mentioned experiment it can be further concluded that the temperature T in the glass transition region can be considered to be the equilibrium transformation temperature. It means that the reversible part of the glass transition ($C_p$ vs. T) may be referred to the temperature dependence of glass ↔ undercooled melt equilibrium. This dependence after normalization represents the temperature dependence of glass → undercooled melt conversion $\alpha(T)$.

If temperature dependent transformation occurs randomly in the material and the fraction dx of glass is transformed to undercooled melt, it is possible to write:

$$d\alpha = (1 - \alpha)dx \tag{9.9}$$

where α varies from zero to unity (the end of the transformation). Integration of this equation yields to:

$$\alpha = 1 - \exp(-x) \tag{9.10}$$

In case of the glass transition, the temperature dependence of the conversion α can be expressed by the measurable values as $\alpha = (C_p(T) - C_{pg})/\Delta C_p$. $C_p(T)$ is the actual value of isobaric heat capacity at temperature T, $\Delta C_p$ is its total change during the glass transition and $C_{pg}$ is the heat capacity of glass at the lowest temperature of the glass transition region. Substitution of x for $(T/T_g')^n$ yields to the following equation:

$$\alpha = 1 - \exp\left(-\left(\frac{T}{T_g'}\right)^n\right) \tag{9.11}$$

which describes well the experimentally obtained data. In this equation $T_g'$ is the temperature of inflexion point of α vs. T dependence, and *n* is a steepness parameter. The α vs. T form is more suitable for computer fitting then $C_p$ vs. T because the use of degree of conversion instead of isobaric heat capacities eliminates their temperature dependencies out of glass transition region.

The applicability of Eq. 9.11 at the description of reversible part of the glass transition was tested on many non-crystalline materials of a different chemical

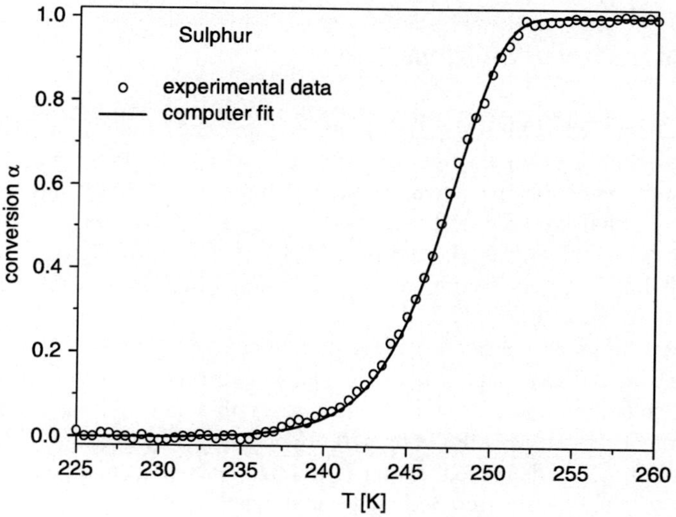


**Fig. 9.14** Normalized reversible part of the glass transition (α) of glassy S and its computer fit, Eq. 9.11

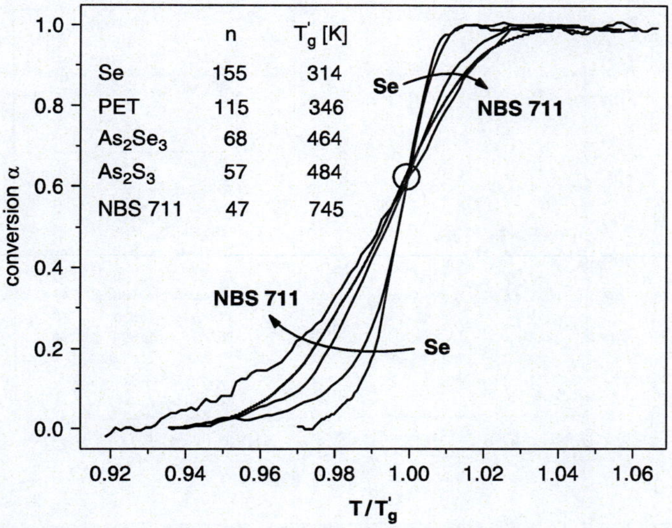


**Fig. 9.15** Normalized reversible parts of the glass transition of some glasses. All curves intersect close to α ~ 0.63

nature mentioned above. The reversible parts of the glass transformation process were normalized and fitted by Eq. 9.11. It was found that this equation fits the glass transition region fairly well in all cases. As an example, see Fig. 9.14.

From the Eq. 9.11 it arises that in the inflection point ($T = T_g'$) the conversion must always be $\alpha = (1 - 1/e) = 0.632$, where e is the base of natural logarithm. Therefore when reduced temperature, $T/T_g'$, is used, the conversion curves of all glasses studied must intersect within an experimental error at $\alpha \sim 0.63$. Experimental results are in the excellent agreement with this assumption, see Fig. 9.15. It is evident that the reversible parts differ in the shape. The steepness of curves is determined by the value of parameter $n$ in Eq. 9.11. By differentiation of this equation for $T = T_g'$ the parameter n is obtained as $n = e^*d\alpha/d(T/T_g')$ and represents the slope of the glass ↔ undercooled melt conversion curve in the inflexion point. Values of n calculated for all materials studied are included into Fig. 9.16.

From it follows that $n$ decreases when $T_g'$ increases regardless of the chemical composition of glass. It means that if glasses of different chemical composition have close values of $T_g'$ than the shape of reversible part of glass transition will be very similar.

The physical meaning of parameter $n$ is not still clear. Its dependence on $T_g'$ is noticeably close to the same dependency of Angell's fragility index m [21, 41]. Dashed straight lines with markedly different slopes crossing at approx. 150°C divide the $n$ vs. $T_g'$ into two regions, see Fig. 9.16. From both $n$ and m dependencies it follows that undercooled melts of glasses with glass transition temperature higher

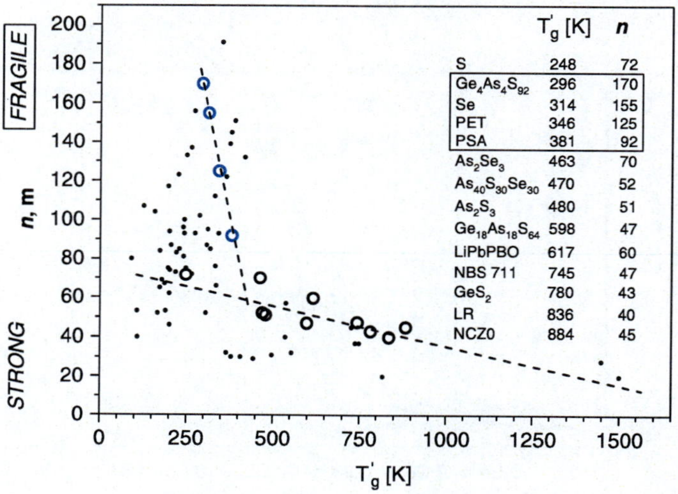


**Fig. 9.16** The dependence of the glass transition steepness parameter $n$ (*circles*) and the fragility index m (*points*) (From [42], on the glass transition temperature, $T_g'$)

then approx. 150°C are "strong" (viscosity close to the glass transition obeys arrhenian temperature dependency). However, further experiments are needed for detailed $n$ explanation.

## 9.3.7 Summary

From experimental results it follows that the glass transition can be regarded as a temperature dependence of a glass $\leftrightarrow$ undercooled melt equilibrium. The semi-empirical two-parametric equation was proposed to describe the temperature dependence of glass $\leftrightarrow$ undercooled melt conversion. One of the parameters, $T_g'$, determines temperature of inflexion point of sigmoidal shaped curve and the second one, $n$, is the steepness at $T_g'$. Applicability of proposed equation was tested using various non-crystalline materials with very different chemical composition. All experimentally obtained reversible parts were fitted fairly well and in agreement with the proposed equation an inflection point at $\alpha \approx 0.63$ was found in all cases. The steepness, $n$, was found to be inversely proportional to the glass transition temperature. The temperature of inflection point, $T_g'$, should be used as unambiguously defined glass transition temperature.

**Acknowledgements** This work was supported continuously by the project of the Ministry of Education of the Czech Republic MSM 0021627501.

## References


1. Schawe JEK (1995) Principles for the interpretation of modulated temperature DSC measurements. Part 1. Glass transition. Thermochim Acta 261:183–194
2. Modulated DSC Compendium (1995) TA Instruments
3. Wunderlich B (1997) The heat capacity of polymers. Thermochim Acta 300:43–65
4. Simon SL (2001) Temperature-modulated differential scanning calorimetry: theory and application. Thermochim Acta 374:55–71
5. Merzyakov M, Schick C (2001) Step response analysis in DSC – a fast way to generate heat capacity spektra. Thermochim Acta 380:5–12
6. Avramov I, Gutzow I (1988) Heating rate and glass transition temperature. J Non-Cryst Solids 104:148–150
7. Grenet J, Larmagnac JP, Michon P, Vautier C (1981) Relaxation structurelle dans les couches minces de sélénium amorphe au dessous de la température de transition vitreuse. Thin Solid Films 76:53–60
8. Kovacs AJ, Hutchinson JM (1979) Isobaric thermal behavior of glasses during uniform cooling and heating: dependence of the characteristic temperatures on the relative contributions of temperature and structure to the rate of recovery. II. A one-parameter model approach. J Polym Sci 17:2031–2058
9. Moynihan CT, Easteal AJ, DeBolt MA, Tucker J (1976) Analysis of structural relaxation in glass using rate heating data. J Am Ceram Soc 59:12–16
10. Grenet J, Saiter JM, Vautier C, Bayard J (1992) The $T_g$ displacement measurements: a way to foresee the physical behaviour and the use of glassy polymers. J Therm Anal 38:557–565
11. Derrey T, Saiter JM, Larmagnac JP, Vautier C (1985) Structural relaxation below $T_g$ of amorphous germanium-selenium alloys. Mater Lett 3:308–310
12. Kovacs AJ (1963) Glass transition in amorphous polymer: phenomenological study. Adv Polym Sci 3:394–505
13. Lasocka M (1976) Thermal stability of Ge–As–Te–In glasses. Mat Sci Eng 23:173–177
14. Černošek Z, Holubová J, Černošková E, Liška M (2002) Enthalpic relaxation and the glass Transition. J Optoelectron Adv Mater 4:489–503
15. Holubová J, Černošek Z, Černošková E (2005) The study of the glass transition by the StepScan DSC technique. J Optoelectron Adv Mater 7:2671–2676
16. Hodge IM (1994) Enthalpy relaxation and recovery in amorphous materials. J Non-Cryst Solids 169:211–266
17. Moynihan CT (1995) Structure, dynamics and properties of silicate melts. In: Stebbins JF, McMillan PF, Dingwell DB (eds) Reviews in mineralogy, vol 32. Mineralogical Society of America, Washington, DC, pp 1–19
18. Tool AQ (1948) Effect of heat-treatment on the density and constitution of high-silica glasses of the borosilicate type. J Am Ceram Soc 31:177–180
19. Suga H (2000) Prospects of materials science: from crystalline to amorphous solids. J Therm Anal Calorim 60:957–974
20. Elliott SR (1990) Physics of amorphous materials, 2nd edn. Longman Scientific and Technical, Essex
21. Angell CA (1991) Relaxation in liquids, polymers and plastic crystals – strong/fragile patterns and problems. J Non-Cryst Solids 131–133:13–31
22. Debenedetti PG, Stillinger FH (2001) Supercooled liquids and the glass transition. Nature 410:259–264
23. Garn PD (1965) Thermoanalytical methods. Academic, New York, p 194
24. Naraynaswamy OS (1971) A model of structural relaxation in glass. J Am Ceram Soc 54:491–498
25. Gutzow I, Schmelzer J (1995) The vitreous state. Springer, Berlin
26. Ediger MD, Angell CA, Nagel SR (1996) Supercooled liquids and glasses. J Phys Chem 100:13200–13212

27. Brunacci A, Cowie JMG, Fergusson R, McEwen IJ (1997) Enthalpy relaxation in glassy polystyrenes – part 1. Polymer 38:751
28. Montserrat S, Gomez Ribelles J, Meseguer JM (1998) The application of a new configurational entropy model to the structural relaxation in an epoxy resin. Polymer 39:3801–3807
29. Ducroux J-P, Rekhson SM, Merat FL (1994) Structural relaxation in thermorheologically complex materials. J Non-Cryst Solids 172–174:541–553
30. Angell CA, Ngai KL, McKenna GB, McMillan PF, Martin SW (2000) Relaxation in glassforming liquids and amorphous solids. J Appl Phys 88:3113–3157
31. Atkinson JR, Hay JN, Jenkins MJ (2002) Enthalpic relaxation in semi-crystalline PEEK. Polymer 43:731–735
32. Bailey NA, Hay JN, Price DM (2001) A study of enthalpic relaxation of poly(ethylene terephthalate) by conventional and modulated temperature DSC. Thermochim Acta 367–368:425–428
33. Scherer GW (1991) Glass formation and relaxation. In: Zarzycki J (ed) Materials science and technology, vol 9. VCH, Cambridge, pp 119–174
34. Richardson MJ, Savill AG (1975) Derivation of accurate glass transition temperatures by differential scanning calorimetry. Polymer 16:753–757
35. Mazurin OV (1977) Relaxation phenomena in glass. J Non-Cryst Solids 25:129–169
36. Mora MT (1997) Chalcogenide glasses. In: Thorpe MF, Mitkova MI (eds) Amorphous insulators and semiconductors. Kluwer Academic, London, pp 45–69
37. Hutchinson JM, Kumar P (2002) Enthalpy relaxation in polyvinyl acetate. Thermochim Acta 391:197–217
38. Pappin AJ, Hutchinson JM, Ingram MD (1994) The appearance of annealing pre-peaks in inorganic glasses: new experimental results and theoretical interpretation. J Non-Cryst Solids 172–174:584–591
39. Saiter JM, Ledru J, Hamou A, Zumailan A (1997) Dependence of the glass transition temperature on the heating rate and structure of chalcogenide glasses. Mater Lett 33:91–96
40. Moynihan CT, Lee SK, Tatsumisago M, Minami T (1996) Estimation of activation energies for structural relaxation and viscous flow from DTA and DSC experiments. Thermochim Acta 280:153–162
41. Angell CA (1988) Perspective on the glass transition. J Phys Chem Solids 49:863–871
42. Böhmer R, Ngai KL, Angell CA, Plazek DJ (1993) Nonexponential relaxations in strong and fragile glass formers. J Chem Phys 99:4201–4209

# Chapter 10
# Viscosity Measurements Applied to Chalcogenide Glass-Forming Systems

**Petr Koštál, Jana Shánělová, and Jiří Málek**

## 10.1 Introduction

Viscosity is an important physical parameter which determines the flow of material. The knowledge of viscous behaviour is important for example for the process of the material production. In the case of glasses and their undercooled melts, viscosity influences also the processes of structural relaxation and crystallization. Structural relaxation is in fact a very slow structural rearrangement of glass. This process can be realized through viscous flow and therefore is influenced by it. Crystallization process which may occur in undercooled melts is also influenced by the diffusion coefficient in the glassy matrix and therefore by its viscosity. This chapter tries to summarize the available viscosity data for chalcogenides and the basic measuring methods which are mostly often used to determine them.

Generally, it should be mentioned that most of amorphous chalcogenides are considered to be Newtonian fluids. Their viscosity coefficients are dependent only on pressure and temperature. Nevertheless, the pressure dependence is often neglected in the case of condensed phase. Measurements of the pressure dependencies are for inorganic glass-formers complicated because the combination of a high-pressure and high-temperature apparatus is needed. This difficulty, together with small influence of pressure on viscosity at common pressures, caused that most works dealing with the viscosity of inorganic glass-formers study only the temperature dependence of viscosity.

P. Koštál (✉), J. Shánělová and J. Málek
Faculty of Chemical Technology
University of Pardubice, Studentská 573, 532 10, Pardubice, Czech Republic
e-mail: Petr.Kostal@upce.cz

J. Šesták et al. (eds.), *Glassy, Amorphous and Nano-Crystalline Materials*,
Hot Topics in Thermal Analysis and Calorimetry 8, DOI 10.1007/978-90-481-2882-2_10,

## 10.2 Measuring Methods

Most important measuring methods for determination of dynamic viscosity of chalcogenide glass-formers are mentioned in Fig. 10.1. The approximate measuring range for each method is also denoted. It is apparent that no method covers whole viscosity interval from melt to glass. It should be mentioned that one method by which is possible to measure this broad interval was introduced by Plazek [1]. But his magnetic bearing torsional creep method is not widespread. The most widespread methods for determination of higher viscosity of chalcogenide glass-formers are penetration method followed by parallel-plate one. The dominance of penetration method is caused by the frequent use of it by Russian authors. Other two methods for high viscosity region, rod elongation method and beam-bending method are used only occasionally. Capillary and rotating cylinder methods are the most popular for determination of dynamic viscosity in the region of lower viscosity. Falling sphere method is used only rarely. Other popular method for melts is torsion oscillating cup method [2] which is used for determination of kinematic viscosity.

### *10.2.1 Capillary Method (C)*

This method is based on measuring of time which a finite volume of liquid needs to go through a narrow tube under a given pressure. The viscosity can be calculated using the Hagen-Poiseuille equation [3] which presumes laminar liquid flow.

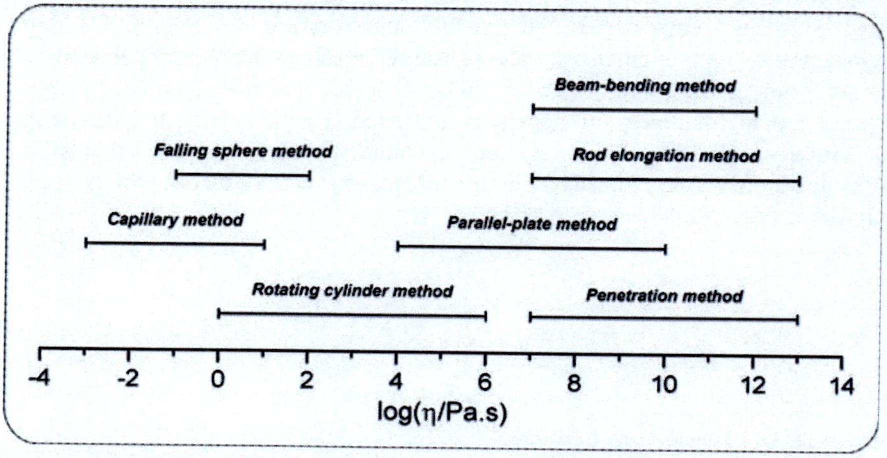


**Fig. 10.1** Overview of the most typical measuring methods for chalcogenide glass-formers with indication of their approximate measurement ranges

$$V = \frac{\pi r^4 \Delta p t}{8\eta l}, \tag{10.1}$$

where $r$ and $l$ stand for the radius and the longitude of the capillary, respectively, $t$ stands for time, $V$ stands for the volume of liquid, $\eta$ stands for viscosity and $\Delta p$ stands for pressure difference which is usually given by the hydrostatical pressure of liquid column. This method is in practice commonly used as a relative one. In this case the viscosity of measured liquid is determined by comparison of its retentive time with retentive time of liquid with known viscosity.

### *10.2.2 Falling Sphere Method (FS)*

This method is based on measuring the velocity of the falling sphere in measured liquid [4]. The viscosity can be calculated by the use of Stokes law.

$$\eta = \frac{2gr^2(\rho_s - \rho_l)}{9v}, \tag{10.2}$$

where $\rho_s$ and $\rho_l$ stand for the density of sphere and density of measured liquid, respectively, $v$ stands for the fall velocity, $g$ stands for the gravitation constant, $r$ stands for the radius of the sphere and $\eta$ stands for viscosity. The non-newtonian or non-transparent liquid can be measured by use of special setting of this viscometer. In this case the sphere is connected by some rigid attachment to the movable arm which can be weighted and the fall of sphere can be observed implicitly by through it.

### *10.2.3 Rotating Cylinder Method (RC)*

In the case of this method the measured liquid is placed between two defined surfaces, usually concentric cylinders (so-called Couette viscometer). One of these cylinders makes rotating motion with constant angular velocity $\omega$, the second one is stationary. The moment of force $M$ produced by viscosity of liquid is measured by the use of torsion spring. Viscosity can be then calculated using following equation [5]:

$$\eta = \left(\frac{1}{r_1^2} - \frac{1}{r_0^2}\right)\frac{M}{4\pi h\omega}, \tag{10.3}$$

where $r_I$ and $r_O$ stands for radius of inner and outer cylinder, respectively and $h$ stands for the height of the column of the sliding liquid. This equation was deduced for laminar liquid flow and for the neglected end and edge effects. This method is commonly used as a relative one.

### 10.2.4 Rod Elongation (RE)

This method is based on measuring of the rod or fibre elongation. The rod is made from measured material. The elongation of rod is caused by constant load or only by self-weight of the rod. Viscosity can be measured using equation:

$$\eta = \frac{Lmg}{3\pi r^2 v}, \tag{10.4}$$

where $L$ stands for the length of the rod, $r$ stands for the radius of the rod, $m$ stands for the weight of load and $v$ stands for the velocity of elongation. The elongation of non loaded rod investigated Littleton [6]. He heated 229 mm long rod with diameter 0.65–1 mm at constant rate (5–10 K/min). The temperature when the velocity of elongation was 1 mm/min is softening point. This point which corresponds to viscosity $10^{6.6}$ Pa.s is often called as Littleton softening point.

### 10.2.5 Torsion Oscillating Cup Method (TOC)

This method is based on measuring of torsion oscillation damping of cylindrical crucible filled with sample. This method was firstly described by Meyer [2]. The viscosity can be determined from following equation:

$$2M\lambda = \sqrt{\eta\rho T}\left[R^4\sqrt{\frac{(\pi-\lambda)^3}{2}} + 2cR^3K\right], \tag{10.5}$$

where $M$ stands for the moment of inertia, $T$ stands for the time of the oscillation, $\lambda$ stands for the damping decrement, $\eta$ stands for viscosity, $\rho$ stands for density, $R$ stands for the semi-diameter of the cylinder, $2c$ stands for the height of the cylinder and $K$ is proportional to the time of oscillation. This method in modified form was widely used by Russian authors for determination of kinematic viscosity. Details can be found for example in Ref. [7].

### 10.2.6 Penetration Method (P)

This method, firstly described by Cox [8], is based on measuring of penetration depth or rate of indenter which is penetrated into sample by constant force. The penetration method can be used to measure viscosity in the range of $10^7$–$10^{13}$ Pa.s. The indenter may have the any basic shape [9] but in most cases only hemispherical and cylindrical indenters are used.

The cylindrical indenter was very popular for Russian scientists, especially for Nemilov [10] and his co-workers. The general description of the penetration of cylinder was not published until the Yang work [9]. He and his co-worker deduced the general equation for calculating of viscosity:

$$\eta = \frac{F}{8R(dh/dt)}, \tag{10.6}$$

where $F$ stands for the applied force, $R$ stands for the radius of indenter and $dh/dt$ stands for the rate of penetration.

The hemisphere is the second one mostly used shape of the indenter. The equation which is needed for calculating viscosity and some theoretical and practical aspects about this method can be found in the work of Exnar et al. [11]:

$$\eta = \frac{9}{32\sqrt{2R}} \frac{Ft}{h^{3/2}}, \tag{10.7}$$

where $h$ stands for penetration depth and other symbols are the same as in previous equation.

### 10.2.7 Parallel-Plate Method (PP)

This method is based on measuring the time dependence of the height of the sample which is squeezed between two parallel plates. The parallel-plate method can be used to measure viscosity in the range of $10^4$–$10^{10}$ Pa.s. The viscosity value can be obtained from the following equation which corresponds to a cylindrical specimen [12]:

$$\eta = \frac{2\pi F d^5}{3V(dd/dt)(2\pi d^3 + V)}, \tag{10.8}$$

where $F$ stands for the applied force, d stands for the height of specimen, V stands for the specimen volume and $t$ stands for time.

### 10.2.8 Beam-Bending Method (BB)

In the case of this method the bar sample is supported by two knife edges. Another rod pushes in the middle of the sample from the upper side. The viscosity can be calculated from the deflection rate of the sample by use of the following equation [13].

$$\eta = \frac{gL^3}{144 \cdot I_c v} \cdot \left[ M + \frac{\rho AL}{1.6} \right], \tag{10.9}$$

where $L$ stands for the support span, $I_c$ stands for cross-section moment of inertia, $v$ stands for midpoint deflection rate of the sample bar, $M$ stands for the applied load, $\rho$ stands for the density of sample and $A$ stands for the cross-section area of bar.

### 10.2.9 Low-Temperature Torsion Viscometer (LV)

This uncommon viscometer is based on measuring of angular velocity $\omega$ of rod from measured material which is screwed by moment $M$. The viscosity can be calculated by use of the following equation [5].

$$\eta = \frac{2LM}{\pi r^4 \omega}, \tag{10.10}$$

where $L$ stands for length of the rod and $r$ stands for radius of the rod.

### 10.2.10 "Eisenberg" Method (EM)

This indirect method to determine viscosity was introduced by Eisenberg and Tobolsky [14]. Their apparatus is constructed to measured stress relaxation of sample and viscosity is determined from Young modulus and relaxation time.

### 10.2.11 Magnetic Bearing Torsional Creep Method (TC)

This method was described by Plazek [1]. The creep experiment starts by fixed stress on relaxed sample. The recoverable portion of deformation is subsequently measured as a function of time after the removal of the stress. The permanent deformation is indirectly proportional to viscosity which can be also determined from this experiment.

## 10.3 Overview of Viscosity of Amorphous Chalcogenides

The overview of measured chalcogenide systems is summarized in the Table 10.1 (dynamic viscosity) and in the Table 10.2 (kinematic viscosity). The kinematic viscosities are published only for melts. This work is primary aimed on the viscosity of glasses and undercooled melts. Hence the kinematic viscosities and the dynamic viscosities of melts are presented only for selected compositions and for completion of general view on amorphous chalcogenides. Both tables contain four columns. The first one includes the chemical formula of measured composition or the schematic formula of measured system. The compositions of substances are expressed in atomic percents but the steady formulas of some compounds are respected. The tables are divided into three parts; each part includes the materials based on one of three chalcogenide glass-formers (S, Se and Te). The substances including more then one glass-former are placed in group of chalcogenide with lower atomic weight. The first in each group is the pure glass-former followed by substances sorted by alphabetical order. The important chalcogenides (as $As_2S_3$, $As_2Se_3$...) have their own entry and are sorted in the front of group which belongs to its alphabetical order. The second column comprises the measured range of viscosity values. In all case only temperature dependence was looked at. The third and fourth column includes measuring method and the source work, respectively.

**Table 10.1** Overview of dynamic viscosity of amorphous chalcogenides

| Substance | Range [Pa.s] | Method | Reference |
|---|---|---|---|
| S | $10^1$–$10^2$ | RC | [15] |
| $As_2S_3$ | $10^8$–$10^{15}$ | RE | [16] |
| $As_2S_3$ | $10^8$–$10^{12}$ | P | [17] |
| $As_2S_3$ | $10^{10}$–$10^{12}$ | BB | [18] |
| $As_2S_3$ | $10^3$–$10^{12}$ | P | [19] |
| $As_2S_3$ | $10^8$–$10^{15}$ | P | [20] |
| $As_2S_3$ | $10^{12}$–$10^{15}$ | RE | [21] |
| $As_2S_3$ | $10^4$–$10^{12}$ | P | [22] |
| $As_2S_3$ | $10^8$–$10^{12}$ | P | [23] |
| $As_2S_3$ | $10^4$–$10^{12}$ | P | [24] |
| $As_2S_3$-$AsI_3$ | $10^4$–$10^{12}$ | P | [24] |
| $As_2Se_3$-S | $10^5$–$10^{10}$ | P | [25] |
| $As_2S_3$-$Tl_2S$ | $10^4$–$10^{12}$ | P | [26] |
| As-S | $10^3$–$10^{12}$ | P | [19] |
| As-S | $10^1$–$10^3$ | RC | [15] |
| As-S | $10^4$–$10^{12}$ | P | [22] |
| As-S | $10^{10}$–$10^{13}$ | BB | [18] |
| As-I-S | $10^{10}$–$10^{13}$ | BB | [18] |
| $AsTlS_2$ | $10^{12}$–$10^{14}$ | RE | [21] |
| $AsTlS_2$ | $10^6$–$10^{12}$ | P | [27] |
| As-Tl-S | $10^{10}$–$10^{12.5}$ | BB | [18] |

(continued)

**Table 10.1** (continued)

| Substance | Range [Pa.s] | Method | Reference |
|---|---|---|---|
| As-Tl-S | $10^5$–$10^{12}$ | P | [23] |
| Ga-La-S | $10^5$–$10^8$ | PP | [28] |
| $GeS_2$ | $10^{10}$–$10^{12}$ | P | [29] |
| $GeS_2$-$Sb_2S_3$ | $10^6$–$10^{13}$ | P<br>PP | [30] |
| $Ge_{30}S_{70}$ | $10^7$–$10^{10.7}$ | P | [31] |
| Ge-S | $10^8$–$10^{13}$ | P | [29] |
| Ge-I-S | $10^7$–$10^{10.5}$ | P | [31] |
| Se | $10^0$–$10^1$<br>$10^8$–$10^{13}$ | RC<br>BB | [32] |
| Se | $10^8$–$10^{12}$ | P | [33] |
| Se | $10^{10}$–$10^{14}$ | RE | [34,35] |
| Se | $10^0$–$10^1$<br>$10^8$–$10^{12}$ | TC | [36] |
| Se | $10^6$–$10^{12}$ | P | [37] |
| Se | $10^{-0.5}$–$10^{0.5}$ | C | [38] |
| Se | $10^{-1}$–$10^0$ | FS | [39] |
| Se | $10^5$–$10^{12}$<br>$10^{-1}$–$10^{0.5}$ | RE<br>C | [40]<br>[41] |
| Se | $10^4$–$10^{17}$ | LV<br>RE<br>RC | [42] |
| Se | $10^0$–$10^2$ | C | [43] |
| Se | $10^5$–$10^{12}$ | EM | [44] |
| Se | $10^{-1}$–$10^0$ | RC | [45] |
| Se | $10^0$–$10^{0.5}$ | C | [46] |
| Se | $10^{-1}$–$10^{0.5}$ | C | [47] |
| Se | $10^{7.5}$–$10^{10}$ | EM | [14] |
| Se | $10^6$–$10^9$ | PP | [48] |
| Se | $10^5$–$10^{10}$ | P | [25] |
| Ag-As-Si-Te-Se | $10^{7.4}$ and $10^{8.8}$ | P | [49] |
| $As_2Se_3$ | $10^8$–$10^{15.5}$ | RE | [16] |
| $As_2Se_3$ | $10^5$–$10^{12}$ | P | [24] |
| $As_2Se_3$ | $10^5$–$10^7$ | PP | [50] |
| $As_2Se_3$ | $10^{7.5}$–$10^{12}$ | PP | [51] |
| $(As_2Se_3)_{99}Ag_1$ | $10^{7.5}$–$10^{14}$ | PP | [51] |
| $As_2Se_3$-$AsI_3$ | $10^5$–$10^{12}$ | P | [24] |
| $As_2Se_3$-Cu | $10^{8.5}$–$10^{13}$ | P | [52] |
| $As_2Se_3$-Te | $10^5$–$10^{10}$ | P | [25] |
| $As_2Se_3$-Tl | $10^5$–$10^{10}$ | P | [25] |
| $As_2Se_3$-$Tl_2Se$ | $10^4$–$10^{12}$ | P | [26] |
| $As_2Se_3$-$Tl_2Se$ | $10^5$–$10^{12}$ | P | [53] |
| AsSe-TlSe | $10^5$–$10^{12}$ | P | [53] |
| As-Se | $10^5$–$10^{10}$ | P | [25] |
| As-Se | $10^3$–$10^{12}$ | P | [37] |

(continued)

**Table 10.1** (continued)

| Substance | Range [Pa.s] | Method | Reference |
|---|---|---|---|
| As-Se | $10^{10}$–$10^{12}$ | See note | [54] |
| As-Se | $10^{2}$–$10^{12}$ | TC | [36] |
| $As_{20}Ge_{22}Se_{58}$ | $10^{10}$–$10^{13}$ | PP | [55] |
| As-Ge-Se | $10^{10}$–$10^{12}$ | See note | [54] |
| As-Ge-Se | $10^{5}$–$10^{7}$ | PP | [50] |
| As-Ge-Se | $10^{3}$–$10^{12}$ | P | [56,57] |
| As-Ge-I-Te-Se | $10^{4}$–$10^{9}$ | PP | [58] |
| As-Ge-Te-Se | $10^{5}$–$10^{7}$ | PP | [50] |
| As-Ge-Te-Se | $10^{4}$–$10^{9}$ | PP | [58] |
| $AsTlSe_2$ | $10^{6}$–$10^{12}$ | P | [27] |
| Cl-Se | $10^{-0.5}$–$10^{0.5}$ | C | [38] |
| Cl-Se | $10^{-1}$–$10^{0}$ | C | [46] |
| $Ga_2S_3$-$La_2S_3$ | $10^{8.5}$–$10^{11}$ | P | [59] |
| $(GeSe_2)$-$(Sb_2Se_3)$ | $10^{6}$–$10^{13}$ | P | [60] |
| | | PP | |
| Ge-Se | $10^{6}$–$10^{9}$ | PP | [48] |
| Ge-Se | $10^{3}$–$10^{12}$ | P | [61] |
| Ge-Se | $10^{10}$–$10^{12}$ | see note | [54] |
| Ge-Sb-Se | $10^{6}$–$10^{9}$ | PP | [48] |
| Ge-Sb-Se | $10^{5.5}$–$10^{12}$ | P | [62] |
| Ge-Sb-Se | $10^{5}$–$10^{7}$ | PP | [50] |
| I-Se | $10^{0}$–$10^{2}$ | C | [43] |
| P-Se | $10^{5}$–$10^{12}$ | P | [63] |
| P-Se | $10^{5}$–$10^{12}$ | P | [64] |
| P-Te-Se | $10^{5}$–$10^{12}$ | P | [64] |
| Sb-Se | $10^{6}$–$10^{12}$ | EM | [44] |
| Tl-Se | $10^{-1}$–$10^{0}$ | RC | [45] |
| ZnSe | $10^{-3}$ | C | [65] |
| As-Si-Te | $10^{7.4}$ and $10^{8.8}$ | P | [49] |
| ZnTe | $10^{-3}$ | C | [65] |
| CdTe | | | |

Note: These data were originally taken from other not so well accessible sources (e.g. Ph.D. thesis or conference)

**Table 10.2** Overview of kinematic viscosity of amorphous chalcogenides

| Substance | Range [$m^2.s^{-1}$] | Method | Reference |
|---|---|---|---|
| $Ag_2S$ | $10^{-6.5}$–$10^{-6}$ | TOC | [66] |
| $As_2S_3$ | $10^{-5.5}$–$10^{-3}$ | TOC | [67] |
| As-S | $10^{-6.5}$–$10^{-3}$ | TOC | [67] |
| $AgSbS_2$ | $10^{-6.5}$–$10^{-6}$ | TOC | [68] |
| $Cu_2S$ | $10^{-5.5}$–$10^{-5}$ | TOC | [69] |
| $Cu_2S$-$Cu_2Se$ | $10^{-6.5}$–$10^{-5.5}$ | TOC | [70–72] |
| $Cu_2S$-$Cu_2Te$ | $10^{-6.5}$–$10^{-5.5}$ | TOC | [70–72] |
| $CuSbS_2$ | $10^{-6.5}$–$10^{-6}$ | TOC | [68] |
| GeS | $10^{-6.5}$–$10^{-5}$ | TOC | [73] |
| PbS | $10^{-7}$–$10^{-6.5}$ | TOC | [74] |
| SnS | $10^{-6.5}$–$10^{-6}$ | TOC | [73] |
| Se | $10^{-6}$–$10^{-3.5}$ | C | [75] |
| Se | $10^{-5.5}$–$10^{-4}$ | TOC | [76] |
| Se | $10^{-5.5}$–$10^{-4}$ | TOC | [77,78] |
| Se | $10^{-5.5}$–$10^{-4.5}$ | TOC | [79] |
| $Ag_2Se$ | $10^{-6.5}$–$10^{-6}$ | TOC | [66] |
| $As_2Se_3$ | $10^{-5.5}$–$10^{-4}$ | TOC | [79] |
| $As_2Se_3$ | $10^{-5}$–$10^{-3}$ | TOC | [7] |
| $As_2Se_3$ | $10^{-5}$–$10^{-2.5}$ | TOC | [80] |
| $As_2Se_3$-$As_2Te_3$ | $10^{-6}$–$10^{-3}$ | TOC | [7] |
| As-Se | $10^{-5.5}$–$10^{-3.5}$ | TOC | [79] |
| As-Se | $10^{-6}$–$10^{-2}$ | C | [75] |
| As-Ge-Se | $10^{-6.5}$–$10^{-2}$ | C | [75] |
| $AgSbSe_2$ | $10^{-6.5}$–$10^{-6}$ | TOC | [68] |
| $AsTlSe_2$ | $10^{-6.5}$–$10^{-3}$ | TOC | [80] |
| $Cu_2Se$ | $10^{-5.5}$–$10^{-5}$ | TOC | [69] |
| $Cu_2Se$-$Cu_2Te$ | $10^{-6.5}$–$10^{-5.5}$ | TOC | [70–72] |
| Cu-Se | $10^{-5.5}$–$10^{-5}$ | TOC | [81] |
| $CuInSe_2$ | $10^{-6.5}$–$10^{-6}$ | TOC | [68] |
| $CuSbSe_2$ | $10^{-6.5}$–$10^{-6}$ | TOC | [68] |
| GeSe | $10^{-6.5}$–$10^{-6}$ | TOC | [73] |
| Ge-Se | $10^{-5.5}$–$10^{-3}$ | TOC | [76] |
| Ge-Se | $10^{-5.5}$–$10^{-3}$. | TOC | [77,78] |
| PbSe | $10^{-7}$–$10^{-6.5}$ | TOC | [74] |
| Pb-Se | $10^{-7}$–$10^{-6.5}$ | TOC | [82] |
| SnSe | $10^{-7}$–$10^{-6.5}$ | TOC | [73] |
| Sn-Se | $10^{-6.5}$–$10^{-6}$ | TOC | [83] |
| $Tl_2Se$ | $10^{-6.5}$–$10^{-6}$ | TOC | [80] |
| TeSe | $10^{-6.5}$–$10^{-4.5}$ | C | [75] |
| Te | $10^{-7}$–$10^{-6.5}$ | TOC | [84] |
| $Ag_2Te$ | $10^{-6.5}$–$10^{-6}$ | TOC | [66] |
| $As_2Te_3$ | $10^{-6.5}$–$10^{-4}$ | TOC | [84] |
| As-Te | $10^{-6.5}$–$10^{-5}$ | TOC | [84] |
| $As_{20}Ge_{10}Te_{70}$ | $10^{-6.5}$–$10^{-5}$ | C | [75] |
| $AgSbTe_2$ | $10^{-7}$–$10^{-6.5}$ | TOC | [68] |

(continued)

**Table 10.2** (continued)

| Substance | Range $[m^2.s^{-1}]$ | Method | Reference |
|---|---|---|---|
| $Cu_2Te$ | $10^{-6.5}$–$10^{-6}$ | TOC | [69] |
| Cu-Te | $10^{-6.5}$–$10^{-6}$ | TOC | [81] |
| $CuGaTe_2$ | $10^{-6.5}$–$10^{-6}$ | TOC | [68] |
| $CuInTe_2$ | $10^{-6.5}$–$10^{-6}$ | TOC | [68] |
| GeTe | $10^{-7}$–$10^{-6.5}$ | TOC | [74] |
| Ge-Te | $10^{-7}$–$10^{-6.5}$ | TOC | [85] |
| PbTe | $10^{-7}$–$10^{-6.5}$ | TOC | [74] |
| Pb-Te | $10^{-7}$–$10^{-6.5}$ | TOC | [85] |
| SnTe | $10^{-7}$–$10^{-6.5}$ | TOC | [74] |
| Sn-Te | $10^{-7}$–$10^{-6.5}$ | TOC | [85] |

**Acknowledgments** This work has been supported by the Ministry of Education Youth and Sports of the Czech Republic under project: LC 523 and the Czech Science Foundation under grant No: 104/08/1021.

## References

1. Plazek DJ (1968) Magnetic bearing torsional creep apparatus. J Polym Sci A2 6:621–638
2. Meyer OE (1891) Ein verfahren zur bestimmung der inneren reibung von flüssigkeiten. Ann Phys Chem 5:1–14
3. Holland FA, Bragg R (1995) Fluid flow for chemical engineers. Elsevier, Oxford
4. Brooks R, Dinsdale A, Quested P (2005) The measurement of viscosity of alloys – a review of methods: data and models. Meas Sci Technol 16:354–362
5. Scholze H, Kreidl NJ (1983) Technological aspects of viscosity. In: Uhlmann R, Kreidel NJ (eds) Glass: science and technology. Academic, New York, pp 233–273
6. Littleton J (1927) A method for measuring the softening temperature of glasses. J Am Ceram Soc 10:259
7. Kolomiets BT, Pozdnev VP (1960) Viscosity of $As_2Se_3$-$As_2Te_3$ glassy semiconductors (Russ.). Sov Phys-Sol State 2:28–34
8. Cox SM (1943) A method of viscosity measurement in the region $10^8$ poises. J Sci Instrum 20:113–114
9. Yang FQ, Li JCM (1997) Newtonian viscosity measured by impression test. J Non-Cryst Solids 212:126–135
10. Nemilov SV, Petrovskii GT (1963) New method for measuring of viscosity of glasses (Russ.). J Appl Chem-USSR 36:222–225
11. Exnar P, Hruba M, Uhlir J, Voldan J (1980) Experience with the penetration viscosimeter. Silikaty 24:169–179
12. Gent AN (1960) Theory of the parallel plate viscometer. Br J Appl Phys 11:85–87
13. Hagy HE (1963) Experimental evaluation of beam-bending method of determining glass viscosities in the range $10^8$ to $10^{15}$ poises. J Am Ceram Soc 46:93–97

14. Eisenberg A, Tobolsky AV (1962) Viscoelastic properties of amophous selenium. J Polym Sci 61:483–495
15. Pohl KD, Bruckner R (1981) Structural viscosity of sulfur and arsenic sulfide melts – density and refractive-index. Phys Chem Glasses 22:150–157
16. Prod'homme M (1956) Viscosité des verres dans le domaine de transformation. Verres Refract 4:208–214
17. Malek J (1998) Dilatometric study of structural relaxation in arsenic sulfide glass. Thermochim Acta 311:183–198
18. Suzuki S, Kamiya Y, Suzuki Y, Kobayashi T (1972) Viscosity of glass in the transition region in the systems arsenic-sulfur, arsenic-sulfur-iodine, and arsenic-sulfur-thallium (Jpn.). J Soc Mater Sci Jpn 21:143–147
19. Nemilov SV (1979) Valence structure, viscous and elastic properties of as-S melts (Russ.). Fiz Chim Stekla 5:398–409
20. Nemilov SV (1964) Relation between free energy of activation for viscous flow and energy of chemical bonds in glasses (Russ.). Solid State Phys 6:1375–1379
21. Kubacki W (1973) Effect of heat treatment on chlacogenide glass – part 1. Verres Refract 27:105–116
22. Vinogradova GZ, Dembovskii SA, Kuzmina TN, Chernov AP (1967) Viscosity and structure of As-S glasses (Russ.). Zh Neorg Chim 12:3240–3247
23. Zhukina NE, Orlova GM, Chalabjan GA (1979) Viscous and elastic properties of As-S-Tl glasses (Russ.). Fiz Chim Stekla 5:223–228
24. Chernov AP, Dembovskii SA, Machova VI (1970) Viscosity and structure of $As_2X_3$-$AsI_3$ glasses (Russ.). Izv Akad Nauk SSSR Neorg Mater 6:823–825
25. Kunugi M, Ota R, Suzuki M (1970) Viscosity of glasses in the system As-Se, As-Se-S, As-Se-Te and As-Se-Tl (Jpn.). J Soc Mater Sci Jpn 19:145–150
26. Vinogradova GZ, Dembovskii SA, Anisimova TN (1970) Study of viscosity of $As_2S_3$-$Tl_2S$ and $As_2Se_3$-$Tl_2Se$ glasses (Russ.). Zh Neorg Chim 15:1949–1952
27. Dembovskii SA (1968) Identification and properties of chemical compounds $TlAsS_2$, $TlAsSe_2$ and $TlAsTe_2$ in glassy and crystalline form (Russ.). Izv Akad Nauk SSSR Neorg Mater 4:1920–1926
28. Brady DJ, Wang J, Hewak DW (1998) Viscosity of chalcogenide glass. In: Hewak D (ed) Properties, processing and applications of glass and rare earth-doped glasses for optical fibres. INSPEC, London, pp 309–312
29. Malek J, Shanelova J (1999) Viscosity of germanium sulfide melts. J Non-Cryst Solids 243:116–122
30. Shanelova J, Kostal P, Malek J (2006) Viscosity of (GeS2)(x)(Sb2S3)(1-x) supercooled melts. J Non-Cryst Solids 352:3952–3955
31. Seddon AB, Hemingway MA (1991) Thermal-properties of chalcogenide-halide glasses in the system – Ge-S-I. J Therm Anal 37:2189–2203
32. Cukierman M, Uhlmann DR (1973) Viscous flow behavior of selenium. J Non-Cryst Solids 12:199–206
33. Yang FQ, Li JCM (1997) Viscosity of selenium measured by impression test. J Non-Cryst Solids 212:136–142
34. Jenckel E (1937) Die Vorgänge bei der Abkühlung von Gläsern und Kunstharzen. Z Elektrochem 43:796–806
35. Jenckel VE (1938) Über Strukturviskosität an erweichendem Selenglas. Kolloid Z 84:266–268
36. Bernatz K, Echeverria I, Simon S, Plazek D (2002) Characterization of the molecular structure of amorphous selenium using recoverable creep compliance measurements. J Non-Cryst Solids 307:790–801
37. Nemilov SV, Petrovskii GT (1963) Study of viscosity of glassy system Se-As (Russ.). J Appl Chem-USSR 36:977–981

38. Hamada S, Yoshida NTS (1969) On the viscosity of liquid selenium. Bull Chem Soc Jpn 42:1025–1029
39. Harrison DE (1964) Effect of pressures (up to 4 kbar) on polymerization of liquid selenium from measurements of viscosity. J Chem Phys 41:844–849
40. Khalilov KM, Kuliev BB (1965) Temperature dependence of viscous properties of amorphous selenium (Russ.). Fiz Tverd Tela 7:2847–2848
41. Khalilov KM (1959) Viscosity of molten selenium (Russ.). Izv Akad Nauk Azerb SSR 6:67–70
42. Ueberreiter K, Orthmann HJ (1951) Viscosity of glassy selenium from 0°C to 100°C. Kolloid Z 123:84–91
43. Krebs H, Morsch W (1950) Die Molekelgröße des amorphen schwarzen Selens. Z Anorg Allg Chem 263:305–309
44. Dzhalilov SU, Orudzheva SO (1966) Viscosity of amorphous selenium doped by antimony (Russ.). Zh Fiz Chim 60:2130–2132
45. Keezer RC, Bailey MW (1967) The structure of liquid selenium from viscosity measurements. Mater Res Bull 2:185–192
46. Shirai T, Hamada S, Kobayaski K (1963) Viscous flow of selenium liquid. J Chem Soc Jpn 84:968–972
47. Dobinski S, Wesolowski J (1937) On the viscosity of liquid selenium. Bull Int Acad Polon Sci Lett Part A 7:7–14
48. Senapati U, Varshneya AK (1996) Viscosity of chalcogenide glass-forming liquids: An anomaly in the 'strong' and 'fragile' classification. J Non-Cryst Solids 197:210–218
49. Anthonis HE, Kreidel NJ, Ratzenboeck WH (1973/1974) Polynary silicon arsenic chalcogenide glasses with high softening temperatures. J Non-Cryst Solids 13:13–36
50. Webber PJ, Savage JA (1981) Measurement of the viscosity of chalcogenide glasses by a parallel plate technique. J Mater Sci 16:763–766
51. Henderson DW, Ast DG (1984) Viscosity and crystallization kinetics of As2Se3. J Non-Cryst Solids 64:43–70
52. Kostál P, Shanelova J, Malek J (2005) Viscosity of Cu-x(AS(2)Se(3))100-(x) supercooled melts. J Non-Cryst Solids 351:3152–3155
53. Orlova GM, Udalov SS, Manachova EN (1985) Elastic and thermal properties of AsSe-TlSe and $As_2Se_3$-$Tl_2Se$ glasses (Russ.). Fiz Chim Stekla 11:215–218
54. Webber PJ, Savage JA (1976) Some physical properties of Ge-As-Se infrared optical glasses. J Non-Cryst Solids 20:271–283
55. Le Bourhis E, Gadaud P, Guin JP, Tournerie N, Zhang XH, Lucas J, Rouxel T (2001) Temperature dependence of the mechanical behaviour of a GeAsSe glass. Scr Mater 45:317–323
56. Nemilov SV (1964) Viscosity and structure of As-Ge-Se glasses in the region of low selenium content (Russ.). J Appl Chem-USSR 37:1699–1708
57. Nemilov SV (1964) Viscosity and structure of As-Ge-Se glasses in the region of high selenium content (Russ.). J Appl Chem-USSR 37:1452–1457
58. Nguyen VQ, Sanghera JS, Aggarwal ID, Lloyd IK (2000) Physical properties of chalcogenide and chalcohalide glasses. J Am Ceram Soc 83:855–859
59. Morgan SP, Furniss D, Seddon AB (1996) Lanthanum-fluoride addition to gallium-lanthanum-sulphide glasses. J Non-Cryst Solids 203:135–142
60. Kostal P, Malek J (2007) Viscosity of (GeSe2)(x)(Sb2Se3)(1-x) undercooled melts. J Non-Cryst Solids 353:2803–2806
61. Nemilov SV (1964) Viscosity and structure of Se-Ge glasses (Russ.). J Appl Chem-USSR 37:1020–1024
62. Korepanova NA, Orlova GM, Pazin AV (1976) Study of the structural-chemical features of Sb-Ge-Se glasses by a viscometric method (Russ.). J Appl Chem-USSR 49:36–40
63. Kim EI, Orlova GM (1974) Viscosity of P-Se glass system (Russ.). J Appl Chem-USSR 47:989–992

64. Orlova GM, Kolomejtseva SE, Timonov AS, Kuznetsova OA (1979) Viscosity and thermal expansion of P-Se-Te glasses (Russ.). Fiz Chim Stekla 5:546–551
65. Kolesnikov NN (1992) Viscosity of ZnSe melt (Russ.). Zh Fiz Chim 66:760–763
66. Glazov VM, Grabchak NM (1978) Change in short-range order structure when silver chalcogenide melts are heated (Russ.). Neorg Mater 14:466–468
67. Chaussemy G, Fornazero J, Mackowski JM (1983) Relationship between viscosity and structure in AsxS1-x molten materials. J Non-Cryst Solids 58:219–234
68. Glazov VM, Mal'sagov AU (1977) Change in short-range order during heating of melts of certain compounds of Type $A^{I}B^{III}C_2{}^{VI}$ and Type $A^{I}B^{V}C_2{}^{VI}$ (Russ.). Neorg Mater 13:1383–1386
69. Glazov VM, Burchanov AS, Krestovnikov AN (1971) Change in Short-range order structure in melts of $Cu_2B^{VI}$ compounds during heating (Russ.). IVZ VUZ Tsvetnaya Metallurgia 14:95–100
70. Glazov VM, Burchanov AS (1994) Viscosity of melts in pseudobinary systems of copper chalcogenides (Russ.). Neorg Mater 30:741–746
71. Glazov VM (2000) Structural Inhomogeneity of melts in quasi-binary systems formed by copper chalcogenides (Russ.). Zh Fiz Chim 74:586–594
72. Glazov VM (2000) Temperature dependence of viscosity and density of the melts of quasi-binary systems formed by copper chalcogenides (Russ.). Teplofiz Vysok Temp 38:557–565
73. Glazov VM, Situlina OV (1970) Temperature variation in short-range structure of molten germanium and tin sulphide and selenide (Russ.). Zh Fiz Chim 44:2480–2485
74. Glagoleva NN, Krestovnikov AN, Glazov VM (1968) Change in short-range order and character of chemical bonds of IV group chalcogenide melts with galenite structure upon heating (Russ.). Neorg Mater 4:1890–1897
75. Fisher VM, Krebs H (1974) Untersuchungen über die Viskosität von Chalcogenidschmelzen. Glass Sci Technol 47:42–51
76. Laugier A, Chaussemy G, Fornazero J (1977) Viscosity of glass-forming Ge-Se liquid solutions. J Non-Cryst Solids 23:419–429
77. Chaussemy G, Fornazero J, Laugier A (1976) Influence of germanium on viscosity of liquid selenium. J Phys Lett Paris 37:L283–L285
78. Chaussemy G, Fornazero J, Laugier A (1977) Viscosity measurements for $Ge_xSe_{1-x}$ liquids in simple vitrification compositions and notion of association. Rev Phys Appl 12:687–690
79. Kadoun A, Chaussemy G, Fornazero J, Mackowski JM (1983) Kinematic viscosity of $As_xSe_{1-x}$ glass forming liquids. J Non-Cryst Solids 57:101–108
80. Tverjanovich AS, Skorobogatova I (1990) Viscosity of $As_2Se_3$-$Tl_2Se$ melts (Russ.). Fiz Chim Stekla 16:369–373
81. Glazov VM, Burchanov AS, Krestovnikov AN (1971) Physico-chemical analysis of Cu-Se and Cu-Te melts (Russ.). Neorg Mater 7:1494–1496
82. Chaussemy G, Fornazero J, Laugier A (1980) Viscosity and local order in $Pb_xSe_{(1-x)}$ melts. J Phys Paris 41:371–373
83. Aleksandrov AA, Andrianova TN, Ochotin VS, Razumeichenko LA, Panina ZI (1977) Study of the density and viscosity of tin-selenium system melts (Russ.). Teplofiz Vysok Temp 15:47–52
84. Tverjanovich AS, Kasatkina EB (1992) Viscosity of As-Te melts (Russ.). Fiz Chim Stekla 18:86–93
85. Glazov VM, Krestovnikov AN, Glagoleva NN (1965) Physico-chemical analysis of binary systems formed by tellurium with germanium subgroup elements in liquid state (Russ.). Dokl Akad Nauk SSSR 162:94–97

# Chapter 11
# Thermal Properties and Related Structural Study of Oxide Glasses

**Marek Liška and Mária Chromčíková**

## 11.1 Introduction

Glass is accompanying people from the early times of mankind. First it was the natural glass generated by the volcanic processes and the glass "produced" by the impact of meteors on the earth. During formation of the earth, highly siliceous melts of rocks froze to natural glasses such as obsidians. After some time the people start the glass melting. Glass was first produced by man about 4,000 years ago in ancient Egypt. From this time the need of the knowledge of glass composition, structure and properties is dated. These are the typical questions answered by the glass chemistry [1–6].

The aim of the present contribution is point out the mutual relationships between the understanding of the glass structure and construction of the thermodynamic models of glass. Finally the method of studying glass structure based on combined statistical treatment of Raman spectra with the thermodynamic model will be presented together with the example of its application on the chalcogenide glasses from the pseudobinary system $As_2S_3$–$As_2Se_3$.

## 11.2 Thermodynamics of Glass Formation [7, 8]

The glass is commonly defined (e.g. [9]) as the non-crystalline solid obtained by cooling the melt without crystallization. Despite the fact that glass can be produced also by other ways, like by the sol–gel method or by the amorphization of crystalline solid [6], it is worth noting the uncommon situation when the method of preparation

M. Liška (✉) and M. Chromčíková
Vitrum Laugaricio, Joint Glass Center of Institute of Inorganic Chemistry
(Slovak Academy of Sciences) and Alexander Dubček University of Trenčín
and RONA, j.s.c. glassfactory, Študentská 2, Trenčín SK-911 50, Slovak Republic
e-mail: liska@tnuni.sk

J. Šesták et al. (eds.), *Glassy, Amorphous and Nano-Crystalline Materials*,
Hot Topics in Thermal Analysis and Calorimetry 8, DOI 10.1007/978-90-481-2882-2_11,

becomes a part of the definition of some kind of substance-material. This indicates the importance of the kinetic factors connected with the glass formation. On the other hand the kinetics – namely the kinetics of nucleation and crystal growth – is determined by the (under-cooled) melt structure. The typical course of the glass formation from the melt can be seen in Fig. 11.1 where the temperature dependence of the (single component) system volume, enthalpy and entropy are plotted against temperature. The melt is in the state of the true thermodynamic equilibrium (corresponding to the global minimum of the Gibbs energy at isothermal-isobaric conditions) at the melting temperature $T_m$. In case of the multi-component system the melting temperature is replaced by the liquidus temperature, $T_{liq}$, and during crystallization the temperature gradually decreases tracing the binary, ternary... eutectic lines of the particular equilibrium phase diagram until the multi-component eutectic is reached at the temperature $T_{eut}$ (obviously $T_{eut} < T_{liq}$). The true thermodynamic equilibrium of the single-component system under the $T_m$ temperature is then represented by the crystalline state. In case of multicomponent system the equilibrium mixture of crystalline phases and the melt corresponds to the true thermodynamic equilibrium in temperature ranging from the liquidus temperature to eutectic temperature. The mixture of crystalline phases (i.e. without melt) represents the equilibrium state under the eutectic temperature.

The equilibrium state represented by the equilibrium mixture of crystalline phases is termed the Crystalline Reference State (CRS) by Conradt [8]. It is worth noting that the knowledge of the particular equilibrium phase diagram is sufficient for the knowledge of the composition of the CRS. Thus mole fractions of individual thermodynamically stable crystalline phases in the CRS can be simply calculated from the particular phase diagram, namely from the coordinates of particular eutectics (non-variant points of the phase diagrams).

When the crystallization is avoided (due to sufficiently high cooling rate) the metastable equilibrium state represented by the under-cooled melt is reached below the $T_m/T_{liq}$ temperature. Depending on the cooling rate the system follows the metastable equilibrium until the glass transition region (the curved part of the dashed line) where the increase of the viscosity and thus the structural relaxation time leads

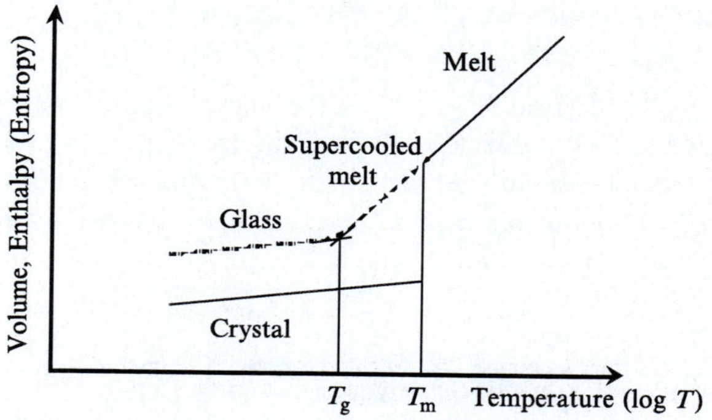


**Fig. 11.1** The temperature dependence of volume during cooling of the single component melt

to freezing in of the structure. Under the glass transition region characterized by the glass transition temperature $T_g$ the glass is obtained (the dotted line). The kinetic character of the glass transition is expressed by the fact that decreasing the cooling rate (but avoiding the crystallization) shifts the glass transition to lower temperatures. Thus the glass structure depends on the cooling rate, i.e. on its thermal history. It is worth noting that the $T_g$ temperature is in fact not strictly defined. On one side it depends on the cooling rate but also on the method of its determination (thermo-dilatometry, DSC, ...). The structure of the glass can be thus characterized by the temperature at which the metastable equilibrium structure was frozen in. This is the fictive or structure temperature $T_f$ introduced by Tool. In the simplified situation of linear volume temperature dependence presented in Fig. 11.1 the glass transition temperature coincides with the Tool's fictive temperature. In fact the *V*(*T*) linearity can be generally assumed with some reasonable accuracy in limited temperature range. This situation may be reached supposing the temperature independent value of the volume thermal expansion coefficient:

$$\beta = \frac{1}{V}\left(\frac{\partial V}{\partial T}\right)_P \tag{11.1}$$

and replacing the exponential volume temperature dependence resulting from the Eq. 11.1 by the linear approximation:

$$V_2 = V_1 \exp[\beta(T_2 - T_1)] \cong V_1[1 + \beta(T_2 - T_1)] \tag{11.2}$$

Thus the slopes of linear dependences plotted in Fig. 11.1 are simply related to the particular thermal expansion coefficients. It is worth noting, that the thermal expansion of glass can be well approximated by the thermal expansion of the CRS. In both cases the volume change induced by temperature increase is connected only with the increase of population of higher vibrational states. The real experimentally obtained temperature dependence of volume (length) obtained by the thermodilatometric measurement is plotted in Fig. 11.2. In the right part of the figure the experimental result of measurement performed on axially loaded sample during zig-zag cyclic time temperature regime is plotted. In this case the thermal expansion is combined with the structural relaxation (represented by the hysteresis) and viscous flow (represented by irreversible shortening of the sample). Such type of experiment can be used for numerical analysis possessing simultaneously the information of thermal expansion, structural relaxation (e.g. the parameters of Tool-Narayanaswamy-Moynihan/Mazurin model) and viscosity [10].

Similar picture can be obtained for the temperature dependence of the system enthalpy, *H*. The slope of lines in the *H*(*T*) line is then given by the isobaric heat capacity $C_P$:

$$C_P = \left(\frac{\partial H}{\partial T}\right)_P \tag{11.3}$$

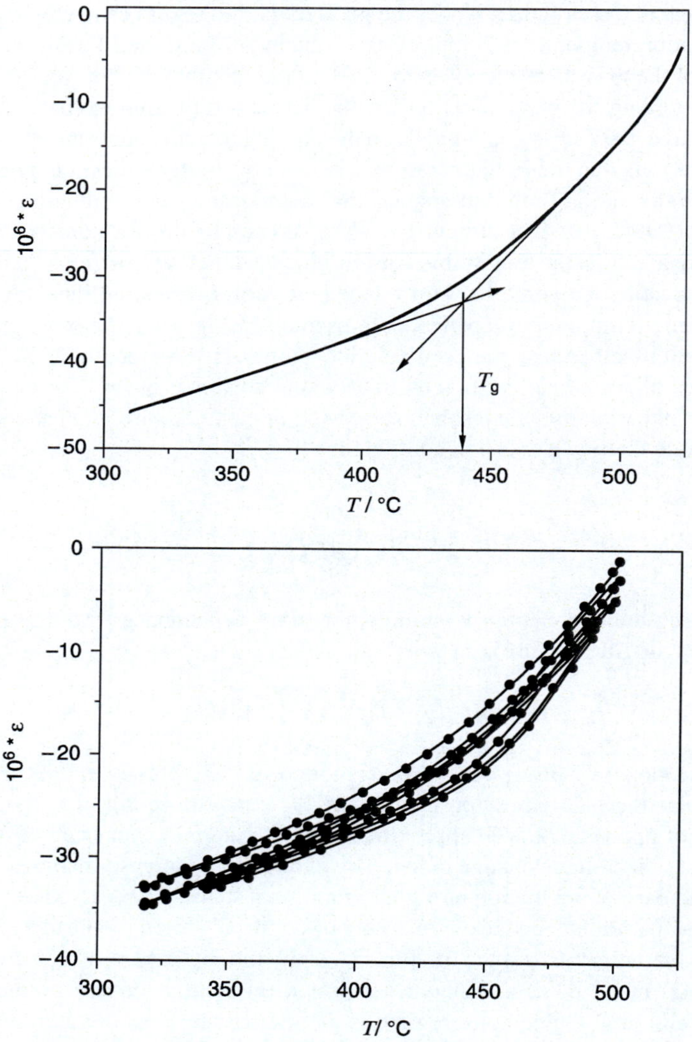


**Fig. 11.2** *Top*: Thermodilatometric cooling curve (cooling rate 5°C/min) of the NBS711 viscosity standard glass. *Bottom*: Results obtained for the same glass during cyclic zig-zag time temperature regime. Points – experimental values, *line* – model accounting for thermal expansion, structural relaxation and viscous flow [10]

Like in the case of the thermal expansion coefficient the temperature independent values of heat capacity can be assumed in the plotted temperature range. Moreover, using the same reasoning as in the case of thermal expansion, the same value of heat capacity of glass and those of the CRS can be assumed. It is also obvious that the value of the heat capacity/thermal expansion coefficient of metastable melt is higher than those of the glass. It is of academic importance only, that assuming the extremely low cooling rate we can reach the hypothetic situation when the volume/enthalpy of the metastable melt will be lower than those of the CRS. Taking into account the logarithmic temperature dependence of the entropy:

$$dS = \frac{dq_{rev}}{T} = C_P d\ln T \tag{11.4}$$

the same picture will result when entropy is plotted against ln$T$. Here the situation when extremely slow cooling would lead to lower entropy of metastable melt than of the CRS is paradoxical. This situation is known as the Kauzmann paradox. The point where the entropy of metastable melt equals to the entropy of the CRS is considered as an ideal glass. The glass close to this state can be obtained by amorphization of some zeolites [6].

To describe the thermodynamics of the glass state we have first to discuss the fact that glass itself is not in the equilibrium state and its structure depends on the kinetics factors, e.g. on the thermal history or the cooling rate when the glass was prepared from the undercooled melt. However, following the reasoning of Conradt [8, 11], we can with some acceptable tolerance accept the glass as a system with definite (mean) values of thermodynamic quantities.

In analogy with melting where we define the melting difference, $\Delta_{fus}$, of any property as the difference between the melt and crystal at the temperature $T_m$ and we define the vitrification change, $\Delta_{vit}$, as the difference between the glass and crystal state at $T$. Once again in many component incongruent systems, the crystal is replaced by CRS and $T_m$ by $T_{liq}$ (or an effective mean temperature between the $T_{liq}$ and $T_{eut}$). Due to the assumption $C_{P,glass} = C_{P,CRS}$ that results in the parallel course of enthalpy temperature dependence the vitrification enthalpy is constant below $T_g$, i.e.:

$$\Delta_{vit}H(T; T \leq T_g) = \Delta_{fus}H(T_g) \tag{11.5}$$

The temperature dependence of the enthalpy of melting can be calculated from the known (from calorimetry experiment) enthalpy of melting at $T_m$ assuming the temperature independent heat capacities:

$$\Delta_{fus}C_P = C_{P,melt} - C_{P,CRS} = C_{P,melt} - C_{P,glass} = \Delta_{vit}C_P = \Delta C_P \tag{11.6}$$

and consequently:

$$\Delta_{fus}H(T) = \Delta_{fus}H(T_m) + \Delta C_P \int_{Tm}^{T} dT' = \Delta_{fus}H(T_m) + \Delta C_P(T - T_m) \tag{11.7}$$

The entropy of melting at $T_m$, $\Delta_{fus}S(T_m)$, can be obtained from the equilibrium condition:

$$\Delta_{\text{fus}}G(T_{\text{m}}) = \Delta_{\text{fus}}H(T_{\text{m}}) - T_{\text{m}}\Delta_{\text{fus}}S(T_{\text{m}}) = 0 \tag{11.8}$$

and for the temperature dependence of the melting entropy we obtain:

$$\Delta_{\text{fus}}S(T) = \Delta_{\text{fus}}S(T_{\text{m}}) + \Delta C_P \int_{T\text{m}}^{T} \mathrm{d}\ln T' = \frac{\Delta_{\text{fus}}H(T_{\text{m}})}{T_{\text{m}}} + \Delta C_P \ln\frac{T}{T_{\text{m}}} \tag{11.9}$$

Using the same reasoning as in the case of vitrification enthalpy, we obtain:

$$\Delta_{vit}S(T;\, T \leq T_{\text{g}}) = \Delta_{\text{fus}}S(T_{\text{g}}) \tag{11.10}$$

Taking into account Eqs. 11.5 and 11.10 we can conclude that we have described the thermodynamics of vitrification by the thermodynamic quantities related to crystallization. Obviously, the driving force for isothermal isobaric devitrification (by crystallization) is represented by the positive value of the vitrification Gibbs energy $\Delta_{\text{vit}}G(\text{T})$:

$$\Delta_{\text{vit}}G(T) = \Delta_{\text{vit}}H(T) - T\,\Delta_{\text{vit}}S(T) \tag{11.11}$$

or

$$\Delta_{\text{vit}}G(T) = \Delta_{\text{fus}}H(T_{\text{m}}) + \Delta C_P(T_{\text{g}} - T_{\text{m}}) - T\left[\frac{\Delta_{\text{fus}}H(T_{\text{m}})}{T_{\text{m}}} + \Delta C_P \ln\frac{T_{\text{g}}}{T_{\text{m}}}\right] \tag{11.12}$$

that can be rewritten as:

$$\Delta_{\text{vit}}G(T) = \Delta_{\text{fus}}H(T_{\text{m}})\left[1 - \frac{T}{T_{\text{m}}}\right] + \Delta C_P\left[(T_{\text{g}} - T_{\text{m}}) - T\ln\frac{T_{\text{g}}}{T_{\text{m}}}\right] \tag{11.13}$$

Because of the positive value of the melting enthalpy and the relationship $T_{\text{g}} < T_{\text{m}}$ it can be seen from the above equation that the thermodynamic driving force towards crystallization (i.e. the opposite to vitrification) increases with decreasing temperature.

The following rules of thumb are frequently used in the glass science [8, 11]

$$\Delta_{\text{vit}}S = \frac{1}{3}\Delta_{\text{fus}}S(T_{\text{m}}) = \frac{1}{3}\frac{\Delta_{\text{fus}}H(T_{\text{m}})}{T_{\text{m}}} \tag{11.14}$$

$$\Delta C_P = \frac{3}{2}\Delta_{\text{fus}}S(T_{\text{m}}) = \frac{3}{2}\frac{\Delta_{\text{fus}}H(T_{\text{m}})}{T_{\text{m}}} \tag{11.15}$$

These rules, although almost never correct in detail, give reasonable estimates of the true values. Inserting the Eqs. 11.14 and 11.15 into the Eqs. 11.9 and 11.10 we obtain the equation:

$$\Delta_{\text{vit}} S = \frac{1}{3} \Delta_{\text{fus}} S(T_{\text{m}}) = \Delta_{\text{fus}} S(T_{\text{m}}) + \frac{3}{2} \Delta_{\text{fus}} S(T_{\text{m}}) \ln \frac{T_{\text{g}}}{T_{\text{m}}} \tag{11.16}$$

leading to the simple relationship between $T_{\text{g}}$ and $T_{\text{m}}$:

$$T_{\text{g}} = T_{\text{m}} \exp(-4/9) = 0.641\, T_{\text{m}} \cong \frac{2}{3} T_{\text{m}} \tag{11.17}$$

Moreover, inserting the Eqs. 11.17 and 11.15 into the Eqs. 11.5 and 11.7 we obtain simple approximate relationship between $\Delta_{\text{vit}} H$ and $\Delta_{\text{fus}} H$:

$$\Delta_{\text{vit}} H = \Delta_{\text{fus}} H(T_{\text{g}}) = \Delta_{\text{fus}} H(T_{\text{m}}) + \frac{3}{2} \frac{\Delta_{\text{fus}} H(T_{\text{m}})}{T_{\text{m}}} \left( \frac{2}{3} T_{\text{m}} - T_{\text{m}} \right) \tag{11.18}$$

or

$$\Delta_{\text{vit}} H = \frac{1}{2} \Delta_{\text{fus}} H(T_{\text{m}}) \tag{11.19}$$

Finally, inserting the Eqs. 11.14 and 11.19 into the Eq. 11.11 we obtain the temperature dependence of the affinity to glass crystallization:

$$\begin{aligned} \Delta_{\text{vit}} G(T_{\text{g}}) &= \frac{1}{2} \Delta_{\text{fus}} H(T_{\text{m}}) - T_{\text{g}} \frac{1}{3} \frac{\Delta_{\text{fus}} H(T_{\text{m}})}{T_{\text{m}}} = \Delta_{\text{fus}} H(T_{\text{m}}) \left( \frac{1}{2} - \frac{2}{9} \right) \\ &= \frac{5}{18} \Delta_{\text{fus}} H(T_{\text{m}}) \end{aligned} \tag{11.20}$$

and

$$\Delta_{\text{vit}} G(0) = \frac{1}{2} \Delta_{\text{fus}} H(T_{\text{m}}) \tag{11.21}$$

The above equations are of especial importance because the enthalpies of melting of crystalline compounds are well experimentally determined and can be obtained from many sources including various thermodynamic databases [12].

## 11.3 Glass Structure

What is the structure [1–5] of glass and how to describe it? In contrary to the crystalline substances where the 3D translational symmetry takes place the structure of glass cannot be quantified giving a small number of numeric data. Due to the 3D translational symmetry the structure of any crystalline substance is given by specifying the basic cell and giving the fractional coordinates of all atoms located in the basic cell. Positions of all other atoms are then simply given by superposition of translations in the three dimensions. Following this concept the structure of glass would be given by specifying Cartesian coordinates of all atoms – i.e. about ~$10^{25}$ data for 1 mol of glass.

On the other side the relatively uniform coordination polyhedra, like $SiO_4$, $AlO_4$, $BO_4$, $BO_3$ etc., can be found in the oxide glass. Therefore the glass structure has to be seen in different scales. The uniform coordination polyhedra correspond to the short-range order (SRO) or the nearest neighbour scale. Going to the next nearest neighbour we can distinguish between various types of coordination polyhedra according to their structural position within the glass network. So the $SiO_4$ tetrahedra can be classified according to the number of bridging oxygen atoms, connecting them with the neighbouring polyhedra. For instance $Q^i$ represents the $SiO_4$ polyhedron with $i$ bridging oxygen atoms, i.e. $Si(\text{-O-X})_i(\text{-O}^-)_{4-i}$, where X stands for any other networkforming atom (Si, Al, B, P,...). Going to the next nearest neighbour (and next-next nearest neighbour) we can reach more detailed structural description corresponding to the medium range order (MRO). The long range order (LRO) structural information concerns the large extent of the polymerized 3D network. Different glass properties are structurally influenced in different extent of structural arrangement. So some optical properties are defined by the SRO, mechanical, electrical and thermal properties depend on the MRO and LRO structure.

On the other hand, the usefulness and applicability of any kind of the structure description/quantification is limited by the experimental methods enabling its quantitative determination. So the nearest neighbour and next nearest neighbour description of networkforming coordination polyhedra is possible mainly due to the experimental techniques like EXAFS, XANES, NMR and Raman vibrational spectroscopy.

The most important structural theory of inorganic glasses was developed by Zachariasen and Waren [13, 14]. According to the *Zachariasen-Warren Continuous Random Network* (CRN) theory the following rules are valid for the formation of glass from simple compounds like $SiO_2$, $B_2O_3$, $P_2O_5$, $GeO_2$, $As_2S_3$, $As_2Se_3$, $BeF_2$, $ZnCl_2$ etc.:

1. An oxide or compound tends to form a glass if it easily forms polyhedral groups as the smallest building units.
2. Polyhedra should not share more than one corner.
3. Anions (like $O^{2-}$, $S^{2-}$, $Se^{2-}$, $F^-$, $Cl^-$, ...) should not bind more than two central atoms (in simple glasses anions form bridges between two adjacent polyhedra).
4. The number of vertices of polyhedron should be less than six.
5. At least three vertices of a polyhedron must be shared with neighbouring polyhedra.

These rules can be well understood when we think about the consequences of their violation. The first rule is in some extent equivalent to the assumption of covalent (or polar covalent) bonding between cations and anions. In another words the covalency implies the sufficient strength of the 3D backbone. The violation of second rule results in more regular (more crystal like) structure and thus the crystallization of the melt can be expected at least when common cooling rates are applied. The same reasoning can be applied in the case of violation of the next rule. The fourth rule can be understood in two different ways. On one side the six (and more) coordinated polyhedra form more regular structures. On the other side the high coordination numbers are typically found in ionic compounds, i.e. in case of electropositive atoms like alkaline and alkaline-earth atoms. The last rule assures the formation of three-dimensional network. In case of sharing only two vertices the chain (or cyclic) structure is formed and the covalent percolation of the melt bulk is not reached.

Zachariasen classified the cations in a glass as follows:

1. Network-formers (NWF, e.g. Si, B, P. . .) with the coordination number 3 or 4.
2. Network-modifiers (NWM, e.g. Li, Na, K, Ca, Sr, Ba, . . .) with coordination numbers generally greater or equal to 6.
3. Intermediates that may either reinforce the 3D network by cross-linking (coordination number 3 or 4) or weaken the network by depolymerization (coordination number 6–8). Moreover the intermediate cations cannot solely form the single component glass.

In case of silicate glasses the NWM oxides (e.g. $Na_2O$) depolymerize the network of $SiO_4$ tetrahedra by disrupting the oxygen bridges (bridging oxygen – BO) and creating the non-bridging oxygen atoms $O^-$ (NBO):

$$\equiv \mathrm{Si} - \mathrm{O} - \mathrm{Si} \equiv + \ \mathrm{Na_2O} \ = 2 \equiv \mathrm{Si} - \mathrm{O}^- + \ 2\mathrm{Na}^+ \tag{11.22}$$

The lack of experimental information about the structure of glass and the difficulties in describing the polymerized structure in the LRO extent can be partially solved by the concept of atom-specific structure elements (ASEs) developed by Sprenger [15–17].

## 11.4 Deducing the Structure from Stoichiometry

### *11.4.1 Silicate Glasses*

The structure of silica glass is formed by mutually 3D interconnected $SiO_4$ tetrahedra. Each tetrahedron is sharing four oxygen atoms with neighbouring tetrahedra, i.e. the structure is on the SRO level composed from $Q^4$ structural units. When alkaline or alkaline earth oxide is added to silica glass the 3D silica network is gradually depolymerized due to the reaction (Eq. 11.22). According to the ratio between the molar amounts ox oxygen and silicon atoms, $R = n(\mathrm{O})/n(\mathrm{Si})$, the structure can be roughly characterized as composed from the mixture of $Q^i$ and $Q^{i+1}$ structural units

(Table 11.1). In case of specific glass compositions listed in Table 11.1, the glass can be considered as composed from single type of Q-units.

It is obvious, that when the assumption of the structure formed by only two successive Q-units (say $Q^i$ and $Q^{i+1}$, where $0 \le i \le 3$) is applied, the relative amount of individual Q-units can be simply calculated from stoichiometry. It is worth noting that this very simple model can provide very important information, e.g. in the field of bioglass [18, 19]. Let us consider the glass of composition given by the formula $x\mathrm{R_2O}\cdot y\mathrm{R'O}\cdot(1\!-\!x\!-\!y)\mathrm{SiO_2}$, where $0 < x,\ y < 1$, and $x + y < 2/3$. Then the $i$ value is determined from Table 11.1 by the inequality:

$$R_i \le \frac{x + y + 2(1 - x - y)}{1 - x - y} = \frac{2 - x - y}{1 - x - y} \le R_{i+1} \qquad (11.23)$$

Then the mole fractions, $x(Q^i)$, of particular Q-units are calculated from the mass balance equations:

$$(1 - x - y)\left[x(\mathrm{Q}^i)\,(i/2 + 4 - i) + x(\mathrm{Q}^{i+1})\,(i/2 + 7/2 - i)\right] = 2 - x - y \quad (11.24)$$

$$x(Q^i) + x(Q^{i+1}) = 1 \qquad (11.25)$$

Unfortunately, the above approach can be used for the rough orientation purposes only. In fact, Q-units disproportionate and the Q-distribution is given by the equilibrium of the disproportionation reactions of the type:

$$2\mathrm{Q}^n \leftrightarrow \mathrm{Q}^{n+1} + \mathrm{Q}^{n-1}, \qquad n = 1,\ 2,\ 3 \qquad (11.26)$$

Nevertheless even the very simple approach considering the coexistence of only two successive Q-units, can successfully explain some experimental property – composition relationships. It is worth noting, that the Q-units distribution represents the SRO only. Going to the next neighbour the interconnectivity between various Q-units (MRO) is taken into account. These level of structure description is especially important for rationalization of the so-called mixed alkali effect, residing in the

**Table 11.1** Structural units of silicate glasses and glass compositions of alkali silicate and alkaline earth silicate corresponding to the *R* value

| Structural unit | $R_i$ | Coordination formula | Glass composition |
|---|---|---|---|
| $Q^i$ | | | $R$ = Li, Na, K, Rb, Cs, Fr<br>R' = Mg, Ca, Sr, Ba, Ra |
| $Q^4$ | 2 | $SiO_{4/2}$ | $SiO_2$ |
| $Q^3$ | 2.5 | $[SiO_{3/2}O_{1/1}]^-$ | $R_2O{\cdot}2SiO_2$, $R'O{\cdot}2SiO_2$ |
| $Q^2$ | 3 | $[SiO_{3/2}O_{1/1}]^{2-}$ | $R_2O{\cdot}SiO_2$, $R'O{\cdot}SiO_2$ |
| $Q^1$ | 3.5 | $[SiO_{3/2}O_{1/1}]^{3-}$ | $3R_2O{\cdot}2SiO_2$, $3R'O{\cdot}2SiO_2$ |
| $Q^0$ | 4 | $[SiO_{4/1}]^{4-}$ | $2R_2O{\cdot}SiO_2$, $2R'O{\cdot}SiO_2$ |

non-monotonous dependence of various physical and chemical glasses and melt properties on the degree of equimolar substitution of one alkaline (or alkaline earth) oxide by the other.

### *11.4.2 Other Oxide Glasses*

When compared with the silicate glasses the structure of borate glasses is more complex namely due to the fact that both trigonal $BO_3$ and tetrahedral $BO_4$ polyhedra form the 3D network. The pure $B_2O_3$ glass is formed from $BO_3$ triangles ($BO_{3/2}$) ordered in six-membered boroxol rings. Due to the equilibria between the three- and four-coordinated boron, the compositional dependence of various physical properties of alkali-borate and alkaline earth-borate glasses is not monotonous. This fact is known as the *boron anomaly*. The structure of borosilicate glasses where the $SiO_4$ Q-units are interconnected with the $BO_3$ triangles and $BO_4$ tetrahedra is even more complex.

Similar complexity is found in the structure of aluminosilicate glasses. The crucial point determining the structure is the ratio between the molar amount of alkali and alkaline earth oxides and alumina, i.e. $[n(R_2O)+n(R'O)]/n(Al_2O_3)]$. In the peraluminous region where this ratio is less than 1, the substantial amount of octahedral $AlO_6$ units is found in the network. In the peralakaline region where sufficient amount of alkalis is present the tetrahedral $[AlO_{4/2}]^{-1}$ units are formed by the reaction:

$$Al_2O_3 + O^{2-} \rightarrow 2[AlO_{4/2}]^- \tag{11.27}$$

The borosilicate an aluminosilicate glasses are presented here as an example illustrating the possible complexity of the SRO glass structure. Obviously the distribution of various types of coordination polyhedra cannot be obtained from the stoichiometry even at the crudest simplifying assumptions. Some results may be obtained by the purely statistical method assuming the purely stochastic formation of connection between various structural units. It is in some respect equivalent to maximizing the configurational entropy of the glass. However the decisive role of the enthalpy (roughly speaking the energy of various bonds) is completely neglected by such treatment.

## 11.5 Thermodynamic Models of Glass

The crucial point of each thermodynamic model is the definition of atomic groupings that will be considered as components of the system. This definition implies the structural range described by the particular model. On the SRO level the components of the system are defined as the particular coordination polyhedra (like Q-units). The system entropy can be then described by regular mixing of system components while the enthalpy can be calculated by summing the bond energies. The equilibrium composition of the system is then obtained by the minimization of the Gibbs energy constrained by the stoichiometry of the system. Such type of models is critically reviewed in the works of Gurman and Conradt [8, 11, 20]. As an example of very simple thermodynamic model of silicate glasses the equations of simultaneous Q-units disproportionation can be presented (Eq. 11.26). The model is represented by the values of the equilibrium constants, $K_n$, of reactions 11.26 – i.e. by the Gibbs energies of different Q-units.

$$K_n = \exp\left(\frac{\Delta_{r,n}G^o}{RT^o}\right) = \frac{a(Q^{n-1})a(Q^{n+1})}{[a(Q^n)]^2} \cong \frac{x(Q^{n-1})x(Q^{n+1})}{[x(Q^n)]^2}, \; n = 1,\, 2,\, 3 \quad (11.28)$$

where $\Delta_{r,n}G^o$ is the standard reaction Gibbs energy of $Q^n$-unit disproportionation, $T^o$ – the standard temperature, $a(Q^n)$ – the activity of $Q^n$ unit in the glass (melt) that is (supposing the standard state of pure substance at standard pressure and temperature) commonly approximated by the mole fraction $x(Q^n)$. The values of equilibrium constants are calculated from the Q-speciation that is commonly determined by the $^{29}Si$ NMR and Raman spectroscopy [4, 5]. However, as far as the energetic and structure of the ionic bonds formed by NWM cations is not included in the model, the $K_n$ values are dependent on the type of NWM cations.

MRO structure is reflected in the thermodynamic models based on the (at least stoichiometric) definition of components as compounds. The computational complexity of the model depends on the number of components. In the model of Conradt [8, 11], the components are defined as having the stoichiometry of crystalline substances forming the CRS. The definition CRS implies that only the coexisting phases are taken into account. Thus the concentrations of all components are straightforwardly defined by the mass balance equations and can be obtained by solving a set of linear equations. As an example we can present the molar volume of $Na_2O$–$SiO_2$ glass at room temperature. According the equilibrium phase diagram the CRS consists – depending on the glass composition – of eutectic mixtures of: $SiO_2 \leftrightarrow 3Na_2O{\cdot}8SiO_2$; $3Na_2O{\cdot}8SiO_2 \leftrightarrow Na_2O{\cdot}2SiO_2$; $Na_2O{\cdot}2SiO_2 \leftrightarrow Na_2O{\cdot}SiO_2$; $Na_2O{\cdot}SiO_2 \leftrightarrow 3Na_2O{\cdot}2SiO_2$; $3Na_2O{\cdot}2SiO_2 \leftrightarrow 2Na_2O{\cdot}SiO_2$; $2Na_2O{\cdot}SiO_2 \leftrightarrow Na_2O$. Taking one mole of $(1-z)Na_2O{\cdot}zSiO_2$ glass and supposing it is formed by the mixture of $iNa_2O{\cdot}jSiO_2 \leftrightarrow kNa_2O{\cdot}lSiO_2$ we can calculate the molar amount, $n$, of both compounds from:

$$n(i\mathrm{Na_2O} \cdot j\mathrm{SiO_2}) = \frac{l}{i.l - j.k} - \frac{k+l}{i.l - j.k} z \quad (11.29)$$

$$n(k\mathrm{Na_2O} \cdot l\mathrm{SiO_2}) = -\frac{j}{i.l - j.k} + \frac{i+j}{i.l - j.k} z \quad (11.30)$$

It was shown by Conradt [21, 22] that the application of his computationally simple model could explain the dependence of chemical durability of glasses including those used for radioactive waste vitrification. The rate of the glass dissolution in aqueous corrosive media has been found correlated with the hydration change of the Gibbs energy calculated as the sum of the hydration energies of all components of the CRS.

Shakhmatkin and Vedishcheva proposed the associated solutions thermodynamic model of glasses and glassforming melts [23–30] that could be considered as the extension of the Conrad's model. This model considers glasses and melts as a solution formed from salt-like products of interaction between the oxide components and the original (unreacted) oxides. These salt-like products (also called associates, groupings or species) have the same stoichiometry as the crystalline compounds, which exist in the equilibrium phase diagram of the system considered. The model does not use adjustable parameters; only the standard Gibbs energies of formation of crystalline compounds and the analytical composition of the system considered are used as input parameters.

Obviously, contrary to Conradt's model [22, 23] the number of system components is greater than the number of oxides. Thus the equilibrium composition cannot be obtained on the basis of mass balance equations. That is why the authors present a calculation of the equilibrium system composition based on the values of particular equilibrium constants that are obtained from the Gibbs formation energies by the standard thermodynamic way. More generally, the constrained minimization of the systems Gibbs energy constrained by the overall system composition has to be performed with respect to the molar amount of each system component to reach the equilibrium system composition [31]. The total Gibbs energy is expressed supposing the state of the ideal solution:

$$G(n_1, n_{2,} \dots n_N) = \sum_{i=1}^{N} \Delta_{\mathrm{f}} G_{\mathrm{m},i} + RT \sum_{i=1}^{N} n_i \ln\left( n_i \middle/ \sum_{j=1}^{N} n_j \right) \qquad (11.31)$$

where $N$ is the numberr of components, $n_i$ is the molar amount of $i$-th component, $T$ is the system temperature (i.e. the glass transition temperature, $T_{\mathrm{g}}$, for particular glass) and $\Delta_{\mathrm{f}}G_{\mathrm{m},i}$ is the molar Gibbs formation energy of pure $i$-th component at the pressure of the system and temperature $T$. The system components are ordered such way that $X_i$ ($i = 1, 2, \dots M < N$) are pure oxides and $X_i$ ($i = M+1, M+2, \dots N$) are compounds formed from oxides by reversible reactions

$$X_i \leftrightarrow \sum_{j=1}^{M} \nu_{i,j} X_j, \qquad i = M+1,\ M+2,\ \dots N \qquad (11.32)$$

Let us suppose the system composition given by the molar amounts of pure unreacted oxides $n_{0,i}$ ($i = 1, 2, \dots M$). Then the mass balance constraints can be written in the form:

$$n_{0,j} = n_j + \sum_{i=M+1}^{N} \nu_{i,j} n_i\,, \qquad j = 1,\ 2, \dots M \qquad (11.33)$$

The molar Gibbs formation energies of the melts of pure components may be used in more advanced version of this model. However these thermodynamic data are relatively scarce that prevents routine application of the model to the study of multicomponent systems. On the other hand the errors caused by substituting the melt by the crystalline state are partially compensated when the reaction Gibbs energy, $\Delta_r G_{m,i}$ is calculated according:

$$\Delta_r G_{m,i} = \Delta_f G_{m,i} - \sum_{j=1}^{M} \nu_{i,j} \Delta_f G_{m,j} = -RT \ln K_i \,, \qquad i = M+1,\, M+2, \ldots N \tag{11.34}$$

where $K_i$ is the equilibrium constant of the $i$-th equilibrium reaction described by the Eq. 11.32.

Recently, the thermodynamic model of Shakhmatkin and Vedishcheva and ab initio molecular dynamics were used for the structural study of nonsilicate binary calcium aluminate glass $x$CaO·(1 − $x$)$Al_2O_3$, where $x = 1/3$ (abbreviated as $CA_2$) 1/2 (CA), and 2/3 ($C_2A$) [32]. Both methods provided mutually comparable quantitative results in a reasonable agreement with the accessible experimental data and previous results of classical MD simulation. The obtained compositional trends were in agreement with the concept of the increase of Al/O coordination number (accompanied with formation of Al-triclusters) with the decreasing CaO content in the per-aluminous region.

When the crystalline state data are used the model can be simply applied to most multicomponent glasses including the nonoxide once. Especially the application of the model to the multicomponent industrially produced glasses can be very important. Taking into account that the common praxis resides in expressing most of the multicomponent glass properties in the form of (mostly) additive functions of the glass composition expressed in percents (even weight and not molar!) of pure oxides [33–35] using the thermodynamic model unambiguously represents the significant progress.

The contemporary databases of thermodynamic properties (like the FACT computer database [12]) enable the routine construction of the Shakhmatkin and Vedishcheva model [31] for most of important multicomponent systems. It is worth noting that other methods of thermodynamic modelling of glasses and glassforming melts (not discussed here, e.g. [5, 36, 37]) do not posses the possibility of routine application to multicomponent systems.

## 11.6 Quantitative Raman Spectroscopy and Thermodynamic Modelling

Raman spectroscopy is commonly used for study of the glass structure allowing the SRO and MRO structural units present to be inferred [4, 5, 38–40]. As a typical example the $Q^n$ distribution in silicate glasses can be mentioned. The Raman spectra are after the background subtraction and thermal correction [39] deconvoluted typically to Gaussian peaks assigned to various structural groupings. From the corresponding peak areas the relative abundance of particular structural units is obtained using the NMR calibration. Sometimes [40] the direct proportionality between the ratio of peak areas and molar amounts of various structural groupings is assumed. Relating purely formally the Raman spectroscopy to the absorption optical spectroscopy the non-validity of the Lambert–Beer's law is the main mathematical difference between both methods.

Recently Zakaznova and Malfait [41–43] proposed a new method of numerical treatment of Raman spectra. The basic assumption of this approach is that the Raman spectra are a sum of partial Raman spectra (generated by individual species) multiplied by the abundance of the species. Thus a set o Raman spectra obtained for series of glasses with different composition spans a linear vector space with the dimensionality given by the number of independent species (i.e. species that independently vary their abundance) with different PRS. Each spectrum is recorded with an arbitrary scaling, i.e. it is known with the exception of any multiplication factors. Principal component analysis (PCA, [44–47]) is then used to find the number of independent components. Following statistical treatment based on proper thermodynamic model results in finding the unknown equilibrium constants. Zakaznova and Malfait use the proposed method for analysis of the binary $K_2O$–$SiO_2$ glassforming melts with the aim to estimate the temperature dependence of the disproportionation constant $K_3$ (see Eqs. 11.26 and 11.28). The resulting values of $K_3$ and reaction enthalpy were in perfect agreement with independent results obtained by $^{29}$Si NMR spectroscopy.

The approach of Zakaznova and Malfait [42] can be mathematically formulated as:

$$\mathbf{AP} = \mathbf{EC} \tag{11.35}$$

where the matrix formalism is used where $\mathbf{A}(N_w/N_s)$ is the matrix of Raman spectra with $N_w$ rows corresponding to $N_w$ wavenumbers and $N_s$ columns corresponding to $N_s$ samples/spectra, $\mathbf{P}(N_s/N_s)$ is the square diagonal matrix with the unknown coefficients each multiplying one particular Raman spectrum, $\mathbf{E}(N_w/N_c)$ is the matrix of PRS stored columnwise ($N_c$ equals to the number of independent components), and $\mathbf{C}(N_c/N_s)$ is the matrix with abundances of $N_c$ individual components in $N_c$ samples. Replacing the product $\mathbf{AP}$ with another matrix $\mathbf{A}^{corr}$ we obtain the equation formally equivalent to the Lambert–Beer's equation of optical absorption spectra – here $\mathbf{A}^{corr}$ is the absorbance matrix, $\mathbf{E}$ – the matrix of molar absorption coefficients and $\mathbf{C}$ the concentration matrix.

Let us suppose that the $\mathbf{C}$ matrix is known from the solution of the thermodynamic model. Then the non-linear least squares problem can be formulated:

$$F(\mathbf{P},\mathbf{E}) = \sum_j^{Ns} \sum_i^{Nw} \left[ A_{i,j} P_{j,j} - \sum_k^{Nc} \mathrm{E}_{i,.k} C_{k,j} \right]^2 = F_{\min} = \min \tag{11.36}$$

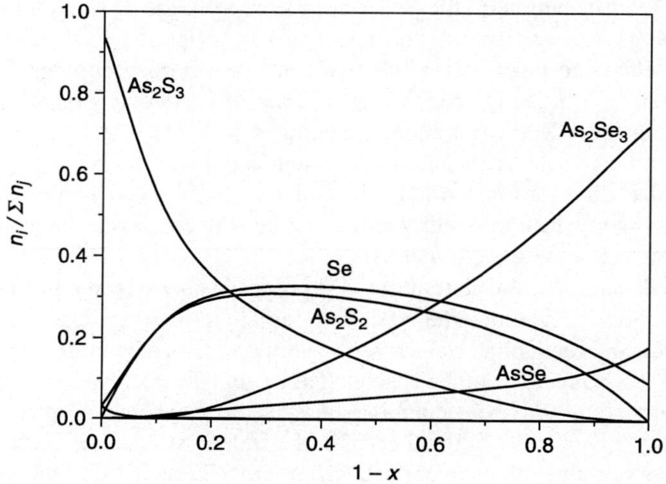


**Fig. 11.3** Results of thermodynamic model of the $xAs_2S_3{\cdot}(1-x)As_2Se_3$ system

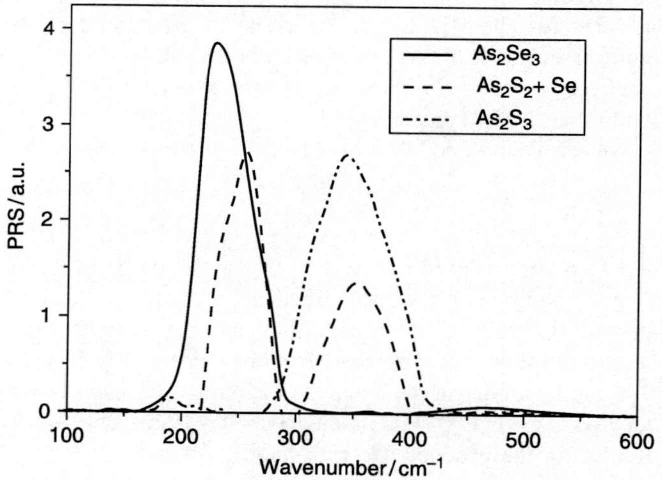


**Fig. 11.4** Partial Raman spectra of the $As_2S_3$–$As_2Se_3$ glasses

Without loss of generality we can set $P_{1,1} = 1$. Then the unknown $P_{i,i}$ multiplication factors are obtained together with the unknown **E** matrix of PRS by solving the set of $(N_s - 1 + N_w N_c)$ linear equations. Minimizing the $F_{min}$ value with respect to some parameters of the thermodynamic model this method can be used for finding some values of missing thermodynamic data.

In our recent paper [48] the structure of chalcogenide glasses of the pseudo-binary system $As_2S_3$–$As_2Se_3$ was investigated by comparison of the results of the thermodynamic model of associated solutions with the results obtained by the above analysis of Raman spectra of $xAs_2S_3{\cdot}(1-x)As_2Se_3$ ($x = 1$; 3/4; 1/2; 2/3; 1/4; 0) glasses. The results of thermodynamic model are presented in Fig. 11.3. On the basis of experimental values of $T_g$ published in [49] the following linear approximation of $T_g$ dependence on the glass composition was obtained

$$T_g/K = 478,85 - 23.087\,x \tag{11.37}$$

reproducing the experimental DTA with the standard deviation of approximation $s_{apr} = 1.6$ K. For each glass composition plotted in Fig. 11.3 the system temperature was set to the $T_g$ value obtained from the Eq. 11.37.

The PCA method identified three independent components in the studied spectral series. On the other hand, the thermodynamic modelling resulted in four components with significant abundance in the studied glasses, i.e. Se, $As_2S_2$, $As_2S_3$ and $As_2Se_3$. Finally, the correlation analysis proved the strong (correlation coefficient of 0.97) linear dependence between the concentrations of Se and $As_2S_2$. Thus the results of Raman spectra analysis were in harmony with the thermodynamic model. Solving the Eq. 11.36 led to the set of PRS plotted in Fig. 11.4.

## References

1. Scholze H (1991) Glass – nature, structure, and properties. Springer, Berlin
2. Vogel W (1992) Glass chemistry, 2nd edn. Springer, Berlin
3. Rao KJ (2002) Structural chemistry of glasses. Elsevier, Amsterdam
4. Mysen BO (1988) Structure and properties of silicate melts. Elsevier, Amsterdam
5. Mysen BO, Richet P (2005) Silicate glasses and melts – properties and structure. Developments in geochemistry, vol 10. Elsevier, Amsterdam
6. Greaves GN, Sen S (2007) Inorganic glasses, glass-forming liquids and amorphizing solids. Adv Phys 56:1–166
7. Gutzow I, Schmelzer J (1995) The vitreous state. Thermodynamics, structure, rheology, and crystallization. Springer, Berlin
8. Conradt R (1999) Thermochemistry and structure of oxide glasses. In: Bach H, Krause D (eds) Analysis of the composition and structure of glass and glass ceramics. Springer, Berlin, pp 232–254
9. ASTM – C162 (1983) Standard terminology of glass and glass products. American Society for Testing and Materials, West Conshohocken, PA
10. Liška M, Štubňa I, Antalík J, Perichta P (1996) Structural relaxation with viscous flow followed by thermodilatometry. Ceramics 40:15–19
11. Conradt R (2004) Chemical structure, medium range order, and crystalline reference state of multicomponent oxide liquids and glasses. J Non-Cryst Solids 345, 346:16–23
12. Bale CW et al (2002) FactSage thermochemical software and databases. Calphad 26:189–228
13. Zachariasen WH (1933) Die Struktur der Gläser. Glastechn Ber 11:120–123
14. Warren BE (1941) Summary of work on atomic arrangement in glass. J Am Ceram Soc 24:256–261

15. Sprenger D (1996) Spektroskopische Untersuchungen und Berechnung zur Struktur anorganischer Gläser. PhD Thesis, Mainz, Germany
16. Sprenger D, Bach H, Meisel W, Gütlich P (1993) Discrete Bond Model (DBM) of sodium-silicate glasses derived from XPS. Raman and NMR measurements. J Non-Cryst Solids 159:187–203
17. Schultz-Münzenberg C (1999) How to describe the topological structure of glasses. In: Bach H, Krause D (eds) Analysis of the composition and structure of glass and glass ceramics. Springer, Berlin, pp 141–152
18. Koga N, Strnad Z, Šesták J, Strnad J (2003) Thermodynamics of non-bridging oxygen in silica bio-compatible glass-ceramics. J Therm Anal Calorim 71:927–937
19. Šesták J, Strnad J, Strnad Z, Koga N, Holeček M (2008) Biomedical thermodynamics of bio-compatible glass-ceramics and otherwise modified inorganic material surfaces. Adv Mat Res 39, 40:329–334
20. Gurman SJ (1990) Bond ordering in silicate glasses: a critique and resolution. J Non-Cryst Solids 125:151–160
21. Conradt R (1999) Predictive modeling of glass corrosion. In: Proceedings of the 5th ESG Conference, B1-2–B1-10, Czech Glass Society, Prague, Czech Republic
22. Conradt R (2001) A proposition for an improved theoretical treatment of the corrosion of multi-component glass. J Nucl Mater 298:19–26
23. Shakhmatkin BA, Vedishcheva NM, Shultz MM, Wright AC (1994) The thermodynamic properties of oxide glasses and glass-forming liquids and their chemical structure. J Non-Cryst Solids 177:249–256
24. Vedishcheva NM, Shakhmatkin BA, Shultz MM, Wright AC (1996) The thermodynamic modelling of glass properties: a practical proposition. J Non-Cryst Solids 196:239–243
25. Shakhmatkin BA, Vedishcheva NM, Wright AC (1997) In: Wright AC, Feller SA, Hannon AC (eds) Borate glasses crystals and melts. Society of Glass Technology, Sheffield, p 189
26. Shakhmatkin BA, Vedishcheva NM, Wright AC (2001) Can thermodynamics relate the properties of melts and glasses to their structure? J Non-Cryst Solids 293–295:220–236
27. Vedishcheva NM, Shakhmatkin BA, Wright AC (2001) Thermodynamic modelling of the structure of glasses and melts: single-component, binary and ternary systems. J Non-Cryst Solids 293–295:312–317
28. Vedishcheva NM, Shakhmatkin BA, Wright AC (2003) Thermodynamic modelling of the structure of sodium borosilicate glasses. Phys Chem Glasses 44:191–196
29. Vedishcheva NM, Shakhmatkin BA, Wright AC (2004) The structure of sodium borosilicate glasses: thermodynamic modelling vs. experiment. J Non-Cryst Solids 345, 346:39–44
30. Shakhmatkin BA, Vedishcheva NM, Wright AC (2004) Thermodynamic modelling of the structure of oxyhalide glasses. J Non-Cryst Solids 345, 346:461–468
31. Vonka P, Leitner J (1995) Calculation of chemical equilibria in heterogeneous multicomponent systems. Calphad 19:25–36
32. Liška M, Macháček J, Perichta P, Gedeon O, Pilát J (2008) Thermochemical modelling and Ab initio molecular dynamics simulations of calcium aluminate glasses. Ceram Silik 52:61–65
33. Seward TP III, Vascott T (eds) (2005) High temperature glass melt property database for process modeling. American Ceramic Society, Westerville, OH
34. Pye LD, Montenero A, Josephs I (eds) (2005) Properties of glass-forming melts. Taylor & Francis, Boca Raton, FL
35. http://www.sciglass.info
36. Pelton AD, Wu P (1999) Thermodynamic modeling in glass-forming melts. J Non-Cryst Solids 253:178–197
37. Stolyarova VL (2008) Thermodynamic properties and structure of ternary silicate glass-forming melts: experimental studies and modeling. J Non-Cryst Solids 354:1373–1377
38. McMillan PF (1984) Structural studies of silicate glasses and melts – applications and limitations of Raman spectroscopy. Am Mineral 69:622–644

39. McMillan PF, Wolf GH (1995) Vibrational spectroscopy of silicate liquids. In: Stebbins JF, McMillan PF, Dingwell DB (eds) Structure, dynamics and properties of silicate melts. Mineralogical Society of America, Washington, DC, pp 247–315
40. Parkinson BG, Holland D, Smith ME, Larson C, Doerr J, Affatigato M, Feller SA, Howes AP, Scales CR (2008) Quantitative measurement of Q3 species in silicate and borosilicate glasses using Raman Spectroscopy. J Non-Cryst Solids 354:1936–1942
41. Zakaznova-Herzog VP, Malfait WJ, Herzog F, Halter WE (2007) Quantitative Raman spectroscopy: principles and application to potassium silicate glasses. J Non-Cryst Solids 353:4015–4028
42. Malfait WJ, Zakaznova-Herzog VP, Halter WE (2007) Quantitative Raman spectroscopy: high-temperature speciation of potassium silicate melts. J Non-Cryst Solids 353:4029–4042
43. Malfait WJ (2009) Quantitative Raman spectroscopy: speciation of cesium silicate glasses. J Raman Spectrosc 40:1895–1901
44. Pelikán P, Čeppan M, Liška M (1994) Computational methods in molecular spectroscopy. CRC Press, Boca Raton, FL
45. Malinowski ER (2002) Factor analysis in chemistry, 3rd edn. Wiley, New York
46. Kramer R (1998) Chemometric techniques for quantitative analysis. Marcel Dekker, New York
47. Factor analysis Toolbox for MATLAB®. Applied Chemimetrics, http://www.chemometrics.com
48. Liška M, Holubová J, Chromčíková M, Černošková E (2008) Raman spectra, structure and thermodynamic model of $As_2S_3$–$As_2Se_3$ glasses. In: 30th International Czech and Slovak Seminar on Calorimetry, University of Pardubice, Pardubice, Czech Republic, pp 3–36
49. Holubová J, Černošek Z, Černošková E, Liška M (2006) Glassy system As-S-Se. In: 28th International Czech and Slovak Seminar on Calorimetry, University of Pardubice, Pardubice, Czech Republic, pp 133–136

# Chapter 12
# Oxide Glass Structure, Non-bridging Oxygen and Feasible Magnetic Properties due to the Addition of Fe/Mn Oxides

**Jaroslav Šesták, Marek Liška, and Pavel Hubík**

## 12.1 Non-bridging Oxygen and the Structure of Oxide Glasses

To a large extent, the physical and thermodynamic properties of glasses are controlled by their inner structural make up within the so called *short range order* (SRO) and its extended viewing as modulated structures identified as *medium range order* (MRO) [1–5] often adjacent to nano-crystalline arrangements. Study of physical nature of nano-scale in-homogeneities in glasses (and in their melts) [6, 8] provides glass inventors with basis for elaboration of glasses of a required state, from extremely homogeneous glasses (optical fibre drawing) to nano-crystallized and/or porous-containing glasses suitable for various application (sorbents, molecular filters, bioglasses, zero-expansion glass-ceramics, matrixes for nano-scale crystals, etc.). Thus structural information is essential for material scientists to predict their thermal, magnetic and other properties.

Over the past 20 years the structural investigation conducted by nuclear magnetic resonance (NMR), Raman infrared spectroscopy (RIS) [4–7, 9] and various other techniques of thermophysical measurements (e.g., methods of thermal analysis) have established that silicate melts and glasses are combination of a relatively small number of building species though the glass can be perceived as a continuous inorganic

J. Šesták (✉)
New Technologies Research Centre, University of West Bohemia, Univerzitní 8, CZ-30614 Plzeň, Czech Republic
e-mail: sestak@fzu.cz

M. Liška
Vitrum Laugaricio, Joint Glass Center of Institute of Inorganic Chemistry (Slovak Academy of Sciences) and Alexander Dubček University of Trenčín and RONA, j.s.c. glassfactory, Študentská 2, Trenčín SK-911 50, Slovak Republic

P. Hubík
Institute of Physics, v.v.i., the Academy of Sciences of ČR, Cukrovarnická 10, CZ-16200 Praha 6, Czech Republic

J. Šesták et al. (eds.), *Glassy, Amorphous and Nano-Crystalline Materials*, Hot Topics in Thermal Analysis and Calorimetry 8, DOI 10.1007/978-90-481-2882-2_12,

*polymeric material.* The basic structural unit (analogous to organic – *mers* in polymers) are regular polyhedrons $AO_n$ (such as $SiO_4$, $GeO_4$, $AlO_3$, $AlO_4$, $BO_4$, $BO_3$, $PO_4$, etc.). By help of Niggli's coordination-substitution formulae [10–12] the thorough state of completely interconnected netting can be described in accordance with the Zachariesen's theory [13] as $AO_{n/2}$ if the bridges are formed by all $\underline{n}$ atoms of oxygen and if oxygen is bridged with two central cations, A- (e.g. $\{SiO_{4/2}\}$). In the case when $\underline{i}$ atoms of oxygen become bridged with A by a double bond (so called *terminal bonds*), such as single component glass composed by phosphorus oxides containing *–mers* of $\{O = P(–O–)_3\}$ the drivable coordination formula ensues as $\{AO_{i/1}O_{(n-i)/2}\}$, i.e., $\{PO_{1/1}O_{3/2}\}$, corresponding thus the stoichiometry of $P_2O_5$. In the analogy with organic polymers the group $\{AO_{i/1}O_{(n-i)/2}\}$ can be considered as $(n\_i)$-functional *–mer*.

Until now the single component glasses were discussed, which can be extended in the sense of the Zachariesen's theory [13] of a randomly interconnected web (*continuous random network* – CRN, see also Fig. 12.1) to more complex systems containing supplements of electropositive elements (such as alkaline oxides, $M_2O$, or oxides of metals and alkaline earths, $M'O$) often called *modifying oxides.* Their additions to a single-component netted glass result in the breakdown of bridging bonds A–O–A and creation of so called *non-bridging oxygen* (NBO), which can be described by the equations:

$$M_2O = 2M^+ + O^{2-} \tag{12.1}$$

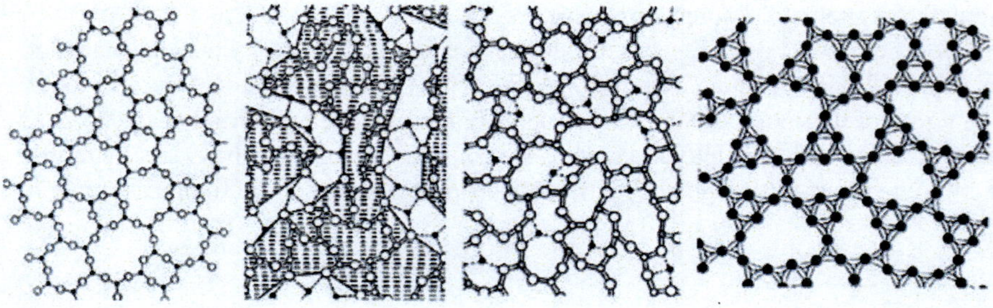

**Fig. 12.1** Two-dimensional illustration of a continuous random network model (CRN) for the corner-sharing tetrahedral oxides (*left*) fitting the structure of $AO_2$-type glass (such as typically $SiO_2$). A-type atoms (Si) are shown as *solid small circles* while the bridging atoms (O) as *open circles.* The modified random network (MRN) correspond to the structure of a modified silicate glass where Si atoms are shown again as small open circles and filled circles refer to modifiers (alkalis or alkaline earth cations). BO atoms are located within the network and NBO atoms along the modifier channels (*left middle*). A compensated continuous random network (CCRN) model is represented by an alumino-silicate glass where all O atoms are bridging (Al atoms are shown as *small bold open circles, right middle*). *Far right* is a portrayal of a two-dimensional depiction of the random network model of a $AO_3$ structure (such as $B_2O_3$ glass consisting of $B_3O_6$ boroxol groups and $BO_3$ *triangles*). B and O atoms are shown as *open* and *filled circles*, respectively.

$$M'O = M'^{2+} + O^{2-} \tag{12.2}$$

$$A - O - A + O^{2-} = 2A - O^{-} \tag{12.3}$$

First two equations are responsible for the dissociation concerning a single molecule of modifying oxides while the third describes breakdown of bridging bond A–O–A. In the case of typically modifying oxides the equation equilibrium 12.1–12.3 is shifted toward the right-sided products and the structural information is straightforwardly accessible from glass initial stoichiometry. Interestingly, there is an apparent analogy with organic polymers mentioning the so called *mean degree of netting* [14, 15]. Structural motives, $Q$, which are appearing throughout glass, is possible to characterize by a coordination formula $\{AO_{(i+j)/1}O_{(n-i-j)/2}\}$, where $j$ is the number of bridging oxygens associated with a single central atom being ripped away by the action of oxygen produced through the dissociation of modifying oxides ($j \leq n\_i$). In the case of silica glasses ($i = 0$, $n = 4$) there is a direct link between the concept of the so called $Q$-notation (i.e., $Q^k$, $k = 0, 1, 2, 3, 4$) a and coordination formula, $Q^k \equiv \{SiO_{(4-k)/1}O_{k/2}\}$.

A most common parameter characterizing quantitatively the degree of netting is the ratio of molar amounts of the oxygen atoms and that of cations in the net-forming oxide. In the case of silica glass with the composition of $\{xM_2O \cdot yM'O \cdot (1 - x - y)$ $SiO_2\}$, where apply both $\{0 \leq x, y\}$ and $\{x + y \leq 1\}$, it reads:

$$r = \frac{n(O)}{n(Si)} = \frac{2 - x - y}{1 - x - y} \tag{12.4}$$

The value of $r = 2$ fully corresponds to 100% networking where (in the terminology of $Q$-motives) the system is composed exclusively by the structural units $Q^4$. On the other hand, when the value reaches 4, the system is composed of isolated anions $[SiO_4]^{4-}$, i.e., by mere units of $Q^0$ matching up to the zero netting. At the fully interconnected web (100% netting) the number of bridging bonds is equal to the double of compound amount of the silica atoms, i.e., $\{2(i - x - y)\}$. Certainly, it is a hypothetical reference state in which the oxygen atoms created by the dissociation of modifying oxides are NOT employed to split the Si–O–Si bridges. In fact, however, the number of bridges (identical to the compound amount of bridging oxygens) is equal $\{2(1 - x - y) - x - y = 2 - 3x - 3y\}$. In this light, the degree of netting can be defined by the percent of maximum networking, i.e.:

$$P\,[\%] = 100\frac{2 - 3x - 3y}{2 - 2x - 2y}, \quad x + y \leq \frac{2}{3} \tag{12.5}$$

The state of zero interconnected web are represented by glasses with an orthosilicate stoichiometry $\{2/3(M_2O,\ M'O) \cdot 1/3SiO_2\}$ or in a more common notation $\{2(M_2O,\ M'O) \cdot SiO_2\}$. From Eq. 12.5 follows that a 100% netting is attainable for pure quartz glass, only, i.e., in the case of $\{x + y = 0\}$.

In contrast to the momentary look of an apparently limited applicability, the above description is, in fact, useful for a generalized description of a variety of glasses containing multiple glass-forming and modifying oxides. The only condition is the internal stability of theirs internal coordination spheres, i.e., values *i* a *n* emerging in coordination formula $\{AO_{i/1}O_{(n-i)/2}\}$ of a glass formed by mere pure glass-forming and thus net-forming oxide. Moreover, the basic significance of netting parameter *r*, gives a possibility to correlate its value with various technologically important properties; for example increasing *r* correlates with the increasing viscosity, chemical durability and/or decrease of $T_g$ while lessening an associated value of thermal expansion. It is clear that such an approach cannot be fully applied to such oxides which do not possess a clear glass-forming and/or modifying status thought it is even applicable for anionically mixed glasses, such a oxynitride glasses [7]. A specificity of alkali aluminosilicate glasses it can be mentioned as the trivalent aluminium cation does not always act as a network former. The structural configuration depends upon the *r* (=$Al_2O_3/R_2O$) ratio. Apparently, if the parameter *r* becomes greater than one the $Al^{3+}$ goes into the network having a tetrahedral coordination. Like ($BO_4$) group, the excess of unit negative charge on ($AlO_4$) group is satisfied by alkali ion in the neighbourhood. Thus, the further addition of an aluminium ion to a silicate glass removes one NBO. At the ratio equal one there is no NBO in the structure. For the further additions of $Al^{3+}$, when the ratio r becomes greater than one, the $Al^{3+}$ cations enters the network as a modifier in the tetrahedral coordination and the formation of ($AlSiO_6$) groups turn out to be unlikely due to the packing difficulties. Presumably three of the oxygens are then non-bridging and three remaining (already) bridging, which favours the formation of triclusters, where an oxygen is shared between three tetrahedrons (one $AlO_4$ and two $SiO_4$ or vice versa). The first type of tricluster is electrically neutral while the second one requires an associated alkali cation. In principle, the triclusters appear equivalent to an $Al^{3+}$ cation in the form of a modifying ion and are bonded to three NBO.

In other melts of the ternary system, such as $CaO–TiO_2–SiO_2$, silica oxide acts always as the network-forming oxide while calcium oxide takes always action as the modifier of the poly-anionic network. The role of $TiO_2$ can be twofold again acting either as a modifier or as a complementary associate undertaking thus the function of the network-forming oxide. Depending on the behaviour of $TiO_2$ in the melt, the structure and thus also the activities of the present components of the melt change revealing variously possessed NBO. The structural role of $Ti^{4+}$ in titanium silicate melts is accordingly a complex function of several variables, namely $TiO_2$ and $SiO_2$ concentrations, type and content of modifying cations [8] as well as temperature. Recently, the structural characteristics of many important glasses and melts (including silicates, borates, aluminosilicates, halides and chalcogenide) are drawn in additional details using results of the recent spectroscopy [7] and scattering experiments by Greaves and Sen [2] and the quantitatively applied Raman spectroscopy as introduced by Malfait et al. [9].

## 12.2 Reflection on Volume of Ionic Sites in Silicate Glasses and the Applicability of NBO Concepts

As mentioned above various physico-chemical properties of silicate glasses can be correlated with the concentration of certainly modelled structural units. Properties, such as molar volume, refractive index, molar refraction, thermal expansion and even some magnetic properties can then be practically described by additive relations. In these relations a specific factor can be used for each structural unit. The factor depends on the type of property under consideration. The obtained factors could successfully be used to calculate the magnitude of those physical properties in binary alkali (or alkaline earth) silicate glasses and in other multi-component glasses. Molar abundant data of alkali and alkaline earth silicate glasses have been recently used to calculate the free volume [14] associated with the presence of BO (bridging oxygen) and NBO as well as modifier ions. The volumes of voids in the structure are of special importance for the transport properties. The mobility of movable ions is affected to a great extent by the openness (nature) of the glass matrix. Such an "open matrix", exhibiting higher concentration of NBO ions, is characterized, e.g., by higher electric conductivity than a closed one. Different spectroscopic studies indicated that up to about 33 mol% of alkali oxide ($R_2O$) so called $Q3$ units are formed at the expense of $Q4$ units where $Q3$ units are $SiO_4$ tetrahedra having one NBO ion per tetrahedron, whereas $Q4$ units are $SiO_4$ tetrahedra without NBOs. For more than 33 mol% of $R_2O$ about 50 mol% of $Q3$ units converts into $Q2$ units the latter representing $SiO_4$ tetrahedra with two NBOs for each tetrahedral [3, 4]. Results of Raman spectroscopy revealed that the conversion process continues along with the increasing $R_2O$ content, from $Q2$ to $Q1$ (for $> 50$ mol% $R_2O$) and finally from $Q1$ to $Q0$ (for $>60$ mol% $R_2O$) where $Q1$ and $Q0$ refer to $SiO_4$ tetrahedra having, respectively, three and four NBOs per each $Si^{4+}$ cation. In the light of these findings some relations could be estimated which describe the concentration of structural units as a function of modifier oxide concentration in binary alkali or alkaline earth silicate glasses.

The relations were modified to calculate the concentration of structural units in mixed alkali silicate glasses on basis of the molar volume ($V_m$) of a glass, which can be given in the form of an additive relation, such as $V_m = N_u V_u$, where $N_u$ is the number of the structural units $u$ per mole of glass and $V_u$ is the volume of this unit [14]. On the basis of the structural unit distribution in alkali silicate glasses and the volumes that the constituting ions occupy in the matrix ($V_{Si}$, $V_\varphi$, $V_R$ and $V_O$), the volumes of individual structural units can be given as follows:

$$\begin{aligned} V_4 &= V_{Si} + 2V_\varphi, V_3 = V_{Si} + 1.5V_\varphi + (V_O + V_R), \\ V_2 &= V_{Si} + V_\varphi + 2(V_O + V_R), \end{aligned} \tag{12.6}$$

$$V_1 = V_{Si} + 0.5V_\varphi + 3(V_O + V_R) \text{ and thus } V_0 = V_{Si} + 4(V_O + V_R). \tag{12.7}$$

Here $V_{Si}$, $V_{\varphi}$, $V_O$ and $V_R$ are, respectively, the volumes occupied by $Si^{4+}$, BO ($\varphi$), NBO (O) and alkali ion (R). The volume occupied by an ion is the volume of ion itself and its associated space (its free volume) in the matrix. These relations can be used to calculate $V_{\varphi}$ and $V_O$. Thus on the basis of above equations we may write $V_m = N_i V_i$ with $N_i$ being the number of the ions $i$ per mole of glass and $V_i$ is the volume occupied by that ion in the given glass. The last above relation can be reformed to $V_m = N_{Si} V_{Si} + N_{\varphi} V_{\varphi} + N_O V_O + N_R V_R$, where $N_{Si}$, $N_{\varphi}$, $N_O$ and $N_R$ are, respectively, the number of $Si^{4+}$, BO, NBO and alkali ions per mole of glass. The radius of $Si^{4+}$ ion in oxides is ~$0.26 \times 10^{-8}$ cm whereas that of $O^{2-}$ ion in silicate units is $1.2 \times 10^{-8}$ cm. Thus the volume of $Si^{4+}$ ions in vitreous $SiO_2$ and then in any modified silicate glass is ~0.5% of the volume of $O^{2-}$ ions. Therefore, in the first approximation, the first term can be neglected with respect to the other terms $V_m \approx N_{\varphi} V_{\varphi} + N_O V_O + N_R V_R$ and further accepting that $N_O = N_R$, we obtain $V_m \approx N_{\varphi} V_{\varphi} + N_O (V_O + V_R)$. Any NBO ion is always bound to an alkali ion (or a half alkaline earth ion) and therefore the term in the brackets can be looked as an undividable quantity that can be labelled by $V_B$, namely $V_B = V_O + V_R$ It follows that $V_m = N_{\varphi} V_{\varphi} + N_O V_B$ which equation can be solved simultaneously for two values of $V_m$ close to each other to get $V_{\varphi}$ and $V_B$.

As the modifier oxide content increases the total contribution of BO ions $(N_{\varphi} V_{\varphi})$ decreases, whereas that of the NBO bindings $(N_O V_B)$ increases. The decrease in $(N_{\varphi} V_{\varphi})$ is related to the decrease in the $SiO_2$ content. Up to 33 mol% of the modifier oxide, the values of $(N_{\varphi} V_{\varphi})$ are mostly the same for all types of modifier ions. This behaviour reveals that $V_{\varphi}$ is not affected by the type of modifier ion up to one NBO ion per $SiO_4$ tetrahedron. Above this concentration $(N_{\varphi} V_{\varphi})$ varies slightly with the type of modifier oxide. On the other hand, $(N_O V_B)$ increases when increasing the modifier oxide content. There is a marked difference between the values from one type of modifier to another showing that $V_m$ of a glass decreases with increasing modifier oxide concentration for $|d(N_{\varphi} V_{\varphi})/d_C| > d(N_O V_B)/d_C$ and vice versa. The increase in $(N_O V_B)$ for a constant modifier concentration and a change of its type (such as from $Li_2O$–$SiO_2$ to $Cs_2O$–$SiO_2$) reveals that the volume $V_B$ becomes greater with the field strength of the modifier.

The cooperatives of the field strength are expressed as $q/r^2$, where $q$ is the charge of modifier ion and $r$ is its radius. It is observed that in all cases the dependence of $V_B$ on field strength can be roughly expressed by linear dependence on $r^2/q$, namely by a relation $V_B = s(r^2/q) + 13.737 \times 10^{-24}$ cm$^3$. It follows that $V_O$ has a constant value and that the increase in $V_B$ with increasing $V_R$ is related to the size of the alkali ion. A greater alkali ion size means lower field strength and weaker bond with the NBO and this leads to a greater $V_B$ value. With reference to the size of the $O^{2-}$ ion in silicate units (~$7.24 \times 10^{-24}$ cm$^3$) it can be said that on average each NBO is associated with a space of ~$6.50 \times 10^{-24}$ cm$^3$. By neglecting the volume of $Si^{4+}$ ion it can be concluded that, in this region, the average associated space is $15.39 \times 10^{-24}$ cm$^3$ per each BO ion.

It was concluded [14] that the free volume associated with the BO is almost constant ($15.39 \times 10^{-24}$ cm$^3$) for all modifier ions below 33.3 mol% of addend oxide. In the case of alkali or alkaline earth silicate glasses it decreases with the increasing number of NBO ions per a structural unit and/or radius of the modifier ion. In all cases the NBO ion is associated with a constant free volume and the modifier ions are associated with a free volume that increases with increasing number of NBO ions per a structural unit and/or radius of the modifier ion. The above model explores the change in the free volume due to changing the concentration of alkali oxides in mixed alkali silicate glasses where the free volume is always related to a certain type of alkali oxide and intensifies with its increasing content.

It is also in agreement with ref. [15], which showed that the partitioning of NBO in silicate melts is not identical to the segregation of Ca, Mn, Mg, and Fe divalent cations [8, 15] between solid (e.g. olivine) and melt, which was revealed as a non-equivalent presence of NBO in the main structural units of *Q*4, *Q*3, and *Q*2 type. For melts with high *Q*3/*Q*2-abundance ratio the increasing ratios of Na/(Na + Ca) and/or Na/(Na + Ca + Mn + Mg + Fe) result in a systematic decrease of the partition coefficients because of ordering of the network-modifying Ca, Mn, Mg, and Fe among NBO in *Q*3 and *Q*2 structural units. This decrease is more pronounced at the smaller value of ionic radius of the cation. With decreasing *Q*3/*Q*2 abundance ratio (e.g., less-polymerized melts) this effect becomes less pronounced. Activity vs. composition relations among network-modifying cations in silicate melts are, therefore, governed by availability of energetically nonequivalent NBO in individual *Qn*-species in the melt. As a result, any composition change that enhances abundance of highly depolymerized *Qn*-species will cause partition coefficients to decrease.

This concept of NBO was applied for various cases and treated under different theories [15–19] as well as for the determination of oxygen states in iron-rich borate glasses [19] (see next paragraph). In the forthcoming text we would like to present two examples of analysis regarding the material properties and the role of NBO with respect to the behaviours of matrix glasses. As for the soda–lime–silica glass system, the correlation between the structural parameters characterized by the anionic constitution and the *bioactivity* has been investigated by Koga et al. [17]. As a simple way for describing the glass composition, *Steevel's* parameters [18], X and Y, were employed, which indicate respectively the mean number of NBO and BO ions per polyhedron in the glass lattice. These Steevel's parameters [17, 18] can be plainly calculated from the molar composition of glass assuming the following equations, $X = 2R - Z$ and $Y = 2Z - 2R$ where Z and R denote the mean number of all types of oxygen per polyhedron and the ratio of the total number of oxygens to the total number of glass-forming cations in glass, respectively. As for the soda–lime–silica system, the pure silica glass $SiO_2$, pseudo-binary $Na_2O–2SiO_2$ glass and pseudo-ternary $CaO–Na_2O–2SiO_2$ glass are characterized by $(X, Y) = (0, 4)$, $(1, 3)$, and $(2, 2)$, respectively. With $X > 2$ and $Y < 2$, the concentration of BO ions is appreciably large and the glasses in this compositional region are commonly called as "invert glasses" (typically $X = Y$). Some further data [20] insinuated a rather good bioactive characteristic for similar borate and phosphate glasses accentuating their potential use as a sympathetic template for analogous strategy in bone-tissue formation.

During the last several decades numerous compositions of alkali borate glasses were studied showing the glasses structure with composition $(B_2O_3)_{12\text{-}x}(R_2O)_x$ where ($R$ = Na, Li, Rb, K) discriminative toward the B–O network, built up from planar three coordinated and tetrahedral four coordinated boron atoms. Pure $B_2O_3$ contains only three coordinated boron atoms, which glass network is almost completely made up of randomly connected boroxol web but if alkali oxide is added some of these units are transformed into four coordinated tetrahedral originally suggested by Krogh-Moe [21, 22]. Adjacent $BO_3$ and $BO_4$ groups are linked by bridging oxygen atoms, building the glassy network. At higher alkali oxide concentration (often greater than 20 mol%), the formation of non-bridging oxygen (NBO) is suggested to amplify providing that the boroxol rings in these glasses are transformed into rings incorporating one or more four coordinated boron atoms.

The presence of magnetic cations such as $Mn^{3+}/Mn^{2+}$ or $Fe^{3+}/Fe^{2+}$ in such glasses involves their interactions with surrounding oxygen, which has been intensively studied for behaviour (redox potential of iron oxides as a particular additive in silicate melts [23–29]). It is widely believed that a similar behaviour can be expected in the other glass-forming mixtures. The details of these interactions depend on the type of oxygen polyhedral around ferric and/or ferrous cations. If, for example, $Fe^{3+}$ is considered in tetrahedral (VI) coordination ($Fe^{3+}(IV)/O_2^{2-}$) and $Fe^{2+}$ is in octahedral (VI), we can compose a following relation ($Fe^{3+}(IV)/O_2^{2-}$) ⇔ $4Fe^{2+}$ + O2 + ($2O_2^{2-}$). However, if the ($2O_2^{2-}$) anions in the melt are under the affect of the other structural units co-existing in the melt, further associated processes, mostly depolymerization, may occur. An alternative is that both $Fe^{3+}$and $Fe^{2+}$ are octahedrally coordinated providing thus optional relation 4($Fe^{3+}(VI)/O_2^{2-}$) + ($2O_2^{2-}$) ⇔ $4Fe^{2+}$ + $O_2$, which is adjacent to opposite process of polymerization. Certainly, it is plausible that both $Fe^{3+}$and $Fe^{2+}$ cations may undergo coordination transformation simultaneously, which would result in structural changes touching all variables as illustrated by the relation ($Fe^{3+}(IV)/O_2^{2-}$) ⇔ ($Fe^{3+}(VI)/O_2^{2-}$) + ($2O_2^{2-}$). Thus any glass property of iron-bearing mixtures can be related to the ratios of $Fe^{3+}/Fe^{2+}$ and/or $Fe^{3+}/\Sigma Fe$ [7] the latter being also controlled by other glass-modifying admixtures such as CaO, $Na_2O$, $K_2O$ with positive and MgO or $Al_2O_3$ with negative correlations towards the redox data and thus affecting the level of NBO. Simply saying, it implies that the $Fe^{3+}/Fe^{2+}$ ratio decreases with increasing NBO in bulk melt.

## 12.3 Magnetic Properties of Non- and Nano-crystalline Splat-Quenched $(Fe,Mn)_2O_3$–$B_2O_3$

Clearly, the action of alkali ions as modifiers in borate glasses is more complex than the one in silicates where the NBO concentration increases almost linearly with the alkali content. On the other hand, in borate glasses the modifier may act according to three specific varieties of mechanisms [21, 22], the one prevailing depends on the modifier content, possibly defined on basis of the stability of polyborate groupings, which may similarly occur in borate crystals. However, the presence of magnetic cations turned into the centre of investigation [29–37].

For example, let us present our experimental work, which was based on nontraditional glasses prepared by melting raw material (mixed from $Fe_2O_3$, $MnO_2$, $SiO_2$, $Al_2O_3$, $B_2O_3$ and $K_2B_4O_7$) and adjusted to the base composition of $31Mn_{1.5}Fe_{1.5}O_4$ + $69B_2O_3$ (marked as 0) and with the addition of 3% $K_2O$ (K), $Al_2O_3$ (Al) and/or $SiO_2$ (Si) as well as the mixture 2% $K_2O$ and 2%$Al_2O_3$ (K + Al). After melt temperature equilibration (~1100°C) and splat quenching between two copper plates, with the estimated cooling rate of about $10^{-2}$ K/s [19, 33–36], a solid specimen (of about 1.5 mm thick) is formed. Its glassiness was examined by XRD and SEM and the given cation content was approved by wet chemical analysis. Furthermore, thermal behaviour was measured by differential thermal analysis (DTA) [35], electric conductivity [37] and magnetic susceptibility (by a modified Faraday method using an inhomogeneous field in the temperature range from 4.2 up to 270 K) [19, 36]. Determination of the hydrostatic sample density $\rho$ and the well-defined knowledge of composition can be further used to provide the average volume per cation, $V_c$, or the average volume $V_{mg}$ per magnetic cation. The cubic roots of volume give us an estimate of the average distances, i.e., $a_c$ ($a_{mg}$) $= \{(M/\rho)\ u\ (1/n)\}^{1/3}$, where $n$ is the number of magnetic cations in the formula units. Similarly we can determine the average volume per an oxygen atom, $V_{ox} = \{(M/\rho)\, u\, (1/n_{ox})\}$ where $n_{ox}$ is the number of oxygen atoms in the formula unit. These data are summarized in Table 12.1.

Basic explorative interest was pointed towards magnetization curves from which typical profile it is apparent that at higher temperatures the reciprocal susceptibility $1/\chi$ is approaching a linear dependence that can be extrapolated to high negative temperatures (of the order of $10^2$ K). Relatively strong interactions among the present magnetic moments can be deduced preferring their antiparallel orientation. However, there is no trace of anomalous behaviour at low temperatures bearing witness on an expected ordering of an antiferromagnetic or ferrimagnetic type. On the other hand, the $1/\chi$ temperature dependence displays a downward curvature at low temperature, which can be explained by a simplified model of coexistence of

**Table 12.1** Properties of unconventional glasses with magnetic moments. $T_g$ is the glass-transformation temperature [°C] and stability indicates a relative ratio ($T_c - T_g$) with respect to that of the undoped sample (0). Similarly is given the relative values for peak area of crystallization and that for related activation energy. The sample density [g $cm^{-3}$] is measured by a hydrostatic method while the average distance between the magnetic cations $a_{mg}$ [in $10^{-8}$ cm] is calculated. Moreover, the average volume of oxygen anions $V_{ox}$ is also determined ($\approx 18\times10^{-24}$ $cm^3$). It worth noting that the oxygen volume is about 5% higher than that often exploited from tables ($\approx 17 \times 10^{-24}$ $cm^3$) and even much higher in the case when assuming the volume of bridging oxygen depicted in the first section ($\approx 15.34 \times 10^{-24}$ $cm^3$), which is likely caused by the specificity of borate matrix involved

| Sample | $T_g$ | stability | Area | Energy | density | dist. $a_{mg}$ | vol. $V_{ox}$ |
|---|---|---|---|---|---|---|---|
| 0 | 528 | (1) | (1) | (1) | 3,251 | 4.033 | 18.42 |
| K | 515 | 0.64 | 1.0 | 1.39 | 3.215 | 4.096 | 18.80 |
| Al | 525 | 1.15 | 1.0 | 0.42 | 3.104 | 4.147 | 18.15 |
| Si | 531 | 1.0 | 1.07 | 0.82 | 3.338 | 4.054 | 17.73 |
| K + Al | 520 | 0.91 | 0.97 | 0.89 | 3.198 | 4.122 | 18.74 |

two kinds of magnetic moments. Certain amount of moments, with a relative concentration $t$ is not exposed to any appreciable interactions from their neighbours and can be treated as paramagnetic whereas the remaining $(1 - t)$ amount behaves like antiferromagnetic nano-clusters characterized by a paramagnetic Neel temperature, $\Theta$, see the Table 12.1.

Introducing an additional assumption that both these kinds of moments have the same average magnitude an expression for the temperature dependence of reciprocal susceptibility can be easily derive in the form

$$1/\chi = \{T + (1-t)\Theta\}/c - t/\{T/(\Theta + t)\}\{(1-t)\Theta/c\} \tag{12.8}$$

Here $c$ is the Curie constant, $T$ is the absolute temperature and the meaning of $t$ and $\Theta$ was shown above. The main features of this dependence reasonably agrees with our experimental data so that we could convincingly determine from them the model parameters such as the Curie constant $c$, the paramagnetic Néel temperature $\Theta$, and the concentration $t$. In order to do that let us first note, that the asymptotic behaviour at high temperatures is only governed by the first term of Eq. 12.8 and the extrapolated $\Theta'$ corresponds to $(1 - t)\Theta$, the slope being equal to $1/c$. Getting an estimate of $t$ from Eq. 12.8, we calculated temperature $T'$ at which the full dependence of $1/\chi$ is equal to three-fourth of the first term linear in $T$. The concentration $t$ of paramagnetic moments is then given by

$$t = \{T'(T' + \Theta')\}/(3\Theta'^2 + T'^2) \tag{12.9}$$

After $t$ is calculated we can determine $\Theta$ from $\Theta'/(1 - t)$. All relevant data are summarized in Table 12.2.

The Curie constant $c$ may be further used to calculate the average magnetic moment, $\mu'$ per an atom of transition metal using the resolved chemical composition of our sample, i.e.,

$$c_g = N_g\mu'^2/(3kT) = N_A a\mu'^2/(MkT) \tag{12.10}$$

Here the subscript in $c_g$ stresses the fact that we are using susceptibilities and consequently Curie constants per gram (g), $N_g$ is the number of magnetic moment $\mu'$ in one gram and $N_A$ is the Avogadro number and $3a$ is equal to the number of magnetic atoms in the formula unit (see Table 12.2).

**Table 12.2** Magnetic moment-related properties of nontraditional glasses. For details see text

| Sample | $\Theta'$ | $t$ | $\Theta$ | $c_g$ | $(\mu'/\mu_B)^2$ | $(\mu'/\mu_t)$ |
|---|---|---|---|---|---|---|
| 0 | 140 | 0.16 | 167 | 2.22 | 22.74 | 0.83 |
| K | 115 | 0.125 | 131 | 2.25 | 24.3 | 0.86 |
| Al | 77 | 0.17 | 93 | 1.89 | 20.1 | 0.78 |
| Si | 130 | 0.105 | 145 | 2.56 | 27.35 | 0.91 |
| K + Al | 121 | 0.15 | 121 | 2.13 | 22.92 | 0.83 |

Magnetic properties consist of an extrapolated temperature $\Theta'$ [in K] and the paramagnetic Neel temperature, $\Theta$ [K] estimated for a given concentration of antiparallel ordered magnetic moments $(1 - t)$. The value $c_g$ is the Curie constant per gram $[10^{-2}]$, $\mu'$ is the average magnetic moment per atom of transition metal yielding the revoking squares of average moments $(\mu'/\mu_B)^2$. Using a theoretical value of $\mu_t$ equal to 5.76 $\mu_B$ we can utilize it for the evaluation and consequent comparison of experimental data in the form of ratio $(\mu'/\mu_t)$.

The squares of the average moments derived from Eq. 12.10 are listed in Table 12.2. In order to calculate the expected average magnetic moment of the present cations we suppose that equilibrium between various valencies of manganese and iron does not deviate in the glass from that stoichiometry found in the standard Mn-Fe spinels with the same $Mn^{3+}/Fe^{3+}$ ratio, i.e., $Fe_{1.5}{}^{3+}Mn_{0.5}{}^{3+}O_4$ [35, 36]. We then ascribe to each cation a single spin-only value of

$$\mu_t'^2 = g^2 s(s+1)\mu_B^2 \tag{12.11}$$

with g = 2, s is the quantum spin number, and $\mu_B$ is Bohr magneton. Assuming squares of the moments in accordance with Eq. 12.11 we arrive at a theoretical value of $\mu_t$ =5.76 $\mu_B$ which is further used for comparison with the experimental values in the last column of Table 12.2.

It is worth noting that DTA was measured on few pieces of glass buried in the powder of fine silver particles placed inside the DTA sample-cell because similar-sized glass pieces were used in all other measurements. Moreover, powdering of glass changed its properties slightly decreasing $T_g$ (for sample 0: → 524°C) but doubling crystallization heat ($A/A_o \approx 2.1$) due to the increased surface (a slight increase of the activation energy $E/E_o = 1.2$ was infertile negligibly enhanced by anticipated surface crystallization).

Glass-forming abilities and associated cation compatibilities with various levels of $B_2O_3$ shown a decrease from $Mn^{3+}$ through $Mn^{2+}$, $Fe^{3+}$ to $Fe^{2+}$ and, consequently, we can expect that the transition metal ions will be incorporated into the glass network along this sequence. This idea was corroborated by our findings about crystallization [35], which for boron glasses containing iron and manganese cations crystallization yields the crystal phase enriched in Fe with the remaining glass matrix augmented of Mn (than that corresponding to the average Mn/Fe ratio [36]). The stability of glasses with respect to the different admixtures can be analyzed on the basis of DTA data, see Fig. 12.2. The ratio of the differences between the temperature of glass crystallization and transformation of the doped glass to the original one (o) may be taken as a measure of the relative glass stability. It is equal to 1 for $SiO_2$ admixture, which is in accordance with the standard concept that $SiO_2$ tends to form a separate network within $B_2O_3$ matrix so that the original matrix remains essentially unaffected. On the other hand $Al_2O_3$ tends to enter $B_2O_3$ matrix and to span its network into three dimensions, which results in an apparent gain in the relative glass stability. Potassium cations with a relatively high ionic radius deform in their vicinity the $B_2O_3$ network, which may thereafter act as

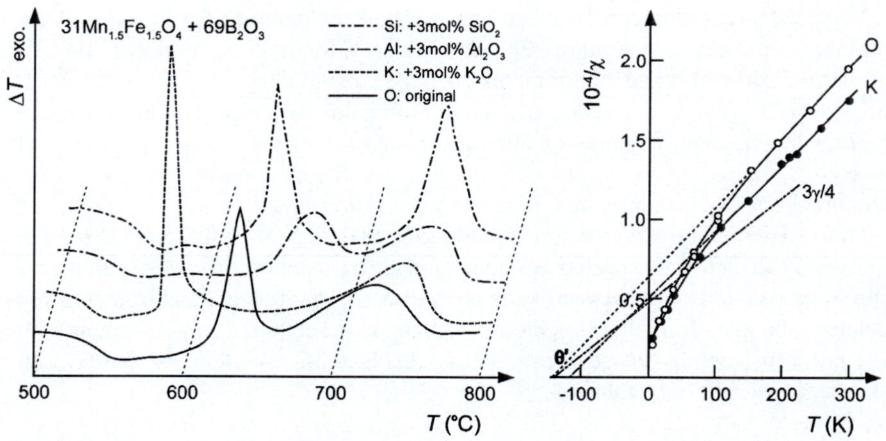


**Fig. 12.2** *Left*: DTA (Netzsch instrument) curves of glasses and resulting data applicable to their stability related to $T_g$ (as the front step-wise change) and $T_c$ (as the onset of the exothermal crystallization peaks), cf. Table 12.1. *Right*: Temperature dependence of the reciprocal magnetic susceptibility, $1/\chi$, extrapolated to negative temperatures of Neel temperature, $\Theta'$, see text and Table 12.2

potential sites for nucleation and crystallization. In consequence the glass stability decreases. All that is also in accordance with the change in activation energy whose high value for $K_2O$ admixture indicates an easier crystallization process started from a large number of ready-to-grow nuclei while its lower value for $Al_2O_3$ indicates rather slower process controlled by diffusion.

Let us turn now our attention to the magnitude of magnetic moments given in Table 12.2. From the last columns it can be seen that for the basic glass (without admixtures) this value is considerably lower than the moment expected for free ions. Various admixtures bearing the magnetic moments are able of further changing the moment per atom of transition metal and we believe that both these phenomena are symptoms of the number and type of incorporation into the network. The reasons may be at three-fold:

1. Increased interaction of incorporated transition metal ions via the intermediating oxygen, which may result in their antiferromagnetic pair (or even small cluster) ordering thus substantially lowering (or even excluding) the contribution of these moments to the Curie constant.
2. The spin moment of covalently bound cations, e.g., $Fe^{3+}$, may be lowered due to the transition to low-spin state. These effects are usually described in terms of strong crystal fields revealed by some compounds supportive of these cations though not so common in oxides.
3. To a large extent, the orbital contribution to the magnetic moment is affected by covalent bonding but in our quantitative estimates we only capture into account the spin moment so that this influence cannot be weighted up in a simplified way.

In any cases it seems that the incorporation of either Mn (or less probably Fe) ions into the glass network can only lower its magnetic moment. The number of ions incorporated will be given by the equilibrium of the reactions discussed in the introductory paragraphs, $Mn^{2+} + O^{2-} = Mn^{\circ} + O^{\circ}$ and/or $2Mn^{3+} + 3O^{2-} = 2Mn^{+} + 2O^{\circ} + O^{2-} = 2Mn + 3O^{\circ}$, where the number of primes denotes the multiplicity of the bond of the given atom in the network. The surface properties, however, may become effective (strongly related to the extent of interfaces in nano-crystalline system [41], which is not subjected herewith).

Concluding on basis of DTA data and magnetic moment determination we suggest the following mechanism for the influence of modifying admixtures. $SiO_2$ enters the $B_2O_3$ matrix as ($Si^{4+} + 2O^{\circ}$), which likely increases the concentration of BO [29]. The NBO equilibrium is thus shifted to the left hand side, i.e., Mn and/or Fe cations are released from the network. The addition of $Al_2O_3$ has the opposite effect [1, 38], because it prefers the incorporation in the form of $AlO^{4-}$ groups whose negative charge is compensated by bonding transition metal ions into the network, which means a shift of equilibrium to their right hand side. The expected effect of $K_2O$ admixture as the donor of $O^{2-}$ should be even stronger than that of $Al_2O_3$ and point in the same direction. Our magnetic measurements show, however, a minor influence though the DTA displays rather significant changes. Similarly, the introduction of $K_2O$ into borate glass would change the $B_2O_3$ matrix itself so that we can expect both the softening of the glass structure and a smaller effect exerted upon the state of the transition metal ions.

The strength of magnetic interactions can be estimated from the value of $\Theta$ as revealed in Fig. 12.2. This quantity $\Theta$ actually gives the lower limit for the entire interactions. In a rough approximation, the effect of various admixtures can be explained by the average distance between magnetic ions presented. In quantitative agreement along our expectations, the value of $\Theta$ smoothly decreases with the increasing distance. Let us finally comment on the magnitude of this concentration, which was calculated for the accepted model on the basis of Eqs. 12.8 and 12.10, and given in Table 12.2. The moments, described here as paramagnetic, do not have a sufficient number of magnetic cations as their nearest neighbours and, for an assumed random distribution, the number dependence on the average concentration of magnetic atoms turn up to be an increasing function for low values attaining a maximum (and again decreasing). The composition of our glasses clearly belongs to the latter part of this dependence, which explains the seemingly contradictory behaviour of the $t$-value with various admixtures. Namely Si releases transition metal atoms from the network, which factually decreases the concentration $t$ and the effect of Al is laying just on the opposite direction. For describing magnetic susceptibility and approving consistency of our model, we plotted in Fig. 12.2 the concentration $t$ against the relative magnetic moment $\mu'/\mu_t$ (see Table 12.2). In agreement with our discussion, admixtures increases the number of transition metal atoms incorporated into the network, which decreases the average magnetic moment and increases the concentration $t$ and vice versa. Antiparallel ordering was also indicated conforming that the temperature of magnetic ordering ($T_N$) is proportional to the square of mean number of interacting neighbours and the Neel temperature ($\Theta$) lies lower than that for analogous polycrystalline ceramics. Particularity of ferromagnetic or speromagnetic properties (of disordered systems with canted spins) were previously identified in metallic glassy alloys [43], which various compositionally different systems found already practical applications [42–44].

## 12.4 Specific Case of Speromagnetic Behaviour of Rapidly Quenched and Laser Melted $(Fe,Mn)_2O_3$–$(B,Bi)_2O_3$

The substitution of $Bi_2O_3$ for $B_2O_3$ [39] increases, however, the upper glass-forming limit for $Fe_2O_3$ to about 20 mol% (such a composition was still vitrifiable by standard splat quenching method at the cooling rate of ~$10^{-2}$ K/s). In the contrast with the simple (Fe, Mn)–B–O system, the substitution of Fe by Mn in the $Bi_2O_3$-rich matrix did not improved the glass-formation but only decreased the Curie temperature. Substitution of B for Bi in Fe–B–Bi–O system changed the phase relation substantially enhancing crystallization tendency of crystalline borates on account of previously precipitated crystalline phases of $Fe_4Bi_2O_6$ and $FeBiO_3$ leaving behind glassy matrix with higher magnetic moments.

However, the $Fe_2O_3$-rich composition (such as the end compound $FeBiO_3$) required for its vitrification much powerful quenching than is a basic splat cooling used above. In some cases it was necessary to exceed rates above $10^{-3}$ K/s. Such a rapid freeze-in of melts could only be attained by a specially developed techniques usually aided by fast laser melting and consequent melted drop pressing in a specially modified coolers, such as twin-rollers [31, 37, 40] being worth of a more detailed description.

These quenching options radically extended the range of vitrified compositions, though the glassy samples were very tiny and difficult to become fully reproducible. From magnetic susceptibility measurements, the temperature dependence of (Bi–B)(Fe–Mn)–O glasses revealed certain evidence of stronger interaction of antiferromagnetic types [19, 32]. The composition containing not only Bi but simultaneously B, exhibited a definite Curie-Weiss behaviour (Fig. 12.3). Samples with Bi and Mn probably contained cations' multivalency as a result of their broader distribution, showing their various degree of incorporation into the glass network. Only difficult to prepare glassy $FeBiO_3$ provided indication of magnetic measurements explainable on the basis of weak ferromagnetism, i.e., so called speromagnetism, probably caused by canted spins in the antiferromagnetic sub-lattice, which was early suggested to take possibly place in the Fe–Bi–O system and which magnetic ordering was already experimentally noted in metallic glasses [43].

The type of incorporation of cations into the Bi–B–glass network and the formation of NBO predictably affects the resulting magnetic moment. Presumably, the addition of $Bi_2O_3$ to the $B_2O_3$ matrix prefers to undertake web incorporations in the form of $BiO_3^{3-}$ groups, their negative charge being compensated by pushing transition metals to bond within the network. Thus the concentration of bridging

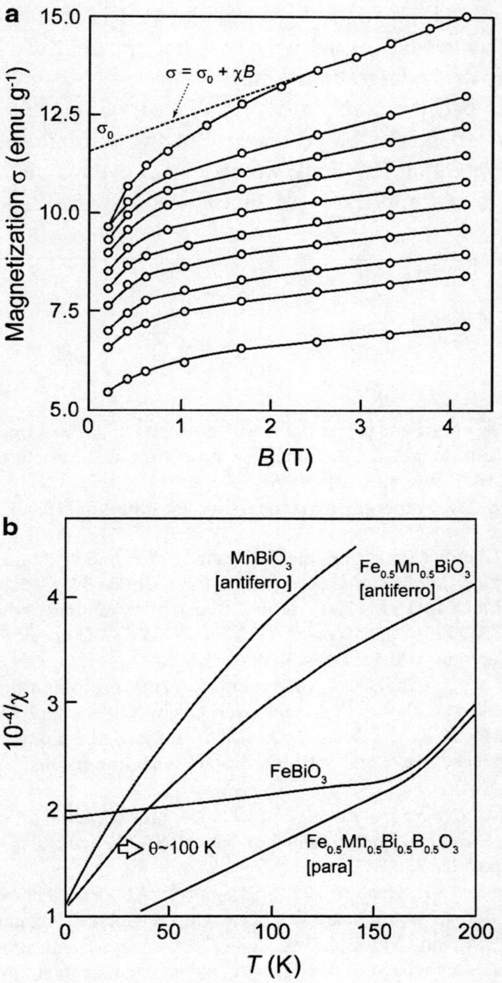


**Fig 12.3** (**a**) Magnetization curves of the antiferromagnetic-like as-quenched Fe–Mn–Bi–O glass viewing the experimental curves and their temperatures of measurements from upper curve at 4.2 K proceeding down with the increasing stepwise temperature changes along 24.5, 50.0, 75.0, 100.0, 125.0, 150.0 175.0, 200.0 up to the final 157.0 K. (**b**) Temperature dependence of the reciprocal magnetic susceptibility, $1/\chi$, portraying various magnetic behaviour, starting from paramagnetic (arbitrary ordering of magnetic moments ↑↖↓↘←) for $Fe_{0.5}Mn_{0.5}Bi_{0.5}B_{0.5}O_3$ to a mixture of paramagnetic and antiferromagnetic (↖↓↑↓↑↗) phases for $Fe_{0.5}Mn_{0.5}BiO_3$ or $MnBiO_3$ exhibiting a Neel temperature $\Theta$ in the order of $-100$ K. A very special (almost extraordinary) behaviour of so called speromagnetism is displayed by the extremely rapid quenched ($\cong 10^6$ K/s) glassy specimen of $FeBiO_3$ showing not completely compensated moments (i.e., not fully antiparallel moment ordering, where *arrow* orientation ↑↓↑↓ is mutually little angled, so called canted spins)

oxygen is thus decreased, i.e., equilibrium of Eq. 12.1 is shifted to the right. This effect may be similar to that caused by $Al_2O_3$ but apparently stronger due to the higher deformability of the larger radius of Bi–ions.

Magnetic glasses became practically relatable in various applications [42–45] recently as a glass – ceramics based on ferroan (wollastonite-like and ε-(Fe,Ca) $SiO_3$-like phases) containing $Fe^{2+}$ ions, which lead to two magnetic phases with the different coercive forces applicable as thermo-seeding for cancer treatment by hyperthermia [46].

## References

1. Mysen BO (1988) Structure and properties of silicate melts. Elsevier, Amsterdam; Mysen BO, Richet P (2005) Silicate glasses and melts – properties and structure. Developments in geochemistry, vol 10. Elsevier, Amsterdam
2. Greaves GN, Sen S (2007) Inorganic glasses, glass-forming liquids and amorphizing solids. Adv Phys 56:1–166
3. Conradt R (2004) Chemical structure, medium range order, and crystalline reference state of multicomponent oxide liquids and glasses. J Non-Cryst Solids 345/346:16–23
4. Macháček J, Gedeon O, Liška M (2006) Group connectivity in binary silicate glasses. J Non-Cryst Sol 352:2173–2179; Macháček J, Gedeon O (2003) Q-species in alkali-disilicate glasses. Silikaty-Ceramics (Prague) 47:45–49
5. Liška M, Macháček J, Gedeon O, Klement R (2004) Molecular dynamics of the $Na_2O$–MgO–CaO–$SiO_2$ glasses. Glastech Ber Glass Sci Technol 77C:267–272; Liška M, Klement R, Macháček J, Gedeon O (2005) Inverse thermodynamic modelling of glass from Raman spectroscopical and molecular dynamics results. Phys Chem Glasses 46:108–111
6. Bogdanov VN, Ananev AV, Golubkov VV, Champagnon B (2007) Micro- and nano-inhomogenities in glasses and their melts studied by optical, SAXS, acoustical and thermodynamic methods. J Phys Conf Ser 93:2033
7. Schneider J, Mastelaro VR, Zanotto ED, Shakhmatkin AB, Vedishcheva NM, Wright AC (2003) $Q^n$ distribution in stoichiometric silicate glasses: thermodynamic calculations, and NMR measurements. J Non-Cryst Solids 325:64–178; Cannas M, Vaccaro L, Boizot B (2006) Spectroscopic parameters related to non-bridging oxygen centers in amorphous-SiO2. J Non-Cryst Solids 352:203–208; Leonova E (2008) Nitrogen-rich La-Si-Al-O-N oxynitride glass structure probed by solid state NMR. J Non-Cryst Solids 354:49–53
8. Peidongz H, Kroger S, Stebbins JF (2000) Non-bridging oxygen sites in barium borosilicate glasses. J Non-Cryst Solids 276:122–131; Allwardt JR, Stebbins JF (2004) Ca–Mg and K–Mg mixing around non-bridging O-atoms in silicate glasses. Am Mineral 89:777–784
9. Malfait WJ, Zajaznova-Herzog VP, Halter WE (2007) Quantitative Raman spectroscopy: principles and application to silicate glasses. J Non-Cryst Solids 353:4015–4028
10. Niggli P (1949) Die vollständige und eindeutige Kennzeichnung der Raumsysteme durch Charakterentafeln. I. Acta Cryst 2:263–270; II. Acta Cryst 3:429–433 (1950)
11. Niggli P (1945) Stereochimija. Inostr Lit, Moscow, translated from Grundlagen der Stereochemie. Birkhäuser, Basel
12. Villars P, Girgis K, Niggli A (1983) Coordination polyhedra and structures of alloys: binary alloys of niobium (and tantalum) with Group IIIb and IVb elements. Acta Cryst B 39:603–606
13. Zachariesen WH (1932) The atomic arrangement in glass. J Am Ceram Soc 54:3841–3851
14. Doweidar H (2007) Volume of ionic sites in silicate glasses. J Phys Condens Matter 19:1561

15. Mysen BO, Shang J (2005) Evidence from melt partitioning that non-bridging oxygen in silicates are not equivalent. Geochim Cosmochim Acta 69:2861
16. Toop GW, Samis CS (1962) Some new ionic concepts of silicate slags. Can Metal Quart 1:129–152; Activities of ions in silica melts. Trans Metal Soc AIME 224:878–887); Masson CR (1977) Anionic constitution of glass-forming melts. J Non-Cryst Solids 25:1–41
17. Koga N, Strnad Z, Šesták J, Strnad J (2003) Thermodynamics of non-bridging oxygen in silica bio-compatible glass-ceramics for bone tissue substitution. J Therm Anal Calorim 71:927
18. Steevels IM (1960) Neue Erkenntnisse über die Struktur des Glases. Philips Tech Rundschau 9/10:337–349
19. Závěta K, Šesták J (1977) Structure and magnetic properties of Fe-rich oxide glasses. In: Proceedings of the International Conference on Glass'77, vol 1, p 399. ČVTS Publication House, Prague
20. Serra J, González P, Liste S, Chiussi S (2002) Influence of the non-bridging oxygen groups on the bioactivity of silicate glasses. J Mater Sci- Mater Med 13:1221–1225; Liang W, Rüssel C, Day DE, Völksch G (2006) Bioactive comparison of a borate, phosphate and silicate glass. J Mater Res 21:125–131
21. Krogh-Moe J (1959) The cation distribution in some crystalline and vitreous cesium borates. Ark Kemi 14:451–459; Crystal structure of lithium diborate, $Li_2O.2B_2O_3$. Acta Cryst 15:190–193 (1962)
22. Chen D, Miyoshi H, Masui H, Akai T, Yazawa T (2004) NMR study of structural changes of alkali borosilicate glasses with heat treatment. J Non-Cryst Solids 345/346:104–107
23. Murdoch JB, Stebins JF, Carmicheal ISE (1985) High-resolution $^{29}Si$ NMR study of silicate and aluminosilicate glasses: the effect of network-modifying cations. Am Mineral 70:332–343
24. Rüssel C (2004) Redox state of glasses. Glastech Ber-Glass 77C:149–159
25. Philips B, Muan A (1959) Phase equilibria in the system CaO-iron oxide-$SiO_2$, in air. J Am Ceram Soc 42:413–423
26. Holmquist S (1966) Ionic formulation of redox equilibria in glass melts. J Am Ceram Soc 49:228–229
27. Mysen BO, Seifert F, Virgo D (1980) Structure and redox equilibria of iron bearing silicate melts. Am Mineral 65:867–884
28. Virgo D, Mysen BO, Seifert FA (1981) Relationship between the oxidation state of iron and structure of silicate melts. Carnegie I Wash 80:308–311
29. Goldman DS (1983) Oxidation equilibrium of iron in borosilicate glass. J Am Ceram Soc 66:205–209
30. Pal M, Chakravorty D (1998) Structural study of iron borate glasses containing NiO and ZnO. J Mater Res 13:3286–3292
31. Horie O, Syono Y, Nakagawa Y, Ito A, Okamura K, Yajima S (1978) Mossbauer study of amorphous Bao-$Fe_2O_3$-$B_2O_3$ system. Solid State Commun 25:423–426
32. Akamatsu H, Tanaka K, Fujita K, Murai S (2008) Magnetic phase transitions in $Fe_2O_3$–$Bi_2O_3$–$B_2O_3$ glasses. J Phys Condens Matter 20:235216(9)
33. Šesták J (1978) Magnetic properties and glass-formation of doped oxide glasses prepared by various methods of rapid quenching: Part I. Sklář a Keramik 28:321; Part II. Sklář a Keramik 28:353 (1978) (both in Czech)
34. Šesták J (1983) Crystallization behavior of rapidly quenched iron oxide containing glasses with regard to thermal and magnetic properties. Wiss Ztschr Friedrich-Schiller-Univ Jena Math-Naturwiss Reihe 32:377
35. Šesták J (1973) On ferrimagnetic glass-ceramics based on $B_2O_3$ and $MnFe_2O_4$. J Therm Anal 5:669–672
36. Závěta K, Šesták J, Roskovec V (1972) Magnetic properties of Mn-Fe-ferrite containing borate glasses. J Am Cer Soc 55:537 and the key lecture at The 3rd Czechoslovak Conf. on Magnetism, Kosice 1971, proceedings in Czech J Phys B 23:837–839 (1973)
37. Šimšová J, Šimša Z, Šesták J (1979) Electrical and structural properties of y($Mn_xFe_{3-x}O_4$)-(1-y) $B_2O_3$ glasses between 400 and 1000 K. J Non-Cryst Solids 30:375–378

38. Kilinc A, Carmichael ISE, Rivers ML, Sack RO (1983) The ferric-ferrous ratio of natural silicate liquids equilibrated in air. Contrib Miner Petrol 83:136–140
39. Šesták J, Šestáková V, Tříska A, Záběta K (1988) Glass-formation, phase relation and magnetic properties of the splat-quenched system of laser melted (Fe, Mn)2O3-(B, Bi)2O3. J Therm Anal 33:789–795
40. Nakamura H, Kishi T, Ohgaki T, Muro Y, Yasumori A (2008) Magnetic properties of phase separated glass and glass ceramics in $Co_3O_4$–$TiO_2$–$SiO_2$ system. J Phys Conf Ser 106:012009(4)
41. Hamad S, Bromley ST (2008) Low reactivity of non-bridging oxygen on stoichiometric silica surfaces. Chem Commun 10:4156–4158
42. Moorjani K, Coey JMD (1984) Magnetic glasses. Elsevier, Amsterdam
43. Lupu N, Chiriac H (2002) Determination of disordered magnetic structures in high-coercivity Nd-Fe-based glassy alloys. Microscopy and Microanal 8:370–371; Bulk amorphous magnetic materials. J Optoelectr Adv Mater 4:207–216 (2002); In: Liu Y, Sellmyer DJ, Shindo D (eds) Handbook of advanced magnetic materials, vol 3: fabrication and processing. Springer, Berlin, pp 1279–1328 (2008)
44. Šesták J (2004) Heat, thermal analysis and society, Chap 19: Modern materials and society. Nucleus, Hradec Kralove, pp. 306–314
45. Zaitsev DD, Kazin PE, Trusov LA, Vishnyakov DA, Tretyakov YuD, Jansen M (2006) Synthesis of magnetic glass-ceramics in the system $SrO$–$Fe_2O_3$–$Al_2O_3$–$B_2O_3$. J Magn Magn Mater 300:473–475
46. del Real RP, Arcos D, Vallet-Regí M (2002) Implantable magnetic glass – ceramics based on (Fe, Ca)SiO3 solid solutions. Chem Mater 14:64–70

# Chapter 13
# New Approach to Viscosity of Glasses

**Isak Avramov**

## 13.1 Apparent Activation Energy

The most widespread assumption is that viscous flow is controlled by activation energy barrier because it changes sharply with temperature:

$$\eta = \eta_o \exp\left(\frac{E(T)}{RT}\right) \tag{13.1}$$

Actually, E(T) is the *apparent* free activation energy. Here, the word *apparent* is very important, because of the existence of direct experimental indication that E(T) is not just a free energy barrier, although it is related to it. This was proven [1] by comparing the equilibrium and the non-equjournaltilibrium viscosities at the break point as shown in Fig. 13.1, where data [2–4] on viscosity of $xNa_2O.yCaO.(100\text{-}x\text{–}y)SiO_2$ glasses are plotted against $10^4/T$.

In the following, we mark with $T_f$ the temperature at which the structure of the system is becoming frozen. We use this notation because it is convenient to choose as $T_g$ the temperature at which viscosity of equilibrium system is $\lg\eta_g = 13.5$ [dPa.s]. In this way $T_g$ is always within the glass transition interval and could serve as thermodynamic definition of the glass transition temperature. With this definition $T_g$ never deviates considerably from $T_o$ of the onset of the changes, i.e. $T_g \approx T_o \pm 1.\%$.

Angell [5, 6] formulated the steepness index (also known as fragility index m) as

$$m = \left.\frac{d \lg \eta}{d(T_f/T)}\right|_{T=T_f}. \tag{13.2}$$

I. Avramov (✉)
Institute of Physical Chemistry, 1113 Sofia, Bulgaria
e-mail: avramov@ipc.bas.bg

J. Šesták et al. (eds.), *Glassy, Amorphous and Nano-Crystalline Materials*, Hot Topics in Thermal Analysis and Calorimetry 8, DOI 10.1007/978-90-481-2882-2_13,

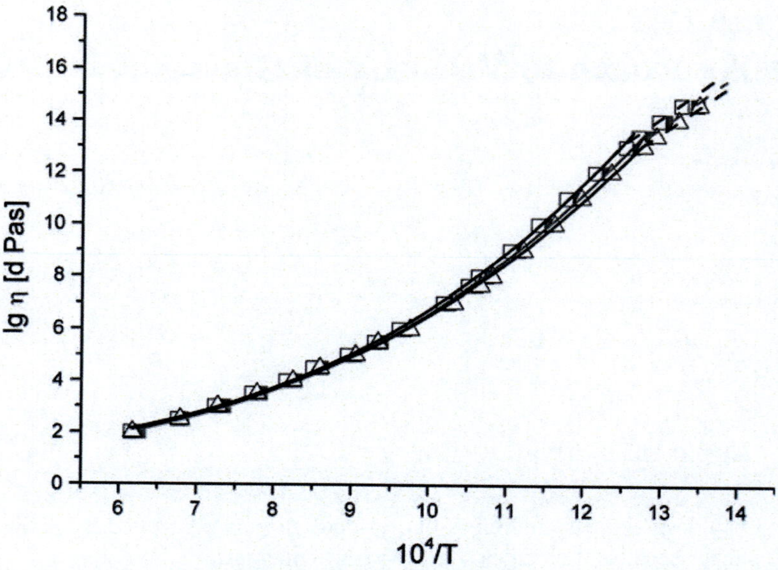


**Fig. 13.1** Viscosity of $xNa_2O.yCaO.(100\text{-}x\text{-}y)SiO_2$. The equilibrium viscosity is given by the solid points. *Lines* are according to Eqs. 13.14, 13.17. □ – $21Na_2O.7CaO.72SiO_2$; ■ – $21Na_2O.9CaO.70SiO_2$; Δ – $19Na_2O.9CaO.72SiO_2$

Here derivative is along the equilibrium viscosity curve. In analogy one can formulate the non-equilibrium steepness index $m_f$ as the same derivative, but along the non-equilibrium curve. Note that Eq. 13.1 leads to the astonishing result that these two steepness indices must be equal, $m = m_f$. Indeed, by definition the activation energy is the difference between the energies of certain active state $E^*$ and the ground state E, so that.

$$\lg \eta = \lg \eta_o + \frac{1}{2.3}\left(\frac{H^* - H(T)}{RT} - \frac{S^* - S(T)}{R}\right). \tag{13.3}$$

Here $H^*$ and $S^*$ are the enthalpy and the entropy of the active state. Under the assumption that non-equilibrium viscosity corresponds to a system with structure fixed at $T = T_f$, i.e. with constant enthalpy $H_f$ and constant entropy $S_f$, the difference m-$m_g$ becomes:

$$m - m_f = \frac{1}{2.3R}\left(\frac{1}{T}\frac{dH(T)}{d(T_f/T)}\bigg|_{T=T_f} - \frac{dS(T)}{d(T_f/T)}\bigg|_{T=T_f}\right)_{equilibrium} \tag{13.4}$$

However, by definition, the term in brackets of Eq. 13.4 must be zero so that $m = m_f$ provided E(T) is either a single free activation free energy or a sum of several free activation energies. At first sign, this result seems absurd, because of the strong experimental evidences that $m > m_f$. This result is obtained for many systems, including organic substances [7], phosphate glasses [2, 8] and silicates (see [2–4, 9]). Thus, it can be shown, that for a standard soda-lime silicate glass NBS710 the experimental data [3, 9] lead to $m_f = 0.36m$.

This is the reason why in the following discussion we discuss only models in which the temperature function E(T) in the exponential term of viscosity equation is not a free energy, although it can be a result of the role of certain energy barriers.

## 13.2 Mean Jump Frequency and Apparent Activation Energy

To move, the molecules have to overcome activation energy barriers [10] created by the resistance of the surrounding molecules. Due to the existing disorder barriers of different height could appear. The jump frequency is thermally activated so that for a given height E of the barrier the frequency is:

$$\nu(E) = \nu_o \exp\left(-\frac{E}{RT}\right). \tag{13.5}$$

If f(E) is the probability distribution function that activation energy barrier of height E will appear, then, in continuous case, the average jump frequency can be determined [11, 12] as:

$$\langle \nu \rangle = \int_0^{E_{max}} \nu(E) f(E)\, dE. \tag{13.6}$$

Since ν(E) decays exponentially with E, of significance for the integral is only the low energy part of the probability distribution function. Therefore, a sufficiently accurate result can be obtained easily. The reason is that most of the tails of the probability distribution functions are getting together away from the maximum, although they could differ near the maximum. If the jumps are considered as a sequence of independent events the probability distribution function is represented by Poissonian law, so that f(E) is a function of the dispersity σ as follows:

$$f(E) = \frac{\exp\left(\frac{E - E_{max}}{\sigma}\right)}{\sigma\left(1 - \exp(-)\frac{E_{max}}{\sigma}\right)}, \quad E \le E_{max}. \tag{13.7}$$

In this way the average jump frequency is

$$<\nu> = \frac{\left[1 - e^{-E_{max}\left(\frac{1}{RT} - \frac{1}{\sigma}\right)}\right]}{\left(\frac{\sigma}{RT} - 1\right)\left(1 - e^{-\frac{E_{max}}{\sigma}}\right)} \nu'_o e^{-\frac{E_{max}}{\sigma}}. \tag{13.8}$$

For RT $< \sigma <$ $E_{max}$, the term to the left of $v_o$, as compared to the exponential term on the right-hand side, is a weak temperature function of the order of unity. Therefore, one can use the approximation:

$$<v> \approx v_\infty e^{-\frac{E_{max}}{\sigma}} \quad , \quad v_\infty = v_o \frac{RT}{\sigma}. \tag{13.9}$$

Equation 13.8 is really useful when the dispersion $\sigma$ is expressed through entropy S.

$$\sigma = \sigma_g \exp\left[\frac{2(S - S_g)}{ZR}\right]. \tag{13.10}$$

Here $\sigma_g$ is the dispersion at the reference state with entropy $S_g$ and Z/2 is the degeneracy of the system, i.e. Z is the number of escape channels available for the moving particle and each channel can be used in two directions. Taking into account that viscosity is inversely proportional to the mean jump frequency the viscosity can be expressed through Eqs. 13.9 and 13.10 as:

$$\lg \eta = \lg \eta_\infty + \left( \lg \frac{\eta_g}{\eta_\infty} \right) \exp\left[-\frac{2(S - S_g)}{ZR}\right], \tag{13.11}$$

where lg $\eta_g$ is viscosity at the reference state. Note that $\lg \eta_\infty$ is larger the preexponential constant $\lg \eta_o$ determined in other models. The reason for this is a term $\frac{\sigma}{RT}$ coming from Eq. 13.9. For this reason it is expected that $\lg \eta_\infty \approx \lg \eta_o + 1.5$.

Equation 13.11 is the main viscosity expression [11, 12]. It permits to follow the temperature and pressure dependencies of viscosity as well as the dependence on how viscosity changes if system is not in equilibrium. It is sufficient to introduce in Eq. 13.11 the dependence of entropy on the corresponding variables. It is quite natural to try to express S as logarithmic function of temperature and/or pressure. Thus, the temperature T dependence of entropy can be presented as:

$$S(T) = S_g + \int_{T_g}^{T} C_p \mathrm{d}\ln \tilde{T}. \tag{13.12}$$

So, we apply the most frequently used approximation that heat capacity is temperature independent, i.e. $C_p$ is the average value for the interval between $T_g$ and T. Under this assumption, the entropy of a melt in metastable equilibrium is becoming:

$$S(T) = S_g + C_p \ln\left(\frac{T}{T_g}\right). \qquad (13.13)$$

With these expressions the temperature dependence of viscosity becomes

$$\lg \eta = \lg \eta_\infty + A \left(\frac{T_g}{T}\right)^\alpha, \qquad (13.14)$$

where the "fragility" parameter α stands for: $\alpha = \frac{2C_p}{ZR}$.

Note that A is not adjustable parameter, instead, it is well defined through $\lg \eta_\infty$ as: $A = \lg \frac{\eta_g}{\eta_\infty} \approx 13.5 \pm 10\%$. When α =1 viscosity gives a straight line in Arrhenian coordinates, indicating that glass is strong. The larger is α the more fragile is the glass. There is a simple relationship between the Angell's steepness parameter m and the fragility parameter α, namely m = A α. Therefore, Eq. 13.14 can be reformulated in terms of m as follows:

$$\lg \eta = \lg \eta_\infty + A \left(\frac{T_g}{T}\right)^{\frac{m}{A}}. \qquad (13.15)$$

Only $\lg \eta_\infty$ and α, play role of adjustable parameters in Eqs. 13.14 and 13.15 because the parameter A is well defined. The best test of applicability of Eqs. 13.14 and 13.15 is to plot experimental data in coordinates lg η against $\left(\frac{T_g}{T}\right)^\alpha$ as this is shown in Fig. 13.2 for different systems (lead-silicate [3, 13], diopside, anortite [14, 15], garnet and basalts [16]).

The experimental data [2, 3, 6, 8, 11] on relationship between the Angell's steepness parameter m and fragility parameter α are shown in Fig. 13.3. It is seen that with sufficient accuracy m = 13.5α.

The pressure dependence of viscosity was already discussed in [12, 17] by introducing into Eq. 13.11 the pressure dependence of entropy.

Note that certain temperature function appears in Eqs. 13.11, 13.14, 13.15, however this function is not just free energy, although it is determined by the presence of many free energy barriers. Therefore, this model is able to describe the break in viscosity curve at the point where system moves out of equilibrium.

## 13.3 Non-Equilibrium Viscosity

Below the glass-transition region the structure is fixed because the relaxation time is too large as compared with the time of observation. In this case the viscosity equation is derived in terms of the temperature $T_f$ at which the system with this fixed structure will be in equilibrium. In terms of Eq. 13.11 this means that the entropy of the non-equilibrium system depends on both the actual temperature T and

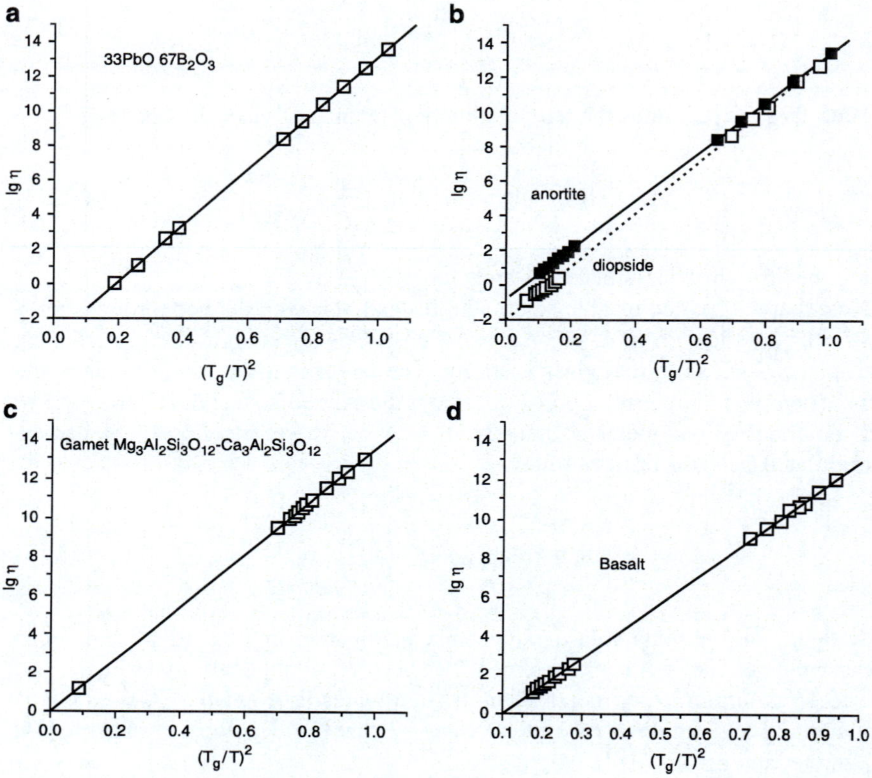


**Fig. 13.2** An illustration how well the existing experimental data on different systems (lead-silicate [3, 13], diopside, anortite [14.15], garnet and basalts [16]) give straight lines in coordinates corresponding to Eqs. 13.14, 13.15

on $T_f$. Basically, the entropy of undercooled melts splits into two parts: the equilibrium entropy $S(T_f)$, and the entropy change while temperature changes from $T_f$ to T with a fixed structure. The corresponding heat capacities are $C_p$ and $C_{gl}$ so that:

$$S(T) = S_g + C_p \ln\left(\frac{T_f}{T_g}\right) + C_{gl} \ln\left(\frac{T}{T_f}\right). \qquad (13.16)$$

Taking into account Eq. 13.16 the nonequilibrium viscosity is given by:

$$\eta = \eta_\infty \exp\left[A\left(\frac{T_g}{T_f}\right)^\alpha \left(\frac{T_f}{T}\right)^\gamma\right]. \qquad (13.17)$$

The dimensionless power $\gamma$ is proportional to the ratio of the heat capacity $C_{gl}$ of the glass and the heat capacity of the undercooled melt $C_p$.

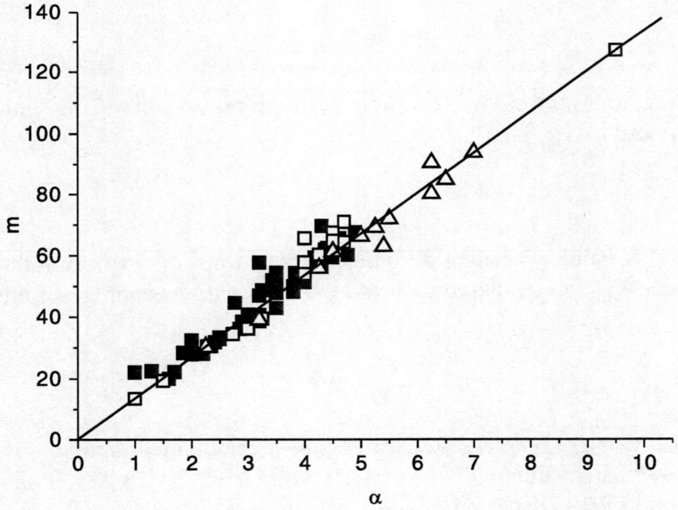


**Fig. 13.3** Relationship between the Angell's steepness parameter m and fragility parameter α. ■ – silicates; □ – borates; Δ – phosphates and organic substances. The straight line is according to m = 13.5 α

$$\gamma = \alpha \frac{C_{gl}}{C_p}. \tag{13.18}$$

Note that usually $T_f \approx T_g$ is a reasonable approximation. The discussed here jump frequency model predicts, in agreement with experimental evidence that

$$\frac{m_g}{m} = \frac{C_{gl}}{C_p}, \tag{13.19}$$

so that the ratio of the two slopes is in agreement with the existing experimental evidences and the activation energy paradox disappears. This is because the role of the jump frequencies instead of activation energies was considered to be of primary importance. In this way a sort of apparent activation energy appeared, accounting in addition to the real activation energies also to the contribution of different frequencies to the overall process.

## 13.4 Preexponential Constant

The preexponential constant is proportional to the reciprocal of the vibration frequency molecules.

$$\eta_o = \frac{G}{\nu_o}. \tag{13.20}$$

According to Maxwell the coefficient of proportionality G is equal to the elasticity modulus $G_\infty$, i.e.

$$G = G_\infty. \tag{13.21}$$

On the other hand, according to Frenkel's equation, G is expressed through the molar volume $V_m$, the temperature T and the ideal gas constant R as follows:

$$G = \frac{RT}{V_m}. \tag{13.22}$$

Experimental data [3] on glasses near the glass-transition temperature give for the elasticity modulus a value of $G_\infty \sim 10$ GPa while from Eq. 13.22 a little bit lower value of $G \sim 1$ GPa is expected.

The main problem is how to define the vibration frequency $\nu_o$. The widespread assumption is that it is determined by the Planck's formula. The alternative is to consider that the vibration frequencies of all atoms constituting the building unit vibrate according to Planck's formula. The vibration frequency of each of them is slightly different. Therefore a 'beat' of the molecule appears. An illustration of this is given in Fig. 13.4 where interactions of two oscillators vibrating with 3%

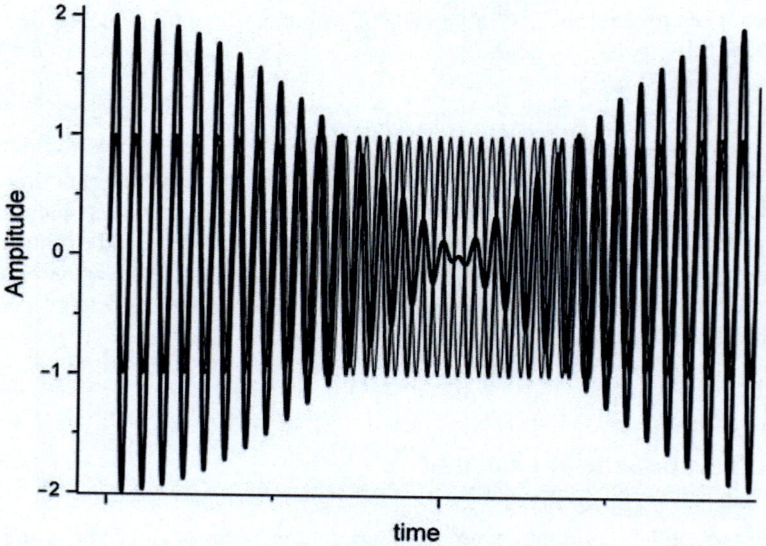


**Fig. 13.4** The 'beat' result (*thick solid line*) when two oscillators with equal amplitude and 3% different frequencies interact

different frequency and equal amplitudes are shown. The resulting 'beat' frequency is shown with a thick solid line. It is seen that the vibration period of the molecule could be considerably larger as compared to the periods of constituting atoms.

## References


1. Avramov I (2009) Non-equilibrium viscosity and activation energy. J Non-Cryst Solids 355:1769–1771
2. Mazurin O, Startsev Yu, Potselueva L (1979) Sov J Phys Chem Glass 5:504 (Engl Transl)
3. Mazurin O, Streltsina M, Shvaiko-Shvaikovskaya T (1985) Handbook of glass data. Elsevier, Amsterdam (1985); SciGlass 6.5 Database and Information System (2005), http://www.sciglass.info/
4. Lillie HR (2006) Viscosity of glass between the strain point and melting temperature. J Am Ceram Soc 14:502–512
5. Angell C (1991) Relaxation in liquids, polymers and plastic crystals—strong/fragile patterns and problems. J Non-Cryst Solids 131–133:13–31
6. Böhmer R, Angell CA (1992) Correlations of the nonexponentiality and state dependence of mechanical relaxations with bond connectivity in Ge-As-Se supercooled liquids. Phys Rev B 45:10091–10094
7. Debendett P (1996) Metastable liquids. Princeton University Press, Princeton
8. Gutzow I, Streltsina M, Popov E (1966) Compt. Rend Acad Bulg Sci 19:15–17
9. Yue Y (2009) The iso-structural viscosity, configurational entropy and fragility of oxide liquids. J Non-Cryst Solids 355:737–744
10. Gladstone S, Laider H, Eiring H (1941) The theory of rate processes. Princeton University, New York, London
11. Avramov I (2005) Viscosity in disordered media. J Non-Cryst Solids 351:3163–3173
12. Avramov I (2007) Pressure and temperature dependence of viscosity of glassforming and of geoscientifically relevant systems. J Volcanol Geoth Res 160:165–174
13. Nemilov S (2007) Structural aspect of possible interrelation between fragility (length) of glass forming melts and Poisson's ratio of glasses. J Non-Cryst Solids 353:4613–4632
14. Taniguchi H (1992) Entropy dependence of viscosity and glass-transition temperature of melts in the system diopside-anortite. Contrib Mineral Petrol 109:295–303
15. Behrens H, Schulze F (2003) Pressure dependence of melt viscosity in the system $NaAlSi_3O_8$-$CaMgSi_2O_6$. Am Mineral 88:1351–1363
16. Neuville D, Richet P (1991) Viscosity of pyroxenes and garnets melts. Geochim Cosmochim Acta 55:1011–1019
17. Avramov I, Grzybowski A, Paluch M (2009) A new approach to description of the pressure dependence of viscosity. J Non-Cryst Solids 355:733–736

# Chapter 14
# Transport Constitutive Relations, Quantum Diffusion and Periodic Reactions

Jiří J. Mareš, Jaroslav Šesták, and Pavel Hubík

## 14.1 Introduction

In this contribution we are discussing a class of linear phenomenological transport equations and in some cases also their relation to microphysical description of corresponding effects. Interestingly enough, in spite of practically identical forms of these constitutive relations there are large differences in their physical content; just such a large diversity of natural processes behind the same mathematical form should serve as a serious warning before making superficial analogies. On the other hand, besides quite obvious analogies there may be found also those much deeper and sometimes quite astonishing. Lesser known or even new aspects of this kind the reader can find especially in paragraphs dealing with Ohm's law and with statistical interpretation of generalized Fick's law. The congruence of the last one with the fundamental equation of quantum mechanics, the Schrödinger equation, opened the possibility to interpret the rather enigmatic "quantum" behaviour of periodic chemical reactions as a special kind of diffusion.

## 14.2 Fourier's Law of Heat Transfer and Analogous Constitutive Relations

There is a class of essentially linear equations describing the transport of substance-like indestructible quantities through the homogeneous medium. Historically the first

J.J. Mareš (✉) and P. Hubík
Institute of Physics ASCR, v.v.i.
Cukrovarnická 10, 162 00 Praha 6, Czech Republic
e-mail: maresjj@fzu.cz

J. Šesták
New Technologies Research Centre, University of West Bohemia,
Univerzitní 8, 30614 Plzeň, Czech Republic

J. Šesták et al. (eds.), *Glassy, Amorphous and Nano-Crystalline Materials*,
Hot Topics in Thermal Analysis and Calorimetry 8, DOI 10.1007/978-90-481-2882-2_14,

known law of this type is famous Fourier's law [1] controlling the transfer of thermal energy through the conductor of heat. Strictly speaking, it is no law of Nature but an approximate linear constitutive relation which can be, with a certain degree of generality, written in a differential form as

$$\mathbf{q} = -\lambda \text{ grad T}, \tag{14.1}$$

where $\mathbf{q}$ is the local current density of thermal energy, $\lambda$ the coefficient of thermal conductivity and T the Kelvin absolute temperature. (We do not consider here the cases where the conductor is anisotropic or where $\lambda$ is not continuous throughout the conductor. The corresponding generalizations are straightforward, nevertheless they are not free from controversies [2].) According to empirical evidence, the terms in Eq. 14.1 have an additional property which may be expressed in the form of constitutive inequality

$$\mathbf{q} \text{ grad T} \leq 0. \tag{14.2}$$

The physical meaning of this condition is obvious; the current vector $\mathbf{q}$ has to have everywhere the component opposite to the direction of grad T representing the local driving force of energy transfer. In order to solve concrete problems, relation 14.1 is, as a rule, completed by a pair of phoronomic equations, namely by restricted equation of continuity

$$\text{div}\, \mathbf{q} = 0 \tag{14.3}$$

which expresses the conservation of substance-like quantity (in this case of thermal energy) during its transfer. Notice that in this case the substance-like quantity is not considered to be only indestructible but that it is not created either. The second phoronomic equation, sometimes called equation of discontinuity then reads

$$q_\nu = 0, \tag{14.4}$$

where $q_\nu$ is the component of current vector which is normal to the surface of heat conductor. This equation is assumed to be valid on all surfaces of the heat conductor except its terminals where the current source and drain are placed. Putting aside the cases where leakage currents or external current sources are present ($q_\nu \neq 0$), Eqs. 14.1, 14.3 and 14.4 may serve as a representative pattern for a wide class of problems of mathematical physics connected with the transport of energy, electricity or matter.

Relations 14.1 and 14.3 describe the steady state of the field represented e.g. by the scalar quantity T. In order to describe also the time evolution of this field the equation of continuity can be no more used in its restricted form 14.3. Instead, it has to involve also a term characterizing the time dependence of accumulation or depletion of the said substance-like quantity which may be represented by the time derivative of the scalar quantity T at a given point, namely

$$\beta\,(\partial T/\partial t) + \operatorname{div} \mathbf{q} = 0\,, \tag{14.5}$$

where $\beta$ is a constant coefficient ensuring the dimensional homogeneity of the equation. Taking then into account Eq. 14.1 and assuming that $\lambda$ is in a certain closed region constant, we obtain immediately

$$\partial T/\partial t \;= (\lambda/\beta)\ \text{div grad T}, \tag{14.6}$$

which is usually called the "second law" conjugated to the "first law" of type 14.1. Returning back to the description of steady state, which may be now characterized by the condition $\partial T/\partial t=0$, we obtain from 14.6 the relation

$$\text{div grad T} = 0, \tag{14.7}$$

i.e. well-known Laplace's equation. The solutions of this equation are called harmonic functions, which are in a particular case of one dimension reduced to the linear change of variable T along the axis x.

It is worth noticing that the most difficult part of the establishment of Fourier's and other similar constitutive relations was not finding out of their mathematical form (which is very simple) but the definition and the physical interpretation of the quantities involved or even the proof of their plain existence [3]. Empirically determined pre-factor in 14.1 is thus very often decomposed into the product of quantities having more straightforward or already known physical interpretation. For example the thermal conductivity is usually written in the form

$$\lambda{=}\alpha/(c_p\,\rho_m), \tag{14.8}$$

where $\alpha$ means the thermal diffusivity introduced by Kelvin as an analogue of diffusion constant, $c_p$ specific heat capacity at constant pressure and $\rho_m$ the density of the material.

## 14.3 Darcy's Law

As another example of linear transport constitutive relation may serve so called Darcy's law [4] describing the flow of fluid through the porous medium brought about by pressure drop. This relation controlling e.g. the movement of groundwater through the aquifer [5] or behaviour of petroleum in oil-deposits thus plays an extraordinary role in geology. It may be written as

$$\mathbf{j} = -(\kappa/\eta)\ \text{grad p}, \tag{14.9}$$

where **j** means the vector of filtration velocity (which differs from the real velocity of liquid in pores) and p the pressure. Notice that the pre-factor is in this case composed from two independent coefficients, one characterizing the porosity and tortuosity of the porous medium, permeability κ, and the second one characterizing the flow of liquid, dynamic viscosity η.

## 14.4 Ohm's Law

The outstanding role among linear relations of type 14.1 plays so called Ohm's law controlling the transport of electricity in metals and semiconductors. This "law" being the cornerstone of modern electronics is probably the most exploited physical relation which has ever been discovered. Its differential form convenient for the description of charge transport in an isotropic conductor reads

$$\mathbf{i} = -\,\gamma\,\mathrm{grad}\,K, \tag{14.10}$$

where **i** is the local current density vector, γ the electrical conductivity and grad K the force driving the electric charge. Of course, Eq. 14.10 has to be accompanied with the phoronomic relations div **i** = 0 and $i_v = 0$, quite analogous to Eqs. 14.3 and 14.4. Originally the driving force was identified by Ohm with the electroscopic force (Elektroskopische Kraft) measured by a mechanical action of the conductor exerted on a small insulated body placed in the vicinity of its surface [6]. Obviously, the direct use of such a definition in the interior of the conductor was impossible. As, however, the electroscopic force was by many savants considered to be a measure of the density of electric charge dwelling on the surface of the body it was quite natural to identify the quantity K simply with the local electric charge density ρ. It seems to be very likely that Ohm was aware that such an assumption results for the limiting case **i**=0 to the condition ρ = const. which is in contradiction with Coulomb's law, the fundamental theorem of electrostatics. He even correctly claimed that [6, Cf.7] "...wenn Gleichgewicht sich hergestellt habe, nach den Versuchen von Coulomb und nach der Theorie, die Elektricität an die Oberfläche der Körper gebunden sei, oder durch eine unmerkliche Tiefe in das Innere eindrige." ("...if equilibrium is established, according to the experiments of Coulomb and according to the theory, the electricity is bound on the surface of the body or penetrates through the tiny depth into its interior".) Nevertheless, he did not provide any definite solution of this puzzling problem which was only due to Kirchhoff who identified the force responsible for the charge transport with the gradient of electrostatic potential φ [8]. Such an assumption reconciled the discrepancy between relation 14.10 and the laws of electrostatics. Indeed according to Kirchhoff's arguments we can compute from Eq. 14.10 div **i** = −γ div gradφ. Taking then into account the fact that at any point in the interior of the conductor steady state condition div **i** = 0 must be valid, we immediately obtain equation div gradφ = 0, which is nothing but the above mentioned Laplace's equation describing behaviour of electrostatic potential φ in the neutral region (ρ ≡ 0). In other words, it means that in the case where K ≡ φ, there is no net space charge in a current-carrying conductor and, vice versa, for any homogeneous conductor where the space charges are present, the application of Kirchhoff's form of Ohm's law cannot be fully justified. Of course, in the cases where the transport is controlled by the regions containing space charge, e.g. in Schottky diodes, p-n junctions and in the vicinity of charge injecting contacts, the violation of Ohm's law is a well-known effect having enormous application impact [9]. It may thus be somewhat astonishing that the space charges must be inevitably present, in contrast to Kirchhoff's proposition, in any transport of electricity via conductors. The indispensable role of the space charges played in conduction was for a long time overlooked in spite of the obvious empirical fact that transport through the wires is to a large extent independent of the arrangement of its surrounding. In order to suppress, namely, the long range electromagnetic interaction between charge carriers in the conductor and external disturbing electric fields, the presence of stable screening charges distributed in the conductor is quite inevitable. Such a necessity of the existence of surface charges on the current-carrying conductor was pointed out, e.g. in Ref. [10]. Accordingly, the surface charges not only prevent the electric flow from escaping from the conductor but also provide, inside the conductor, an appropriate field distribution ensuring the constancy of the total current flow throughout any cross-section of the conductor regardless of its complicated topology (e.g. in knots on the wire) and external electric fields. The quantitative theory of this effect is practically lacking, however, a relatively simple approximate formula for local charge density $\sigma$ dwelling on a free-standing conductor having no loops may be written as:

$$\sigma = i_\tau\,(\varepsilon\varepsilon_0/\gamma)\,b(1/d_1 + 1/d_2), \qquad (14.11)$$

where $i_\tau$ is the local tangential component of the current density, $\varepsilon\varepsilon_0$ the permittivity of outer space, b the linear distance from the ground and $d_1$ and $d_2$ the principal radii of curvature at a given point of the surface. For example, for the straight vertical copper wire of diameter $d_1 = 10^{-3}$ m with one end earthed, carrying the total current of 1 A and for a point lying on it at the distance $b = 10$ m from the ground (other needed parameters are: $\varepsilon\varepsilon_0 \approx 8.85 \times 10^{-12}$ F/m, $\gamma \approx 6.4 \times 10^7$ S/m, $d_2 \to \infty$) we obtain from formula 14.11 that $\sigma \approx 1.76 \times 10^{-9}$ C/m$^2$. It is an extremely low value of surface charge density, especially because of high value of $\gamma$. Nevertheless, regardless of the smallness of this effect we can conclude that in conductors, the presence of surface charges is a non-separable part of the steady state electric current flow. It is further clear that the concept of finite charge strictly confined to the geometric surface of the conductor is only an abstraction, while in a real case, this charge must be deposited somewhere in the bulk. Interestingly, essentially the same conjecture was expressed by Ohm in his quotation above. An adequate mathematical formulation of this idea may be achieved by the following slight modification of Ohm's law [11]. Taking first into account the fact that for the characterisation of the local electric state we have only two independent intensive quantities at our disposal, i.e. potential $\varphi$ and charge density $\rho$, we suggest completing of quantity K containing purely electrostatic term $\varphi$ by a chemical (or diffusion related) term linearly depending on charge density ρ, resembling original Ohm's concept of "Elektroskopische Kraft". In such a case it should be

$$K = \varphi + \rho\,\delta^2/\varepsilon\varepsilon_0, \tag{14.12}$$

where δ is a length parameter not specified yet, added in order to ensure the dimensional homogeneity of both terms. The resulting form of Ohm's law thus reads

$$\mathbf{i} = -\gamma\,\mathrm{grad}\,(\varphi + \rho\,\delta^2/\varepsilon\varepsilon_0). \tag{14.13}$$

Solving this equation with respect to the boundary condition $i_\nu = 0$, we find that the net charge density should decrease exponentially with increasing depth ν under the surface of current-carrying conductor, i.e. as $\rho(\nu) = \rho_0 \exp(-\nu/\delta)$ where the pre-factor can be obviously expressed as $\rho_0 = \sigma/\delta$. Putting, moreover, $\mathbf{i} = 0$ identically, we obtain from Eq. 14.13 a usual electrostatic screening formula, namely $\delta^2\rho/\varepsilon\varepsilon_0 = \varphi_0 - \varphi$, playing an important role in the description of numerous contact phenomena near the equilibrium. From all these relations, it is evident that the parameter δ has a physical meaning of the screening length, in metals, particularly, of the Thomas–Fermi screening length [12]. In order to have a more specific idea of the significance of the chemical term we can estimate its upper bound for the above mentioned case of straight wire using Eq. 14.11. Accordingly it follows $\delta^2\ \rho_0/\varepsilon\varepsilon_0 = \sigma\delta/\varepsilon\varepsilon_0 \approx (i_\tau/\gamma)\ (b\delta/d_1)$. Simultaneously we can determine, by means of Davy's formula, the potential drop along this wire, $\varphi = ib/\gamma$. As $i_\tau \approx i$, we thus immediately obtain for the ratio of chemical and electrostatic term φ an estimate $\approx \delta/d_1$. In ordinary metallic conductors used in electronics (e.g. Cu wire where $\delta = 5.5 \times 10^{-11}$ m, and $d_1 \approx 10^{-4}$ m) this parameter attains the value $\approx 5.5 \times 10^{-7}$. Evidently, because of the smallness of this parameter the correction to Ohm's law due to the chemical term is absolutely negligible there having only theoretical significance. In contrast to bulk metals, however, the situation may be rather different in semiconductor-based structures in which δ typically ranges from $10^{-6}$ to $10^{-9}$ m and where one of the dimensions of the conductor is confined to nanometre scale. In such a case the original current-carrying neutral bulk region which was perfectly screened against the influence of external fields by a surface charge layer is appreciably reduced giving thus rise to qualitatively new effects generally, but not quite correctly, connected with the so called "quantum confinement" [11].

## 14.5 Fick's Law

The spontaneous transport of dispersed substance from the region of high concentration to the one of lower concentration through a material medium is called diffusion. There are a lot of combinations of species which may take part in such a process.

The pioneering systematic studies in this field were performed by Scottish professor of chemistry T. Graham who formulated the first quantitative principles controlling the diffusion of gases into another gas and the diffusion of salts in aqueous solutions [13]. Similar experiments were made somewhat later by a young Swiss pathologist A. E. Fick who in his fundamental article [14] established the constitutive relations satisfactorily describing the process of diffusion in general. For the analytical form of his law Fick used with awareness Fourier's law of heat conduction (and very likely also Ohm's law) as a pattern, while for its physical interpretation he exploited consequently the molecular hypothesis. The resulting first Fick's law in its simplest form may thus be written as

$$\mathbf{J} = -\mathrm{D}\ \mathrm{grad}\ \mathrm{n}, \tag{14.14}$$

where **J** is the density of diffusion current (flux), D the coefficient of diffusion and n is the local concentration of the diffusing species. As was later recognized, such a form of this constitutive relation is related only to the case where we have to do with rarefied gases or with the so called ideal solutions. For more complicated compound systems the driving force of the diffusion is not simply proportional to the gradient of concentration n but to the gradient of chemical potential $\mu$ of particular species. The relation 14.14 must then be rewritten as

$$\mathbf{J} = -(\mathrm{Dn}/\mathrm{RT})\ \mathrm{grad}\,\mu, \tag{14.15}$$

where R is the universal gas constant and T the Kelvin absolute temperature. It is necessary to note that there was in fact only a rather vague connection between the phenomenological relation 14.14 and Fick's molecular model. In accordance with the obsolete modification of molecular hypothesis currently used in the middle of the nineteenth century, namely, the molecules consist of ponderable atoms surrounded by a dense aether atmosphere. It was believed that the thermal agitation of these two components was responsible for thermal dilation of bodies and for movement of molecules within the bodies as well. Due to the high abstractness of such ideas, however, the arguments are somewhat teleological and the causal derivation of relation 14.14 was practically impossible. It is therefore a remarkable fact that both these ingredients appearing in Fick's approach i.e. molecular structure and thermal agitation became essential for the construction of modern microscopic (statistical) models describing diffusion and allied phenomena. The first model of this kind was developed by W. Sutherland [15] and another one, very similar, a year later by A. Einstein [16]. Accordingly, the diffusion constant must be given by the formula

$$\mathrm{D} = \mathrm{k_B T}/\,\xi, \tag{14.16}$$

where $k_B$ is the Boltzmann universal constant ($k_B = 1.38 \times 10^{-23}$ J/K) and $\xi$ the coefficient of friction affecting the movement of an individual molecule through the medium. Useful approximation of $\xi$ is that using the Stokes formula, namely,

$\xi = 3\pi\eta a$, where a is the diameter of the molecule [15, 16]. Very soon after the appearance of these theories it became quite clear that there was a close relation between the diffusion and the so called Brownian motion. Within the frame of the mathematical theory of Brownian motion and other fluctuation phenomena worked out by M. von Smoluchowski [17] and R. Fürth [18] it was possible to show that the process of diffusion may be described as follows. Every diffusing molecule is simultaneously submitted to the influence of force of friction characterized by the coefficient $\xi$ and to the random impacts due to the neighbouring thermally agitated molecules. Under such conditions it can be rigorously proved that the statistical shift of ensemble of diffusing molecules must be controlled by Eq. 14.14 (or 14.15) with the diffusion constant given by Eq. 14.16. Careful consideration of the second form of Fick's law conjugated to Eq. 14.14 (cf. also Eq. 14.6), i.e. of the equation

$$\partial n/\partial t = D \, \mathrm{div} \, \mathrm{grad} \, n, \tag{14.17}$$

defining the relation between continuous phenomenological quantity n and its time derivative, may lead to paradoxes which may be removed only by careful reconsideration of concepts used. For example, for continuous $n(\mathbf{x},t)$ the "speed of diffusion", which is intuitively associated with the speed of molecules, is not well defined or it has to be assumed to be infinite. This odd statement can be elucidated by the following argument [19]. If the concentration n of a substance at time $t = 0$ is finite only in a certain bounded region (e. g., $n = \text{const.}$), being identically zero out of this region, the equation implies that after an arbitrarily small time interval dt the concentration at any point of the whole space is non-zero, so that the transport of matter to any distant point would be instantaneous, i.e., with an infinite speed. On the other side, taking into account the discrete molecular structure of ordinary matter, the concentration at a given point should remain zero till it is reached by the first diffusing particle (molecule), i.e., the velocity should be finite in any case. Very similar unpleasant discordance between mathematical solution and common sense arguments was much earlier discovered also by J. Stephan [20] who studied the time dependent transfer of heat controlled by Fourier's law 14.1. This paradox which was shown to be intrinsic to the mathematical assumptions used by the derivation of relations 14.14 and 14.1 may, as will be shown below, serve as an important key to physically meaningful definitions of "speed of diffusion" or "speed of heat transfer".

## 14.6 Diffusion and Stochastic Quantum Mechanics

Second Fick's law in the form 14.17 describes the process of diffusion only in the case where the external forces are absent. According to laws of classical mechanics the diffusion in an external field can be formally decomposed into a superposition of movement of the centre of mass of the diffusing swarm of particles under the influence of external force and of the diffusion without the external field. Therefore the resulting formula has to contain beside the right side of Eq. 14.17 extra additive terms corresponding to various types of external forces. As was pointed out by Fürth [21], such a generalized formula is fully congruent with the fundamental equation of quantum mechanics, Schrödinger equation[1]. These equations may be mapped one onto another by substituting for the diffusion coefficient the value (Fürth's Ansatz)

$$D = iD_Q = i\hbar/2M, \tag{14.18}$$

where i is the imaginary unit $i=\sqrt{(-1)}$ and $\hbar \approx 1.05\times10^{-34}$ Js Planck's universal constant. This very fact together with the feeling of dissatisfaction with some basic quantum concepts going against the common sense leads to the attempts to construct unified statistical theory treating the quantum motion as a kind of a stochastic process (see e.g. [22–24] and references therein). Accordingly, the source behind the assumed stochastic quantum process was tentatively identified with the universal noise generally known as "zero-point" fluctuations of vacuum (more correctly "temperature independent" fluctuations of vacuum). As can be shown this assumption is consistent with the structure of Fürth's diffusion constant $D_Q$ which differs essentially from that of Sutherland and Einstein 14.16. Indeed, it is clear at first glance that the coefficient $D_Q$ is temperature independent so that the diffusion of quantum particle can be really considered to be a "zero-point" effect. Moreover, we cannot attribute any finite friction coefficient analogous to $\xi$ to the movement of particle through the vacuum space because it would be then possible, in contrast to the Principle of Relativity, to experimentally distinguish coordinate systems in absolute rest from that in uniform motion. This requirement is also evidently fulfilled; $D_Q$ does not really contain explicitly any friction coefficient. Taking now into account the fact that the large macroscopic bodies do not appear to exhibit "quantum" behaviour, we can speculate that the diffusion coefficient $D_Q$ is inversely proportional to the mass M of the body. A special interest deserves the presence of imaginary unit in Fürth's Ansatz 14.18. The necessity to use the imaginary unit there is connected with the fact that the quantum mechanics similarly to the analytical mechanics is formulated in the phase space while von Smoluchowski's theory describes the Brownian motion and diffusion in the configuration space. As the cardinality of phase space is two times higher than that of the configuration space related to the same physical problem, the existing one-to one correspondence between solutions of generalized diffusion equation and Schrödinger equation is ensured just by the implementation of complex numbers effectively doubling the cardinality of the real space. The attempts to interpret quantum mechanics as a stochastic theory, however, bring about further serious problems. The classical Einstein–von Smoluchowski description of diffusion (i.e. of a special case of Brownian motion of a small particle) is essentially a description of Markovian stochastic process, i.e. a process in which the following steps are absolutely non-correlated. Any stochastic process in the phase space assumed to underlie motion of a small quantum particle cannot be, however, Markovian, because, by definition, its state at $t < 0$ determines the probability of states at any later time t [25]. In other words, the quantum particle has "memory". That is the reason why e.g. advanced Uhlenbeck-Ornstein-Wang theory [26, 27] which is based on stationary Markovian stochastic process characterized by measure-preserving flow in the phase space is inevitably mathematically incompatible with the formalism of quantum mechanics.

A famous controversial proof of von Neumann's theorem [28] on the impossibility of entering of "verborgene Parameter" (hidden variables) into description of the quantum processes, is also based on practically the same grounds, i.e. on the discrepancy between cardinality of various spaces of variables used for the description of quantum object. On the other side, as will be shown below, the "naive" theory of von Smoluchowski and Fürth constructed in the configuration space and operating with the concept of intermittent measurement can provide a satisfactory description of the quantum process. Intermittent measurements, namely, performed in the configuration space reduce repeatedly the wave packet of a particle and may thus cause the "memory loss" of the particle so that the resulting process can be treated as Markovian. Such a compatibility of von Smoluchowski's approach and quantum mechanics is in very convincing way manifested below by the fact that Hausdorff's fractal dimensions of Brownian and quantum motions in the configuration space are identical (d=2).

## 14.7 Periodic Reactions and Quantum Diffusion

As an interesting example, on which the continuous transition from classical diffusion to the domain of quantum physics can be demonstrated, may serve so called periodic chemical reactions. These reactions, known from the second half of the nineteenth century [29], perform a curious class of reactions generating marvellous macroscopic patterns periodic both in space and time (see Fig. 14.1). They are mostly considered to be spectacular manifestations of self-organization due to the non-equilibrium nature of thermodynamic processes involved. As these reactions violate traditional view on chemical kinetics characterized by the natural tendency to reach the equilibrium by the shortest way, they have been interpreted as a precursor of life processes [30]. In the 1930s, besides the fact that the kinetics of periodic reactions is controlled by diffusion (so called Nernst–Brunner kinetics), another peculiar and somewhat enigmatic feature of these reactions was discovered [31, 32] which may be concisely expressed as

$$M v \ell \approx \hbar/2, \tag{14.19}$$

where M is the molecular weight of precipitate, v the speed of spreading of reaction fronts and $\ell$ the length parameter of reaction patterns. For a particular configuration of the system in which reaction takes place the left-hand side of Eq. 14.19 should be

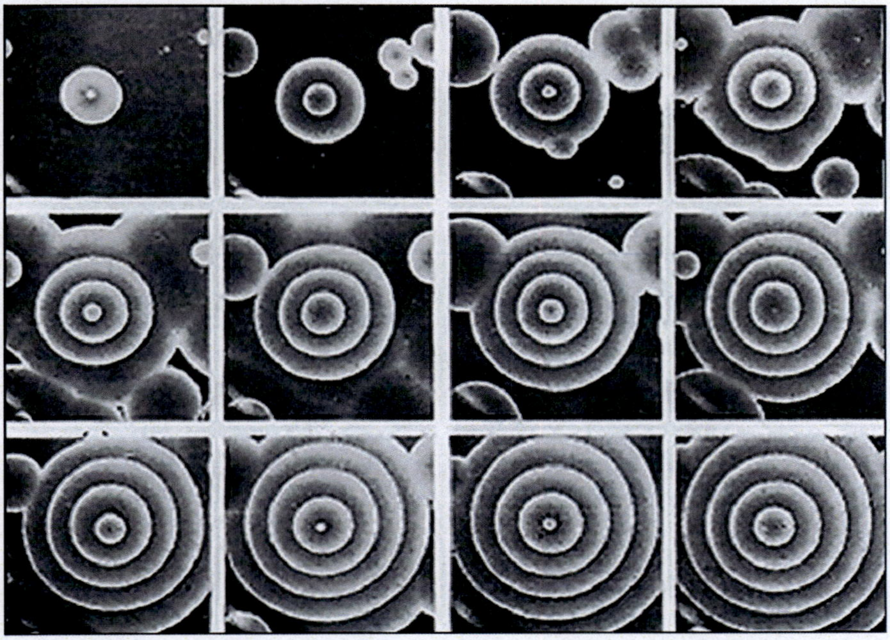

**Fig. 14.1** The record of evolution of the so called Liesegang's rings on gelatinous substrate. An example of a regular pattern resulting from the periodic reaction and revealing the features of quantum diffusion

completed by a geometric factor (e.g. 2 for three- and $1/\pi$ for two-dimensional case) and by tortuosity characterizing the detailed topology of the system. As a rule, the resulting factor is of order unity.

## 14.8 Speed of Diffusion

Taking now into account Sommerfeld's criterion, according to which any effect belongs to the scope of quantum physics just if the corresponding relevant quantity of action is comparable with quantum of action, ħ [33], we can claim that the diffusion-controlled periodic reactions fulfilling Eq. 14.19 may be interpreted as quantum effect. What we, however, urgently need for constructing the corresponding "relevant quantity of type action" is the definition of something like the instant "speed of diffusion", already mentioned above as a rather controversial concept. If we, for the sake of simplicity, confine ourselves only to one dimension, the diffusion can be described by the second Fick's law 14.17 in the form

$$\partial n/\partial t = D\,(\partial^2 n/\partial x^2). \tag{14.20}$$

The meaningful definition of “speed of diffusion” can be then made even under the assumption that the concentration n (x,t) is continuous. Indeed, let us assume that at time t=0 there are N particles concentrated in the plane $x = 0$ (source layer). In this case the solution of Eq. 14.20 reads:

$$n(x,t) = N/(2\sqrt{\pi Dt})\exp(-x^2/4Dt). \tag{14.21}$$

Since the times of Fourier it has been a well-known property of solution 14.21 (known as a source–integral) that the time record of concentration taken in a neighbourhood of a certain fixed point x has a local maximum. The mathematical condition for this maximum reads $\partial n/\partial t = 0$. This is, however, according to Eq. 14.20, equivalent to the condition $(\partial^2 n/\partial x^2) = 0$ (for constant $D \neq 0$). The second space derivative of solution 14.21 gives then the expression:

$$\partial^2 n/\partial x^2 = \{N/4\sqrt{\pi}(Dt)^3\}\exp(-x^2/4Dt)\{(x^2/2Dt) - 1\}. \tag{14.22}$$

Using now the above-mentioned condition for extreme, we obtain immediately relation (Einstein-von Smoluchowski’s relation)

$$x^2 = 2Dt, \tag{14.23}$$

the time derivative of which provides

$$x\,u = D, \tag{14.24}$$

where $u = \partial x/\partial t$ has evidently physical meaning of the instant speed of transfer of concentration maximum. As the quantity u, in fact, represents the movement of the most numerous swarm of diffusing molecules, it is quite reasonable just to call u the “instant speed of diffusion”. It is a remarkable circumstance that Eqs. 14.23 and 14.24 are practically the same as the equations describing random walk of a single Brownian particle (molecule) [16, 17]. The only differences are that there x is no more the position of the concentration maximum but the mean-square-root $\sqrt{\langle x^2 \rangle}$ of the position of a particular Brownian particle at time t and u has a meaning of its mean-square root of stochastic speed $\sqrt{\langle U^2 \rangle}$. If we start, namely, from the one-dimensional Fick’s law 14.14 in a probabilistic notation, i.e., from

$$Uw = -D\,(\partial w/\partial x), \tag{14.25}$$

where the probability density is defined as $w = n/N$, U is the stochastic speed and (Uw) has the meaning of the probability flux, we obtain for the mean-square of the stochastic speed the expression (the integrals here are taken over the range from $-\infty$ to $+\infty$)

$$\sqrt{\langle U^2 \rangle} = \int (U^2 w)dx = D^2 \int (1/w)(\partial w/\partial x)^2\,dx. \tag{14.26}$$

In consequence of evident inequality $\{(\partial w/\partial x)/w + x/\langle x^2\rangle\}^2 \geq 0$, one obtains after a simple algebra that $\int(1/w)\,(\partial w/\partial x)^2\,dx \geq 1/\langle x^2\rangle$ and therefore eventually

$$\sqrt{\langle x^2\rangle}\sqrt{\langle U^2\rangle} \geq D, \tag{14.27}$$

where equality in this "uncertainty relation" takes place for the probability distribution corresponding just to the source-integral 14.21. Because Eqs. 14.23 and 14.24 describe the same physical process of diffusion, the diffusion constants D must naturally be identical for the microscopic as well as for the macroscopic cases and simultaneously the relations

$$u = \sqrt{\langle U^2\rangle} \text{ and } x = \sqrt{\langle x^2\rangle} \tag{14.28}$$

must be valid. We can thus conclude that a typical "average" Brownian particle follows the position of the concentration maximum or in other words that the most significant packet of diffusing molecules consists of "average" Brownian particles.

As we believe, this is just the way how Planck's universal constant can in principle enter essentially macroscopic Eq. 14.24. Namely, if the microscopic movement of a Brownian particle of mass M is controlled by a purely quantum process, where the diffusion constant in three dimensions should have Fürth's limiting value of $D_Q = \hbar/2M$, then Eq. 14.24 will formally attain the same form as empirical Eq. 14.19, i.e.

$$M\,u\,x = \hbar/2, \tag{14.29}$$

provided that the experimentally observed quantities v and $\ell$ are identified with $u = \sqrt{\langle U^2\rangle}$ (speed of diffusion) and $x = \sqrt{\langle x^2\rangle}$ (distance spanned by diffusion), respectively.

## 14.9 Resemblance of Quantum and Brownian Motion in a Configuration Space

It seems thus plausible that to prove the quantum nature of Eq. 14.19 it is enough to make clear conditions under which the numerical value of diffusion constant D attains the Fürth's value $D_Q$. Unfortunately, realization of this task is by no means trivial, mainly because the frequently stressed analogy between quantum and Brownian motions [34–36] is rather incomplete. On the other side, there is an important common characteristic of these two types of stochastic processes which makes them identical in a certain sense, namely their Hausdorff's dimension in the configuration space [37, 38]. To make clearer the relation of this useful mathematical concept to the present problem we must go behind the continuous approximation of diffusion represented e.g. by Eq. 14.23. Extrapolating the validity of this formula from the experimentally verified range to arbitrarily small scales, we can use it, however, for the construction of a more realistic model of movement of a Brownian particle in the following way.

Let us make intermittent observations of total duration $\tau$, sampling every $(\tau/k)$ seconds the position of a Brownian particle, where k is an arbitrary integer. Every time interval $(\tau/k)$ involves some delay and time necessary to determine the position of the particle. Such a measurement will provide in one dimension a sequence of intervals $l_1, l_2,.., l_k$, which define the apparent length of the path passed through as:

$$L_k = l_1 + l_2 + \ldots + l_k \tag{14.30}$$

As the movement along every of these intervals is, according to our assumption, controlled by law 14.23, for a sufficiently large k instead of 14.30 we can write:

$$L_k = k\sqrt{\langle l^2 \rangle_k}, \tag{14.31}$$

where index k means the averaging over all k intervals in 14.30. It is simultaneously clear that the length $\sqrt{\langle l^2 \rangle_k}$ defines for a given k also the length resolution $\Delta l_k = \sqrt{\langle l^2 \rangle_k}$ of the measurement, because the sampling by time intervals $(\tau/k)$ evidently ignores the details of the actual path of the Brownian particle finer than $\Delta l_k$. Average speed determined from the same experiment as $k\,\Delta l_k/\tau$ must, moreover, fulfil the relation 14.27 in the form (reduced to equality):

$$\Delta l_k\,(k\,\Delta l_k/\tau) = D. \tag{14.32}$$

It is obvious that the more frequent measurement with a better resolution $\Delta l_k$ will reveal more details of motion of the Brownian particle. As a result, with increasing k the number of recorded abrupt changes on the Brownian path will increase and also $L_k$ will increase simultaneously. For very large k's (theoretically for $k \to +\infty$) the shape of the Brownian path will resemble a continuous, at every point non-differentiable curve. As was shown by Hausdorff [39] such a complicated mathematical object may be, without ambiguity, characterized by introducing its (Hausdorff's) measure $\Lambda$ and generalized dimension d as follows:

$$\Lambda = L_k\,(\Delta l_k)^{d-1}, \tag{14.33}$$

where dimension d should (and can) be chosen in such a way that $\Lambda$ is independent of k. Putting then relations 14.32 and 14.33 together, we obtain:

$$\Lambda = (D\tau/\Delta l_k)(\Delta l_k)^{d-1}, \tag{14.34}$$

from which, due to the said independence of $\Lambda$ on k, it immediately follows that Hausdorff's dimension of the Brownian motion is $d = 2$ (cf. e.g. [37]).

As was shown by L.F. Abbott and M.B. Wise [38], just the same Hausdorff's dimension in the configuration space (i.e. $d = 2$) has to be ascribed to the jiggling movement ('Zitterbewegung') of a quantum particle. This can be proved very easily using, instead of Eq. 14.32, common Heisenberg's uncertainty relation in a (reduced) form

$$\Delta l_k \,(M\, k\, \Delta l_k/\tau) = \hbar/2, \tag{14.35}$$

preserving the previous meanings of all quantities involved. The corresponding "quantum" expression for Hausdorff's measure then reads:

$$\Lambda = (\hbar\ \tau/2M\,\Delta l_k)(\Delta l_k)^{d-1}, \tag{14.36}$$

This formula gives the same value for $d = 2$ as Eq. 14.34 and can be obtained from it by a formal substitution of $D_Q$ for D. Because both quantum and Brownian motions have in the configuration space exactly the same Hausdorff's dimension revealed, e.g., by the intermittent measurement described above, the quantum jiggling can be, from the phenomenological point of view, considered as a continuation of the Brownian motion down the smaller scales. This circumstance enables us to treat the diffusion of a particular molecule together with its quantum jiggling movement as a single stochastic process, formally described by a convenient combination of classical stochastic and quantum diffusion constants, $D_S$ and $D_Q$, respectively. Using analogy with the composition rule for independent mobilities well known from electrochemistry, we can tentatively write:

$$D\ = D_S D_Q/(D_S + D_Q). \tag{14.37}$$

Then quantum limit is represented by the inequality $D_S >> D_Q$, which ensures that $D \approx D_Q$. The quantum effects should prevail in the case, where $D_S$ attains a very high value, or in other words, if the diffusing particle is decoupled from all stochastic sources in environment which are not of quantum nature. We claim that just this inequality, i.e.

$$D_S >> D_Q \tag{14.38}$$

with the accompanying physical meaning mentioned above is the condition we are looking for. If this takes place, namely, Eq. 14.29 and consequently Eq. 14.9 are valid, satisfying simultaneously the quantum criterion.

To assess the range of validity of inequality 14.38 it is necessary to evaluate stochastic diffusion constant $D_S$. Somewhat crude estimate for ball-like particles (molecules) without observable persistency in motion can be provided by a classical Einstein-von Smoluchowski's formula [16, 17] the use of which is justified rather by practical than physical reasons. Then condition 14.38 reads:

$$k_B T/3\pi\eta a >> \hbar/2M, \tag{14.39}$$

where $\eta$ is the dynamic viscosity of environment, T the absolute temperature and a the characteristic dimension of diffusing particle (molecule). To find out among various aqueous solutions at room temperature molecular systems for which the "quantum" behaviour is to be expected, we put in 14.39 $T = 300$ K and $\eta = 10^{-3}$ kg m$^{-1}$ s$^{-1}$. The resulting condition is

$$8.4 \times 10^{15} >> a/M, \tag{14.40}$$

where a is in meters and $M = (1.67 \times 10^{-27} \times$ molecular weight) in kilograms. Taking for a the ionic diameter [40], we can see that, e. g., sodium, calcium and silver, for which ratios in 14.40 have values $6.0 \times 10^{15}$, $3.4 \times 10^{15}$ and $9.0 \times 10^{14}$, respectively, are good candidates for "quantum" Brownian particles. For $H^+$ ions (i.e. protons) which are the most mobile ions in aqueous solutions condition 14.38 is also valid in spite of the fact that estimate 14.38 (with the ionic diameter $a \approx 10^{-11}$ m) provides rather a high value $\sim 1.2 \times 10^{16}$. It can be accounted for by the fact that the system of protons in water being a Fermionic system requires a consequent quantum-theoretical treatment (involving e.g. the Pauli Exclusion Principle), which is essentially non compatible with classical formulae 14.39 and 14.40.

Summarizing, different linear transport constitutive relations controlled by the equations congruent with Eq. 14.1 were compared. Among the most interesting achievements a new form of force driving the electric current in Ohm's law was introduced. Another interesting item was the statistical interpretation of classical diffusion. Basing further on the so called Fürth Ansatz enabling one-to-one mapping between the Schrödinger and generalized Fick's law, it appear to be possible to extent the statistical approach also on quantum phenomena. As an example, the case of diffusion-controlled periodic chemical reactions was analyzed in detail. It has been shown that in the configuration space and by intermittent observations, which resemble the actual observations performed on periodic patterns, the classical and quantum stochastic processes are practically indistinguishable. This remarkable fact enabled us to formulate in terms of the classical and quantum diffusion constants the condition where the quantum stochastic process should prevail. It has been shown on the basis of this criterion that it is very probable that ions in aqueous solutions can possess macroscopically observable quantum behaviour.

**Acknowledgments** This work was supported by Institutional Research Plan of Institute of Physics No AV0Z10100521.

# References

1. Fourier J-BJ (1822) Théorie analytique de la chaleur. Paris (1822), English transl.: The analytical theory of heat. Dover, Mineola – New York (2003)
2. Truesdell C (1969) Rational thermodynamics. McGraw-Hill, New York

3. Burr AC (1933) Notes on the history of the concept of thermal conductivity. Isis 20:246–259
4. Darcy H (1856) Les fontaines publiques de la ville de Dijon. Dalmont, Paris
5. Forchheimer P (1914) Hydraulik. B.G. Teubner, Leipzig/Berlin
6. Ohm GS (1827) Die galvanische Kette, mathematisch bearbeitet. T.H. Riemann, Berlin
7. Von der Mühll K (1892) Ueber die theoretischen Vorstellungen von Georg Simon Ohm. Ann Phys-Leipzig 283:163–168
8. Kirchhoff RG (1849) Ueber eine Ableitung der Ohm'schen Gesetze, welche sich an die Theorie der Elektrostatik anschliesst. Ann Phys-Leipzig 154:506–513
9. Henisch HK (1984) Semiconductor contacts. Clarendon, Oxford
10. Rosser WGV (1997) Interpretation of classical electromagnetism. Kluwer Academic, Dordrecht
11. Mareš JJ, Krištofik J, Hubík P (2002) Ohm-Kirchhoff's law and screening in two-dimensional electron liquid. Physica E 12:340–343
12. Kittel C (1976) Introduction to solid state physics, 5th edn. Wiley, New York
13. Graham T (1850) On the diffusion of liquids. Philos T Roy Soc London 140:1–46
14. Fick AE (1855) Ueber Diffusion. Ann d Phys (Leipzig) 170:59–86, English transl.: On liquid diffusion. London, Edinburgh and Dublin Phil Mag and J Sci 10:30–51 (1855)
15. Sutherland W (1905) A dynamical theory of diffusion for non-electrolytes and the molecular mass of albumin. Phil Mag 9:781–785
16. Einstein A (1956) In: Fürth R (ed) Investigations on the theory of the Brownian movement. Dover, New York
17. von Smoluchowski M (1923) In: Fürth R (ed) Abhandlungen über die Brownsche Bewegung und verwandte Erscheinungen. Akademische Verlagsgesellschaft, Leipzig
18. Fürth R (1920) Schwankungserscheinungen in der Physik. Vieweg und Sohn, Braunschweig
19. Frank P (1918) Über die Fortpflanzungsgeschwindigkeit der Diffusion. Phys Z 19:516–520
20. Stephan J (1863) Über die Fortpflanzung der Wärme. Sitz Akad Wiss Wien 47:326–344
21. Fürth R (1933) Über einige Beziehungen zwischen klassicher Statistik und Quantenmechanik. Z f Physik 81:143–162
22. Fényes I (1952) Eine wahrscheilichkeitstheoretische Begründung und Interpretation der Quantenmechanik. Z f Physik 132:81–106
23. Comisar GG (1965) Brownian-motion model of nonrelativistic quantum mechanics. Phys Rev 138:B1332–B1337
24. Nelson E (1966) Derivation of Schrödinger equation from Newtonian mechanics. Phys Rev 150:1079–1085
25. Gillspie DT (1994) Why quantum mechanics cannot be formulated as a Markov process. Phys Rev A 49:1607–1612
26. Uhlenbeck GE, Ornstein LS (1930) On the theory of the Brownian motion. Phys Rev 36:823–841
27. Wang MC, Uhlenbeck GE (1945) On the theory of the Brownian motion II. Rev Mod Phys 17:323–341
28. von Neumann J (1932) Mathematische Grundlagen der Quantenmechanik. Springer, Berlin, English transl.: Mathematical foundations of quantum mechanics. Princeton University Press, New Jersey (1955)
29. Runge FF, Liesegang RE, Belousov BP, Zhabotinsky AM (1987) Selbsorganisation chemischer Strukturen. In: Kuhnert L, Niedersen U (eds) Ostwald's Klassiker, vol 272. Verlag H. Deutsch, Frankfurt am Main
30. Runge FF (1855) Der Bildungstrieb der Stoffe veranschaulicht in selbständig gewachsenen Bildern. Selbstverlag, Oranienburg
31. Michaleff P, Nikiforoff W, Schemjakin FM (1934) Über eine neue Gesetzmässigkeit für periodische Reaktionen in Gelen. Kolloid Z 66:197–200
32. Christiansen JA, Wulff I (1934) Untersuchungen über das Liesegang-Phänomenon. Z Phys Chem B 26:187–194
33. Lévy-Leblond J-M, Balibar F (1990) Quantics, rudiments of quantum physics. North-Holland, Amsterdam

34. de la Peña L, Cetto AM (1996) The quantum dice – an introduction to stochastic electrodynamics. Academic, Dordrecht
35. Mareš JJ, Stávek J, Šesták J (2004) Quantum aspects of self-organized periodic chemical reactions. J Chem Phys 121:1499–1503
36. Mareš JJ, Šesták J, Stávek J, Ševčíková H, Krištofik J, Hubík P (2005) Do periodic chemical reactions reveal Fürth's quantum diffusion limit? Physica E 29:145–149
37. Mandelbrot BB (1977) The fractal geometry of nature. W.H. Freeman, New York
38. Abbott LF, Wise MB (1981) Dimension of a quantum-mechanical path. Am J Phys 49:37–39
39. Hausdorff F (1919) Dimension und äusseres Mass. Math Ann 79:157–179
40. Douglas B, McDaniel DH, Alexander JJ (1976) Concepts and models of inorganic chemistry, 2nd edn. Wiley, New York

# Chapter 15
# In-Situ Investigation of the Fast Lattice Recovery during Electropulse Treatment of Heavily Cold Drawn Nanocrystalline Ni-Ti Wires

**Petr Šittner, Jan Pilch, Benoit Malard, Remi Delville, and Caroline Curfs**

## 15.1 Introduction

Shape memory alloys (SMA) such as the near equiatomic Ni-Ti alloy [1] have attracted considerable attention for their unique functional thermomechanical properties as superelasticity or shape memory effect deriving from the martensitic transformation. Ni-Ti wires are being produced from extruded bars by multiple hot working passes finished by a final cold drawing. In this so called "cold worked" (as-drawn, hard, etc.) state, the alloy possesses a heavily deformed microstructure resulting from severe plastic deformation [2] consisting of mixture of austenite, martensite, and amorphous phases with defects and internal strain [3]. In this state, the wires do not show any functional property (Fig. 15.1 left)

As-drawn Ni-Ti wires need to be heat treated so their cold worked microstructure changes into an annealed austenite microstructure for which the wire shows the desired functional properties. At the same time, if the shape of a SMA element is constrained during this final heat treatment, it acquires a new "parent shape". The final thermomechanical treatment thus has two purposes – setting the functional properties and setting the new shape of the wire.

P. Šittner (✉) and J. Pilch
Institute of Physics, Na Slovance 2, 182 21 Praha, Czech Republic
e-mail: sittner@fzu.cz

B. Malard
Institute of Physics, Na Slovance 2, 182 21 Praha, Czech Republic
and
Now at SIMaP, Domaine Universitaire, BP 75 38402 Saint Martin d'Hères, France

R. Delville
EMAT, University of Antwerp, Groenenborgerlaan 171, B-2020 Antwerp, Belgium

C. Curfs
ESRF, 6 rue Jules, Horowitz 38043, Grenoble, France

J. Šesták et al. (eds.), *Glassy, Amorphous and Nano-Crystalline Materials*,
Hot Topics in Thermal Analysis and Calorimetry 8, DOI 10.1007/978-90-481-2882-2_15,

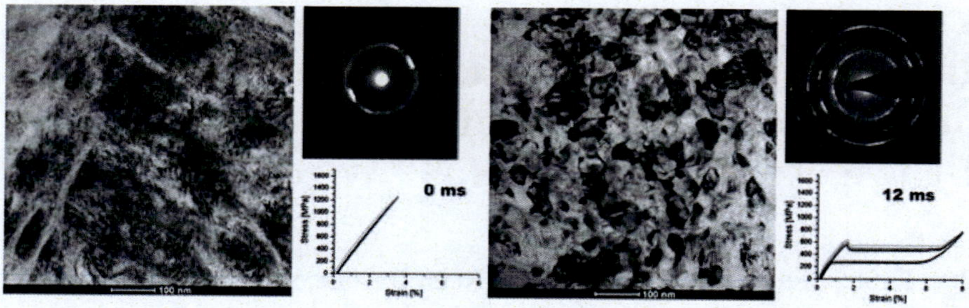


**Fig. 15.1** Microstructures, electron diffraction patterns and tensile stress–strain response of thin Ni-Ti wires just after cold drawing (*left*) and after electropulse heat treatment 125 W/12 ms [10]

Conventionally, this final heat treatment is performed in an environmental furnace [4–6]. Depending on the alloy (chemical composition, cold work) and desired application, heat treatment conditions in the range of temperatures 400–500°C and time 10–60 min [4] are applied. If the shape of the wire is constrained during the heat treatment [5], the stress generated in the wire affects the functional properties. This is the reason why the tensile force applied during the commercial straight annealing treatment of superelastic Ni-Ti wires is a very important technological parameter of the Ni-Ti wire production process.

Recently, the relatively long time of the conventional heat treatment of Ni-Ti wires became an obstacle for emerging technology of textile fabrics knitted or woven using continuous Ni-Ti filaments. Maximum respooling speed of ~1 m/min achievable with straight annealing treatment in conventional ~6 m long tubular electrical furnaces is still painfully slow for this purpose. To solve this problem the possibility to treat continuously long thin Ni-Ti filaments by passing electric current through it during respooling has been investigated as an alternative method to the conventional treatment. The method has been developed based on a series of dedicated studies [7–10] and called "Final Thermo Mechanical Treatment by Electric Current"/FTMT-EC/. Further experimental details can be found in Refs. [8, 9].

As reported in related works [7–10], thin superelastic Ni-Ti filaments heat treated by the FTMT-EC method display excellent functional properties due to the specific nanosized microstructures found by TEM in the treated wires. This suggests that the recovery processes responsible for the functional property setting (polygonisation, crystallization form amorphous, recrystallization, grain growth, plastic deformation, etc.) are capable to change the cold worked microstructure (Fig. 15.1 left) into the annealed one (Fig. 15.1 right) very fast .

Major difference with respect to conventional furnace heat treatment is that the heat comes from inside the thin wire and that the temperature can rise and fall very quickly. The time of heat treatment is in the order of milliseconds instead of minutes. The opportunity to control the fast evolving sample temperature and tensile force by the FTMT-EC method (Fig. 15.2b) has opened the way for detailed investigations of the progress of the recovery processes closer to the rate at which they naturally proceed at high temperatures. This motivated us to perform dedicated

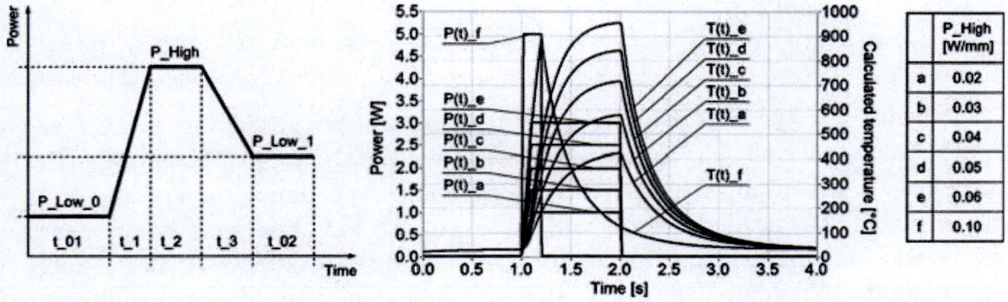


**Fig. 15.2** (**a**) Parameters describing the DC power pulse applied in the FTMT-EC treatment, (**b**) temperature profiles T = T(t) achieved in pulses with parameters $P = 2, 3, 4, 5, 6$ W; $t_2 = 0.9$ s and $P = 10$ W, $t_2 = 0.18$ s

experimental studies during the electropulse treatments by combination of in-situ tensile force measurement, electrical resistance measurement and high speed synchrotron X-ray diffraction with the aim to obtain direct experimental information on the progress of the recovery processes forming nanocrystalline microstructures at heating rates of the order of ~5,000°C/s.

## 15.2 Experiment and Method

### *15.2.1 Heat Treatment*

All experiments were performed on Fort Wayne Metals #1 superelastic as-drawn Ni-Ti wires (56.0 wt.% Ni) having a diameter d = 0.1 mm. The wire is first mounted on the miniature deformation rig especially designed and built for efficient testing of functional thermomechanical properties of thin Ni-Ti filaments in tension. The rig consists of a stepping motor, a 100 N load cell, electrically isolated grips, a Peltier furnace, a laser micrometer for strain measurement and special electronics allowing to send controlled electric power pulse to heat the Ni-Ti wire up to the melting point and perform simultaneously electric resistance measurement.

The initial length $l_0$ (~50 mm) and electrical resistance $\rho_0$ of the wire at room temperature were first evaluated. The wire was then preloaded to a reach starting values of the mechanical constraint $\sigma_0$ and $\varepsilon_0$. In the present experiments, the wire was preloaded to 400 MPa and its length was fixed. The wire was than heated by a controlled DC power pulse P (Fig. 15.2a). It is essential that the electronics system is capable of controlling the desired power P(t), even if the electric resistance of the wire drastically changes during the Joule heating event. Evolution of the wire temperature, T(t), as a function of time (Fig. 15.2b) is calculated according to Eq. 15.1, taking into account the Joule heat supply *P* and the ambient temperature losses.

$$\frac{d}{dt}(T(t) \cdot C) = P - h \cdot (T(t) - T_{ext}) - \varepsilon \cdot \sigma \cdot A \cdot T^4 \quad (15.1)$$

The heat capacity, C, is assumed to be temperature independent, $\sigma$ is a Stefan-Boltzman constant and A is the surface area of the wire. The specific heat transfer coefficient, h, describing the heat dissipation into air per unit time and the NiTi wire emissivity, $\varepsilon$, needed for the calculation of the radiation heat loss per unit time were indentified from series of calibration experiments. Other effects influencing the wire temperature such as heat conduction losses into grips and latent heats were neglected. In some experiments described below, two parameters of the power pulse (maximum power P and pulse duration t_2) are used to represent the temperature history T = T(t) the wire is exposed to. In other experiments, the temperature profile T(t) is directly set by controlling the P(t) and in accordance with Eq. 15.1.

As the temperature of the wire rises and falls during the treatment (Fig. 15.2b), thermally activated recovery processes are triggered and proceed with their characteristic intrinsic kinetics which changes with temperature. The electric resistance of the wire, macroscopic tensile force and X-ray diffraction signal varying in response to the progress of the recovery processes are evaluated. Finally, after the treatment is finished, superelastic response of the treated wire in 10 tensile cycles at room temperature is evaluated using the same stress rig. The parameters determined from the superelastic stress–strain curves and their stability during cyclic tensile loading serve as measures of the achieved functional properties of the treated wires.

### *15.2.2 Synchrotron X-Ray Diffraction*

Synchrotron radiation enables to extend the X-ray measurements from the static to the dynamic regime, thanks to its unique time structure and very high flux even at high energy, which reduces considerably the data collection time. Very fast time resolution down to the picosecond regime is achievable [11] with stroboscopic studies of reversible phenomena. In case of the irreversible microstructure evolution, the X-ray data have to be acquired on a single shot basis [12] requiring high flux source, very fast and sensitive data acquisition system and a high speed data transfer.

The present experiment was performed on ID11 diffractometer at the European Synchrotron Radiation Facility (ESRF) in Grenoble. A monochromatic X-ray beam with energy of 45 keV ($\lambda \sim 0.27552$ Å) is obtained with a Laue monochromator. The beam is focussed down to 1 mm in vertical direction by bending the Laue crystal and down to 100 μm in horizontal direction. Frelon2K 2D camera with taper has been chosen as a fast detection system [12]. In order to speed up the data acquisition for the targeted 10 ms time resolution only a radial slice of the 2D detector (Fig. 15.3a) is used.

The thin Ni-Ti wire mounted between the grips of the stress rig is positioned vertically in the beam (Fig. 15.3a). The heat treatment experiment is performed

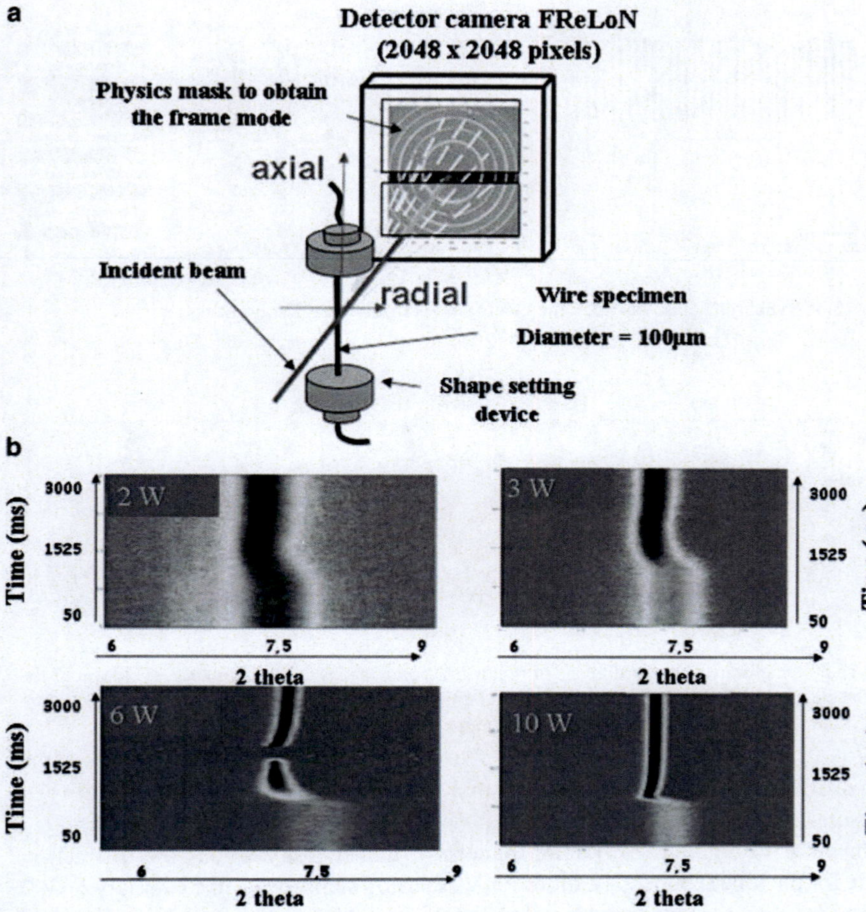


**Fig. 15.3** **(a)** Setup of the in-situ high speed synchrotron X-ray experiment during the heat treatment by electric current. The FReLON camera reading was optimized for high speed data acquisition (only part of the detector – strip of 2048 × 64 pixels – was used to increase the time resolution to ~10 ms); **(b)** Evolution of the {110} austenitic peak measured during heat treatment experiments with $P = 2, 3, 6$ W/0.9 s and $P = 10$ W/0.18 s

with synchronized pulse, force, electrical resistivity and diffraction measurements (Fig. 15.4). A strip of 2048 × 64 pixels is acquired and 64-binned in order to obtain a line of 2048 × 64 pixels. Then, this line is read out of the CCD and stored before next image is acquired. Afterwards, 300 lines are put together into a single 2D binary image which represents the evolution of the diffraction patterns versus time, as illustrated in Figure 15.3b. To minimise the time needed to read out the line, the 64 active pixels have to be chosen at the bottom edge of the camera. The part of the detection area, which is not active, is masked by a lead mask. By using such kinetic mode of the data acquisition by the FReLON camera, it has been possible to acquire

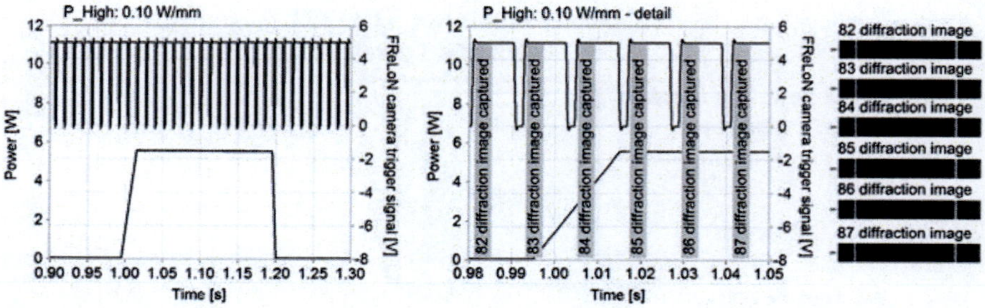


**Fig. 15.4** Synchronisation between the electric power pulse and X-ray measurements

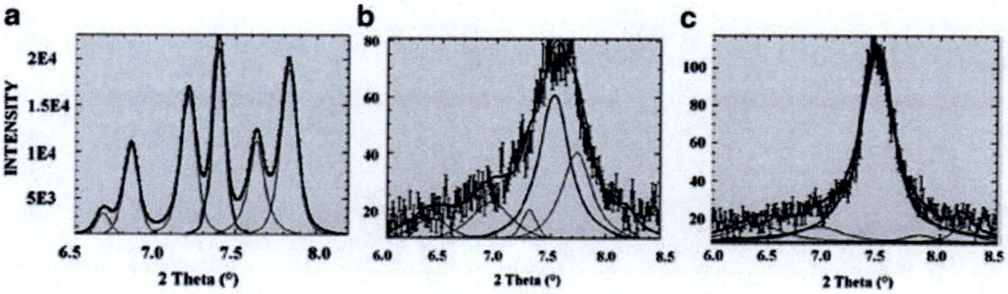


**Fig. 15.5** Quantitative phase analysis of the diffraction patterns to determine the evolution of the height, position and FWHM of the {110} austenitic peak (**a**) Ni-Ti powder; (**b**) Ni-Ti wire before heat treatment; (**c**) Ni-Ti wire after heat treatment

100 diffraction patterns per second with a readout time of less than 1 ms between 2 images.

In order to obtain a diffraction pattern of the intensity versus the diffraction angle for the kinetic mode, a calibration was performed. A very fine capillary of 300 μm with $LaB_6$ powder was placed at the Ni-Ti wire position and 2D diffraction pattern was acquired. The position of each $LaB_6$ peak in pixels was measured using DASH [13] and the corresponding diffraction angle was calculated. The calculated calibration function relating the diffraction angle to the pixel number was used to transform each line of the experimental pattern (intensity versus pixel) into a diffraction pattern (intensity versus diffraction angle). Since the beam centre is in the middle of the detector, the right side of the diffraction pattern is added to its left side to increase the statistics.

In order to obtain an etalon for stress-free Ni-Ti diffraction pattern, Ni-Ti powder of similar composition as that of the wire was placed in the $LaB_6$ container and diffraction pattern was acquired. Since the Ni-Ti powder existed in a mixture of austenite and martensite phases at room temperature, the diffraction pattern contained both austenite and martensite diffraction lines. The Ni-Ti powder gives diffraction pattern with much better resolution (Fig. 15.5a) than the thin as-drawn Ni-Ti wires (Fig. 15.5b). This is due to the complex microstructure of the thin Ni-Ti wires consisting of a mixture of heavily deformed austenite, R phase, martensite, amorphous phase with frozen in internal stresses. After heat treatment yielding ideal superelastic properties, the microstructure of the wire is essentially austenitic (Fig. 15.5c). The diffraction lines were analyzed for height position and FWHM of the {110} austenite peak using the LAMP software [14] as well as the GSAS software [15] considering the structure of the austenite phase Pm-3m, R phase P-3 and martensitic phase P2$_1$/m [1].

## 15.3 Experimental Results

### *15.3.1 Heat Treatment Experiments – Macroscopic Results*

The heat treatment experiments were performed using different power $P = 2, 3, 4, 5, 6$ W ($t_2 = 0.9$ s) and 10 W ($t_2 = 0.18$ s). Same experiments were performed twice with identical results – once with simultaneous *in-situ* X-ray diffraction once without. The results of 3 selected experiments are presented in Fig. 15.6 showing the variations of the supplied electric power, temperature, tensile stress and electric resistivity during the FTMT-EC pulse on the left side and the obtained superelastic

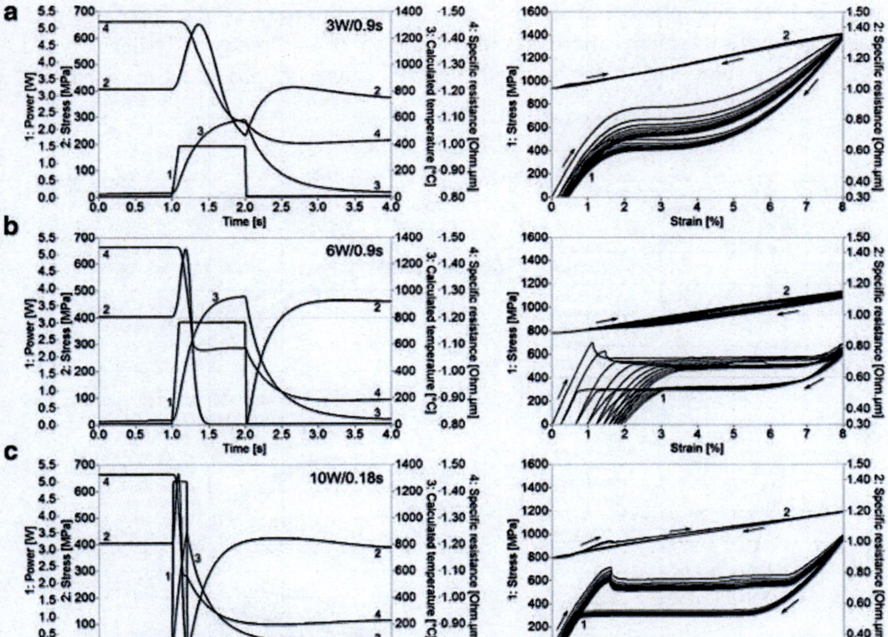


**Fig. 15.6** Evolution of power, temperature, tensile stress and electric resistance (*left*) of the Ni-Ti wire during FTMT-EC treatments with maximum power $P = 3, 6, 10$ W/100 mm and resulting superelastic response and electric resistivity records of the treated wires (*right*)

stress–strain response with superimposed variation of the electric resistivity of the wire on the right side. The observed evolution of tensile stress and electric resistivity in response to the prescribed temperature evolution during the treatment (Fig. 15.2b) is analyzed below.

The tensile stress evolves with time (temperature) due to the several mutually competing processes: (1) relaxation of residual stress, (2) thermal expansion, (3) plastic deformation processes, (4) reversible phase transformations. The tensile stress starts to increase with increasing temperature right after the onset of heating (Fig. 15.6). This is due to the unlocking of elastic deformation held by internal stresses in the heavily cold worked microstructure of the as-drawn Ni-Ti wire coupled with reverse transformation of small fraction of the martensite phase which retransforms back to the austenite phase upon heating. The stress increases with the increasing temperature up to ~450°C, where it reaches a maximum of ~690 MPa (Figs. 15.6 and 15.7b) and falls ultimately down to zero if the maximum temperature reached in the pulse is high enough. This is due the plastic deformation and thermal expansion processes prevailing at high stress and temperature. Upon cooling, the stress varies simply due to thermal expansion.

The electric resistivity starts to decrease when the temperature reaches ~200°C (Figs. 15.6 and 15.7) and decreases monotonically during heating until it reaches a plateau at ~75% of its starting value. The electric resistivity changes with the temperature due to: (1) intrinsic thermal dependence of the resistivity of a metallic wire, (2) reversible phase transformations and (3) progress of the recovery processes causing irreversible microstructure changes. If an already fully heat treated Ni–Ti wire is subjected to the same FTMT-EC pulse [7], the electric resistivity

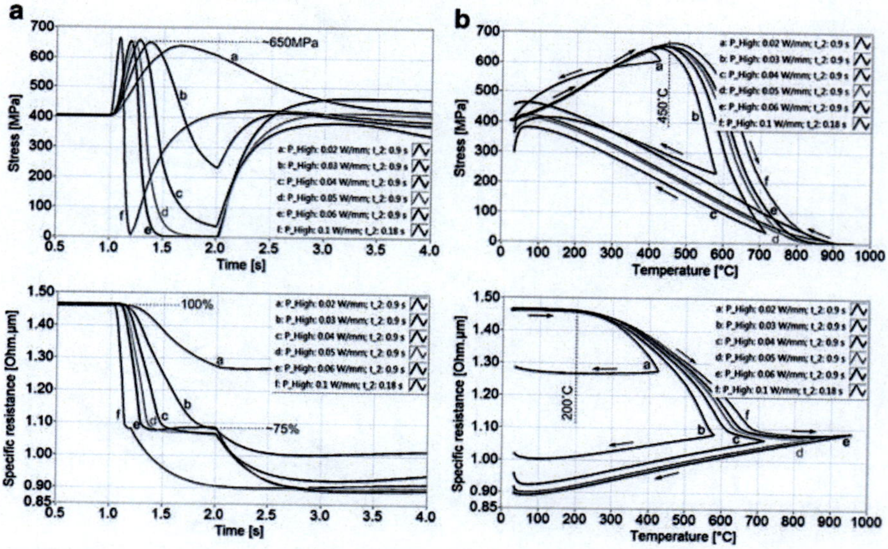


**Fig. 15.7** Comparison in experiments with maximum power $P = 2, 3, 4, 5, 6$ W/0.9 s and 10 W/0.18 s of **(a)** the evolution of the tensile stress (*left*) and the electrical resistivity (*right*) with time **(b)** the evolution of tensile stress (*left*) and electrical resistivity (*right*) with temperature

monotonically increases with increasing temperature and returns to its original value after cooling back to room temperature. This is a simple proof that the electric resistance decrease is caused by the recovery processes (3). Upon cooling, the electrical resistivity decreases due to intrinsic thermal dependence of the electric resistance of a metal (Fig. 15.7). Overall, the total decrease of the wire electric resistance due to microstructure recovery reaches ~40% – from 1.45 to 0.88 Ωμm for the 6 W/0.9 s treatment.

### *15.3.2 Heat Treatment Experiments - In-Situ X-Ray Diffraction Results*

Figure 15.3b shows the evolution of the diffraction pattern near the {110} austenitic peak as measured by the 2D detector during the experiments using pulses with power $P = 2$, 3, 6 and 10 W (see also the synchronization in Fig. 15.4). It evolves from the broad diffraction pattern corresponding to the complex microstructure of the as-drawn wire to the well-defined narrow peak of the heat treated wire. The intensity of the peak increases, the peak shifts left and right and its FWHM decreases during the electric pulse treatment. Figure 15.8a shows the evolution of the height of the {110} austenitic peak during the experiments. The higher the temperature reached in the treatment, the larger the peak height is. The shift in the onset of the increase of the peak height to shorter time with increasing P is due to the fact that the temperature rises faster for more energetic pulses (Fig. 15.2b).

Figure 15.8b shows the results of the analysis of the evolution of {110} austenite peak position recalculated into radial strain (Eqs. 15.2–15.4) of the {110} austenitic peak for treatment with $P = 2$, 3, 6 and 10 W pulses. Radial strain means elastic strain in transverse direction. The axial strain would also be of interest but it could not be evaluated due to the geometry of the diffraction experiment (Fig. 15.3a). Using the Bragg law (Eq. 15.2) and the definition of strain (Eq. 15.3) an expressionfor the radial strain of the {110} austenitic peak $\varepsilon_{110}$ is given by Eq. 15.4.

$$2d_{hkl}.\sin\theta_{hkl} = n.\lambda \tag{15.2}$$

$$\varepsilon_{hkl} = \frac{d_{hkl} - d_{0,hkl}}{d_{0,hkl}} = \frac{\Delta d_{hkl}}{d_{0,hkl}} \tag{15.3}$$

$$\varepsilon_{110} = 1 - \frac{\sin\theta_{110}}{\sin\theta_{0,110}} \tag{15.4}$$

$\theta_{110}$ is the value of the Bragg angle of the {110} austenitic peak in the wire during the heat treatment and $\theta_{0;110}$ is the value of the Bragg angle {110} austenitic peak measured on the powder. Because the radial strain is strongly negative before the heat treatment experiment (Table 15.1), it means that the {110} austenite lattice of the

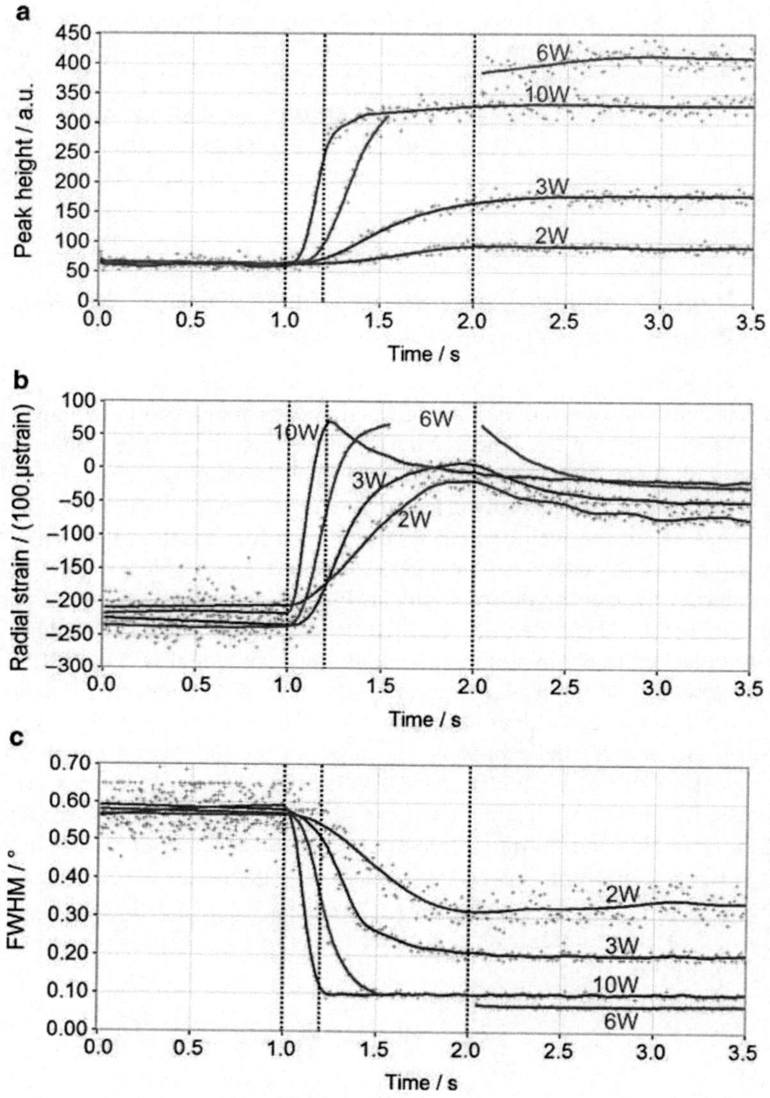


**Fig. 15.8** Evolution of (**a**) the height, (**b**) the position (radial strain), (**c**) the FWHM of the austenite given by the {110} austenitic peak during experiments at $P = 2, 3, 6$ W/0.9 s and 10 W/0.18 s

as-drawn wires experienced a tensile stress along its axis. Assuming a Young modulus E = 60 GPa and a Poison ratio of 0.33, the stress is evaluated to 2,290 MPa. Subtracting the 400 MPa external tensile stress, the tensile internal stress (corresponds to 3.15% tensile elastic strain) existing in the as-drawn wire prior the heat treatment comes down to 1,890 MPa. If the as-drawn wire is heated in stress free conditions,

**Table 15.1** Key features of the evolution of the peak position in heat treatment experiments

| Power | 2 W | 3 W | 6 W | 10 W |
|---|---|---|---|---|
| Onset of the peak position change observed at (ms) | 1,260 | 1,224 | 1,116 | 1,020 |
| Maximum radial strain observed (ms) | 1,920 | 1,716 | | 1,188 |
| Radial strain value before treatment (μstrain) | −12,600 | −12,600 | −12,600 | −12,600 |
| Maximum radial strain value (μstrain) | 1,140 | 1,055 | | 7,290 |
| Radial strain after treatment (μstrain) | −6,225 | −5,175 | −2,755 | −2,145 |

it becomes ~3% shorter and the radial strain equals to zero after the treatment. As the temperature increases during the heat treatment, the austenite lattice expands in radial direction due to two processes (1) isotropic thermal expansion and (2) directional decrease of the internal tensile stress. Recall that the wire is exposed to oscillating external tensile stress (Fig. 15.7a) which affects the radial strain as well. The radial strain increases and decreases but does not return to its original starting value $-12600.10^{-6}$ μstrain in any of the treatments (Table 15.1).

Figure 15.8b evidences how internal tensile stress frozen in the wire after the final cold drawing reduction becomes relaxed during the FTMT-EC treatment. The radial strain $-2145.10^{-6}$ μstrain measured after the 10 W/0.18 s treatment (Table 15.1) gives a tensile stress of 390 MPa which nearly equals the external tensile stress. This suggests that the internal tensile stress does not exist in the 10 W/0.18 s and 6 W/0.9 s treated wires but it still partially remains in the wires treated with 2 W/0.9 s and 3 W/0.9 s pulses.

Figure 15.8c shows the results of the analysis of the evolution of the width of the {110} austenitic peak (FWHM) for experiments with $P = 2, 3, 6$ and 10 W pulses. FWHM is equal to 0.35° before the heat treatment. It decreases with increasing temperature mainly due to the recovery of the crystal defects and related strain fields existing in the cold worked microstructure, crystallization of amorphous and/or due to the increase of the subgrain size. While in case of the $P = 10$ W and 6 W heat treatments, the decrease of the FWHM from 0.35° to ~0.1° corresponds to well annealed microstructure, the decrease to only 0.2° in case of the 3 W treatment, suggests that this microstructure probably still contains many defects and related strain fields. In case of the $P = 2$ W/0.9 s treatment, the FWHM almost does not change suggesting that the defects and strain fields essentially remained in the microstructure.

## 15.4 Discussion

The recovery of the cold worked microstructure of the Ni-Ti wire takes place as the temperature increases from 200°C to 700°C during the heat treatment. The question is whether it is possible to control the progress of the lattice recovery processes with a sufficient precision. In conventional furnace treatment, the lattice recovery processes occur in uncontrolled manner as soon as the wire enters the pre-heated furnace.

Obviously, precise control of the progress of the lattice recovery processes having fast kinetics (~ms) is not possible by environmental furnace treatment. In the FTMT-EC treatment, the chance is better since the temperature profile T(t) is prescribed.

In addition to the recovery processes, there are diffusion processes characterised by much slower kinetics (min) such as precipitation or dissolution of precipitates which may proceed during the exposure of the sample to elevated temperatures and affect the functional properties of the wire as well [6]. Hence, in conventional furnace treatments, it is not possible to find out whether the change of functional properties is due to recovery processes or/and to diffusion processes. Since the activity of the diffusional processes is largely suppressed by the short time FTMT-EC treatments, it is possible to establish the *heat treatment parameters – microstructure – functional property* relationship for heat treated Ni-Ti wires [7, 10].

Obviously, the principal technological goal was to find out the parameters of the FTMT-EC treatment which leads to optimum superelastic functional properties of the treated Ni-Ti wires. It comes out that, among the performed treatments, the optimal ones would be the 5 W/0.9 s or 10 W/0.18 s. The 3 W/0.9 s treatment is not sufficient and the 6 W/0.9 s treatment results in a loss of stability of the superelastic behaviour due to plastic deformation accompanying the stress induced phase transformation during tensile cycling. Much more detailed results concerning the functional property setting of Ni-Ti wires were reported in [8]. Detailed TEM investigation of microstructures in FTMT-EC treated wires confirmed that the functional properties of Ni-Ti wires can be very well correlated with the microstructures [10] and that slip deformation leading to unstable cyclic response takes place during tensile cycling of overtreated wires [16]. The microstructure corresponding to optimum superelastic response of the wire is a partially polygonized/recrystallized microstructure with grain size 20–50 nm (Fig. 15.1 right). Such microstructure was found in a wire treated by 125 W/0.012 s pulse [10]. The 5 W/0.9 s and 10 W/0.18 s treated wire are assumed to contain a very similar microstructure although there was no direct observation.

Five recovery processes (Table 15.2) are expected to become subsequently active during the heat treatment of as-drawn Ni-Ti wires. Though the same processes take place also during conventional heat treatments in a furnace, there are following differences that need to be emphasized: (1) fast rate of the FTMT-EC treatment possibly leading to overheating which differently affects the recovery processes with different kinetics, (2) mechanical stress action on the recovery processes, (3) the possible direct action of passing electrons on the recovery processes. It is assumed that these three FTMT-EC specific features assist the formation of the desired polygonized/recrystallized nanograin microstructure [7, 10] in the FTMT-EC treated Ni-Ti wires.

The goal of this work was to obtain more detailed information on the lattice recovery processes with the help of the high speed synchrotron X-ray diffraction experiments. Particularly, the goal was to interpret the apparently curious *in-situ* electric resistance-time and stress-time responses (Figs. 15.6 and 15.7) in terms of the activity of various recovery processes subsequently taking place during the heat

**Table 15.2** Recovery processes expected to occur during electropulse treatment of Ni-Ti wires take place in sequence I–V as the temperature increases and those dominating in given temperature range are in *bold*

| | Recovery process | Temperature range |
|---|---|---|
| I | **Residual stress relaxation** and reverse martensitic transformation | 20–500°C |
| II | **Polygonization** and crystallization from amorphous phase and subgrain growth | 200–700°C |
| III | **Recrystallization** and grain growth and plastic deformation | 500–900°C |
| IV | **Plastic deformation** and grain growth | > 800°C |
| V | **Grain growth** | > 1,000°C |

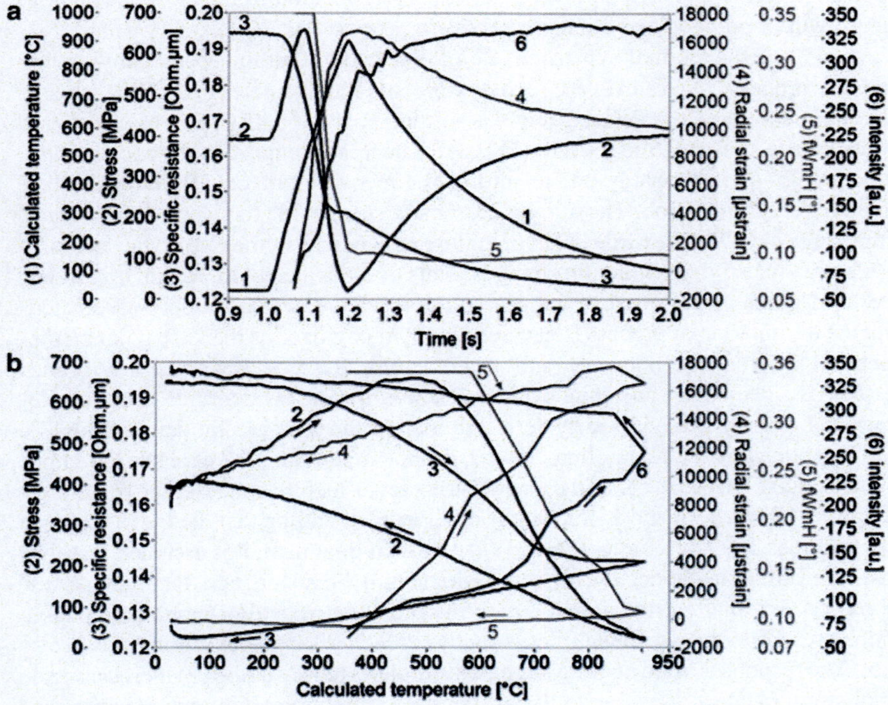


**Fig. 15.9** In-situ macroscopic and diffraction results of the $P$ =10 W/0.18 s experiment plot together in dependence on (**a**) time, (**b**) temperature

treatment. To synthesize the obtained results, the macroscopic and diffraction results from 10 W/0.18 s treatment were plotted together in Fig. 15.9.

The tensile stress (curve 2) increases from the onset of heating and electric resistance (3) starts to fall at 200°C before any change of the diffraction signal can be detected. However, one has to admit that the experimental error related to {110} peak analysis at temperatures up to 300°C (Fig. 15.5b) is too large to determine the peak parameters with sufficient precision. The increase of the tensile stress is thus taken as a main evidence for the recovery process I. This process is largely reversible if the maximum achieved temperature remains relatively low (see the stress-temperature response during the 2 W/0.9 s treatment – curve a in Fig. 15.7). It can be deduced that the process I dominates over the linear thermal expansion up to ~450°C, otherwise the stress would have to decrease with increasing temperature due to thermal expansion. Upon further heating, the electrical resistivity (3) falls, peak height (6) and radial strain (4) starts to increase at ~360°C. It is assumed that the decrease of electrical resistivity starting at ~200°C evidences the beginning of the irreversible microstructure change due to recovery of lattice defects through migration, redistribution and annihilation of point defects and migration of dislocation dipoles into cell walls called polygonization, crystallization from amorphous phase and growth of polygonized domains (recovery process II).

At ~450°C, the thermal expansion and/or plastic deformation accompanying the recovery process II takes over the recovery process I and tensile stress (2) reaches a maximum and starts to decrease. Right after that, at about 600°C, together with the falling tensile stress (2), the FWHM (5) starts to decrease suggesting the accelerating defect annihilation probably due to start of the recovery process III dominated by dynamic recrystallisation. The alloy is extremely susceptible to plastic deformation at this stage. At ~700°C, while the temperature still increases, the rate of the decrease of electric resistivity suddenly changes. This is considered to be an important threshold point in the microstructure evolution which can be roughly associated with the termination of process II. It can be easily recognized as the knee point on the electrical resistivity response. Upon further heating, the recovery processes III and IV (FWHM decreases, austenite volume fraction increases) continue. The tensile stress had already reached nearly zero terminating the wire plastic deformation. It shall be noticed that, after cooling down to room temperature, the wire has the same length as it had before the test (the tensile stress levels before and after the treatment are similar) (Figs. 15.6 and 15.7). On the other hand, recalling that the free as-drawn wire becomes 3% shorter after the 10 W/0.18 s heat treatment, it is assumed that the plastic deformation did occur even if the wire length did not change. It compensated the relaxed internal elastic strains due to internal stresses and/or martensite phase strains present in the as-drawn microstructure which disappeared during the treatment. The responses upon cooling are much simpler – tensile stress (2) increases and radial strain (4) decreases due to linear thermal contraction taking place under the condition of the constant sample length, electric resistivity (3) decreases linearly due to conventional linear dependence on temperature. As the temperature decreases below ~70°C, the stress (electrical resistivity) starts to decrease (increase) again resulting in local maxima (minima) recorded at temperature 50°C on the respective responses in Figs. 15.7 and 15.9. This is due to the B2-R transformation taking place upon cooling in the already heat treated wires [8]. The situation upon cooling is somehow more complicated in case of lowest power treatments which did not result in complete relaxation of internal stresses (Fig. 15.7).

Although it was possible to follow the progress of the sequential lattice recovery processes at heating rates reaching up to 5,000°C/s, it is difficult to discuss the kinetics of individual lattice recovery processes, since these are still fast enough to nearly follow the prescribed T(t) profile. Nonetheless, the fact that individual responses in Fig. 15.7 do not follow same path upon heating is a clear experimental evidence that the intrinsic kinetics of individual processes matters. In case of the 125 W/0.012 s treatment [10], in which the heating rate is ~80,000°C/s, the intrinsic kinetics of the lattice recovery processes plays a much more important role.

## 15.5 Conclusion

Superelastic functional properties of Ni-Ti wires can be precisely set by the recently developed nonconventional electropulse treatment. In order to learn about the recovery processes responsible for that, variations of tensile force, electrical resistance and synchrotron X-ray diffraction signal from 0.1 mm thin as-drawn Ni-Ti wire prestrained in tension were recorded simultaneously during the short time electric pulse treatment. The data were used to obtain direct experimental information on the phase fractions, internal stresses and defects in the microstructure fast evolving in response to the prescribed temperature and tensile stress in the treated wire.

**Acknowledgments** The authors acknowledge the support of ESRF for performing the in-situ synchrotron experiment (MA-358) and support from projects AV0Z10100520, IAA200100627.

## References

1. Otsuka K, Ren X (2005) Physical metallurgy of Ti-Ni based shape memory alloys. Prog Mater Sci 50:511–678
2. Inaekyan K, Brailovski V, Prokoshkin S, Korotitskiy A, Glezer A (2009) Characterization of amorphous and nanocrystalline Ti-Ni-based shape memory alloys. J Alloy Compd 473:71–78
3. Schaffer JE (2009) Structure–property relationships in conventional and nanocrystalline NiTi intermetallic alloy wire. J Mater Eng Perform 18:582–587
4. Duerig TW, Melton KN, Stöckel D, Wayman CM (1990) Engineering aspects of shape memory alloys. Butterworth-Heinemann, London
5. Liu X, Wang Y, Yang D, Qi M (2008) The effect of ageing treatment on shape-setting and superelasticity of a nitinol stent. Mater Charact 59:402–406
6. Undisz A, Fink M, Rettenmayr M (2008) Response of austenite finish temperature and phase transformation characteristics of thin medical-grade Ni–Ti wire to short-time annealing. Scripta Mater 59:979–982
7. Malard B, Pilch J, Šittner P, Gartnerova V, Delville R, Schryvers D, Curfs C (2009) Microstructure and functional property changes in thin Ni–Ti wires heat treated by electric current – high energy X-ray and TEM investigations. Funct Mater Lett 2:45–54
8. Pilch J, Heller L, Šittner P (2009) Final thermomechanical treatment of thin NiTi filaments for textile applications by electric current. Proceedings of the ESOMAT09, EDP Sciences, 05024

9. patent application PCT/CZ2010/000058
10. Delville R, Malard B, Pilch J, Sittner P, Schryvers D (2009) Microstructure changes during non conventional heat treatment of thin Ni-Ti wires by pulsed electric current studied by transmission electron microscopy. Acta Mater 58:4503–4515
11. Plech A,Wulff M, Bratos S, Mirloup F, Vuilleumier R, Schotte F, Anfinrud PA (2004) Visualizing chemical reactions in solution by picosecond x-ray diffraction. Phys Rev Lett 92:125505(4)
12. Labiche J, Mathon O, Pascarelli S, Newton M, Ferre G, Curfs C, Vaughan G, Homs A, Carreiras D (2007) Invited article: the fast readout low noise camera as a versatile x-ray detector for time resolved dispersive extended x-ray absorption fine structure and diffraction studies of dynamic problems in materials science, chemistry, and catalysis. Rev Sci Instr 78:091301(11)
13. David WIF, Shankland K, Van de Streek J, Pidcock E, Motherwell WDS, Cole JC (2006) DASH: a program for crystal structure determination from powder diffraction data. J Appl Crystallogr 39:910–915
14. http://www.ill.fr/data_treat/lamp/front.html/. Accessed 3 January 2010
15. http://www.ccp14.ac.uk/solution/gsas/. Accessed 3 January 2010
16. Delville R, Malard B, Pilch J, Sittner P, Schryvers D (2010) Transmission electron microscopy study of dislocation slip activity during superelastic cycling of NiTi. Int J Plasticity, doi: 10.1016/j.ijplas.2010.05.005

# Chapter 16
# Emanation Thermal Analysis as a Method for Diffusion Structural Diagnostics of Zircon and Brannerite Minerals

**Vladimír Balek, Iraida M. Bountseva, and Igor von Beckman**

## 16.1 Emanation Thermal Analysis of Solids

Emanation thermal analysis (ETA) [1–3] based on the measurement of the radon release from samples, is one of the methods used in the diffusion structure diagnostics of solids. Changes in surface morphology and microstructure of solids during their thermal treatments and changes due to chemical, mechanical or radiation interactions can be studied by the emanation thermal analysis method.

As most of the solids to be investigated do not naturally contain atoms of radon it is necessary to introduce the radon atoms in the samples prior to the ETA measurements.

To introduce the radioactive trace $^{220}$Rn into solids, the samples are labelled by parent radio-nuclides $^{228}$Th and $^{224}$Ra, serving as a quasi-permanent source of radon atoms $^{220}$Rn. The used specific activity of the parent radionuclide $^{228}$Th is in the order $10^5$ Bq g$^{-1}$ of the sample.

### *16.1.1 Radon Atoms Implantation into Solids by the Recoil Energy of α-Decay*

Atoms of the $^{220}$Rn radon radionuclide are formed by a spontaneous α-decay of $^{228}$Th and $^{224}$Ra radio-nuclides according to scheme Eq. 16.1

V. Balek (✉)
Nuclear Research Institute Řež, plc, 250 68 Řež, Czech Republic
e-mail: bal@ujv.cz

I.M. Bountseva and I. von Beckman
Faculty of Chemistry
Moscow MV Lomonosov State University, Moscow 199234, Russia

J. Šesták et al. (eds.), *Glassy, Amorphous and Nano-Crystalline Materials*, Hot Topics in Thermal Analysis and Calorimetry 8, DOI 10.1007/978-90-481-2882-2_16,

$$^{228}\text{Th} \xrightarrow[T_{1/2}=1.9\text{y}]{\alpha} {}^{224}\text{Ra} \xrightarrow[T_{1/2}=3.8\text{y}]{\alpha} {}^{220}\text{Rn} \xrightarrow[T_{1/2}=55\text{s}]{\alpha} \tag{16.1}$$

and the $^{220}$Rn atoms can be introduced into the solid owing to the recoil energy of the spontaneous $\alpha$-decay (85 keV/atom). The samples can be labelled by using an adsorption of traces of $^{228}$Th as nitrate from a solution.

Due to the energy of the spontaneous $\alpha$-decay of $^{228}$Th and $^{224}$Ra radio-nuclides the atoms of $^{220}$Rn can penetrate into the sample several tens of nanometres from the surface depending on the composition of the materials.The values of the maximum penetration depths of $^{220}$Rn were determined by the Monte Carlo method using TRIM code [4], e.g. for $SiO_2$: 65.4 nm, for zircon ($ZrSiO_4$): 60 nm and brannerite mineral ($U_{1-x}Ti_{2+x}O_6$ ): 60 nm.

Radon atoms can be trapped in solids at structure defects such as vacancy clusters, grain boundaries and pores. The structure defects in the solids can serve both as traps and as diffusion paths for radon atoms.

### *16.1.2 Mechanisms of the Radon Release from Solids*

The radon formed by the spontaneous $\alpha$-decay of $^{224}$Ra may escape from the solid either by recoil energy ejection or by diffusion. The term emanation rate, *E*, has been used to express the release of radon from solids. It is defined as the ratio of the radon release rate to the rate of radon formation by the spontaneous $\alpha$-decay of $^{228}$Th and $^{224}$Ra in the investigated solids.

It has been determined experimentally (in relative units) as $E = A_\alpha/A_{\text{total}}$, where $A_\alpha$ is the $\alpha$-radioactivity of radon released in unit time from the labelled sample and $A_{\text{total}}$ is the total $\gamma$ -radioactivity of the labelled sample. The $A_{\text{total}}$ value is proportional to the rate of radon formation in the sample.

In the evaluation of the radon release from solids several mechanisms have been supposed, namely the radon release by recoil mechanism, the diffusion in open pores, and the volume diffusion mechanism.

The experimentally obtained values of the emanation rate, E, can be considered as:

$$\text{E} = \text{E(recoil)} + \text{E(pores)} + \text{E(solid)} \tag{16.2}$$

The emanation rate due to recoil, E (recoil), can be expressed as

$$\text{E(recoil)} = \text{K}_1\text{S}_1 \tag{16.3}$$

where $K_1$ is a temperature independent constant, proportional to the penetration depth of recoiled radon atoms in solids investigated and $S_1$ is external surface area of sample particles. The path of recoiled atoms of radon is dependent on the "nuclear stopping power" of the sample material.

The emanation rate due to diffusion in pores, E (pores), is expressed as

$$E(\text{pores}) = K_2 S_2 \quad (16.4)$$

where $K_2$ is a constant that depends on temperature and $S_2$ is internal surface area of the sample depending on the surface of open pores, cracks and intergranular space.

The emanation rate due to volume diffusion mechanism, E (solid), is expressed as

$$E(\text{solid}) = K_3 \exp(-Q/2\mathbf{R}T) S_3 \quad (16.5)$$

where $K_3$ is a constant related to the atomic properties of the lattice, Q is the activation energy of Rn diffusion in the solid, $S_3$ is surface area, **R** is molar gas constant, and T is temperature.

The growth of the emanation rate values, E (T), may characterize an increase of the surface area of interfaces, whereas a decrease in the E(T) may reflect processes like closing up structure irregularities that serve as paths for the radon migration, closing pores and/or a decrease in the surface area of the interfaces [3–6].

## 16.2 Application of the Emanation Thermal Analysis in the Diffusion Structural Diagnostics of Solids

The equipment for the emanation thermal analysis (ETA) was developed in the 1960s [1, 2]. Since that time the ETA method was used in various investigations, e.g. the re-crystallization of solids, annealing of structure defects and changes in the defect state of both crystalline and amorphous solids, sintering, phase changes, the characterization of surface and morphology changes accompanying chemical reactions in solids and on their surfaces, including the thermal degradation, solid–gas, solid–liquid, and solid–solid interactions [7–11]. The ETA made it possible to reveal even fine changes in poorly crystalline or amorphous solids. Differences in the morphology and behaviour of samples prepared by the sol–gel technique under different conditions were revealed by the ETA.

Changes in defects annealing and pore sintering of the samples were characterized by using the ETA results under in situ conditions of their heat treatments. The determination of optimized conditions for the preparation and thermal treatments of advanced ceramic materials was achieved [10, 11]. By this way the ETA results contributed to the solution of practical tasks in the materials technology.

Recently, this method made it possible to characterize the thermal stability of ceramic materials designed for the immobilization/encapsulation of high level radioactive waste [12]. Moreover, the thermal stability of self- irradiated amorphous minerals that serve as natural analogues of the ceramic matrices was evaluated by using the emanation thermal analysis.

In the study of the self- irradiated metamict materials the ETA measurements were carried out by using modified NETZSCH DTA-ETA equipment, Type 404. Details of the measurements and the data treatments are described elsewhere [2–6]. During the ETA measurements the samples were heated at the rate of 6 K $min^{-1}$ in air (zircon sample) and argon (brannerite sample). The specific activity of the labelled samples was $10^5$ Bq $g^{-1}$.The used amounts of the samples were 0.02 and 0.1 g respectively. During the ETA measurements, a constant flow of the carrier gas (air, nitrogen, or another gas) has been used to take the radionuclide of radon of $^{220}Rn$ released by the sample into the detector of α-activity of radon (semiconductor detectors).

### *16.2.1 Thermal Behaviour of Natural Zircon Mineral*

Natural zircon mineral (general formula $ZrSiO_4$), containing an average concentration up to 0.4% of uranium and 0.2% of thorium, has attracted much interest from both fundamental and technological view points. The α-particles and heavy recoil nuclei released during the decay of radioactive impurities (typically $^{238}U$, $^{235}U$ and $^{232}Th$) interact with the surrounding crystalline matrix displacing atoms from their equilibrium positions [13]. Over geological periods of time this process disrupts the crystalline order to such a point that specimens covering all the stages from fully crystalline to amorphous can be found, depending on the uranium/thorium content.

Understanding the radiation effects in crystalline zircon and the determination of the structure of the aperiodic state are essential to ensure the reliability of zircon based ceramics for nuclear waste disposition [13, 14]. During nuclear disintegration, the emission of the α-particle is accompanied by a recoil nucleus. Amorphization taking place in natural zircon is called metamictization. The α-particles have energy of 4–6 MeV, and almost all the energy is dissipated by the ionization processes. It is believed that various isolated defects, such as Frenkel pairs, are formed along their paths. A number of studies have been devoted to structural changes of zircon under irradiation, in particular to understanding the amorphization and/or metamictization process [15]. This process can lead to an increased solubility and fracturing [16].

Ceramic forms used in the encapsulation of nuclear waste are subjected to a similar transformation, with the corresponding variation of their physical and chemical properties. An understanding of radiation effects in crystalline zircon and a determination of the structure of the aperiodic state are essential to ensure the reliability of zircon and related ceramics for nuclear waste disposition [13, 14].

Zircon ceramics can incorporate significant amount of $UO_2$, $PuO_2$ or $ThO_2$ in a solid solution with $ZrO_2$. The zircon undergoes an amorphization promoted by α-decay events of radiogenic elements. During the nuclear disintegration, the emission of an α-particle is accompanied by a recoil nucleus and ballistic collisions of the recoil nucleus cause displacement cascades. A number of studies have been devoted to the evolution of amorphous zircon under irradiation, in particular to understanding of the metamictization process [15].

Natural zircon mineral sample characterized by ETA was from the locality Sri Lanka. The sample was X-ray amorphous [16].

Figure 16.1 shows ETA results of the zircon mineral sample measured on heating (curve 1a) in air flow in the temperature range 20–1,100°C and subsequent cooling (curve 1b). The increase of the emanation rate, *E*, observed in the temperature range of 170–250°C characterized the diffusion mobility of radon atoms along surface cracks and other subsurface defects, the subsequent decrease of the *E* values in the range 250–420°C can be ascribed to healing the surface and subsurface defects.

We supposed that the increase of the emanation rate, *E*, in the range 420–750°C is due to the radon diffusion along structure irregularities in the amorphous zircon. The phase transformation of initially amorphous zircon was characterized by the decrease of the emanation rate values *E* in the range 750–950°C. From ETA results of the amorphous zircon mineral sample measured on heating to 1,200°C and subsequent cooling to room temperature it followed that the microstructure changes taking place in the sample on heating were irreversible.

The results of DSC measured on a parallel sample of amorphous/metamict zircon are demonstrated in Fig. 16.1 as the full line curve. The transformation of amorphous zircon to the crystalline zircon was characterized by a DSC exothermal effect with the maximum at 918°C.

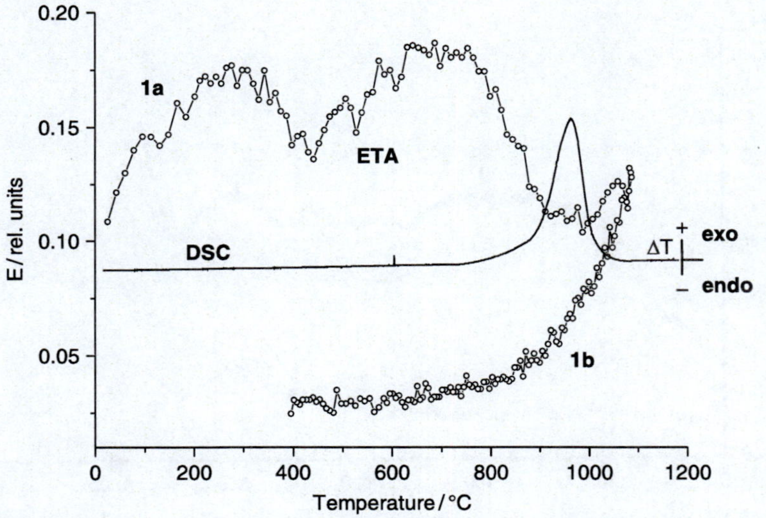


**Fig. 16.1** ETA results (points) of natural zircon mineral sample measured on heating (curve 1a) and subsequent cooling (curve 1b) in air in the range 20–1,100°C. The DSC results measured on the sample heating are depicted as the *full line curve*

### 16.2.2 Thermal Behaviour of Natural Brannerite Mineral

Brannerite mineral (general formula $U_{1-x}Ti_{2+x}O_6$) has been found in nature as amorphous due to α-decay damage caused by high content of U, Th. The formula of natural brannerite can also be written $(U, Th)_{1-x}Ti_{2+x}O_6$. The natural brannerite generally contains impurity elements like Pb, Ca, Th, Y and rare earth elements (REE) on the U-site and Si, Al and Fe on the Ti-site. The brannerite is a minor phase in titanate-based ceramics designed for the geological immobilization of surplus Pu [17, 18]. Therefore, it was of interest to investigate the thermal behaviour of the metamict brannerite mineral as a natural analogue of the brannerite ceramics to be used for immobilization of hazardous radioactive elements.

The diffusion structural diagnostics based on the results of emanation thermal analysis (ETA) made it possible to characterize the annealing of the structure irregularities in the brannerite mineral sample on heating to various temperatures up to 1,200°C. Natural brannerite mineral was from the locality El Cabril mine near Cordoba, Spain. The sample was X-ray amorphous and contained Ca, Pb and other impurity elements [17].

Figure 16.2 shows the ETA results of the metamict brannerite mineral measured during heating in argon in the range 20–1,200°C and subsequent cooling. The increase of emanation rate, *E*, observed on the sample heating in the range of 40–300°C characterized the diffusion mobility of radon atoms along surface cracks and other subsurface defects to depth of 60 nm.

The slight decrease of *E(T)* observed in the temperature range of 400–500°C (curve 1a, Fig. 16.2) was ascribed to healing surface cracks and voids. The decrease

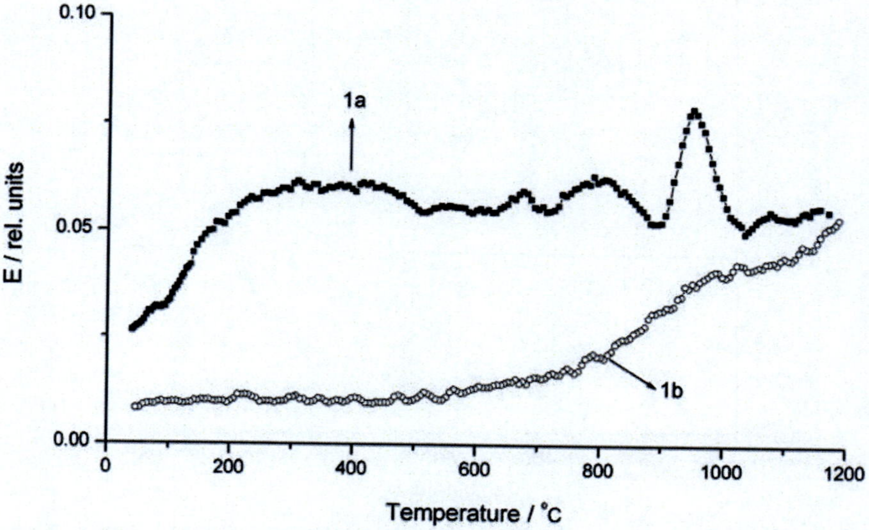


**Fig. 16.2** ETA results of natural brannerite mineral sample measured on heating (curve 1a) and subsequent cooling (1b) in argon in the range 2–1,200°C

of the emanation rate *E(T)* observed on the ETA curve in the range 800–880°C, corresponding to the healing microstructure irregularities, was considered as a first step of the formation of crystalline brannerite. The increase of *E(T)* observed in range 900–965°C followed by the sharp decrease of *E(T)* in the range 970–1,020°C indicated the formation of the crystalline brannerite phase, as confirmed by XRD spectroscopy [17, 19]. From ETA results of the brannerite mineral sample measured on heating up to 1,200°C and subsequent cooling it followed that the microstructure changes on sample heating are irreversible.

The release of $CO_2$ was detected by mass spectrometry of evolved gases in the temperature range 700–800°C [17] due to the thermal degradation of minor carbonate containing components of the sample. The release of $CO_2$ gave rise to the sample porosity [17]. It was of interest to investigate the self-irradiated metamict brannerite mineral during "step by step" heating and subsequent cooling of the sample to the temperatures of 300, 550, 750, 880, 1,020 and 1,150°C, respectively. Results of ETA measured by the "step by step" heating runs (Fig. 16.3) made it possible to compare the annealing of microstructure irregularities of the sample in the selected temperature intervals.

As it follows from the ETA results in Fig. 16.3, the "step by step" heating of the sample to these temperatures caused a decrease of the amount of structure irregularities serving as radon diffusion paths.

A good reproducibility of the ETA results measured on heating from 20°C to 300°C is obvious from the comparison of the results in Fig. 16.3, curves 2a and 1a.

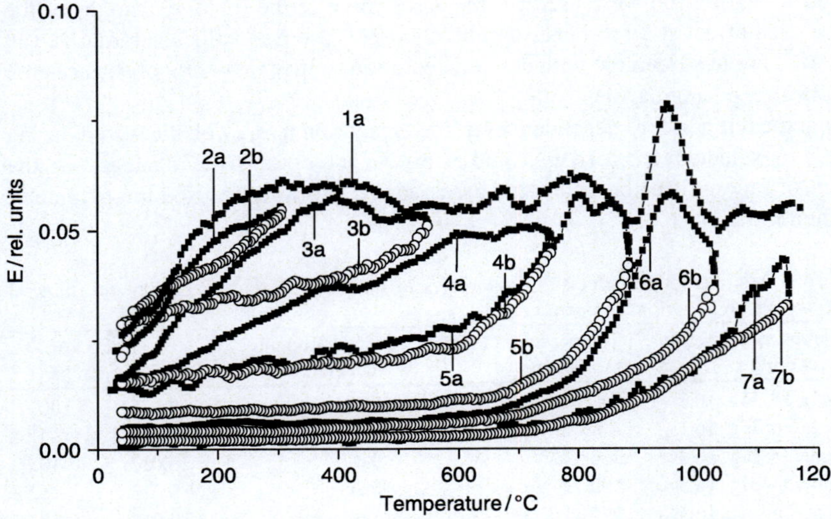


**Fig. 16.3** ETA results of natural brannerite mineral sample measured on heating and subsequent cooling in argon in the range 20–1,150°C: curve 1a corresponds to the "as received" sample measured during heating from 20°C to 1,150°C, curves 2a/2b, 3a/3b, 4a/4b, 5a/5b, 6a/6b and 7a/7b were measured with a parallel samples pre-heated to the temperatures of 300, 550, 750, 880 and 1,020°C, respectively

The ETA curves 3a/3b, 4a/4b and 5a/5b characterized the thermal behaviour of the metamict brannerite sample pre-heated to 300 and 750°C, respectively. The increase of the emanation rate, $E$, in the temperature range of 20–360°C, due to the diffusion of radon along micropores in the sample, was followed by the decrease of $E$, characterizing the partial healing of voids and structure irregularities that served as diffusion pathways for radon.

The ETA curves 6a/6b in Fig. 16.3 characterized the thermal behaviour of the sample pre-heated to 880°C. As already observed by curve 1a the amount of structure irregularities serving as radon diffusion paths further diminished in the sample pre-heated to 880°C. The decrease of the emanation rate on sample observed on heating in the range of 970–1,020°C indicated the next step of the formation of crystalline brannerite. A good reproducibility of the ETA measurements can be seen from the temperature coincidence of the effects on the curve 1a and curve 6a in Fig. 16.3.

From curves 7a/7b characterizing the thermal behaviour of the sample pre-heated to 1,020°C, it is obvious that after the pre-heating the sample to this temperature an irreversible crystallization of amorphous self-irradiated brannerite mineral took place.

From Fig. 16.3 it is obvious that the amount of structure irregularities serving as radon diffusion paths further diminished and the radon permeability in the pre-heated brannerite samples decreased with the temperature used for pre-heating of the samples.

Values of the emanation rate, $E_{RT}$, measured at room temperature before and after each heating run were used for the assessment of the relative changes of the surface area affected by the heat treatments used. The $E_{RT}$ values summarized in Table 16.1 are in agreement with our considerations of the annealing of surface area and subsurface irregularities.

From the temperature dependences of the emanation rate, $E(T)$, measured during heating to selected temperatures and subsequent cooling, the decrease in the amount of radon diffusion paths was assessed. To this aim we used the parameter $\xi$ defined in Eq. 16.6 as:

**Table 16.1** Microstructure defects characteristics of natural self-irradiated brannerite mineral sample pre-heated to various temperatures

| ETA curves measured on heating/cooling | Temperature of sample pre-heating | Defect amount characteristics $\xi^*$ | $E_{RT}$ [rel. units] | $\Delta\xi^{**}$ [%] |
|---|---|---|---|---|
| Curves 1a/1b, Fig. 16.2 | As received | 38.1 | 0.026 | 100 |
| Curves 2a/2b, Fig. 16.3 | As received | 0.41 | 0.023 | 1.08 |
| Curves 3a/3b, Fig. 16.3 | 300°C | 3.82 | 0.017 | 10.02 |
| Curves 4a/4b, Fig. 16.3 | 550°C | 10.26 | 0.015 | 26.93 |
| Curves 5a/5b, Fig. 16.3 | 750°C | 13.62 | 0.014 | 35.75 |
| Curves 6a/6b, Fig. 16.3 | 880°C | 6.30 | 0.005 | 16.54 |
| Curves 7a/7b, Fig. 16.3 | 1,020°C | 0.98 | 0.001 | 2.57 |

* $\xi(T_{max}) = \int_{T_{min}}^{T_{max}} E(T)_{heating} dT - \int_{T_{min}}^{T_{max}} E(T)_{cooling} dT$ ** $\Delta\xi = \frac{\xi_n}{\xi_1} \times 100$ [%]

$$\xi(T_{max}) = \int_{T_{min}}^{T_{max}} E(T)_{heating} dT - \int_{T_{min}}^{T_{max}} E(T)_{cooling} dT \qquad (16.6)$$

Moreover, values of $\Delta\xi$ (see Eq. 16.7) were calculated with the aim to compare the amounts of the annealed microstructure defects during the "step by step" heating of the sample. The difference of integrals used for the assessment of the amount of the microstructure defects can be expressed as $\Delta\xi$ defined as

$$\Delta\xi = \frac{\xi_n}{\xi_1} \times 100[\%] \qquad (16.7)$$

As it followed from values of $\xi$ and $\Delta\xi$ summarized in Table 16.1, the most significant decrease of the structure irregularities serving as diffusion paths for radon diffusion was annealed prior to the crystallization of the sample in the range of 970–1,020°C.

Figure 16.4 depicts a comparison of the relative amount of structure irregularities, expressed by parameter $\xi$, that were annealed during heat treatments to the selected temperatures.

It was shown that the emanation thermal analysis revealed differences in the amount of structure irregularities that served as radon diffusion paths in the brannerite

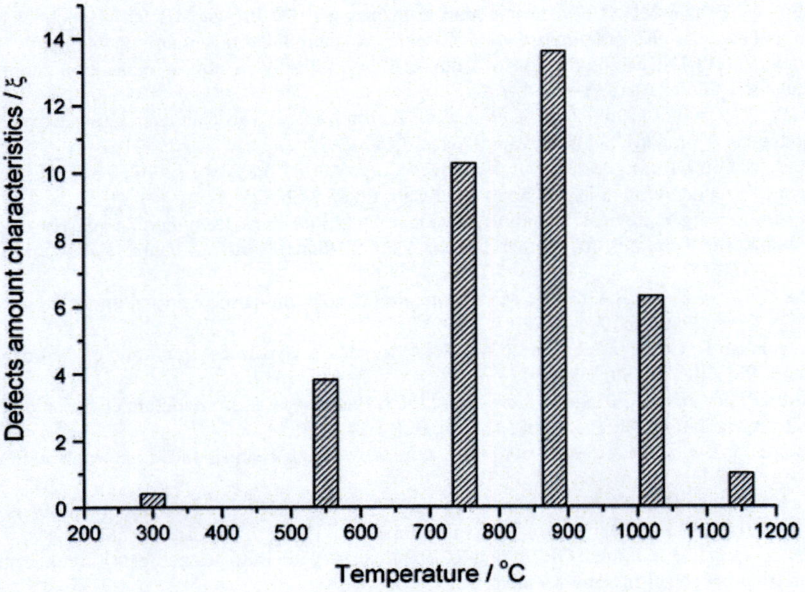


**Fig. 16.4** Relative amounts of micro structure irregularities in natural brannerite sample healed in the heating runs to temperatures 20–300, 20–550, 20–750, 20–880, 20–1,020 and 20–1,150°C. Parameter $\xi$ was used to characterize the amount of structure irregularities of the following samples: preheated to 300, 550, 750, 880 and 1,020°C respectively

samples. Additional information about thermal behaviour of self-irradiated metamict minerals was obtained by using the diffusion structural diagnostics.

**Acknowledgments** Authors thank to the Ministry of Education, Youth and Sports of the Czech Republic for the support (Project MSM 2672244501).

## References

1. Balek V (1978) Emanation thermal analysis. Thermochim Acta 22:1–156
2. Balek V (1991) Emanation thermal analysis and its application potential. Thermochim Acta 192:1–17
3. Balek V, Šubrt J, Mitsuhashi T, Beckman IN, Györyová K (2002) Emanation thermal analysis: ready to fulfil the future needs of materials characterization. J Therm Anal Calorim 67:15–35
4. Ziegler JF, Biersack JP, Littmark U (1985) The stopping and range of ions in solids. Pergamon, New York
5. Beckman IN, Balek V (2002) Theory of emanation thermal analysis XI. Radon diffusion as the probe of microstructure changes in solids. J Therm Anal Calorim 67:49–61
6. Emmerich WD, Balek V (1973) Simultaneous application of DTA, TG, DTG, and emanation thermal analysis. High Temp-High Press 5:67
7. Balek V, Tölgyessy J (1984) Emanation thermal analysis and other radiometric emanation methods. In: Svehla G (ed) Wilson and Wilson's comprehensive analytical chemistry, Part XIIC. Elsevier Science, Amsterdam
8. Balek V, Brown ME (1998) Less common techniques. In: Brown ME (ed) Handbook on thermal analysis and calorimetry, vol 1. Elsevier Science BV, Amsterdam, pp 445–471
9. Balek V (1989) Characterization of high-tech materials by means of emanation thermal analysis. J Therm Anal 35:405–427
10. Balek V, Šesták J (1988) Use of emanation thermal analysis in characterization of superconducting $YBa_2Cu_3O_x$. Thermochim Acta 133:23–26
11. Balek V, Pérez-Rodriguez JL, Pérez-Maqueda LA, Šubrt J, Poyato J (2007) Thermal behaviour of ground vermiculite. J Therm Anal Calorim 88:819–823
12. Balek V, Zhang Y, Zeleňák V, Šubrt J, Beckman IN (2008) Emanation thermal analysis study of brannerite ceramics for immobilization of hazardous waste. J Therm Anal Calorim 92:155–160
13. Ríos S, Boffa-Ballaran T (2003) Microstructure of radiation-damage zircon under pressure. J Appl Crystallogr 36:1006–1012
14. Devanathan R, Corrales LR, Weber WJ (2004) Molecular dynamics simulation of disordered zircon. Phys Rev B 69:064115(9)
15. Carrez P, Forterre Ch, Braga D, Leroux H (2003) Phase separation in metamict zircon under electron irradiation. Nucl Instrum Methods B 211:549–555
16. Wang LM, Ewing RC (1992) Detailed in situ study of ion beam-induced amorphization of zircon. Nucl Instrum Methods B 65:324–329
17. Balek V, Vance ER, Zeleňák V, Málek Z, Šubrt J (2007) Use of emanation thermal analysis to characterize thermal reactivity of brannerite mineral. J Therm Anal Calorim 88:93–98
18. Lian J, Wang LM, Lumpkin GR, Ewing RC (2002) Heavy ion irradiation effects of brannerite-type ceramics. Nucl Instrum Methods B 191:565–570
19. Zhang Y, Lumpkin GR, Li H, Blackford MG, Colella M, Carter ML, Vance ER (2006) Recrystallisation of amorphous natural brannerite through annealing: The effect of radiation damage on the chemical durability of brannerite. J Nucl Mater 350:293–300

# Chapter 17
# Scanning Transitiometry and Its Application in Petroleum Industry and in Polymer and Food Science

**Jean-Pierre E. Grolier**

## 17.1 Introduction

Liquid–solid phase equilibria in asymmetric binary mixtures are not only of general interest to explore phase equilibria in three-phase (gas, liquid, solid) systems but they play a major role in understanding and monitoring the $pT$-behaviour of petroleum fluids. Such fluids present a vast variety of compositions in terms of their respective constituents from light gases and liquids of various molecular sizes to macromolecular solids. Nowadays, the lack of thermodynamic data on asphaltenic fluids prevents the large scale exploitation of heavy oils in deep deposits. The main concern is the uncontrolled precipitation/flocculation of heavy fractions (asphaltenes, waxes) which causes obstruction and plugging of underground as well as surface installations and pipes. Research in polymer science continues to develop actively while the concepts of thermodynamics and kinetics together with polymer chain structure enhance the domain of polymer development and transformation. In many industrial applications, during extrusion processing or as all purpose materials, polymers are usually submitted to extreme conditions of temperature and pressure. Furthermore, most of the time they are also in contact with gases and fluids, either as on-duty materials (containers, pipes) or as process intermediates (foaming, molding). Since such materials are often used in special environments or under extreme conditions of temperature and pressure, their careful characterization must be done not only at the early stage of their development but also all along their life cycle. In addition, their properties as functions of temperature and pressure must be well established for the optimal control of their processability. This also stands for phase transitions; ignorance of a phase diagram, particularly at extreme conditions of pressure, temperature, and of chemical reactivity, is a limiting factor to the

J.-P.E. Grolier (✉)
Laboratoire de Thermodynamique des Solutions et des Polymères Université Blaise Pascal, 24, Avenue des Landais, Clermont Ferrand 63177, Aubière Cedex, France
e-mail: j-pierre.grolier@univ-bpclermont.fr

J. Šesták et al. (eds.), *Glassy, Amorphous and Nano-Crystalline Materials*,
Hot Topics in Thermal Analysis and Calorimetry 8, DOI 10.1007/978-90-481-2882-2_17,

development of an industrial process, e.g., sol–gel transitions, polymerization under solvent near supercritical conditions, micro- and nano-foaming processes. Natural and bio-polymers constitute an important class of components largely used in food science. Among the numerous such polymers, starch serves to illustrate the complexity of state equilibria of systems containing other species like fibers, fat, proteins, and extended ranges of water percentages. In food science, industrial processing of such systems, for example during cooking extrusion, requires in depth thermodynamic as well as thermophysical characterization of the systems to process. All above fields to cite a few, in oil industry and in polymer and food applications, necessitate the acquisition of key data.

Undoubtedly, thermal and calorimetric techniques are essential in this respect. In relating thermal as well as mechanical behaviour to materials' structures these techniques are perfectly adapted to provide accurate data in wide ranges of temperature and pressure. Typically, thermophysical properties feature the most important information expected when dealing with materials submitted to thermal variations and/or mechanical constraints. The properties of interest are of two types, bulk properties and phase transition properties. The bulk properties are either caloric properties, like heat capacities $C_P$, or mechanical properties, like isobaric thermal expansivities $\alpha_P$, isothermal compressibilities $\kappa_T$, and isochoric thermal pressure coefficients $\beta_V$. Two main thermal properties concern the first order transitions, fusion and crystallization, and the glass transition. All these properties are now accessible thanks to recent progress in various technologies which allow measurements in the three physical states over extended ranges of $p$ and $T$, including in the vicinity of the critical point. In this respect, knowledge, i.e. measurements, of the thermophysical properties of polymers over extended ranges of temperature and pressures and in different gaseous environments is absolutely necessary to improve the use and life-time of end products made of such polymers. The purpose of this chapter is to demonstrate the contribution of the new technique, *scanning transitiometry*, in providing accurate information to meet the demand for the different data pointed out. Examples have been selected in three main domains, oil exploitation and transport, polymer foaming and modification, and starch-water systems. As a matter of fact, these examples are directly connected to industrial activities: the petroleum industry, the insulating material industry, and the food industry. In many cases, gases and polymers of different types and from different origins (synthetic, natural) are intimately interacting under external conditions of temperature ($T$) and pressure ($p$). In the subsequent examples the gas/polymer systems are either selected for a targeted industrial purpose i.e. foaming materials and materials processing, or are polymeric materials in contact with gas/liquid systems, i.e. pipes or tanks in gas and petroleum industry. The foaming materials industry is a rapidly growing area where constant innovation and added value products are key factors for economic success where international competition is high. The mastering of polymer degradation (typically blistering) by high pressure dissolved gases is another key issue. In what follows, in a first section the newly developed technique will be described. In a second section selected examples will illustrate how such technique is providing valuable data for significant progresses in different fields.

## 17.2 Scanning Transitiometry

Certainly, calorimetry is a major technique to measure thermodynamic properties of substances and to follow phase change phenomena. In most applications, calorimetry is carried out at constant pressure while the tracked phenomenon is observed on increasing or decreasing the temperature. The possibility of controlling the three most important thermodynamic variables ($p$, $V$ and $T$) during calorimetric measurements makes it possible to perform simultaneous measurements of both thermal and mechanical contributions to the thermodynamic potential changes caused by the perturbation. Calorimetric techniques provide valuable additional information on transitions in complex systems. Their contributions to the total change of thermodynamic potential not only leads to the complete thermodynamic description of the system under study, but also permits investigation of systems with limited stability or systems with irreversible transitions. By a proper external change of the controlling variable the course of a transition under investigation can be accelerated, impeded or even stopped at any degree of its advancement and then taken back to the beginning, all with simultaneous recording of the heat and mechanical variable variations. In what follows the main characteristics of scanning transitiometry are reviewed.

The seminal presentation by Randzio [1] of thermodynamic fundamentals for the use of state variables ($p$,$V$,$T$) in scanning calorimetric measurements has open the path [2–4] from $p$,$V$,$T$-calorimetry to the now well established scanning transitiometry technique [5]. With this technique the simultaneous determination of thermal and mechanical responses of the investigated system, perturbed by a variation of an independent thermodynamic variable while the other independent variable is kept automatically constant, allows the determination of thermodynamic derivatives over extended ranges of pressure and temperature, impossible to obtain by other known techniques. Four thermodynamic situations are thus possible to realize in the instruments based on such technique, namely, $pVT$-controlled scanning calorimeters or simply *scanning transitiometers*, since they are particularly adapted to investigate transitions by scanning one the three thermodynamic variables. The four possible thermodynamic situations (Fig. 17.1) are obtained by simultaneous recording of both

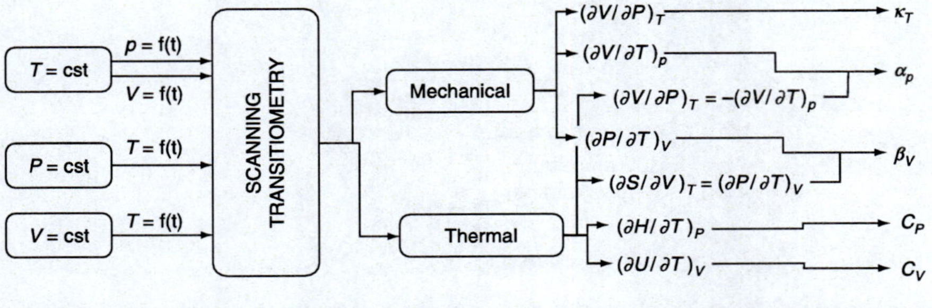


**Fig. 17.1** Thermodynamic scheme of scanning transitiometry showing the four possible modes of scanning. Each of these modes delivers two output derivatives (mechanical and thermal) which in turn lead to four pairs of the different thermomechanical coefficients, namely: $\alpha_P$, $\kappa_T$, $\beta_V$, $C_P$, and $C_V$.

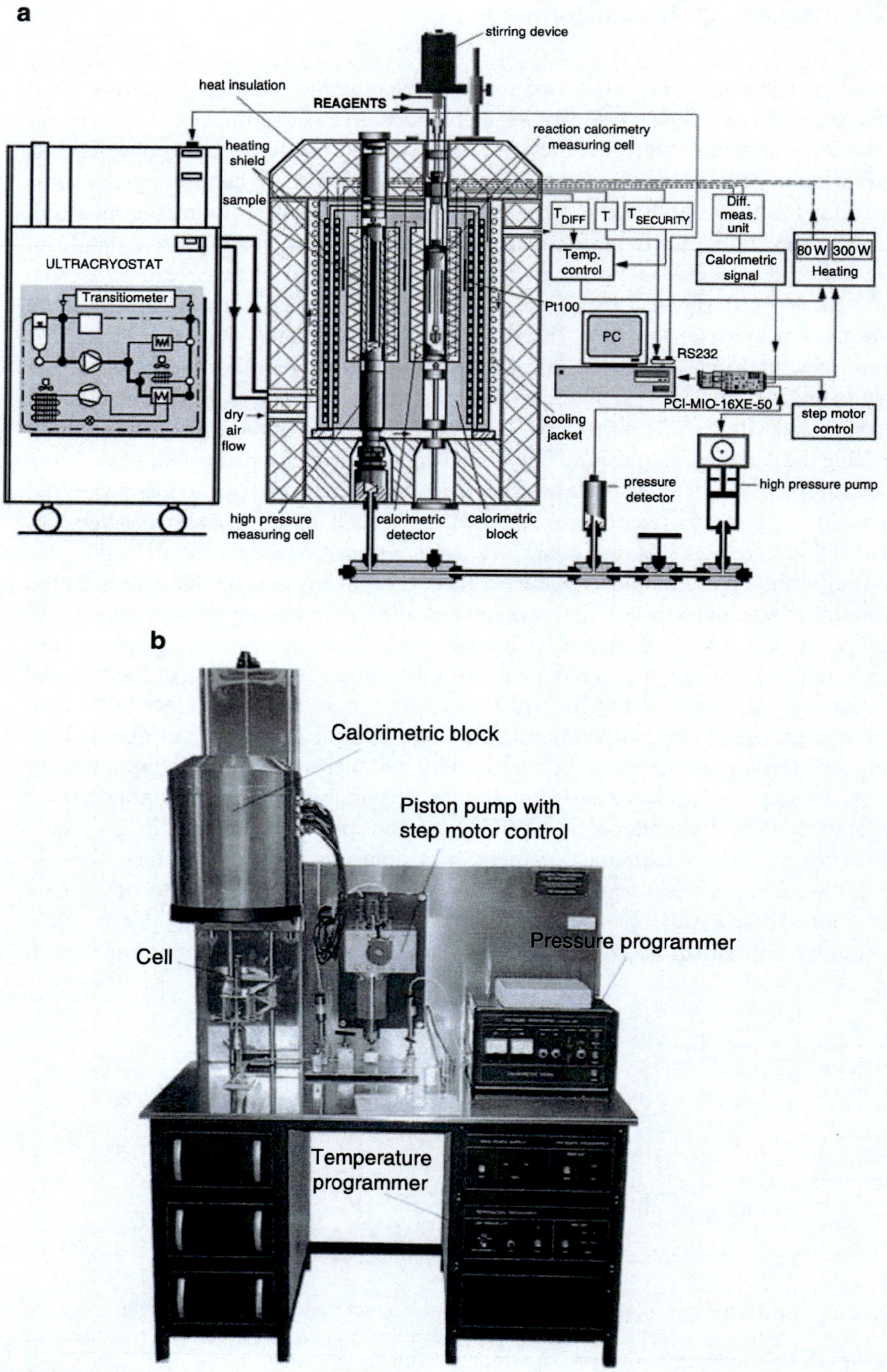


**Fig. 17.2** (**a**) Detailed schematic representation of a scanning transitiometer setup for in situ simultaneous determination of the thermal and mechanical derivatives. For convenience, two

heat flow (thermal output) and the change of the dependent variable (mechanical output).

Then, making use of the respective related Maxwell relations one readily obtains the main thermophysical properties as follows: (a) scanning pressure under isothermal conditions yields the isobaric thermal expansivity $\alpha_P$ and the isothermal compressibility $\kappa_T$ as functions of pressure at a given temperature; (b) scanning volume under isothermal conditions yields the isochoric thermal pressure coefficient $\beta_V$ and the isothermal compressibility $\kappa_T$ as functions of volume at a given temperature; (c) scanning temperature under isobaric conditions yields the isobaric heat capacity $C_P$ and the isobaric thermal expansivity $\alpha_P$; (d) scanning temperature under isochoric conditions yields the isochoric heat capacity $C_V$ and the isochoric thermal pressure coefficient $\beta_V$.

A detailed description of a basic scanning transitiometer is given elsewhere [6]. A schematic representation of the instruments (from BGR TECH, Warsaw, Poland) used in the present applications to polymers, and constructed according to the principle of scanning transitiometry, is presented in Fig. 17.2. It consists of a calorimeter equipped with high-pressure vessels, a *pVT* system, and a LabVIEW based virtual instrument (VI) software. Two cylindrical calorimetric detectors (ext. diameter 17 mm, length 80 mm) made from 622 thermocouples chromel-alumel each are mounted differentially and connected to a nanovolt amplifier. The calorimetric detectors are placed in a calorimetric metallic block, the temperature of which is directly controlled with an entirely digital feedback loop of 22-bit resolution (~$10^{-4}$ K), being part of the transitiometer software. The calorimetric block is surrounded by a heating-cooling shield connected to an ultracryostat (Unistat 385 from Huber, Germany) and the temperature difference between the block and the heating-cooling shield is set constant (5, 10, 20 or 30 K) as controlled by an analogue additional controller. The whole assembly is placed in a thermal insulation enclosed in a stainless steel body and placed on a stand, which permits to move the calorimeter up and down over the calorimetric vessels. The actual operating ranges of scanning transitiometry are respectively 173 K $< T <$ 673 K and 0.1 MPa $< p <$ 200 MPa (or 400 MPa).

Actually, transitiometry is at the centre of different types of utilization since with such technique, bulk properties, transitions as well as reactions (for example polymerization) can be advantageously studied. In the case of a non-reacting system which remains in a homogeneous state, both the mechanical and thermal outputs as explained before give straightforward access to pairs of the thermomechanical coefficients. When the system or the material sample evolves through a chemical reaction or a phase change, the recorded information yields the corresponding heat

**Fig. 17.2** (continued) types of cells are shown: on the left-hand side is the standard high pressure cell and on the right-hand side is a reaction type cell which can accommodate various accessories (stirrer, reagents feeding, capillaries, optical fibers/probes for UV/Vis/near IR spectroscopic analysis). (**b**) Photography of a standard scanning transitiometer (from BGR TECH, Warsaw). The calorimetric detector which can be moved up and down over the measuring and reference cells (in twin differential arrangement) is shown in the upper position. In this position the cells which are firmly fixed on the stand table are then accessible for loading and $pVT$ characteristics. In the case of polymer synthesis, a scanning transitiometer was used as an isothermal reaction calorimeter, the advancement of a polymerization reaction being accurately monitored through the rigorous control of the thermodynamic parameters [7, 8]. The thermocouples composing fluxmeters serve as a heat conduction path between cell walls and the block. To gather additional information on a reaction, the reaction can be coupled with other analytical devices (e.g., on-line FTIR, particle sizing probes, turbidity probes, pH or other ion selective probes) [9]. A photographic presentation of a transitiometer is given in Fig. 17.2.

## 17.3 Selected Results

Performances and advantages of scanning transitiometry are well demonstrated by typical applications in several important fields: (1) asymmetric fluid mixtures and petroleum fluids; (2) transitions of polymer systems under various constraints (temperature, pressure, gas sorption) including first-order phase transitions [10, 11] and biopolymer gelatinization [12–14]; (3) polymer thermophysical properties [15] and influence of gas sorption [11, 16]. In what follows illustrative examples have been selected, namely: in the petroleum industry, in polymer science and in food science.

### *17.3.1 Petroleum Industry*

The oil industry, where petroleum products and associated fluids of different nature with multicomponent and complex compositions present a large variety of phases and phase equilibria, is certainly the domain by excellence for scanning transitiometry applications. The ongoing determinations of the thermophysical properties ($C_P$, $\alpha_P$, $\kappa_T$) over extended $p$ and $T$ ranges of newly developed fuels (including biofuels) will not be reported here. Instead, two other completed studies are worth to report: (1) appearance of the solid phase in asymmetric binary systems and (2) precipitation of heavy cuts in asphaltenic fluids under $p$ and $T$ conditions of deep underground reservoirs.

The investigation of asymmetric systems i.e. binary mixtures of two components having large difference in the molecular size is of interest in relation with the solid precipitation or flocculation of heavy components in high pressure reservoir fluids containing large concentrations of light components like methane. Typically these systems exhibit a three-phase equilibrium curve (solid–liquid-vapor) with a high temperature segment. Scanning transitiometry allows to precisely detecting up to high pressures the transition through the three-phase line in both directions by varying any of the state variables ($p$, $V$, $T$). The method was tested [6] with the system {tetracosane + methane} by comparing with the reference data measured

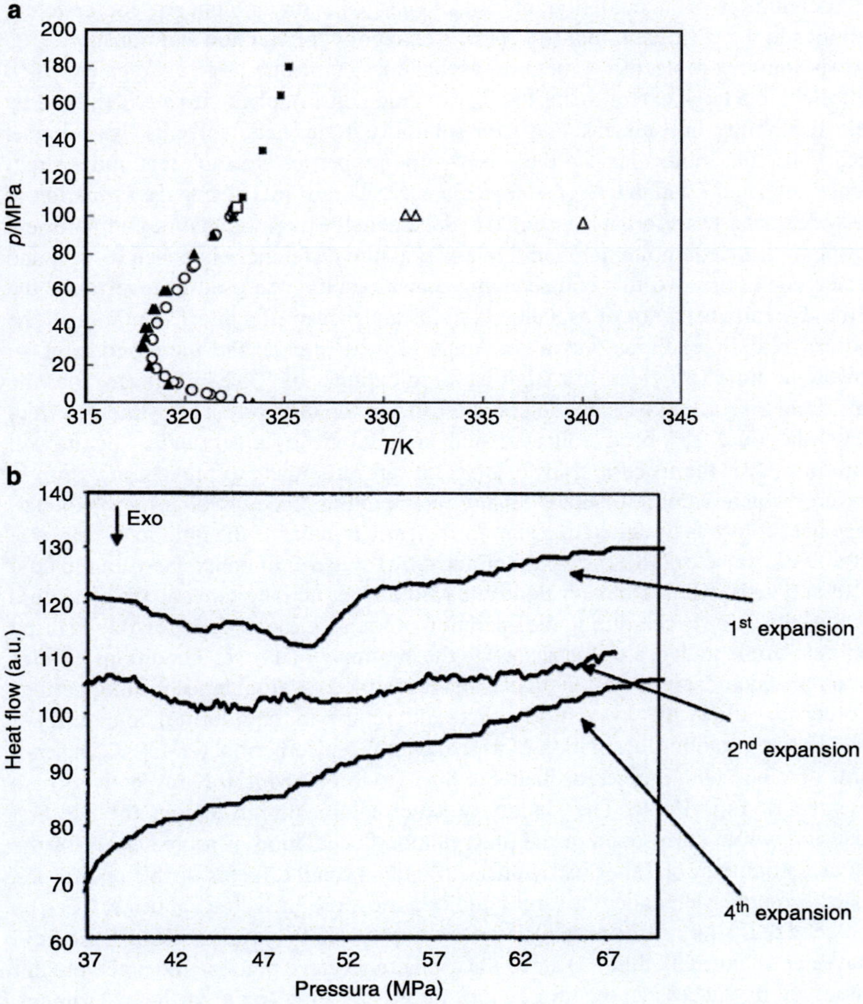


**Fig. 17.3** (a) Phase equilibria in the system {tetracosane + methane}.▲, Our work on the three-phase equilibrium [6]. ○, Three-phase equilibria data of Flöter et al. [17]. □, Second critical endpoint. ■, Our work for the high pressure equilibrium (solid + fluid + liquid) [6]. Note that the three (*open triangles*) points correspond to a near critical liquid–vapor isopleth at 0.04 mole fraction of tetracosane. (**b**) Thermograms obtained after cycling successive decompressions and recom-pressions. During the first expansion the exothermic effect which would correspond to the precipitation/flocculation of asphaltenes clearly appears; it is slightly visible for the second expansion and completely disappears after the third one. The shape of the forth expansion thermogram is similar to what is observed for a simple fluid [4]

using the conventional visual method [17]. The Fig. 17.3a shows the very good agreement of data obtained with scanning transitiometry and the indirect method based on the visual observation of phase boundaries.

Precipitation or flocculation of asphaltenes is a major concern for different activities in the petroleum industry such as extraction, production and transport. Due to their complex molecular structures, asphaltenes constitute large aggregates which contribute to a large extent to the stability of crude oils. Asphaltenes are identified by their insolubility in n-alkanes and their solubility in aromatic solvents like toluene. Their particular molecular and thermodynamic properties are such that under slight changes of $p$ and $T$ asphaltenes can flocculate in the crude oil causing the formation of heavy deposits. Therefore it is essential to document the thermodynamic comportment of asphaltenes undergoing temperature and pressure differences between in-well and surface conditions. To this end scanning transitiometry was used to investigate the thermodynamic behavior of asphaltenic fluids under in-well $p$ and $T$ conditions. The study of real (live) fluids was a challenge [18] as regards the introduction of an asphaltenic fluid (provided in a high pressure cylinder by TOTAL France) into the transitiometric cell under isobaric conditions from the high pressure cylinder. Effectively, the fluid has been collected and kept in the cylinder under the in-well conditions, thus the isobaric transfer must be carefully made to prevent any drop in pressure which would inevitably cause precipitation/flocculation of asphaltenes. A special setup was designed to insure the isobaric transfer of the fluid into the active calorimetric detector [18]; this was facilitated by the use of mercury as hydraulic fluid to pressurize the whole (transfer setup and calorimetric cell) system and push the fluid into the calorimetric cell during the operation of transfer. It was then possible to bring the fluid sample in the calorimetric cell to the nominal $p$, $T$ in-well conditions. Cycles of compression/decompression could be performed over extended time periods in order to allow the system to relax and return to equilibrium and observe possible precipitation/flocculation of asphaltenes. The isothermal (at 430 K) thermograms obtained with a given asphaltenic fluid (61 MPa and 430 K respectively) are presented in Fig. 17.3b. The shallow exothermal maximum shown for the first expansion would correspond to the precipitation/flocculation of asphaltenes; for the second decompression (after recompression) only a small effect is visible and for the fourth decompression under the same conditions no more such effect appears. It can be concluded that after few decompression/re-compression the original asphaltenic fluid behaves as a "normal" fluid [4] since the asphaltenes have precipitated and were not redissolved by successive recompressions during the time frame of the experiment. These preliminary investigations have shown the advantage of scanning transitiometry to characterize real (live) oils and systematic studies are underway on heavy oils from different geological origins.

### *17.3.2 Polymer Science*

As multipurpose technological materials, polymers must be perfectly characterized. In particular, among the thermophysical properties the isobaric thermal expansivities, $\alpha_P$, and the isothermal compressibilities, $\kappa_T$, are key properties to document over extended ranges of $T$, $p$, and crystallinities. Also, when dealing with polymer materials, the glass transition must be unambiguously established in particular with respect to its dependence with temperature, pressure and plasticizers, especially high pressure gases. Although the following examples might appear as simply relevant of "applied" science, they imply fundamental approaches in terms of developing specific instruments and associated methodologies to determine thermophysical properties and to investigate the glass transition temperature $T_g$ which is the pivotal transition to understand and control for the proper processing of polymeric materials. Herein, an account of the influence of the crystallinity and $p$ on the isobaric thermal expansivities $\alpha_P$ is reported; also the influence on $T_g$ of $p$, $T$ and gas solubility is illustrated.

The thermodynamic principle of scanning transitiometry is based on the Maxwell equality

$$\left(\frac{\partial V}{\partial T}\right)_p = -\left(\frac{\partial S}{\partial p}\right)_T \cong -\frac{1}{T}\left(\frac{dQ}{dp}\right)_T. \tag{17.1}$$

Thus, if a sample of a polymer of a mass $m_s$ is placed in the calorimetric vessel and is submitted to a pressure variation $dp$, then the heat exchanged in such a process, $dQ_{pl}$, is defined by the following equation

$$dQ_{pl} = m_s T\left(\frac{\partial V_s}{\partial T}\right)_p dp. \tag{17.2}$$

On the other hand, if the substance contained in the calorimetric vessel is in the fluid state and the pressure is exerted through the fluid itself then the real mass of the fluid sample contained in the calorimetric vessel is changing along with the pressure variation and is equal to $V_E/V_s$, where $V_E$ is the active internal volume of the calorimetric vessel accessible for the fluid and $V_s$ is the specific (or molar) volume of that fluid. Thus, in this case the thermal effect, $dQ_{fl}$, associated with the pressure variation is defined by the following equation

$$dQ_{fl} = -\frac{V_E}{V_s}T\left(\frac{\partial V_s}{\partial T}\right)_p dp = V_E T\alpha_p dp, \tag{17.3}$$

where $\alpha_p = 1/V_s(\partial V_s/\partial T)_p$ is the isobaric thermal expansivity of the fluid. Then, if a polymer sample is placed in the calorimetric vessel and is compressed or decompressed by intermediary of a fluid like a gas or mercury, the total thermal effect will be composed of the main three following terms: compression of the solid (Eq. 17.2), compression of the fluid phase (Eq. 17.3) and "pure" interaction of the fluid with the polymer. Of course it is assumed that there are no state or phase transitions in both the fluid and the polymer within the experimental $p$, $T$ range.

The thermal calibration (energy and temperature) was done with the use of benzoic acid. The thermomechanical calibration, especially the determination of the internal active volume $V_E$ (Eq. 17.3) was done with the use of nitrogen, for which the thermal expansion is known from its equation of state.

#### 17.3.2.1 Isobaric Thermal Expansivities of Polyethylenes with Various Crystallinities over Extended $p$ and $T$ Ranges

As a matter of fact thermal expansivity $\alpha_P$ as well as heat capacity $C_P$, both second derivatives of the thermodynamic potential $G$ (the free enthalpy) are key thermophysical properties. The knowledge of $C_P$ as a function of $T$ allows to evaluate the temperature effects on thermophysical properties, similarly, the knowledge of $\alpha_P$ as a function of $p$ permits to access the effects of pressure on thermophysical properties. However, direct measurements of $\alpha_P$, especially of polymers, are rather scarce. Contrary to fluids (gases and liquids) the thermal expansivity of polymers presents a specific (theoretical) aspect because like in solid substances thermal expansion results from anharmonic lattice vibrations. The particularity of polymers is that they usually are constituted of partially crystal phases. Therefore, this property of both fundamental and engineering interests needs to be documented by experimental measurements taken over extended $p$ and $T$ ranges for various crystallinities.

We conducted the first systematic series of direct measurements along this line [15, 19]. For this, we used a pressure controlled scanning calorimeter (PCSC), actually later called Scanning Transitiometer (ST). Remarkably, scanning transitiometry is perfectly designed to measure $\alpha_P$, in the same way temperature controlled scanning calorimetry (TCSC), usually named DSC, is the technique by excellence to measure $C_P$. In our measurements [15, 19] mercury was used as inert hydraulic fluid enveloping the polymer sample. Isothermal linear pressure scans were performed from where $\alpha_P$'s of polymer samples were determined through the procedure detailed in Refs. [15, 19]. Data for two different types of low and high density polyethylenes (respectively LDPE and HDPE) are presented on Fig. 17.4 as smooth isotherms of $\alpha_P$'s versus pressure. One observes that there are small differences between the four sets of data which show the characteristic converging trend at elevated pressures.

#### 17.3.2.2 Glass Transition Temperatures of Elastomers Under High Pressure

The glass transition temperature is affected by pressure since an increase of pressure causes a decrease in the total volume then an increase of $T_g$ is expected due to the decrease of free volume. This result is important in engineering operations such as molding or extrusion, when operations too close to $T_g$ can result in a stiffening of the material. Investigation of the glass-transitions of polymers under pressure is not a simple problem, especially in the case of elastomers of which $T_g$'s are usually well-below the ambient temperature. In the case of scanning transitiometry the traditional pressure-transmitting fluid, mercury, must be replaced since its crystallization temperature is relatively high, i.e. 235.45 K. Then the choice of the replacement fluid is again a challenge [20] because it should be chemically inert with respect to the investigated sample. Also, values of its thermomechanical coefficients, compressibility, $\kappa_T$, and thermal expansivity, $\alpha_p$, should be smaller than those of the investigated sample. An additional difficulty in the investigation of second order

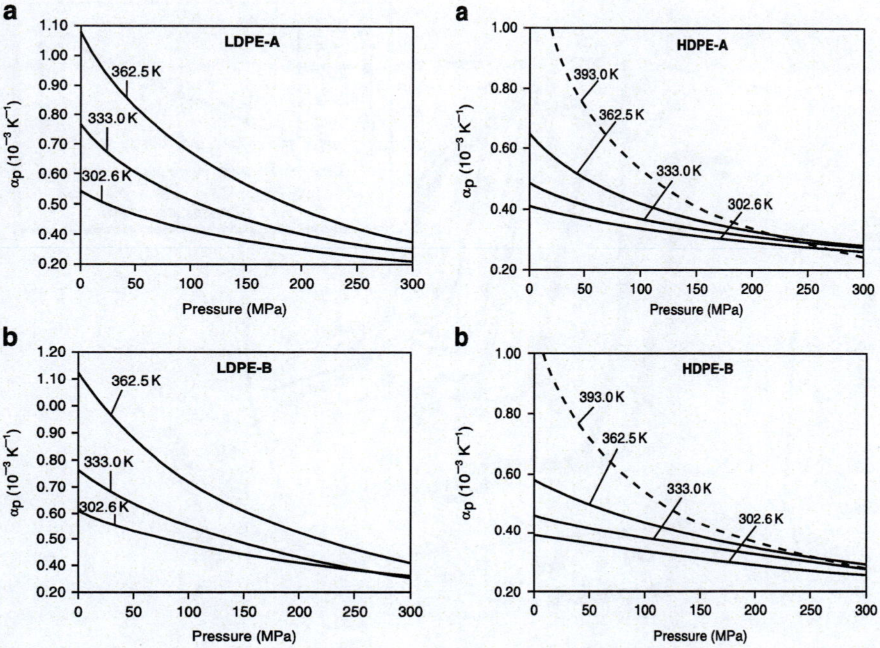


**Fig. 17.4** Isobaric thermal expansivities as functions of pressure at various temperatures for four polyethylenes (PE) with different densities ($\rho$ in g cm$^{-1}$), crystal phase volume fractions ($\Phi_c$), and crystal phase mass fractions ($\chi_c$). On left-hand side low density PE: LDPE-A ($\rho = 0.921$, $\Phi_c = 0.46$, $\chi_c = 0.42$), LDPE-B ($\rho = 0.936$, $\Phi_c = 0.51$, $\chi_c = 0.55$). On the right-hand side high density PE: HDPE-A (0.73$^{¥}$), HDPE-B ($\rho = 0.957$, $\Phi_c = 0.64$, $\chi_c = 0.80^{¥}$). $^{¥}$*N.B.* from enthalpimetric measurements; all other values from volumetric measurements

type transitions is the relatively weak effect measured. It is well-known that the amplitude of the heat flow at the glass transition $T_g$ increases with the temperature scanning rate while the time-constant of heat flow type calorimeters, like scanning transitiometers, imposes temperature scan rates which are slow compared to typical DSC scan rates.

For the present measurements silicon oil was used instead of mercury as the hydraulic pressurizing fluid, and the polymer sample was placed in a lead (soft metal) ampoule. Test measurements were made on polyvinyl acetate for which the $\Delta T_g/\Delta p$ coefficient was found to be $0.212 \pm 0.002$ K MPa$^{-1}$ in good agreement with the literature value 0.22 K MPa$^{-1}$ [21]. The calorimetric traces obtained with the same method for a poly(butadiene-*co*-styrene) vulcanized rubber [22] during isobaric scans of temperatures ranging from 218.15 to 278.15 K at 0.4 K min$^{-1}$ are shown in Fig. 17.5. This Figure shows also the evolution of $T_g$ at different pressures, at 0.25, 10, 30, 50, and 90 MPa, respectively: $T_g$ increases linearly with pressure with a $\Delta T_g/\Delta p$ coefficient of $0.193 \pm 0.002$ K MPa$^{-1}$. It should be noted that $T_g$ is expressed as the temperature corresponding to the peak of the first derivative of the heat flux (i.e. the inflexion point of the heat flow).

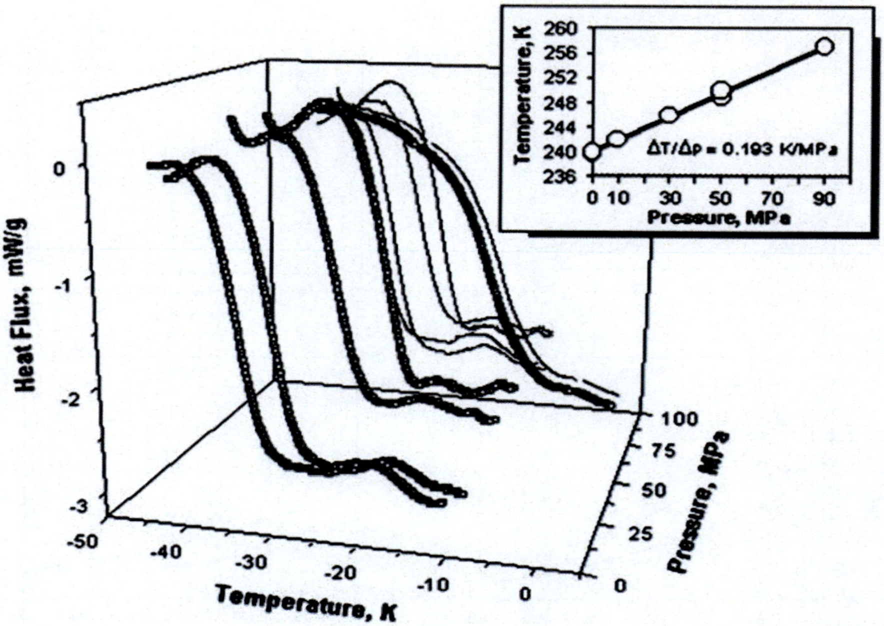


**Fig. 17.5** Illustration of the scanning transitiometry technique for the investigation of polymers glass transition temperature at low-temperature and high pressure. The typical thermograms (heat flow vs $T$) for the transition domain of a vulcanized rubber are shown for different pressures. The pressure coefficient of the glass transition temperature is given in the inset

#### 17.3.2.3 Glass Transition Temperature of Polystyrene Modified by High Pressure Methane

There is not much information available in the literature on calorimetric study of plasticization of polymers at high pressures, above say 50 MPa, induced by gases. Plasticization is well characterized by the shift of the temperature of the glass transition, $T_g$. Actually, when pressure is induced by a gas, both plasticization and hydrostatic effects contribute to the shift of $T_g$. If plasticization tends to lower $T_g$ because of the gain of mobility of the polymeric chains, the hydrostatic effect raises it in diminishing the free volume. Methane ($CH_4$) is assumed to be a non-plasticizing gas but, our results show that, at higher pressures, plasticization overtakes again the hydrostatic effect, due to a probably higher solubility of the gas in polystyrene (PS) at higher pressures; this kind of behavior has been suggested for high enough pressures [23]. The plasticization of PS using $CH_4$ seems to be possible but it is necessary for this to apply high pressure, i.e. 200 MPa, in order to obtain approximately the same shift of the $T_g$ as with ethylene ($C_2H_4$) under 9,0 MPa! In this respect $CH_4$ cannot really be considered as a good plasticizing gas.

#### 17.3.2.4 Shift of the Glass Transition Temperature by High Pressure Gases

An important aspect of polymer foaming is certainly the "easiness" of the blowing agent to enter, to dissolve and to diffuse into the polymer matrix. Two parameters, $T$ and $p$, are essential to control these phenomena. The nature and properties of the polymer and of the fluid play evidently a major role. In this context, the physical state of the polymer must be appropriately modified to undergo plasticization; this optimal condition for having the "blowing" effect taking place depends upon the glass transition temperature $T_g$. Plasticization depends on all the thermodynamic variables and parameters listed above. In particular, it is necessary to know to what extent $T_g$ is advantageously decreased in order to optimise the foaming process. From a practical point of view the $\Delta T_g$ shift should be accurately determined or predicted. Moreover, many properties can be correlated with the glass transition temperature depression $\Delta T_g$ due to plasticization. In order to predict the $\Delta T_g$ the model of Chow [24] was selected. The calculations using the model of Chow were made using experimental data of solubilities directly measured with a new technique combining a vibrating wire (VW) weight sensor and a $pVT$ setup [25].

Chow has proposed a relation based on Gibbs and Di Marzio principle (the entropy of the glassy state is zero) to account for the change in $T_g$ due to the sorbed component as follows:

$$\ln\left(\frac{T_g}{T_{go}}\right) = \beta\,[(1-\theta)\ln(1-\theta) + \theta\ln\theta], \tag{17.4}$$

where

$$\beta = \frac{z\,R}{M_p \Delta C_p}, \theta = \frac{M_p}{z\,M_d}\ \frac{\omega}{1-\omega},$$

$T_g$ and $T_{go}$ are the glass transition temperatures for the polymer-gas system and the pure polymer respectively, $M_p$ is the molar mass of the polymer repeat unit, $M_d$ is the molar mass of the (diluent) gas, $R$ is the gas constant, $\omega$ the mass fraction of the gas in the polymer, $\Delta C_p$ is the heat capacity change associated with the glass transition of the pure polymer and $z$ is the lattice coordination number. All parameters of the model have physical meanings, except the number $z$. The value of this parameter may change according to the state of the diluent: $z = 2$ when the diluent is in the liquid state and $z = 1$ when it is gas.

In order to compare the model calculations with experimental calorimetric data, polystyrene (PS) samples were modified in a transitiometer used in that case as a small reactor to modify under equilibrium conditions the polystyrene in presence of a chosen fluid. Modifications of polystyrene have been done in presence of $N_2$ and $CO_2$, along isotherms at a given pressure. For these two fluids, a final temperature of 398.15 K and a final pressure of 80 MPa have been attained. The glass transition temperatures $T_g$'s of modified and non modified PS samples were determined by Temperature Modulated DSC (TMDSC). The solubilities of the different gases were measured using the VW-*pVT* sorption technique [25, 26] along different isotherms and the mass fraction of the gas in the polymer was then determined with the following equation:

$$\omega = \frac{s}{s+1}, \qquad (17.5)$$

*s* being the solubility of the fluid in the polymer, in mg of fluid/mg of polymer.

Using the values of $\omega$ determined for each system PS/gas, the Chow equation (Eq. 17.4) has allowed to estimate the variation with pressure, $\Delta T_g$, of the temperature of the glass transition this, along the different isotherms of the sorption measurements. The use of the Chow model is then rather delicate since the choice of the value of *z*, that is to say in fact, the state of the diluent influences significantly the results. The *Tg*-shift under $CO_2$ pressures is spectacular showing the high plasticizing effect of $CO_2$. The good agreement of the literature data for PS/$CO_2$ with the calculated values [23] as seen in Fig. 17.6 can certainly be explained by the state of the diluent which is most likely in the critical state in the ranges of *T* and *p* considered.

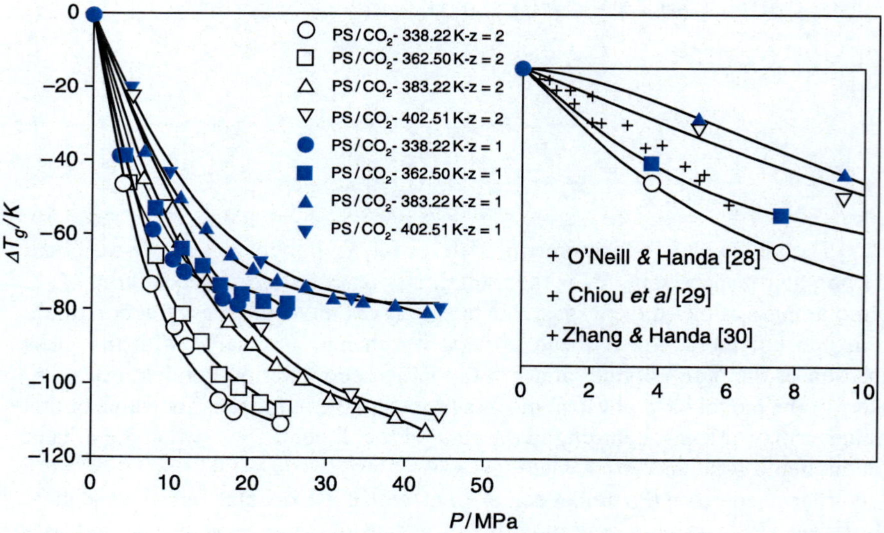


**Fig. 17.6** Variation of the glass transition temperature with pressure for the system polystyrene–$CO_2$. Calculations have been made for 338.22, 362.50, 383.22 and 402.51 K. Full symbols represent results for z = 1 and empty symbols for z = 2. Literature values are represented by crosses in the zoom of the graph (the same scale of temperature being kept). Lines are hand drawn through the points

Effectively, the critical temperature $T_c$ and critical pressure $p_c$ of $CO_2$ support the hypothesis of the gas being in the near critical region. As a matter of fact, depending on the experimental conditions in the vicinity of the critical point, the fluid can exist in one or the other state (gas or liquid) or even in both. In the present case, literature data [28–30] for the PS/$CO_2$ system have been obtained under a pressure $p \leq p_c$ and at a temperature $T \geq T_c$ for $CO_2$; then two phases of the diluent can coexist in different proportions. Despite the difficulty to determine exactly the variation of $T_g$, particularly under supercritical conditions of a diluent fluid, the model of Chow is actually a useful guide in order to predict the variation of the glass transition of a polymer modified by a high pressure fluid. However, the exact determination of the glass transition depression, $\Delta T_g$, becomes more difficult when the pressure increases, specially near and above the critical point of the diluent fluid. This means that when plotting $\Delta T_g$ as a function of pressure, the temperature of measurement plays a major role. If we do not take into account this temperature, it is preferable to represent $\Delta T_g$ as a function of the mass fraction of the fluid in the polymer.

Compared to polar $CO_2$ and because of its non-polarizability, $N_2$ should be a weaker plasticizing agent although, as shown in Fig. 17.7, it induces significant shifts of $T_g$ with increasing pressures [27]. However, $N_2$ which should be also a good foaming agent is not used in the foaming industry because of the need of too high a pressure to attain the desired depression in $T_g$.

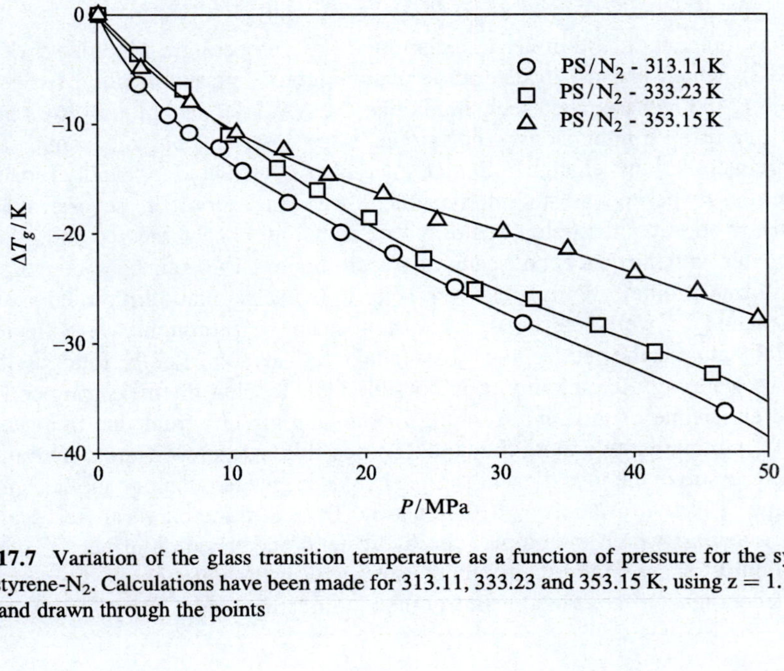


**Fig. 17.7** Variation of the glass transition temperature as a function of pressure for the system polystyrene-$N_2$. Calculations have been made for 313.11, 333.23 and 353.15 K, using z = 1. *Lines* are hand drawn through the points

### *17.3.3 Food Science*

Recent and fast developments in food science result from the large scale use of extruders to process different types of products from biscuits, crackers, breakfast cereals, flours, to more elaborate culinary components and pet food. Actually, food materials are extruded as other materials like polymers. Extrusion processes shape the final products in terms of structural organization (fibre like crystallisation, foam, soft or hard species) in combining elevated temperatures and pressures.

Optimization of extrusion processes requires a detailed knowledge of the properties of the starting ingredients. In this respect, the properties of starch which is a major component of food systems are essential to document, particularly over extended pressure and temperature ranges as well as water content. In this context, the contribution of scanning transitiometry to investigate starch-water systems has been recently demonstrated [14, 31]. Figure 17.8 presents an example of a transitiometric study of a 56% water suspension of wheat starch. Temperature scans have been performed at various pressures while recording simultaneously two output signals, heat flow and the thermal expansion. Interestingly, the endothermic main transition (positive enthalpy) corresponding to gelatinization is associated with a negative volume change, in complete agreement with the Clapeyron equation characterizing the transition.

## 17.4 Concluding Remarks

New developments in calorimetric techniques like Temperature Modulated DSC, (TMDSC), which permits to determine unambiguously glass transition temperatures [32], and in associated techniques like the VW-*pVT* technique [26] which allows measuring simultaneously the amount of gas entering a polymer sample and the subsequent volume change of the polymer, have contributed to notably broaden the area of experimental thermodynamics, with incursions in several major domains. In addition, the field of calorimetric techniques has witnessed an impressive impetus with the concept of scanning transitiometry. The main characteristic of scanning transitiometry is its great versatility in the sense that all physical states, homogeneous as well heterogeneous systems can be thoroughly investigated. A special feature is the possibility it gives to take full advantage of the four possible modes of scanning. Remarkably, it is possible after loading the measuring cell to activate either one of the four scanning modes and to shift from one to another without removing or reloading the cell. The possibility to use different hydraulic fluids to pressurize the investigated sample or system is undoubtedly another great advantage. The hydrostatic effect of a neutral fluid like mercury can serve as a reference against which to compare the hydrostatic/plasticization effects of other fluids (liquids or gases). Furthermore, a scanning transitiometer can be used as an instrumented reactor to perfectly control the modifications induced by $T$, $p$ and the

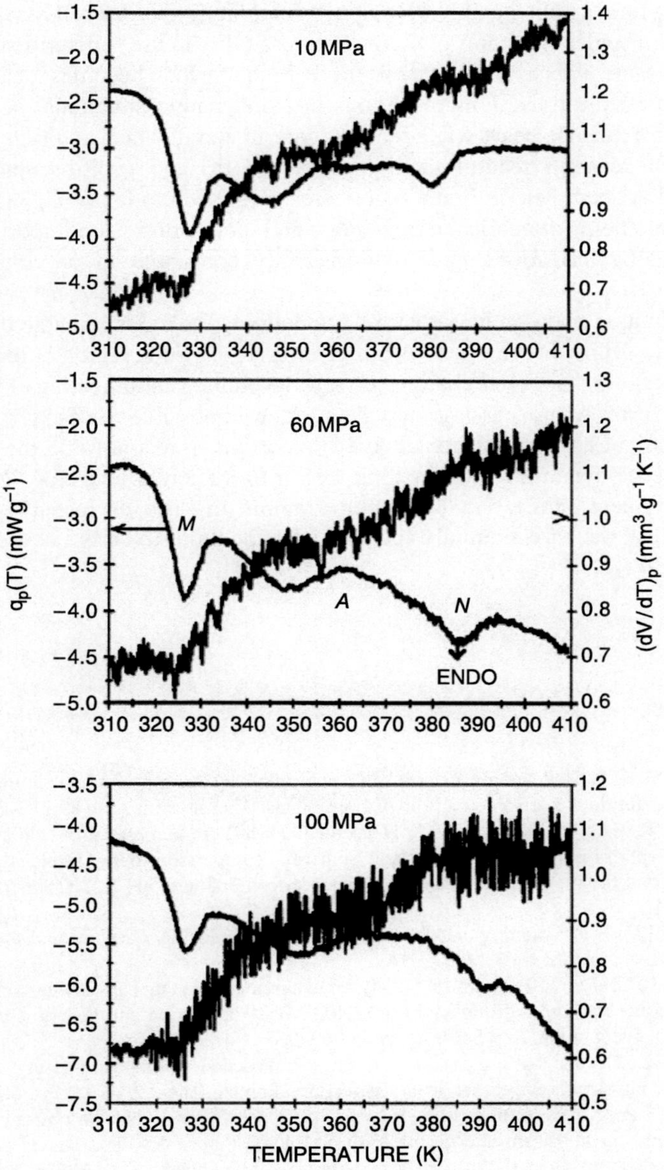


**Fig. 17.8** Transitiometric traces (thermal and mechanical outputs per gram of dry starch) obtained simultaneously in situ by scanning temperature at the rate 2.5 mK s$^{-1}$ at three different pressures for a starch-water suspension (56 mass% total water content). The middle graph allows distinguishing between the heat flow $q_p$ (thermal) thermogram and the volume change $(dV/dT)_p$ (mechanical) trace

pressurizing fluid as for example for polymer foaming [33] or for self assembling of nanoscale molecular structures [34–36]. The possibility to use supercritical fluids to transmit pressure adds a new feature to the technique which becomes ideally suited to investigate state transitions in various thermodynamic situations. A scanning transitiometer can also be used as a reaction calorimeter in which additional devices (for example reagents feeding capillaries, stirrer, pH and spectroscopic probes) allow to collect complete information on a reacting medium [8, 9, 37, 38]. Among other possible fields of applications, biology and biochemistry are important fields where scanning transitiometry should certainly contribute to provide original results.

This technique plays an important role in defining the most probable thermodynamic path to follow in order to in depth investigate a special region of interest in a system undergoing imposed changes or self developing modifications kinetically dependent. In making accessible new data, often impossible to obtain with other known techniques, scanning transitiometry contributes to reinforce the place of rigorous thermodynamics in providing key information to develop theoretical models; a typical illustration being for example the use of recently obtained thermophysical (isobaric thermal expansion) data on fluids to design new equations of state [39, 40].

## References

1. Randzio SL (1985) Scanning calorimeters controlled by an independent thermodynamic variable: definitions and some metrological problems. Thermochim Acta 89:215–241
2. Randzio SL, Eatough DJ, Lewis EA, Hansen LD (1988) An automated calorimeter for the measurement of isobaric expansivities and isothermal compressibilities of liquids by scanning pressure from 0.1 to 400 MPa at temperatures between 303 and 503 K. J Chem Thermodyn 20:937–948
3. Randzio SL (1991) Scanning calorimetry with various inducing variables and multi-output signals. Pure Appl Chem 63:1409–1414
4. Randzio SL, Grolier J-PE, Quint JR (1994) An isothermal scanning calorimeter controlled by linear pressure variations from 0.1 MPa to 400 MPa. Calibration and comparison with the piezothermal technique. Rev Sci Instrum 65:960–965
5. Randzio SL, Grolier J-PE, Zaslona J, Quint JR, Procédé et dispositif pour l'étude des transitions physicochimiques et leur application. French Patent 91-09227, Polish Patent P-295285; Randzio SL, Grolier J-PE, Procédé et dispositif pour l'étude de l'effet d'un fluide supercritique sur la transition d'un matériau de l'une à l'autre des deux phases condensées et leur application au cas d'un matériau polymère. French Patent 97-15521. http://www.transitiometry.com
6. Randzio SL, Stachowiak Ch, Grolier J-PE (2003) Transitiometric determination of the three-phase curve in asymmetric binary systems. J Chem Thermodyn 35:639–648
7. Dan F, Grolier J-PE (2002) Isothermal fluxmetry and isoperibolic calorimetry in anionic lactames polymerization in organic media. Setaram News 7:13–14
8. Grolier J-PE, Dan F (2007) Advanced calorimetric techniques in polymer engineering. In: Moritz H-U, Pauer W (eds) Polymer reaction engineering macromolecular symposia 259. Wiley VCH, Weinheim, pp 371–380

9. Dan F, Grolier J-PE (2004) Spectrocalorimetric screening for complex process optimization. In: Letcher T (ed) Chemical thermodynamics for industry. The Royal Society of Chemistry, Cambridge, pp 88–103
10. Randzio SL (1999) Transitiometric analysis of pressure effects on various phase transitions. J Therm Anal Calorim 57:165–170
11. Grolier J-PE, Dan F, Boyer SAE, Orlowska M, Randzio SL (2004) The use of scanning transitiometry to investigate thermodynamic properties of polymeric systems over extended T and p ranges. Int J Thermophys 25:297–318
12. Randzio SL, Flis-Kabulska I, Grolier J-PE (2002) Re-examination of phase transformations in the starch-water systems. Macromolecules 35:8852–8859
13. Randzio SL, Flis-Kabulska I, Grolier J-PE (2003) Influence of fiber on the phase transformations in the starch-water system. Biomacromolecules 4:937–943
14. Orlowska M, Randzio SL, Grolier J-PE (2003) Transitiometric in situ measurements of pressure effects on the phase transitions during starch gelatinization. In: Winter R (ed) Advances in high pressure bioscience and biotechnology. Springer, Berlin, pp 393–398
15. Rodier-Renaud L, Randzio SL, Grolier J-PE, Quint JR, Jarrin J (1996) Isobaric thermal expansivities of polyethylenes with crystallinities over the pressure range from 0.1 MPa to 300 MPa and over the temperature range from 303 K to 393 K. J Polym Sci B Poly Phys 34:1229–1242
16. Randzio SL, Grolier J-PE (1998) Supercritical transitiometry of polymers. Anal Chem 70:2327
17. Flöter E, De Loos ThW, de Swan Arons J (1997) High pressure solid-fluid and vapour-liquid equilibria in the system (methane + tetracosane). Fluid Phase Equilib 127:129–146
18. Stachowiak Ch, Grolier J-PE, Randzio SL (2001) Transitiometric investigation of asphaltenic fluids under in-well temperature and pressure conditions. Energy Fuels 15:1033–1037
19. Rodier-Renaud L (1994) Doctoral dissertation. Blaise Pascal University, Clermont-Ferrand
20. Dan F, Grolier J-PE (2006) High pressure-low temperature calorimetry. I. Application to the phase change of mercury under pressure. Thermochim Acta 446:73–83
21. O'Reilly JM (1962) The effect of pressure on glass temperature and dielectric relaxation time of polyvinyl acetate. J Polym Sci 57:429–444
22. Grolier J-PE, Dan F (2004) Calorimetric measurements of thermophysical properties for industry. In: Letcher T (ed) Chemical thermodynamics for industry. The Royal Society of Chemistry, Cambridge, pp 144–158
23. Ribeiro M, Pison L, Grolier J-PE (2001) Modification of polystyrene glass transition by high pressure methane. Polymer 42:1653–1661
24. Chow TS (1980) Molecular interpretation of the glass transition temperature of polymer-diluent systems. Macromolecules 13:362–364
25. Boyer SAE, Grolier J-PE (2005) Modification of the glass transitions of polymers by high pressure gas solubility. Pure Appl Chem 77:593–603
26. Boyer SAE, Grolier J-PE (2005) Simultaneous measurement of the concentration of a supercritical gas absorbed in a polymer and of the concomitant change in volume of the polymer. The coupled VW-pVT technique revisited. Polymer 46:3737–3747
27. Grolier J-PE, Unpublished results
28. O'Neill ML, Handa YP (1994) Plasticization of polystyrene by high pressure gases: A calorimetric study. In: Seyler RJ (ed) Assignment of the glass transition. ASTM, Philadelphia, pp 165–173
29. Chiou JS, Barkow JW, Paul DR (1985) Plasticization of glassy polymers by $CO_2$. J Appl Polym Sci 30:2633–2642
30. Zhang Z, Handa YP (1998) An in situ study of plasticization of polymers by high pressure gases. J Polym Sci B Pol Phys 36:977–982
31. Randzio SL, Orlowska M (2005) Simultaneous and in situ analysis of thermal and volumetric properties of starch gelatinization over wide pressure and temperature ranges. Biomacromolecules 6:3045–3050

32. Ribeiro M, Grolier J-PE (1999) Temperature modulated DSC for the investigation of polymer materials: a brief account of recent studies. J Therm Anal Calorim 57:253–263
33. Boyer SAE, Klopffer M-H, Martin J, Grolier J-PE, Grolier (2006) Supercritical {gas-polymer} interactions with applications in the petroleum industry. Determination of thermophysical properties. J Appl Polym Sci 103:1706–1722
34. Boyer SAE, Grolier J-PE, Pison L, Iwamoto C, Yoshida H, Iyoda T (2006) Isotropic transition behavior of an amphiphilic di-block copolymer under pressure using carbon dioxide and mercury as pressure medium. J Therm Anal Calorim 85:699–706
35. Boyer SAE, Grolier J-PE, Yoshida H, Iyoda T (2007) Effect of interface on thermodynamic behavior of liquid crystalline type amphiphilic di-block copolymers. J Polym Sci B Pol Phys 45:1354–1364
36. Yamada T, Boyer SAE, Iyoda T, Yoshida H, Grolier J-PE (2007) Effect of $CO_2$ pressure on isotropic transition of amphiphilic side-chain type liquid crystalline di-block copolymers. J Therm Anal Calorim 89:717–721
37. Dan F, Hamedi MH, Grolier J-PE (2006) New developments and applications in titration calorimetry and reaction calorimetry. J Therm Anal Calorim 85:531–540
38. Grolier J-PE, Dan F (2006) The use of advanced calorimetric techniques in polymer synthesis and characterization. Thermochim Acta 450:47–55
39. Randzio SL, Deiters UK (1995) Thermodynamic testing of equations of state of dense simple liquids. Ber Bunsen Phys Chem 99:1179–1186
40. Deiters UK, Randzio SL (2007) A combined determination of phase diagrams of asymmetric binary mixtures by equations of state and transitiometry. Fluid Phase Equilib 260:87–97

# Chapter 18
# Constrained States Occurring in Plants Cryo-Processing and the Role of Biological Glasses

**Jiří Zámečník and Jaroslav Šesták**

## 18.1 Unique Properties of Water Affecting Plant Life

The freezing temperatures well below 0°C [1] are common and there are several mechanisms to assure life survival. Processes associated with water freezing (particularly in conjunction with its supercooling) have been intensively studied [2–4] for many years because their significance in the bionetwork of both plants [5] and living. In human activity they play an important role in various production from a plain ice making to the complex foodstuffs freezing [6, 7] and pharmacy finishing. Even more important role they play in the viability of plants in natural overwintering and in controlled cryopreservation of plants.

Water is a compound, which exhibits some extraordinary uncommon and amazing properties and strange behaviour. It concerns a relatively high boiling-point, exceptionally large specific heat capacity and surface tension, and anomalous ability to dissolve both ionic and polar compounds. When water is left standing for a longer period, it tends to develop thixotropic properties, which implies a fragile but charge-containing (pH-variation) macrostructure ($OH^-/H^+$ assembling), curiously capable to even store and release amounts of preintroduced charge [8]. Impurities dissolved in water decrease both its melting and boiling temperatures. Recent studies, done via molecular dynamics, identified experimentally water clusters [2–4, 9], which grow in size and become more compact as temperature decreases. Their size can be characterized by a fractal dimension consistent with patterns common in natural world, having

J. Zámečník (✉)
Molecular Biology Department, Plant Physiology and Cryobiology Laboratory
Crop Research Institute, Drnovská 507, CZ-161 06 Prague 6, Czech Republic
e-mail: zamecnik@vurv.cz

J. Šesták (✉)
Division of Solid-State Physics, Institute of Physics, v.v.i., Academy of Sciences of CR, Cukrovarnická 10, CZ-16200 Prague 6, Czech Republic
e-mail: sestak@fzu.cz

J. Šesták et al. (eds.), *Glassy, Amorphous and Nano-Crystalline Materials*, Hot Topics in Thermal Analysis and Calorimetry 8, DOI 10.1007/978-90-481-2882-2_18,

the correlation length proportional to the configurational entropy [10]. Upon cooling, ice-like crystals were more easily transformed from the smaller clusters of $(H_2O)_n$ with n up to 20, more likely existing at temperatures >5°C. The larger clusters ($n$>20) can undergo supercooling more readily to alter to glassy structures at low temperatures. The decrease in the region of glass transition temperature with the increasing cluster size was experimentally observed to be much less operative than the corresponding change of melting temperatures. The mutual order of the melting and glass-transition temperatures were found to be reversed compared with that observed for bulk water.

Plants, tolerating ice crystals in their tissues possess several overall strategies to remain alive during temperatures below zero (Fig. 18.1). The first living strategy is to avoid nucleation while supercooled. Many vegetation and plant tissues are capable to overcome freezing by adjusting certain supercooled status [4] where the level of supercooling may go down to below −40°C [11]. There are multiple examples in different plant organs, such as parenchymatic cells [12], whole organs as generative buds of *Ribes* [13], *Malus* [14] or vegetative buds of *Abies* [15].

The second living strategy of plants is based on tolerating the extracellular freezing of water [13, 15–18]. Survival of such plant cells and/or tissues is based on their acceptability of excessive dehydration of the protoplast (Fig. 18.2). It is worth noting that quite a few of plant tissues can survive naturally temperatures down to −40°C, however, with the attuned techniques used for cryopreservation of plant genetic material they can survive even the liquid nitrogen temperatures [19].

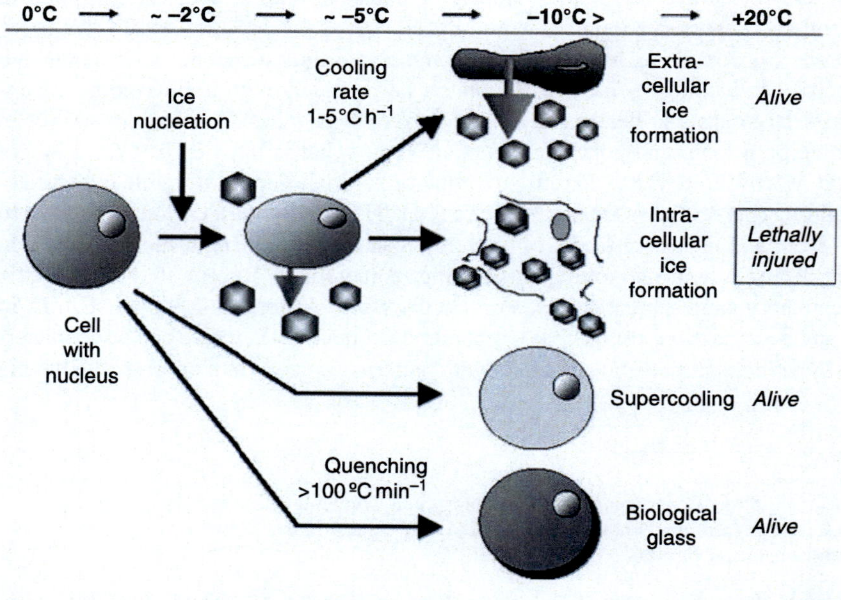


**Fig. 18.1** Natural plant behaviour while exposed to cooling explained by diagram of the freezing process in plant tissue. Freezing process is sorted according to the cooling rate and the degree of the ice crystal formation. Ice crystal inside the cell formation is predominantly lethal for plant

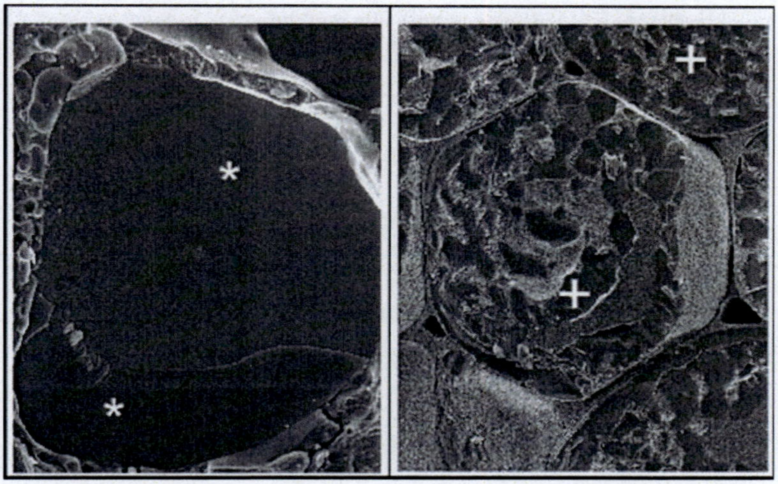


**Fig. 18.2** Cells from garlic shoot tip of dormant bulbils. Samples were quenched directly in liquid nitrogen, fractured and etched. Left is a cross-section of the fractured vitrified cells (*) after the process of plant vitrification (Plant Vitrification Solution No. 2 – treatment, magnification 10,000×) Right is non-vitrified cell with the visible ice-crystal (+) structure inside the protoplast (magnification 3,000×) [24]

The third strategy towards an enhanced survival during sudden cooling is called as an "extra-organ" freezing [14, 20] when the whole plant organ (for example bud) stays protected against nucleation and ice spreading inside the organ. Extra-organ ice formation causes frost dehydration of inner bud meristematic tissues resulted in their deep supercooling [21, 22]. Above mentioned survival strategies of plant species in their tissues are commonly witnessed such as the woody plants can exhibit extra-organ freezing in their buds, extracellular freezing in their bark tissues and supercooling in their parenchymatic cells of xylem rays [23].

However, a question remains what is the state of water or the state of protoplasm in the plant cells subjected to subzero and even to lower temperatures. Supercooled liquid can eventually switch into the form of non-crystalline solid, called glass. Such a water vitrification was treated within the concept of polymeric-polymorphic structure [4, 25]. The inherent glass formation takes place at a narrow temperature interval called "glass transition" (and abbreviated $T_g$) [26], which, fortunately, does not possess the volume change as it is associated with the transformation of ice. For example, the glassy state found in dormant twig of a poplar tree [2, 23, 27] offers the justification of what a curious state of solid water can be found in the frozen tissues.

The process of vitrification is [28], however, closely associated with the anomalous behaviour of liquid water [29] and its apparent tendency to pentameric netting at falling temperatures. Worth noting is possible interaction of liquid water with plant cells, which structures possess fractal and self-affinity architecture [15] yielding thus focal sites for compulsory nucleation enhancing and/or inhibiting ice formation. Therefore, it seems to be essential to reiterate some basic data about the structure of liquid water mainly of its configurationally quasi-periodicity, which seems be crucial for the formation conditions of quasi-crystalline nuclei and their capability to outline the state of either biologically optimal glass or hazardous ice.

## 18.2 Glass Forming Components in Plants

### *18.2.1 Supercooled Water*

Any commence of a state diagram for water relies, at least, on two sources of data:

- Variation of glass transition temperature [26] as a function of water content;
- Curve-based determination of a freezing point depression [30] (often completed up to the intersection with the $T_g$ curve).

Temperature of both curves for homogenous and heterogeneous nucleation is depending on water concentration. A state diagram cannot be created without a precise knowledge of water content and the corresponding value of glass transition temperature, $T_g$, which must be defined experimentally – $T_g$ for hyper-quenched water is generally believed to lay at about −133°C to −138°C [31]. Note the highest heat capacity water has and the same applies for ice, at temperatures below zero (Fig. 18.3), in comparison with other substances occurring in plants [32].

Water supercooling and ice formation have been theoretically analysed [12] within a traditional nucleation theory and the ice formation can be conventionally described by homogeneous nucleation if no foreign (nucleation extraneous) centres

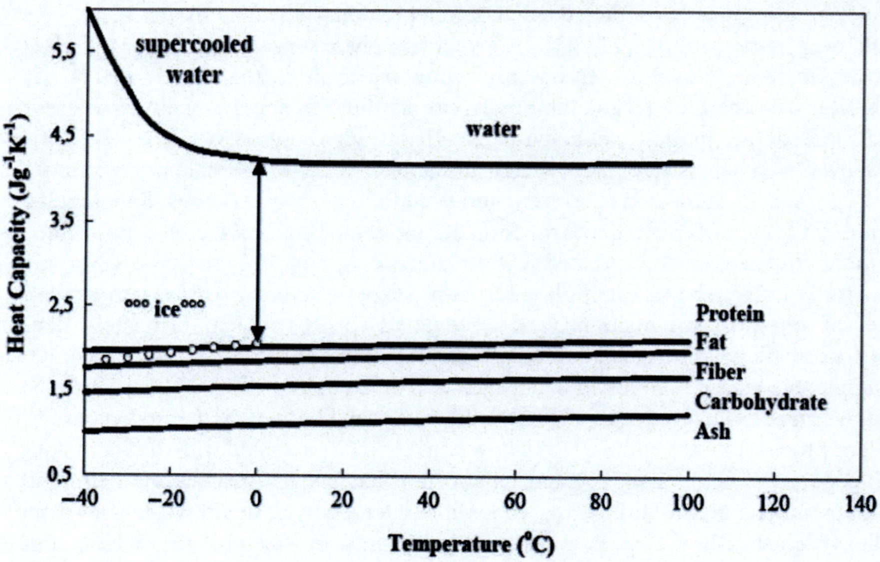


**Fig. 18.3** Heat capacity ($Jg^{-1}$ $K^{-1}$) of water and ice. The main organic substances and ash from plants. Data calculated according published equations [40]. Heat capacity line for protein and fat represents the same line. Note the abrupt change of heat capacity at 0°C of ice

are present. However, this is a rare case in nature, because microscopic inhomogeneities or biological interfaces (e.g. bacteria) are always present in water (together with the outer-bordering surfaces) acting thus as nucleation sites and triggering irresistible heterogeneous nucleation. Under normal atmospheric pressure, the ordinary liquid water can be easily supercooled down to about −25°C, with some further requirements (purity) as low as −38°C and with an enhanced supercooling (such as in small droplets of ~5 μm diameter) down to lowest −40°C. The bottom temperature limit for water supercooling (also known as the homogeneous freezing point) is achieved for water activity equal one and the associated freezing point depression [30] is close to 0°C. Where salts or hydrophilic solutes are present, the homogeneous freezing point reduces about twice (as much as the melting point). However, some sugar solutions have very low glass transition temperatures and the associated time for ice formation [26] becomes almost immeasurable.

The intracellular ice-crystal formation is in every case lethal for plant survival due to the inherent volume expansion of newly formed ice in comparison with the original volume of liquid (Fig. 18.2). One way how plants can withstand the low temperatures in the nature is their tolerance to movement of intracellular water to extracellular domains, which, in result, brings a higher tolerance to cell dehydration [4, 33–35]. However, the supercooled states frequently possess no dehydration effects or ice-crystal formation, but this state is unstable. The exceedingly low viscosity at rapidly cooling conditions of 'freeze-in' states give often birth to a new non-crystalline state of a rigid state of glasses, and remaining once again, such a process of vitrification is not associated with any dangerous volume changes. The freeze-in states can nearly hinder any diffusion of molecules, preventing thus biochemical transformations and, accordingly, excluding also any genetic changes. From a comparison of the elements and inorganic compounds forming glasses [36–38] in plants it is not clear to derive a strict relationship of glass formability in plants.

### *18.2.2 Carbohydrates*

Organic compounds, such as sucrose, glucose, glycerol fructose and proteins occur in higher concentration in plants and it is not clear how to derive a strict relationship of glass formability in a wider organic disposition. A certain exception can be seen in carbohydrates, mainly in the region of high concentration of sucrose. The levels of various organic and inorganic solutes in stem tracheal (xylem) sap and fruit tip [39] sap of legume *Lupinus albus* is about a thousand times higher than elements content in the same plants [41]. Sucrose and specific proteins are the main compounds of plants, which are synthesised during an acclimation procedure toward cold.

Sugars are commonly used as cryoprotectants. Their concentrations and the individual types of sugars must be optimized toward the developing cryopreservation methods [34], those based on glass involvement. The cryobiologists tested various carbohydrates in an attempt to improve the percentage of explants surviving with their regeneration. Unlike sugar alcohols such as sorbitol and manitol, which are not readily metabolised in the cells of many plant species, sucrose is readily metabolized and is thought to have several protective functions during dehydration and subsequent freezing and thawing. Most of these sugar alcohols are not as effective cryoprotectants as sucrose is. Sucrose has been found to be the most effective cryoprotectant in many other studies [34, 42].

Carbohydrate activities are an important intermediary for cryopreservation:

- Can decrease water content by osmotic dehydration;
- After influx of carbohydrate into the cells, the osmotic potential of plant tissues often decreased;
- Osmoprotectants (sugars like sucrose) are thought to contribute maintaining the membrane integrity during the dehydration and freezing processes. It is intermediated by the formation of hydrogen bonds between sugars and the hydrophilic components of cellular membranes [43];
- Act as a glue sticking for shoot tips to the carrier (for example aluminum foil) after dehydration;
- Accomplish the first aid during the energy starvation throughout the plant regeneration after warming.

Sugars are seen as the most usable cryoprotectants involved in plants as a reaction to abiotic stress (mainly in drought and cold conditions). The sucrose is used also as the most acceptable cryoprotectant added to the plant during their cryopreservation at ultra-low temperatures. Using sugar as a cryoprotective example, it can be shown how difficult it is to interpret a more complex behaviour even for such a relatively simple substance and how difficult it can become to define its mixtures in the plant tissue (Fig. 18.4). None explicit expression derived from theoretical models is able to predict the behaviour satisfactorily enough.

The glass transition temperature for a binary mixture of water and sucrose seems to follow the semi-empirical Couchman and Karasz expression [35], which is only partially profitable in the description of aqueous solutions. Although empirical, this equation [35] seems to be the best model to describe varying $T_g$ along with the composition. Modified Gordon–Taylor equation (MGT, Eq. 18.1) can be used to estimate the $T_g$ values of single-phase sucrose solutions (when no ice is present). Glass transition temperature data for sucrose with various water contents were collected from the literature (Fig. 18.4) and then fitted by the following Eq. 18.1

$$\begin{aligned} T_g = \left(\left(kT_{g2} - C_w\left(kT_{g2} - T_{g1}\right)\right)/\left(C_w\left(1 - kT_g\right) + kT_g\right)\right) \\ + \alpha_1 C_w(1 - C_w) + \alpha_2 C^2(1 - C_w), \end{aligned} \tag{18.1}$$

where $C_w$ is the concentration of water at $T_g$ (weight fraction), $T_{g1}$ and $T_{g2}$ are the mid-points of the glass transitions for pure water and sucrose, while k, $\alpha_1$ and $\alpha_2$ are constants. Here we use 348.2 K for $T_{g2}$, the glass transition temperature of

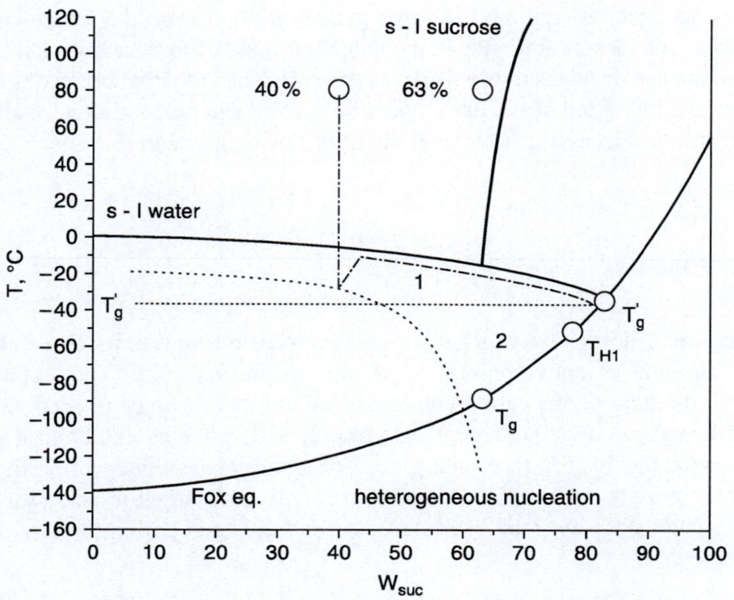


**Fig. 18.4** Schematic state diagram of aqueous solution of sucrose [30], an example of 40% concentrated sucrose. Curve 1 is for equilibrium freezing, curve 2 is for thawing of 63% frozen solution. Glass transitions corresponding to the nominal composition of the initial solution of sucrose $T_g$, glass transition of maximum freeze concentrated phase $T_{g1}$, glass transition domains around sucrose crystals $T_g{'}$. Solid curves are for water solid–liquid boundary (s–l water) and sucrose solid–liquid boundary (s–l sucrose)

pure sucrose, and $T_{g1}$ for water is taken as 135 K. The "best fit values" for the parameters are $k = 0.092$, $\alpha_1 = 481$ and $\alpha_2 = -1{,}225$ [31].

No attempts are made to apply this equation to multi-component plant samples, but it could become feasible under further credentials. However, the equation allows considering a mixture of several components as one component and water as another one having chiefly some practical values only. The problem still remains because it does not recognize the definite $\Delta Cp$ even for pure water, which the Couchman-Karasz equation uses for predicting the $T_g$ data of multi-component systems. Nevertheless, both equations were satisfactorily used for description of water content and glass transition for such multi-component biological material as the fruit powders are [44].

From the above mentioned results, it follows that diluted and semi-concentrated aqueous solutions of sucrose ($w_{suc}$ weight fraction of sucrose in Fig. 5.4, Table 5.1) freeze under a non-equilibrium state. The system is heterogeneous and contains crystals of ice and sucrose. These exhibit two types of domains of amorphous solids [4], which composition corresponds to that of maximum freeze concentrated phase (MFCP) characterised by glass transition temperature $T'_g$. The solution inside the domains around sucrose crystals is characterised by glass transition temperature $T_{g1}$. At very low temperatures, the resulting structure of the solution is heterogeneous containing ice, sucrose crystals, and two glasses formed from the freeze-concentrated phase and the solution lying in the neighbourhood of sucrose crystals. During plant freezing and drying, sucrose works as an explicit and inherent protecting agent, but a mere accumulation of sucrose alone is not a sufficient precondition for all plant species toward a successful survival at the ultra-low temperatures.

### *18.2.3 Proteins*

The formation of intracellular glass is proposed to be relevant to protein stabilization and survival of anhydrobiotic organisms in the dry state. The stability of proteins in the amorphous carbohydrate matrix and its relevance to seed survival has been investigated [45]. Thermal stability of seed proteins exhibited a strong dependence on the $T_g$ of intracellular glass. These results indicate an important role of the glassy state in protein stabilization. This data suggest an association between protein stability in intracellular glass and seed and pollen survival during storage [45–48].

## 18.3 Artificial State of Biological Glasses Contributing Plants' Survival at Ultra-Low Temperature

A new approach how to store plant samples at ultra-low temperature is called the 'ice free cryogenic storage' [1, 50], which is based on the instigated formation of biological glasses. Such a strategy how to reach the glass transition in plant samples for their long-term storage by cryopreservation methods is prospective in the three following ways of an external control of:

- The rate of cooling and warming;
- Dehydration removing excess of water from the tissue, safeguarding enough water volume in the tissue to keep the plants alive. The dehydration can be done often by three ways (see below);
- Infiltration of plant samples with substances shielding the unwanted crystallization, so-called 'cryoprotective' fixatives or cryoprotectants, which are usually efficacious glass formers.

### *18.3.1 Cooling and Warming Rate*

An important factor is the rate of freezing, which is crucial for the ice growth and sizing. The ice formation in plants meets a certain temperature stress. By application of a well guarded rate of cooling/warming, the cryopreservation method subsists as the oldest method, which was ever used [51, 52]. For the first time, it was demonstrated that the willow twigs collected during very strong winter, additionally prefrozen at −30°C, were successfully cryopreserved in liquid nitrogen for 1 year with subsequent plant regeneration [51]. The principle of this method is to give enough time to precede the extracellular freezing of water by slow cooling to −30°C. In principle, it is like the natural freezing occurring in wildlife.

The two-step freezing method based on extracellular freezing was successfully applied to hardy tree buds, which were recovered by grafting [53, 54]. Controlled rate of cooling is efficient for storing suspension and callus cultures, embryogenic cultures and in vitro shoot tips from temperate and subtropical plants. In addition, cryopreservation of cultured cells and meristems was achieved by controlled slow pre-freezing to about −40°C in the presence of a relevant cryoprotectant [55] where single chemicals or their combination in various ratios (mixed in a suitable cryoprotective cocktails) can be employed. These cryoprotectants help the plant samples to stabilize their membranes and proteins after deep dehydration, alternatively acting as an energy pool during the recovery of plants. For a common use, dimethylsulfoxide (DMSO) is utilized as a representative cryoprotectant of colligative group capable of penetrating inside the cells. The polysaccharides (mainly sucrose and polyethylene glycols) are representative of non-penetrating cryoprotectants [56]. The volume of water inside the cell is decreasing while the concentration of solutes increases until the equilibrium of water activity outside and inside the cells is reached.

A control of glass formation in the tissue during this researching method was invented by authors [52, 57] but not proved yet. After freeze dehydration it is speculated that plants can adapt to survival due to inherent glass transition ongoing inside the plant's cells. It is anticipated by following facts. The intracellular matrix occurs highly concentrated owing to extracellular freezing so the intracellular ice nucleation cannot take place during the slow cooling and, therefore, the plants turn out to be capable to survive even at ultra-low temperatures. When the intracellular ('freezable') water is reduced to that level, which is not dangerous for the plants plunging into liquid nitrogen, its remaining amount probably caught in the glass is, however, still high enough to keep viable function of cellular compartments inside the cells. Usually, after the dehydration the cells are able to recover and/or repair their full function indispensable for living.

Warming rates are rather critical for the plant survival when taken from liquid nitrogen. The controlled rate of warming is mainly important for the two-step freezing method. In controlled rate of cooling taking place in the presence of cryoprotectant, the consequent higher warming rates are frequently applied. The highest rate of warming can be reached by quench-immersing cryovials into the hot water held at about 40°C. Upon such a higher rate of warming it is believed that the recrystallization becomes avoidable. However, immediately after warming, the potentially toxic cryoprotectant must be washed out from the plants. The high rates of warming minimize cold crystallization often proceeding after the glass transition.

In the course of a slow cooling, ice crystals start to grow inside the cells, which is mostly lethal. Extracellular artificially-induced ice nucleation can, however, avoid the random intracellular nucleation. Moreover, slow cooling rate may result in colligative concentration of the solution inside the cell where the ions in the highly concentrated solution can interact with the main components of membranes [58], causing thus non-reversible injury.

When the rate of cooling is too rapid (>3°C $min^{-1}$), cells will not likely remain in an equilibrium with extracellular ice, and the intracellular solution would become supercooled. Such a supercooled state of the intracellular solution is lacking ice crystal formation and bears alongside a sensitive state of plant called metastable state, which could be dangerous, likewise. When the ice formation occurs in a supercooled state at ultra-low temperatures, it usually leads to intracellular ice formation from either intracellular ice nucleation [59] or through seeding via extracellular ice [28, 60]. Mostly, this metastable state becomes noxious for plants. At ultra-high cooling rates (applied down to ultralow temperatures taking place during rapid quenching under ~500°C $s^{-1}$) [52], the supercooled water solution is often transited into an amorphous/glassy state. Plant cells exploit such a non-extensible glassy state and its diminishable possibility of ice crystal formation.

The survival of seeds exposed to liquid nitrogen temperatures is also influenced by an interaction between cooling rate and moisture content. Rapid cooling of seeds with higher moisture contents (where 'freezable' water is clearly displayed) has beneficial effects, while rapid cooling of dry seeds with high lipid contents is detrimental. It is suggested that glass transitions are associated with the two effects of the water and lipid components of the seed [61].

### *18.3.2 Dehydration*

Plant cells and tissues that are supposed to survive impact of liquid nitrogen need to be dehydrated to a certain level ahead of their immersion into liquid nitrogen. The aim of all following dehydration approaches is to increase cell-inner viscosity to the level at which the ice crystallization is inhibited and the intracellular matter becomes present in glassy state.

Three main methods for dehydration of plant tissue used in cryopreservation of plants are the following:

- Evaporative dehydration in the air – the driving force for dehydration is a lower vapour tension than that over the plant sample.
- Osmotic dehydration – the driving force for dehydration is a lower osmotic potential in solution than that of the plant sample.
- Freezing dehydration – the driving force for dehydration is a difference in water vapour over the supercooled state of water in plants and over the ice in the surrounding space.

During dehydration, the viscosity of plant protoplast increases to a level required for glass formation, avoiding accordingly the dangerous voluminosity of ice crystal formation.

### *18.3.3 Desiccation*

It is worth reminding that the water activity depends on the inherent temperature, water content and on the associated glass transition. If we know them we can assume a certain critical water activity corresponding to the relation $T_g = T$. The relationship between water activity and the state diagram was described by Slade and Levine [65]. Water remains fairly mobile within glassy plant parts and the water activity in glassy state can thus be determined.

The relationship between glass transition temperature, $T_g$, and the water content, can be measured. This relationship is useful for finding at what a low level we need to dry/desiccate the plants to incorporate the glass transition inside plant tissues. The subsequent importance for such a desiccation is the threshold of water content up to which the plant is able to fully regenerate its recovering to original, new plant. Between these two limits of water content the cryopreservation of plants is feasible for enough long time. A high probability of plant liveability and associated plant crucial recovery is reaching after warming them from the conservation state at ultra-low temperatures.

Glass transition temperature must be determined under the number of gradually changed water contents. This must be accomplished in such a way as to become sure that the transition measured is really a glass transition relevant to the water and not, for example, an associated glass transition for lipids and other substances. Since the plant matrix is complex, a phase separation often occurs. The local water content in sub-localities (micro-regions, micro-domains) may become very important. The driving force for desiccation of the plants in the air is related to the difference between water vapour in the tissue and the surrounding air (Table 18.1). The greater the difference in water vapour the greater the strength for removing water from plant tissue.

Osmotic dehydration is one of the most common ways how to remove water from plant samples not only providing less water but also a possible incorporation of some osmotic compounds for a direct protection. In principle, during the osmotic dehydration a plant part is factually bathed in solution, which has a higher concentration of solutes than that inside the cells. The driving force for the osmotic dehydration is the difference in water potential between cells and the osmotic potential of surrounding solution (Table 18.1).

Fully turgid[1] plant cells have water potential close to zero like a very low concentrated solution. The highly concentrated solution inside the cells (representing low osmotic potential) is compensated by turgor potential with opposite signs. For an

---

**Table 18.1** Temperature and dehydration conditions for glass transition involvement as a basis for plant cryopreservation methods

| Dehydration method | Cryopreservation method | Dehydration driving force | Dehydration at temperature | Glass transition temperature |
|---|---|---|---|---|
| Air dehydration | Encapsulation-dehydration | $\Delta p = p_p - p_a$ | T>0 | $T_g$<T |
| Osmotic dehydration | Plant vitrification solution (PVSx, MPVS) | $\Delta\psi = \psi_{\pi s} - (\psi_{\pi p} + \psi_T)$ | $T>T_m$ | $T_g$ <T |
| Freezing dehydration | Two-step freezing, Controlled rate cooling | $\Delta p = p_{sw} - p_i$ | ~ −40<T<0 | $T_g < T_m$ |

Note: Δp – difference in water vapour tension; $p_p$ – water vapour tension in plants; $p_a$ – water vapour tension in air; Δ ψ – difference in water potential; $\psi_{\pi s}$ – osmotic potential of plant vitrification solution; $\psi_{\pi p}$ – osmotic potential in plants; $\psi_T$ -turgor potential (turgor potential at plasmolysis is zero); $p_{sw}$ -water vapour tension of suppercooled solution in plant cells; $p_i$ – water vapour tension over ice; T – ambient temperature; $T_g$ – glass transition temperature; $T_m$ – temperature of freezing point depression. PVSx Plant Vitrification Solution (PVS2 [62], PVS3 [63], MPVS [64])

involved glass transition, the osmotic dehydration must be very deep. In the case of an excessive volume of osmotic solution, the changes in osmotic potential of the surrounding solution can be constant, taking in account rather low volume of water removed from the tissue. The samples are gradually dehydrated by water removal until equilibrium with the surrounding osmotic solution is reached.

## 18.4 Distinctiveness of Measurements of Biological Glass – DTA/DSC

Different types of methods are commonly used to determine the temperature span of glass transition, preferably based on:

- Thermodynamics, that is, using conventional differential scanning calorimeter (DSC) or differential thermal analysis (DTA) technique
- Other dynamic measurements of properties such as mechanical spectroscopy
- Characteristic times can then be associated with enthalpy and mechanical relaxations, which may be different because of certain “decoupling” between different relaxation modes. It should be considered there are differences induced by the given method of experimental data analysis
- Definition of $T_g$ on the DSC, DTA curves
- Characterization of relaxation times
- Mechanical measurements often determined in terms of modulus, such as dielectric measurements in terms of compliance

One of the main advantages of DSC, DTA is that the measurements can be performed in sealed pans. The size of plant samples is usually small enough to fit to the hermetic pan. One to three pieces of *Allium* shoot tips weight 3–5 mg and one to three 3–8 mg vegetative apple buds weight of 1–3, as an example.

Although DSC is not the most sensitive method to measure $T_g$ the main advantage of DSC is that it allows repeated and relatively reproducible measurements easy to carry out along a wide temperature range.

Glassy state and associated phase appraisal during cryopreservation can be determined by DSC, which is a sensitive method for the determination of various crossing phenomena occurring towards the vitrification (Fig. 18.5, as an example). Second consequent DSC run with the same biological sample could, however, be substantially different from *the first run*, because the plant samples could be injured or completely destroyed after the first run completion. Starting temperature is important to hold plant samples in equilibrium allied with the starting temperature. Room temperature as a starting temperature for plant investigation is often for their non-acclimated state.

Starting temperature for cold-hardened plants should be at least close to 0°C and should be accepted after a long-time equilibrium. At extremely slow cooling rates, degrading biological samples could occur frequently. The maximum temperature of warming cannot cause a change of substances, e.g., protein denaturation, sugar caramelization and starch gelatinization. The temperature condition of samples in an autosampler is important for long-running measurement.

A new technique, called dynamic cooling protocols, is using a conventional DSC and has been developed to obtain chart of dynamic and quantitative water transport data in cell suspensions during freezing [66].

## 18.5 Stability of Biological Glasses

The quality of glass is possible to classify. In simplicity it is possible to express as the distance of glass transition and freezing onset of melting. The greater difference between glass transition and freezing onset of melting the higher stability of glass is [27].

Glassy status as a common solid, rigid state occurring in biological samples habitually results from a suitable circumvention of crystallization. The quality of glassy state is possible to classify simply by ratio ($T_g$, $T_m$) – the lower ratio reveals the higher difference of $T_g$ and $T_m$, thus showing a greater stability of glassy state. A more sensitive interrelation to the glass formation peculiarities can be found based on Hruby coefficient $K_{g1}$:

$$K_{g1} = (T_c - T_g)/(T_m - T_g), \qquad (18.2)$$

where $T_c$, $T_g$, $T_m$ are temperatures of crystallization, glass transition and melting temperature, respectively, the values of which, however, can be found only by

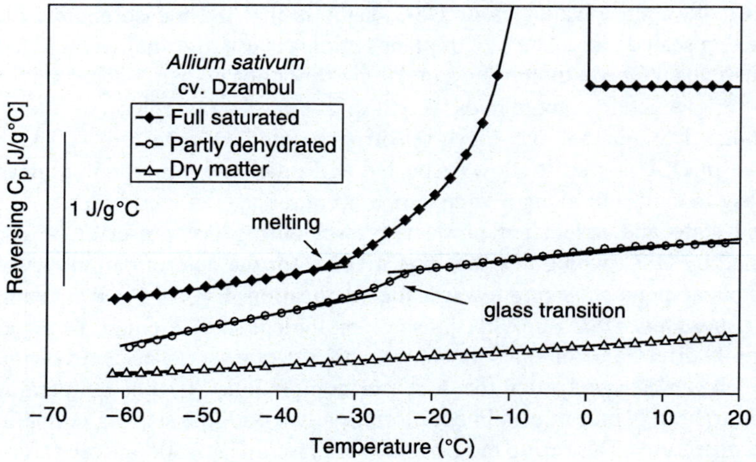


**Fig. 18.5** An example of the glass transition (*middle curve*) in *Allium* shoot tips measured by quasi-isothermal modulated differential scanning calorimetry (Zámečník et al. submitted)

suitable experiment using the ready-to measure sample (pre-prepared glasses) [15]. It is clear that the greater the value of $K_{gl}$ the better is the glass-forming. The best utilization of $K_{gl}$ is in the comparison of glass-forming ability and glassy state stability of different biological materials under different conditions and thermal treatments. Hruby coefficient ($K_{gl}$) indicates the glass stability against crystallization on heating and could be used to estimate the vitrification ability of glass forming liquids. These both characteristics ($T_g$,$T_m$) are easily measured by conventional DTA, by DSC samples in glassy state. The correlation between glass stability characterized by Hruby parameter and the glass forming tendency characterized by critical rate of cooling for glass formation was confirmed [67].

The kinetic property of biological objects in glassy state is poorly understood yet. Dynamic processes including the membrane fusion and solute leakage of frozen and dry liposomes in carbohydrate glasses, and ice formation haemolysis of frozen human red blood cells, were recently studied [68]. The kinetic stability of biological glasses is clearly $T_g$-dependent. The reaction rate constants deviated notably from the traditional Arrhenius behaviours can be well fitted to the Williams-Landel-Ferry (WLF) [26, 69] equation

$$\log_a(T) = -C_1(T - T_g)/\{C_2 + (T - T_g)\}, \tag{18.3}$$

where $C_1$ and $C_2$ are universal WLF constants. It was found that different polymers exhibit similar WLF constants. The universal values of constants in the WLF equation, used for abroad spectrum of amorphous materials ($C_1 = 17.44$, $C_2 = 52.1$) do not match, however, to the dynamic processes of biological material. The derived C1 and C2 constants of the WLF equation for seed survival of three species at $T>T_g$ are: for Glycine ~ 37.5 and 269.8, for *Pisum sativum* 22.3 and 179.2, and for *Phaseolus vulgaris* 22.1 and 183.1, respectively [26]

Angell [70] defined fragility via the diagrams of log (η) vs. $T_g/T$, where η is viscosity. When the temperature dependency of viscosity is high, such glass is called fragile. At temperatures higher than the glass transition temperature, a different behaviour of glasses can be commonly distinguished according to the ratio of the crystalline melting point [30] and glass transition temperature [26]. Consequently according to Slade and Levine [65], it is possible to define three classes of amorphous systems:

- The ratio $T_m/T_g \approx 1.5$ is characteristic for the first-class. The amorphous system in this class of biologic-like materials is represented by glucose, for example, where the viscosity decrease is moderate. Description is typical to comply with some 'well-behaved' polymers or 'fragile' liquid [35]. The universal constants in WFL equation are 17.44 for $C_1$ and 51.6 for $C_2$.
- The ratio $T_m/T_g >> 1.5$ is often categorized for the second-class. The viscosity decrease is very small. Such glasses are typical products of 'poorly behaved' polymers or strong liquids. This 'typical but not well-behaved' class is a readily crystallisable matter (e.g., water) and the WLF constants are about 20 and 155 for $C_1$ and $C_2$, respectively.
- The ratio $T_m/T_g << 1.5$ (when the temperature of glass transition is close to the temperature of the melting) is distinguished for the third-class. The WLF constants of $C_1 = 12.3$ and $C_2 = 22.3$ apply [7] and the glass-forming systems are described as 'a typical and poorly behaved' class represented by, e.g., native starch, gelatine, fructose or galactose and/or other very fragile liquids.

The kinetics of ice formation also deviates from the Arrhenius relationship, following the WLF kinetics. The WLF constants for the process of ice formation were calculated to be 2.1 and 11.2 for $C_1$ and $C_2$, respectively. Therefore, the ice formation in frozen cells may be distinguished by WFL constants from glass transition in the above mentioned three classes.

For some supercooled liquids, typically water, a fragile-strong transition exists [71]. The simple model system has been reported by Angel et al. [72]. For electrolyte solutions that are very fragile, and become stronger as concentration increases due to the replacement of weaker hydrogen bonding with stronger ion-dipole interaction. In contrast, any sugar solution becomes more fragile as concentration increases. High molecular weight proteins such as gluten are relatively strong.

The prediction of the stability of biological materials preserved in the amorphous matrix is of a very particular interest. From the WLF Eq. 18.3, one can easily derive that the stability of biological materials in the amorphous matrix will decrease at a rate proportional to $(10^{T-Tg})$ if held in the rubbery state at temperatures $T > T_g$.

However, more accurate estimation is needed in practice. A conceptual illustration has been provided by Slade and Levine, who used the constants of WLF equation to characterize the three classes of amorphous systems (see previous discussion). For the systems of the first class with the ratio $T_m/T_g \approx 1.5$, a change by factor 10 for every 3°C is recognizable. However, for both, for the second ratio $T_m/T_g >> 1.5$ [73] and for the third class with the ratio $T_m/T_g << 1.5$, this stability can show the decrease by a factor of 10 for any increase of every 6°C [25]. Sun [26] has done a similar calculation using the derived WLF constants knowing that the rate of ice formation in frozen blood cells (prepared during the devitrification) increases approximately by the factor $10^{(T-Tg)/6}$ (this is a 10-fold increase over a 6°C interval above the $T_g$). This value matches wonderfully to the second-class of amorphous systems, which includes water [7]. However, the stability of frozen and dehydrated biological cells and membranes, as measured by haemolysis and solute leakage, turns up to decrease much slower than expected, roughly at a rate of $10^{(T-Tg)/15}$ at temperatures $T > T_g$, (i.e., a 10-fold decrease over 15°C above the $T_g$).

## 18.6 Biodiversity Long-Term Storage Based on Glassy State Involvement

Detrimental influence of man-affected environment bears, however, an injurious impact on plant growth and endurance. In some cases, the hostile environmental conditions may even lead to a serious depression, and in limit, extinction of the population density of certain plant specimens on our planet. On the other hand, there are also many plant species close to the edge of their disappearance under the current deterioration effect of changing environment or even under a routine fatality of species vanishing. There is a supplementary call for the urgent need to store plants with their entire genetic information, using preferably the cultural varieties, land-races and original wild-plant genetic resources. Fortunately, it is not necessary to store a whole plant but it is enough to have its buds, even just protoplast, because the plant cells are totipotent. Therefore it is possible to store only a small part of plants, mostly a part of meristematic tissues of vegetatively propagated plants [33] or a seed axis (mostly of recalcitrant seeds) of generatively propagated plants [49]. The plant storage at low or even at ultra low temperatures (cryopreservation) is one of the appropriate ways how to keep their viability in a long term prospect.

From the view of cryopreservation, the glassy state, which possesses the same volume as liquid, occurs as the most suitable state for a long-term biological storage of germplasm. This strategy of 'being in the glassy state', is repeatedly exploited by plants in the nature in order to withstand unfavourable freezing conditions of surrounding environment, which can be seen as a sort of miracle. For example, plant pollen flying in the air has a potential to be in glassy state, keeping its viability after stigma pollination. Dormant poplar twigs can survive the low temperatures being in the glassy state in midwinter. It is possible to add other examples of glassy state occurrence in plant species in the nature. However, the conditions at which the glassy state is induced are not fully understood. Through the glass formation in plants can be of help to control the 'cryogenic status' in which the optimal cooling conditions achieve a suitable cryopreservation of plant germplasm.

**Acknowledgments** This work was supported by MSMT grant 0002700604

## References


1. Sakai A (2004) Plant cryopreservation. In: Fuller B, Lane N, Benson EE (eds) Life in the frozen state. CRC Press, London, pp 329–345
2. Bartell LS (1997) On possible interpretations of the anomalous properties of supercooled water. J Phys Chem B 101:7573–7583
3. Nitsch K (2009) Thermal analysis study on water freezing and supercooling. J Therm Anal Calorim 95:11–14
4. Sestak J, Zamecnik J (2007) Can clustering of liquid water and thermal analysis be of assistance for better understanding of biological germplasm exposed to ultra-low temperatures. J Therm Anal Calorim 88:411–416
5. Burke D, Kaufman P, Mcneil M, Albershe P (1974) Structure of plant-cell walls. 6. Survey of walls of suspension-cultured monocots. Plant Physiol 54:109–115
6. Goff HD, Sahagian ME (1996) Glass transitions in aqueous carbohydrate solutions and their relevance to frozen food stability. Thermochim Acta 280:449–464
7. Slade L, Levine H (1991) A food polymer science approach to structure-property relationships in aqueous food systems: non-equilibrium behavior of carbohydrate-water systems. Adv Exp Med Biol 302:29–101
8. Ovchinnikova K, Pollack GH (2009) Can water store charge? Langmuir 25:542–547
9. Giovambattista N, Buldyrev SV, Stanley HE, Starr FW (2005) Clusters of mobile molecules in supercooled water. Phys Rev E 72:11202
10. Chvoj Z, Šesták J, Tříska A (1991) Kinetic phase diagrams: non-equilibrium phase transitions. Elsevier, Amsterdam
11. Šesták J (2004) Heat, thermal analysis and society. Nucleus, Hradec Kralove
12. Ristic Z, Ashworth EN (1996) Response of xylem ray parenchyma cells of red osier dogwood (*Cornus sericea* L.) to field freezing stress, and to freeze-thaw cycle. J Plant Physiol 149:735–745
13. Ashworth EN, Wisniewski ME (1991) Response of fruit tree tissues to freezing temperatures. Hortscience 26:501–504
14. Bilavčík A, Zámečník J (1996) Localization of endogenous ice nuclei in flowering apple shoots. Biologia 51:62
15. Sakai A (1982) Freezing tolerance of shoot and flower primordia of coniferous buds by extra-organ freezing. Plant Cell Physiol 23:1219–1227
16. Beurroies I, Denoyel R, Llewellyn P, Rouquerol J (2004) A comparison between melting-solidification and capillary condensation hysteresis in mesoporous materials: application to the interpretation of thermoporometry data. Thermochim Acta 421:11–18
17. Malone SR, Ashworth EN (1991) Freezing stress response in woody tissues observed using low-temperature scanning electron-microscopy and freeze substitution techniques. Plant Physiol 95:871–881
18. Warmund MR, George MF, Cumbie BG (1988) Supercooling in darrow blackberry buds. J Am Soc Hortic Sci 113:418–422
19. Sakai A, Kobayashi S, Oiyama I (1991) Survival by vitrification of nucellar cells of navel orange (Citrus-sinensis var brasiliensis Tanaka) cooled to 196-degrees-C. J Plant Physiol 137:465–470
20. Sakai A (1979) Freezing avoidance mechanism of primordial shoots of conifer buds. Plant Cell Physiol 20:1381–1390
21. Chalkerscott L (1992) Disruption of an ice-nucleation barrier in cold hardy azalea buds by sublethal heat-stress. Ann Bot 70:409–418
22. Kang SK, Motosugi H, Yonemori K, Sugiura A (1998) Supercooling characteristics of some deciduous fruit trees as related to water movement within the bud. J Hortic Sci Biotech 73:165–172
23. Hirsh AG, Williams RJ, Meryman HT (1985) A novel method of natural cryoprotection – intracellular glass-formation in deeply frozen populus. Plant Physiol 79:41–56

24. Bilavčík A, Faltus M, Zámečník J (2005) Nové teoretické poznatky o vlivu ultranízkých teplot na rostliny (in Czech) In: Sborník příspěvků a laboratorních postupů z workshopu "Metody kryoprezervace vegetativně množených rostlin". VÚRV Praha-Ruzyně, pp 8–15
25. Minaev VS, Timoshenkov SP, Chernykh SP (2004) Polymorphous-crystalloid nature of vitreous and liquid $H_2O$. J Optoelectron Adv M 6:103–112
26. Sun WQ (1997) Glassy state and seed storage stability: the WLF kinetics of seed viability loss at $T>T_g$ and the plasticization effect of water on storage stability. Ann Bot 79:291–297
27. Šesták J (2005) Science of heat and thermophysical studies: a generalized approach to thermal analysis. Elsevier, Amsterdam
28. Šesták J (2000) Miracle of reinforced states of matter: glasses as innovative materials for the third millennium. J Therm Anal Calorim 61:305–323
29. Buitink J, Leprince O (2004) Glass formation in plant anhydrobiotes: survival in the dry state. Cryobiology 48:215–228
30. Sikora A, Dupanov VO, Kratochvil J, Zamecnik J (2007) Transitions in aqueous solutions of sucrose at subzero temperatures. J Macromol Sci B Phy 46:71–85
31. Chang L, Milton N, Rigsbee D, Mishra DS, Tang X, Thomas LC, Pikala MJ (2006) Using modulated DSC water to investigate the origin of multiple thermal transitions in frozen 10% sucrose solutions. Thermochim Acta 444:141–147
32. Shamblin SL, Tang XL, Chang LQ, Hancock BC, Pikal MJ (1999) Characterization of the time scales of molecular motion in pharmaceutically important glasses. J Phys Chem B 103:4113–4121
33. Zámečník J, Bilavčík A, Faltus M, Šesták J (2003) Water state in plants at low and ultra-low temperatures. Cryo Lett 24:412
34. Paul H, Daigny G, Sangwan-Norreel BS (2000) Cryopreservation of apple (Malus x domestica Borkh.) shoot tips following encapsulation-dehydration or encapsulation-vitrification. Plant Cell Rep 19:768–774
35. Couchman PR, Karasz FE (1978) Classical thermodynamic discussion of effect of composition on glass-transition temperatures. Macromolecules 11:117–119
36. Zanotto ED, Countinho FAB (2004) How many non-crystalline solids can be made from all the elements of the periodic table. J Non-Cryst Solids 347:285–288
37. Debenedetti PG (1996) Metastable liquids: concepts and principles. Cambridge University Press, Chichester
38. Zarzycki J (1991) Glasses and the vitreous state. Cambridge University Press, Cambridge
39. Pate JS, Atkins CA, Hamel K, McNeil DL, Layzell DB (1979) Transport of organic solutes in phloem and xylem of a nodulated legume. Plant Physiol 63:1082–1088
40. Choi Y, Okos MR (1987) Effects of temperature and composition on the thermal properties of foods. In: Maguer M, Jelen P (eds) Food engineering and process applications. Elsevier Applied Science, New York, pp 93–102
41. Pate JS, Sharkey PJ, Lewis OAM (1975) Xylem to phloem transfer of solutes in fruiting shoots of legumes, studied by a phloem bleeding technique. Planta 122:11–26
42. Martinez D, Arroyo-Garcia R, Revilla MA (1999) Cryopreservation of in vitro grown shoot-tips of Olea europaea L-var. Arbequina Cryo-Lett 20:29–36
43. Taylor MJ (1987) Physico-chemical principles in low temperature biology. In: Grout BWW, Morris GJ (eds) The effects of low temperature on biological systems. Edward Arnold, London, pp 3–70
44. Khalloufi S, El-Maslouhi Y, Ratti C (2000) Mathematical model for prediction of glass transition temperature of fruit powders. J Food Sci 65:842–848
45. Sun WQ, Davidson P, Chan HSO (1998) Protein stability in the amorphous carbohydrate matrix: relevance to anhydrobiosis. Biochim Biophys Acta: Gen Sub 1425:245–254
46. Leprince O, Waltersvertucci C (1995) A calorimetric study of the glass-transition behaviors in sxes of bean-seeds with relevance to storage stability. Plant Physiol 109:1471–1481

47. Wolkers WF, Hoekstra FA (1997) Heat stability of proteins in desiccation-tolerant cattail (*Typha latifolia* L) pollen – A Fourier transform infrared spectroscopic study. Comp Biochem Phys A 117:349–355
48. Matveev YI (1997) Determination of the temperatures of transition into the state of viscous flow, denaturation, and the onset of intensive destruction of proteins with various structures. Polym Sci Series A 39:476–484
49. Makeen MA, Noor NM, Dussert S, Clyde MM (2005) Cryopreservation of whole seeds and excised embryonic axes of Citrus suhuiensis cv. limau langkat in accordance to their desiccation sensitivity. Cryo Lett 26:259–268
50. Benson EE (2008) Cryopreservation of phytodiversity: a critical appraisal of theory practice. Crit Rev Plant Sci 27:141–219
51. Sakai A (1960) Survival of the Twig of Woody plants at -196 degrees-C. Nature 185:393–394
52. Sakai A (1966) Survival of plant tissue at super-low temperatures. 4. Cell survival with rapid cooling and rewarming. Plant Physiol 41:1050–1054
53. Forsline PL, Towill LE, Waddell JW, Stushnoff C, Lamboy WF, McFerson JR (1998) Recovery and longevity of cryopreserved dormant apple buds. J Am Soc Hortic Sci 123:365–370
54. Sakai A, Nishiyama Y (1978) Cryopreservation of winter vegetative buds of hardy fruit-trees in liquid-nitrogen. Hortscience 13:225–227
55. Sugawara Y, Sakai A (1974) Survival of suspension-cultured sycamore cells cooled to temperature of liquid-nitrogen. Plant Physiol 54:722–724
56. Tao DL, Li PH (1986) Classification of plant-cell cryoprotectants. J Theor Biol 123:305–310
57. Reed BB, Uchendu E (2008) Controlled rate cooling. Plant cryopreservation: a practical guide., pp 77–92
58. Iljin WS (1933) Über den Kältetod der Pflanzen und seine Ursachen. Protoplasma 20:105–124
59. Levitt L (1980) Responses of plants to environmental stresses. Academic, New York
60. Morris GJ, McGrath JJ (1981) Intracellular ice nucleation and gas bubble formation in Spirogyra. Cryo Lett 2:341–352
61. Vertucci CW (1989) Effects of cooling rate on seeds exposed to liquid-nitrogen temperatures. Plant Physiol 90:1478–1485
62. Sakai A, Kobayashi S, Oiyama I (1990) Cryopreservation of nuclear cells of navel orange (citrus sinensis osb. var. brasiliensis tanaka) by vitrification. Plant Cell Rep 9:30–33
63. Nishizawa S, Sakai A, Amano Y, Matsuzawa T (1993) Cryopreservation of asparagus (Asparagus-Officinalis L) embryogenic suspension cells and subsequent plant-regeneration by vitrification. Plant Sci 91:67–73
64. Seuferheld MJ, Fitzpatrickm J, Walsh T, Stushnoff C (1991) Cryopreservation of dormant buds from cold tender taxa using a modified vitrification procedure. Cryobiology 28:576
65. Slade L, Levine H (1991) Beyond water activity – recent advances based on an alternative approach to the assessment of food quality and safety. Crit Rev Food Sci Nutr 30:115–360
66. Devireddy RV, Raha D, Bischof JC (1998) Measurement of water transport during freezing in cell suspensions using a differential scanning calorimeter. Cryobiology 36:124–155
67. Cabral AA, Fredericci C, Zanotto ED (1997) A test of the Hruby parameter to estimate glass-forming ability. J Non-Crystal Solids 219:182–186
68. Sun WQ (2008) Stability of frozen and dehydrated cells and membranes in the amorphous carbohydrate matrices: The Williams-Landel-Ferry kinetics. Cryo Lett 19:105–114
69. Williams ML, Landel RF, Ferry JD (1955) The temperature dependence of relaxation mechanisms in amorphous polymers and other glass-forming liquids. J Am Chem Soc 77:3701–3707
70. Angell CA, Bressel RD, Green JL, Kanno H, Oguni M, Sare EJ (1994) Liquid fragility and the glass-transition in water and aqueous-solutions. J Food Eng 22:115–142
71. Bergman R, Swenson J, Borjesson L, Jacobsson P (2000) Dielectric study of supercooled 2D water in vermiculite clay. J Chem Phys 113:357–363

72. Angell CA, Poole PH, Shao J (1994) Glass-forming liquids, anomalous liquids, and polyamorphism in liquids and biopolymers. Nuovo Cimento D 16:993–1025
73. Sakai A, Hirai D, Niino T (2008) Development of PVS-based vitrification and encapsulation-vitrification protocols. In: Reed BM (ed) Plant cryopreservation: a practical guide. Springer, Berlin, pp 33–59

# Chapter 19
# Thermophysical Properties of Natural Glasses at the Extremes of the Thermal History Profile

**Paul Thomas, Jaroslav Šesták, Klaus Heide, Ekkehard Füglein, and Peter Šimon**

## 19.1 Introduction

Natural amorphous glassy silicates are widely distributed and are found in quantities that range from micrograms to kilo tonnes and, hence, their occurrence is from microscopic glassy inclusions to "glassy mountains" [1]. These natural glasses have two generic origins which may be generalised as vitreous glasses, formed from the melt state by relatively rapid cooling at cooling rates that inhibit crystal formation, or diagenetic glasses, formed by a dissolution-precipitation mechanism where

P. Thomas (✉)
Department of Chemistry and Forensic Science, University of Technology, Sydney, PO Box 123, Broadway NSW 2007, Australia
e-mail: paul.thomas@uts.edu.au

J. Šesták
Academy of Sciences, Institute of Physics, v.v.i., Solid-State Physics Section, Cukrovarnická 10, CZ-162 00 Praha 6, Czech Republic
e-mail: sestak@fzu.cz

K. Heide
Chemisch-Geowissenschaftliche Fakultat, Universitat Jena, Burgweg 11, 07749 Jena, Germany
e-mail: Klaus.Heide@uni-jena.de

E. Füglein,
Netzsch-Gerätebau GmbH, Wittelsbacherstraß e 42, 95100 Selb, Germany
e-mail: Ekkehard.Fueglein@netzsch.com

P. Šimon
Faculty of Chemical and Food Technology, Department of Physical Chemistry, Slovak University of Technology, Radlinského 9, SK-812 37 Bratislava, Slovak Republic
e-mail: peter.simon@stuba.sk

J. Šesták et al. (eds.), *Glassy, Amorphous and Nano-Crystalline Materials*, Hot Topics in Thermal Analysis and Calorimetry 8, DOI 10.1007/978-90-481-2882-2_19, 

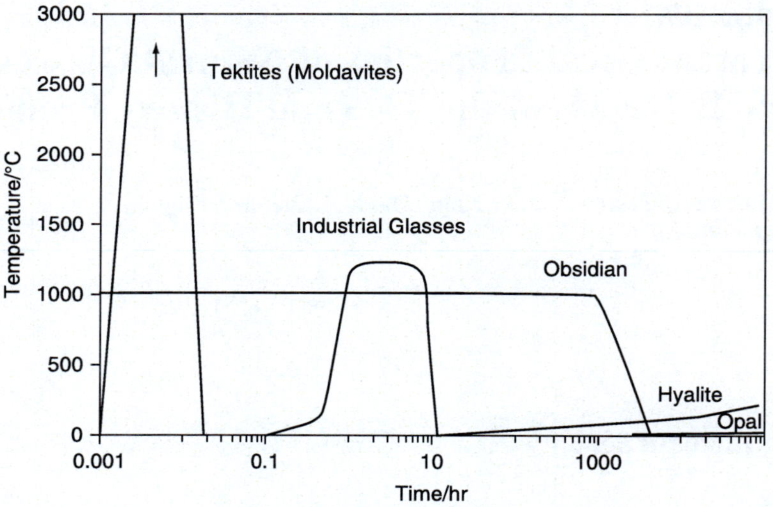


**Fig. 19.1** Schematic T–t plot of glass-forming processes in nature and industry (Adapted from Ref. [3])

crystallisation is inhibited by the Ostwald's rule of stepwise petrogenesis [2]. The thermal histories of a range of natural glasses are depicted in the schematic of Fig. 19.1 and vary significantly from the typical conditions used in the glass industry which are optimised between processing speed and energy conservation. In the extremes, tektites like moldavites are formed by extremely fast heating and melting at very high temperatures ($> 3,000$ K) followed by quenching at extreme cooling rates ($\geq 10$ K/s). By contrast the formation of amorphous glasses from mineral diagenesis or biotic processes occurs at much lower temperatures and over longer time periods; the formation of sedimentary opal, for example, occurs at ambient temperatures, it is essentially isothermal, and takes place over long periods of time of the order of months to years.

The chemical composition of natural glasses also differs significantly from the industrial glasses. The composition of industrial glasses is tailored to the optimisation of the thermo-rheological properties for processing and durability during application; as a consequence, industrial glasses tend to have low alumina contents and their softening temperatures are reduced by the addition of alkali oxides which reduces the proportion of network forming ions. The majority of natural vitreous glasses, on the other hand, are network forming ion rich ($>70\%$ $SiO_2$, S $\Sigma(SiO_2 + Al_2O_3) > 80\%$) and per aluminous (i.e. $Al_2O_3 \geq \Sigma(Na_2O + K_2O + CaO)$) and are principally characterised by a rhyolitic composition reflecting the elemental compositions of their origins (Table 19.1). The diagenetic or biotic glasses tend to have even higher silica contents which can approach 99% of the anhydrous composition and are also peraluminous.

The influence of the chemical composition is reflected in the significant durability of the natural glasses. The glassy state is a frozen in metastable and, hence, there is a

**Table 19.1** Typical chemical compositions of natural and industrial glasses (%)

| Oxide | Moldavites [4] | Coober Pedy sedimentary opal [5] | Obsidians utah [6] | Pitchstone [7] | Soda-lime industrial glass |
|---|---|---|---|---|---|
| $SiO_2$ | 79.0 | 90.8 | 75.0 | 70.3 | 70–75 |
| $TiO_2$ | 0.3 | – | 0.11 | 0.1 | 0.01 |
| $Al_2O_3$ | 10.3 | 0.8 | 12.5 | 13.3 | 0–1 |
| $Fe_2O_3$ | – | 0.05 | – | 1.28 | 0–0.1 |
| FeO | 1.7 | – | 1.0 | – | – |
| MgO | 2.1 | 0.02 | 0.03 | 0.1 | 0–4 |
| CaO | 2.5 | 0.3 | 0.6 | 1.2 | 8–10 |
| $Na_2O$ | 0.4 | 0.1 | 2.9 | 4.4 | 10–15 |
| $K_2O$ | 3.4 | 0.1 | 5.1 | 2.1 | 0–0.3 |
| $P_2O_5$ | 0.06[a] | – | 0.1 | – | – |
| $H_2O$ | <0.02[b] | 7.8 | 0.3 | 7.9 | 0.2 |

[a][8], [b][9]

tendency towards crystallisation or devitrification as well as the potential for weathering. Industrial glasses are relatively susceptible to weathering and degradation at relatively short periods of time (<1,000 years) due to the high alkali content. Specimens of natural vitreous or diagenetic glasses, however, are known to have significant durability. For example, in relatively inert environments, such as on the surface of the moon, vitreous solids have existed for more than a billion years. This inherent stability may be attributed to the low alkali and high network forming ion ($\Sigma(Si + Al)$) content of natural glasses (Table 19.1).

The water contents of the natural glasses also vary significantly and ranges from <0.01% in tektites to between 3% and 13% (or greater) in opaline glasses which compares with 0.2% in typical soda-lime industrial glasses. An example of the effect of the water content is given in the thermal mechanical analysis (TMA) data shown in Fig. 19.2 which is a comparison of the expansion properties of a series of siliceous-aluminous glasses which are discussed in more detail below. The expansion curves for the as received opal and pitchstone, which contain significant proportions of water (Table 19.1), are complex, compared to the natural silica glasses of low water content, and expansion and contraction segments attributable to the water removal are observed. Once dehydration has been completed, the thermal expansion behaviour reverts to that typical of amorphous siliceous-aluminous glasses.

Although a variety of natural glasses exist, this Chapter limits its scope to a discussion of the thermophysical properties of tektites and opal as examples of vitreous and digenetic natural glasses formed at the extremes of the thermal histories shown in Fig. 19.1. Both tektites and opal are rare non-crystalline rocks which are prized for their aesthetic value and both have an important role to play in the understanding of the environments in which they are found. Reference is, however, made to industrial and rhyolitic glasses formed by magmatic processes in order to provide a context for the discussion.

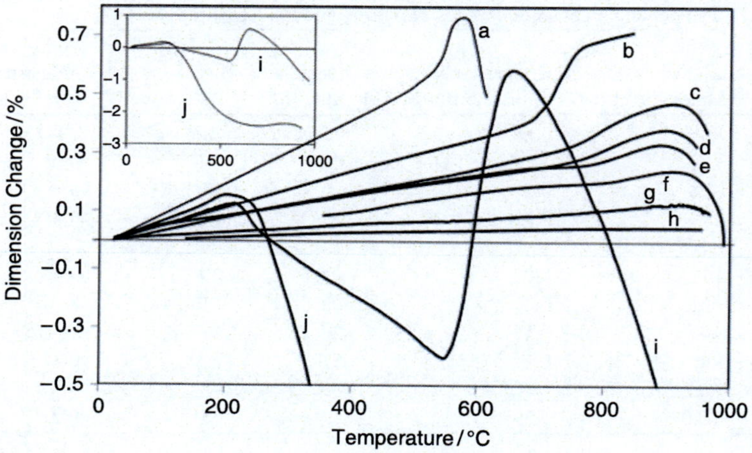


**Fig. 19.2** Thermal expansion curves of (**a**) soda-lime glass, (**b**) obsidian, (**c**) pitchstone – second heating, (**d**) Bohemian moldavite – second heating, (**e**) Bohemian moldavite – first heating, (**f**) Mahren moldavite, (**g**) Coober Pedy white play-of-colour opal – second heating, (**h**) fused silica, (**i**) Coober Pedy white play-of-colour opal – first heating and (**j**) pitchstone – first heating. Inset shows and expanded scale for curves of (**i**) and (**j**) (Adapted from the data in Refs. [3, 10])

## 19.2 Tektites – Moldavites

Tektites are centimetre to decimetre-sized bottle green to blackish glassy "bodies" found in gravels ranging in age from upper tertiary to alluvial in four main strewn fields; the European strewn field associated with the Nördlinger Ries (which includes the moldavites), and Australasian strewn field (no associated crater), the North American strewn field associated with the Chesapeake Bay crater and the Ivory Coast strewn field associated with the Lake Bosumtwi crater. Tektites are also known originate in east Asia and microtektites (vitreous spherules <500 μm in diameter) have been found in deep-sea deposits in the Gulf of Mexico and north-western Atlantic Ocean [11].

The origin of tektites was intensively studied in the latter decades of the twentieth century, particularly, with respect to the mass extinction during the short time event known as the Cretaceous – Tertiary boundary. The discussion was dominated by two mechanisms for the formation of tektite material: the Terrestrial Impact Theory (TIT) and the Lunar Volcanic Theory (LVT) [12, 13]. As Izokh [12] noted the TIT "won a complete victory over O'Keefe's Lunar Volcanic Theory ([13])" as, based on trace element ratios of geochemically well known pairs which show a clear distinction of origin between different bodies of the solar system (e.g. K/Sc; Ba/Zr, La/Yb), the chemical evidence demonstrated that tektites were chemically indistinguishable from terrestrial rocks and soils and, hence, tektites were attributed a terrestrial origin [14–17].

According to the TIT, tektites were formed during the significant energy transfer from asteroids impacting with the earth's surface where high pressures and temperatures in the impact zone resulted in the fusion and vaporisation of the terrestrial rocks. The energy of these impacts was such that the fused and vaporised material was catapulted out to near space. As this material returned to earth, it rapidly cooled and solidified with an absence of gaseous inclusions (although vacuous inclusions are observed). During the quenching process the solidifying phase was generally shaped aerodynamically and, as the molten droplets were spinning, a range of bizarre shapes; plate, spheroid, rod, dumbbell, teardrop, star, curly, winkled and bubble, were produced (Fig. 19.3). As these particles were returned to earth, the particles were distributed by aerodynamic transport over large areas of the Earth-surface in strewn fields and, as such, tektites from a particular strewn field have a high degree of compositional homogeneity [4].

Tektites have a similar composition to the terrestrial rhyolitic volcanic glasses, in particular, to the obsidians (Table 19.1). Despite the similar elemental composition, a remarkable difference in the water content (tektites $< 0.01$ wt%, obsidians $> 0.1$ wt%) and in the redox state (in tektites $Fe^{2+} >>> Fe^{3+}$, in obsidians $Fe^{2+} \geq Fe^{3+}$) exists between tektites and obsidians. These differences are a direct consequence of the formation history; the high temperature and vacuous formation environment of the tektites removed the volatile content, including water, and the shifted the redox equilibrium from $Fe^{3+}$ to $Fe^{2+}$.

The rhyolitic composition of the tektites is also responsible for the significant durability with ages determined to be between 700,000 years (australites) and 34 million years (bediasites, North American tektites). In the more inert lunar environment, durability of these metamorphic glasses has been demonstrated by glass spherules which have been aged at more than 2 billion years. In general, recovered specimens of tektite have been observed to be resistant to hydration or

**Fig. 19.3** Images of typical Moldavites specimens conforming to the aerodynamic shapes expected from the quenching of the molten glass at high velocities. The characteristic surface topography of the moldavites is a modification and a result of a 'characteristic corrosion' of the surface (see text)

devitrification of the bulk [15]. Only a characteristic corrosion of the surface is observed in the Bohemian tektites – the moldavites (Fig. 19.3) [18].

## 19.3 Thermophysical Properties of Tektites

As the volatile contents of the tektites are very low (<0.02 wt%), the thermal properties of tektites are most easily characterised by the thermal expansion (Fig. 19.2). The thermal expansion coefficient, α, for moldavite ($3.7 \times 10^{-6}\ K^{-1}$) is greater than that of fused silica ($0.5 \times 10^{-6}\ K^{-1}$), but less than that of obsidian ($6.3 \times 10^{-6}\ K^{-1}$) and soda lime glass ($7.5 \times 10^{-6}\ K^{-1}$) [3, 15]. The comparison with obsidian is apt as both glasses have a similar rhyolitic composition (Table 19.1). The difference in the expansion coefficient is most likely due to the relative proportions of sodium and potassium with respect to calcium and magnesium within the silica network. The higher CaO and MgO and lower $Na_2O$ content of moldavites results in a more tightly bound structure and a lower expansion coefficient. This difference is also reflected in the glass transition temperature ($T_g$ ~ 780°C) and is approximately 100 K higher for moldavite than obsidian.

The fictive temperature, which is indicative of the frozen in temperature and the rate of quenching, tends to be higher for moldavites as the high melting temperature produces a lower density melt which is rapidly cooled producing a lower density or higher fictive temperature glass. Relaxation of this thermal history is observed at ~600°C as the moldavite is heated (Fig. 19.2). The differences in the fictive temperature have their origins in the relative cooling rates which have been estimated from expansion coefficient measurements [19] and, more recently, by the geospeedometric double-density technique developed by Klöss [20]. The geospeedometric method is based on the difference in density between a quenched and thermally annealed glass specimen. A linear relationship is produced that can be extrapolated by regression of the experimental cooling rates to an unknown natural cooling rate (Fig. 19.4). Based on this model, the cooling rates of magmatic glasses such as obsidian have been determined to be of the order of K/annum while for tektite glasses cooling rates of the order of 10 K/s have been estimated.

Further evidence of the extreme thermal history of the tektites may be gleaned from the characteristic intermediate rage order which, in glasses, may be defined by the magnitudes of two parameters *a* and *f* [21]. The "*a*-value" is a measure of the average diameter of the typical space-filling polyhedra, which in tektites are composed of $SiO_4$ and $AlO_4$ tetrahedra. This parameter results from the bond length and the ring size; for low-quartz $a = 0.436$ nm and for high-cristobalite $a = 0.497$ nm. Moldavite has a cristobalite-like intermediate structure with $a \sim 0.485$ nm; for industrial silica glass $a \sim 0.510$ nm; this high value may be attributed to the more open structure resulting from a higher alkali and lower network forming ion ($\Sigma$(Si + Al)) content. The "*f*-value" characterises the fluctuations of the intermediate range order. It monitors the "quality" of glass structure and can be calculated from the ratio of the "*a*-value" to the "correlation length". In a homogenous and

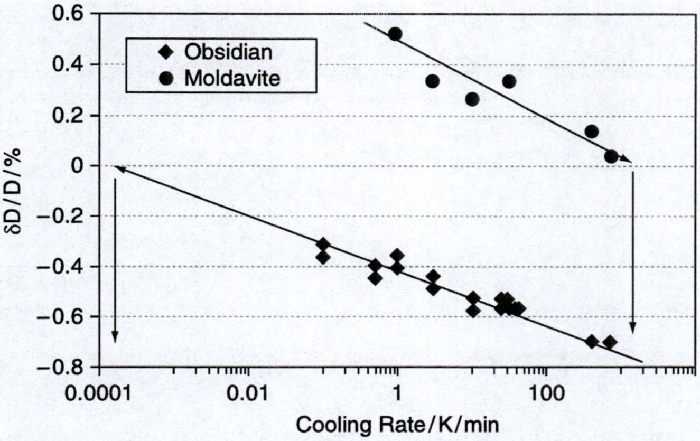


**Fig. 19.4** Density changes as a function of the cooling rates for known thermal histories are plotted to allow estimation of the cooling rates of the natural glasses during formation (Adapted from Ref. [3])

relaxed silicate glass $f$ is ~ 0.35. Short fusion processes and fast cooling increase the "$f$-value". For moldavites $f$ ~ 0.46 and is greater than that of a relaxed glass. The value of $a$ tending toward that of cristobalite correlates with the quenching of the glass from high temperature. The high value of the "$f$-value" correlates with a rapid cooling regime. The values of these parameters for moldavite suggest a short fusion process leading to high degree of structural disorder. Annealing of moldavites by heating up to 1,000°C reduces the fluctuations and especially the mechanical stress.

The origins of tektites as rapidly quenched glasses from high temperature melts is further supported by the low degree of dissolved gasses found in tektites. Mass loss measurements at atmospheric pressure show little mass loss up to 1,600°C for Bohemian moldavites [10]. Additionally, due to the high alumina content, the glass is inhibited from devitrifying and, hence, no crystallisation takes place. The degassing under vacuum (using a special high-vacuum-hot-extraction method DEGAS combined with a quadrupol mass spectrometer) shows even lower gas content and water content of the order of few ppm (0.0004 wt%) [10, 22–24]. The low volatile content of the moldavites correlates with the micrograph shown in Fig. 19.5 where a number of bubbles and striations are observed as a consequence of rapid gas release during solidification.

## 19.4 Precious Opal

Precious or noble opal is a post depositional mineral formed in voids in the host rock by the diagenesis and transport of inorganic silica and is prized for its play-of-colour, the origins of which are based on the diffraction of visible light off ordered arrays of monodispersed silica spheres [25, 26]. Precious opal is composed of amorphous hydrous silica with the general formula $SiO_2.nH_2O$ and is precipitated as

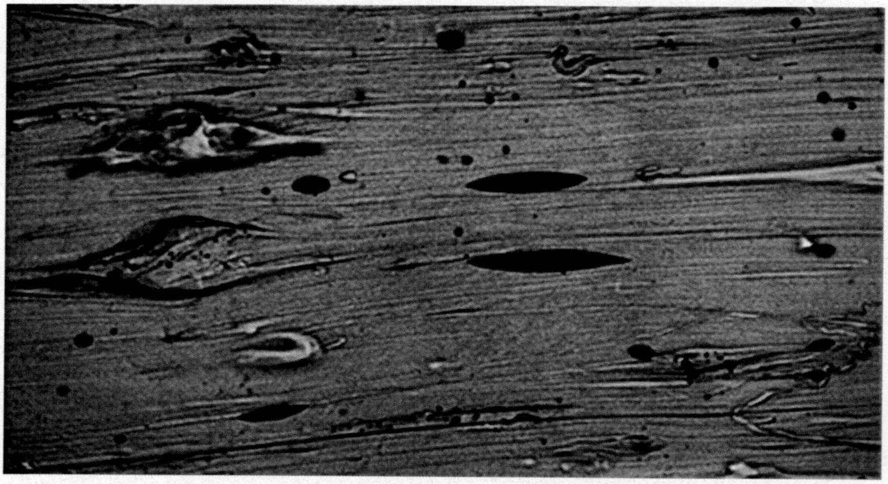

**Fig. 19.5** Micrograph of a moldavite specimen "thick-section" (ca. 3 mm) in transmitted light showing bubbles and striations resulting from rapid gas release [3]

**Fig. 19.6** SEM micrographs of hydrofluoric acid vapour etched fracture surface of a Coober Pedy white play-of-colour (POC) opal (width of field 25.4 µm). Note the 'grain boundary' in the *top left* to *bottom right* diagonal. Random orientation of the ordered arrays provides the play-of-colour

monodispersed colloidal particles from 150 to 400 nm in diameter which are concentrated by sedimentation, evaporation or by filtration under pressure to form the ordered arrays of that are responsible for the diffraction of light observed (Fig. 19.6).

A likely source of the monodispersed silica colloid is the weathering of silicates such as feldspars [2, 26]. In the weathering model, the source of the silica is the sandstones associated with the GAB where the chemical weathering of relatively soluble silicates such as the feldspars contained in these sediments results in the formation of an alkaline silica solution. An example of such a mineral is potassium feldspar which weathers through the idealised stoichiometry:

$$2KAlSi_3O_8 + 3H_2O \rightarrow Al_2Si_2O_5(OH)_4 + 2KOH + 4SiO_2 \qquad (19.1)$$

by the permeation of ground water through the sediments resulting in kaolinite, dissolved silica and an increase in pH through the release of potassium hydroxide. Reported trace element distributions in opal from a wide variety of sources are consistent with such a weathering model [5, 27, 28]. Once the silica is in solution, enrichment of the solution by evaporation or filtration can occur. Increasing the concentration of the silica solution coupled with a lowering of the pH through alkali ion exchange with the surrounding clays allows the nucleation of primary silica spheres and, subsequent, sphere growth as more silica is supplied to the system. The supply of silica appears to be a cyclic process as generations of growth rings are commonly observed in the silica spheres [25, 29]. The monodispersed silica colloid is then concentrated and in certain cases, 'crystallisation' occurs to form the ordered arrays which result in the prized play-of-colour. The interstices are subsequently infilled with a silica cement completing the formation of the opal producing a hard material which, despite the gel-like structure, has a Berkovich hardness of 5.7–6.2 GPa which is similar to that of soda-lime glass (6.4 GPa), but, as might be expected for a gel like material, is significantly less than that of fused silica (10.8 GPa) [30].

Although the weathering model is important in aiding the understanding of the formation of opal, a number of other models have been proposed to account for specific observations in specimens acquired by the authors of these models. Examples of such models are the microbial model where microbes have been observed in Lightning Ridge matrix specimens suggesting that microbial action is responsible for the source of the silica [31], the syntechtonic fluid model where high pressure, warm hydraulic silica rich fluids are the silica source [32] and the mound spring model where the silica rich waters derived from the artesian basin well up through the mound springs supplying the silica rich solutions required for opal formation [33]. Although these models source the silica from different origins, the formation of the monodispersed colloid must occur by homogenous precipitation followed by a concentration mechanism such as evaporation, filtration or sedimentation [5, 34]. This 'sol-gel' mechanism for the formation of opal is responsible for the high water contents (6–10% for Australian sedimentary opal) and the high proportion of molecular water; in Australian sedimentary opal as much as 90% has been reported to be molecular, trapped in silica cages in the bulk of the opal, with the remainder as silanol water [35–39].

The environment in which opal is formed is important as can be discerned from the differences in physical properties of opal derived from the volcanic and sedimentary environments [38–42]. In general, hydrous opaline silica has been found to occur with a range of morphologies which, based on x-ray diffraction (XRD) measurements [43], were originally divided into three categories; crystalline opal predominantly composed of cristobalite, opal-C, partially crystalline opal containing XRD characteristics of cristobalite and tridymite, opal-CT, and amorphous opal, opal-A. Subsequent analysis by Langer and Flörke [36] resulted in a sub-division of opal-A into two further categories, amorphous network silica, opal-AN (e.g. hyalite), and an amorphous gel silica, opal-AG (e.g. precious opal displaying play-of-colour), based on their relative proportions of water and appearance. These forms of silica are related by their relative solubility in water and their thermodynamic stability resulting in an Ostwald like succession of formation [2, 44, 45]:

$$\text{Opal - A} \rightarrow \text{Opal - CT} \rightarrow \text{Opal - C} \rightarrow \text{microcrystalline quartz} \qquad (19.2)$$

Australian sedimentary opal, opal-AG, may be considered as a first generation of silica formed by these low temperature aqueous processes as it is the first of the 'pure' silica phases that are produced from the weathering of basic aluminosilicates.

Opal derived from volcanic environments is generally found to have the opal-CT morphology. This significant difference in morphology demonstrates the significance of the environment on the formation of opal. Although volcanic opal may be categorised as opal-CT based on the XRD data, the designation of volcanic opal as a second generation silica polymorph is contentious. As volcanic opal displays play-of-colour, it contains a monodispersed ordered array of silica spheres. This fact suggests that a monodispersed colloid is the precursor of the opal which suggests homogeneous precipitation and, hence, a first generation silica [38, 39]. The first generation silica in Eq. 19.2, Opal-A, should, therefore, be extended to include precious play-of-colour volcanic opal-CT, being formed by homogeneous precipitation, as later generations are formed through a heterogeneous nucleation and growth mechanism.

## 19.5 Thermophysical Properties of Precious Opal

Opal is a hydrous silica containing circa 8% water (depending on the origin) [46] and due to the formation mechanism, 90% of the water in sedimentary opal is molecular water trapped in silica cages [35–39]. The thermophysical properties of opal are, therefore, dominated by the irreversible process of removing the water. An example of the effect of the water content is given in the TMA data shown in Fig 19.2 [10]. The opal initially has a high rate of expansion ($7.4 \times 10^{-6}\,\mathrm{K}^{-1}$) as it contains 7.5% water (based on the TG mass loss in Fig. 19.7) and compares well with the expansion of pitchstone ($8.5 \times 10^{-6}\,\mathrm{K}^{-1}$) which contains a similar proportion of water. After a period of dehydration a peak in the expansion curve is observed (210°C, circa 200°C for pitchstone). Contraction of the specimen is then observed until the remaining

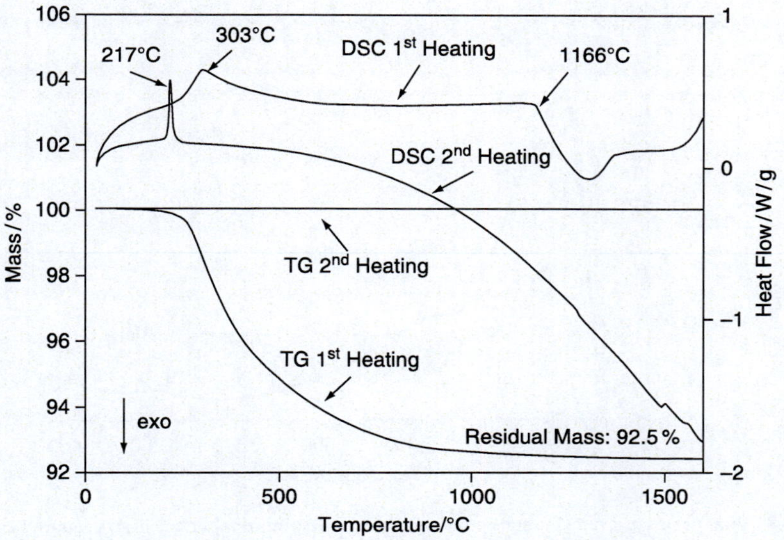


**Fig. 19.7** TG data of the CP white play-of-colour opal heated to 1,640°C at 20 K/min in an air atmosphere. The second heating run shows the α−β cristobalite transition at 217°C suggesting crystallisation has occurred corresponding to the exotherm observed above 1,166°C in the first heating run [10]

trapped water exerts such a high hydrostatic pressure that expansion is once again observed above 540°C. This type of behaviour is relatively common for opal as well as other natural glasses which contain a significant proportion of water such as the pitchstone specimen shown in Fig. 19.2 where contraction occurs at similar temperature although, for this pitchstone specimen, no significant secondary expansion is observed under the heating rates used (1°C/min). Once all the water has been removed from the opal, the expansion coefficient on reheating the sample is low ($1.4 \times 10^{-6}\,K^{-1}$) and much closer to that of a fused silica ($0.5 \times 10^{-6}\,K^{-1}$) due to the high silica content. As the pitchstone has a rhyolitic composition, the expansion coefficient for the dehydrated material ($4.3 \times 10^{-6}\,K^{-1}$) is much closer to that of moldavite ($3.7 \times 10^{-6}\,K^{-1}$) and obsidian ($6.3 \times 10^{-6}\,K^{-1}$).

Although opal has been observed to be relatively stable to thermal treatment, the amorphous character, as, indeed, the play-of-colour, of the opal has been observed to be retained even after prolonged heating up to 1,000°C, crystallisation of this metastable silica polymorph may be achieved at higher temperatures and is observed in the exothermic behaviour above 1,166°C and is confirmed in the second heating run, where a sharp endothermic peak is observed at 217°C corresponding to the α − β cristobalite transition (Fig. 19.7). Crystallisation of the amorphous opal is facilitated by the high silica and low alumina content which is in contrast to the rhyolitic composition glasses such as the moldavite which are observed to be resistant crystallisation even at high temperature due to the high aluminium content.

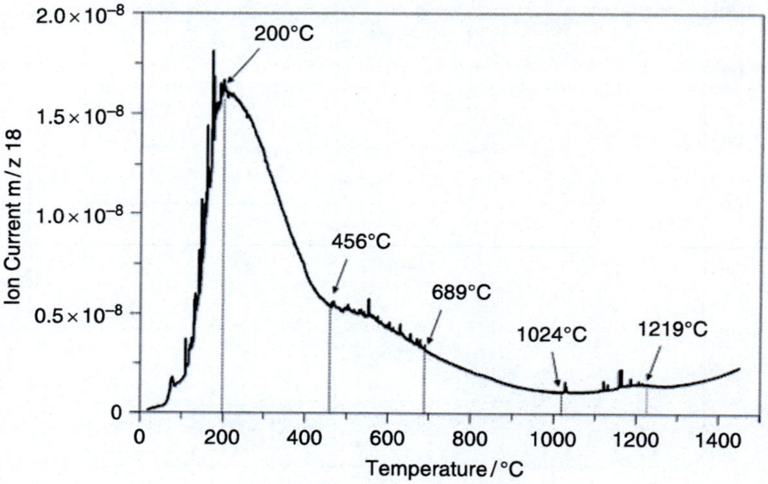


**Fig. 19.8** Water release from Coober Pedy white play-of-colour opal under high vacuum heated to 1,440°C at a heating rate of 10 K/min. The water loss corresponds to 99.63 wt% of the total mass loss. The spiky release of water between 456°C and 689°C is 0.29 wt% and between 1,024°C and 1,219°C is 0.08 wt% is as a result of the sudden release of water from enclosed pores or fluid inclusions [10]

Characterisation of the water evolution under high vacuum using the DEGAS technique has also yielded further understanding of the character of the water and the morphology of opal (Fig. 19.8) [10]. The typical water loss is observed around 200°C which corresponds to the TG data [38, 47]. A shoulder peak between 400°C and 700°C and continued water mass loss up to and beyond 1,000°C is also observed. It is also interesting that spikes in the data in Fig. 19.8 are observed and correspond to sudden losses and may be ascribed to the sudden release of water from inclusions or entrapped micropores in which the pressure escalates beyond the breaking stress of the opal. In addition to water, hydrogen and methane have also been observed to be released. The hydrogen released around 880°C is likely to be derived from micellar decomposition of silanols [48], but the methane evolved around 100°C is a little more difficult to explain as the temperatures are too low for the micellar reactions. It is possible that very small proportions of molecular methane are trapped in the gel structure as the opal network is formed giving a further indication of the nature of the waters in which precious opal is formed.

## 19.6 Final Remarks

An analysis of the thermophysical properties of tektites, as an extreme example of the vitreous glasses, and precious opal, as an extreme example of the diagenetic glasses, yields the formational and chemical characteristics expected of these precious natural glasses. The high silica and alumina content of the tektites results in significant durability and thermal stability and the high concentrations of trace elements results in the higher expansion coefficients observed. The high water content of opal, by contrast, is the dominant characteristic controlling its thermophysical properties. The thermophysical properties observed for the tektites and the opal are typical of their origins and, as these observed properties are the properties of the natural glasses at the extremes of the thermal history profiles, the glasses with intermediate thermal histories and compositions will exhibit intermediate combinations of these properties, as was briefly noted for obsidian and pitchstone.

Although tektites and sedimentary opal are rare natural non crystalline solids, whose primary value is aesthetic, these natural glasses that are formed at the extremes of the thermal history profiles are also important as a record of the environments in which they are found. Both of these material types are important in their as found shapes, in the case of moldavite – the aerodynamic forms and in the case of opal – the infilling of voids, seams and cracks in the host rock and, in particular, in fossil replacement (opalised fossils). Both of these natural minerals have an important part to play in the understanding of geological processes as their formation is a record of change in the natural environment [2, 49]. The tektites, based on their models of formation have given an insight into the impact processes that have remodelled the Earth's fauna and flora. The opal is the first step to understanding the diagenesis of silica and its mobility. The characteristics of these rare gems are exemplified by their thermal properties and, even though both of these non-crystalline solids are formed at the extremes of forming processes, both tektites and opals, as forms of silica, help to enrich our lives.

## References

1. Heide K (2007) Die Geheimnisse des "gläsernen Bergs". Forschung 32
2. Iler RK (1979) The chemistry of silica, solubility, polymerisation, colloid and surface properties, and biochemistry. Wiley, New York
3. Heide K, Heide G, Kloess G (2001) Glass chemistry of tektites. Planet Space Sci 49:839–844
4. Lange JM (1995) Lausitzer Moldavite und ihre Fundschichten. Schriftenr f Geowiss 3:7–138
5. Brown LD, Ray AS, Thomas PS (2004) Elemental analysis of Australian amorphous Opals by laser-ablation ICPMS. Neues Jb Miner Monat 2004(9):411–424
6. Hölzle-Vuynovich A (1992) Petrographische und geochemische Untersuchungen an Schneeflockenobsidianen und verwandtem Material aus den USA, Island und der Osterinsel. Heidelberger geowissenschaftliche Abhandlungen, vol 56. Ruprecht-Karls-Universität Heidelberg, Heidelberg
7. Ray A, Sriravindrarajah R, Guerbois J-P, Thomas PS, Border S, Ray HN, Haggman J, Joyce P (2007) Evaluation of waste perlite fines in the production of construction materials. J Therm Anal Calorim 88:279–283
8. Luft E (1983) Zur Bildung der Moldavite beim Ries-Impakt aus tertiären Sedimenten. Enke Verlag, Stuttgart, p 57
9. Rost R (1972) Vltaviny a tektity. Academia, Prague, p 241
10. Thomas PS, Šesták J, Heide K, Fueglein E, Šimon P (2010) Thermal properties of Australian sedimentary opals and Czech moldavites. J Therm Anal Calorim 99:861–867

11. Glass BP (1984) Tektites. J Non-Cryst Solids 67:333–344
12. Izokh EP (1996) Origin of tektites: an alternative to terrestrial impact theory. Chem Erde 56:458–474
13. O'Keefe JA (1976) Tektites and their origin. Elsevier, Amsterdam
14. O'Keefe JA (1984) Natural glasses. J Non-Cryst Solids 67:1–17
15. Heide K, Kletti H (2003) Resistance of natural glass. Glass Sci Technol 76:118–124
16. Koeberl C, Kluger F, Kiesl W (1986) Trace element correlations as clues to the origin of tektites and impactites. Chem Erde 45:1–21, 1
17. Schnetzler CC, Pinson WH Jr (1963) The chemical composition of tektites. In: O'Keefe JA (ed) Tektites. Chicago Press, Chicago, p 101
18. Bouská V (1993) Natural glasses. Academia, Praha
19. Arndt J, Rombach N (1976) Derivation of the thermal history of tektites and lunar glasses from their thermal expansions characteristics. In: Proceedings of the 7th Conference on Lunar Science. pp 1123–1141, Houston GCA Supplement 7
20. Kloess G (2000) Dichtfluktuationen natürlicher Gläser. Habilitation, Jena
21. Heide G, Müller B (1999) Zur Struktur von Moldavitglas. Schriften Staatl. Museen Min. Geol. Dresden 10:30–33
22. Kloess G, Heide G (1999) $\rho$-$\rho_t$ Geospeedometrie an Tektiten. Schriften Staatl. Museen Min. Geol. Dresden 10:52–54
23. Heide K, Gerth K, Hartmann E (2000) The detection of an inorganic hydrocarbon formation in silicate melts by means of a direct-coupled-evolved-gas-analysis-system (DEGAS). Thermochim Acta 354:165–172
24. Heide K, Schmidt Ch (2003) Volatiles in vitreous basaltic rims, HSDP 2, Big Island, Hawaii. J Non-Cryst Solids 323:97–103
25. Jones JB, Sanders JV, Segnit ER (1964) The structure of opal. Nature 4962:991
26. Darragh PJ, Gaskin AJ, Sanders JV (1976) Opals. Sci Am 234(4):84
27. McOrist GD, Smallwood A (1997) Trace elements in precious and common opals using neutron activation analysis. J Radioanal Nucl Chem 223:9–15
28. Erel E, Aubriet F, Finqueneisel G, Muller JF (2003) Capabilities of laser ablation mass spectrometry in the differentiation of natural and artificial opal gemstones. Anal Chem 75:6422–6429
29. Brown LD, Ray AS, Thomas PS (2003) $^{29}$Si and $^{27}$Al NMR study of amorphous and paracrystalline opals from Australia. J Non-Cryst Solids 332:242–248
30. Thomas PS, Smallwood AS, Ray AS, Briscoe BJ, Parsonage D (2008) Nanoindentation hardness of banded Australian sedimentary opal. J Phys D Appl Phys 41:074028
31. Behr HJ, Behr K, Watkins JJ (2000) Cretaceous microbes–producer of black opal at Lightning Ridge, NSW, Australia. Geological Abstracts No. 59. 15th Australian Geological Convention. Sydney
32. Pecover SR (1996) A new genetic model for the origin of precious opal. Extended abstracts No. 43. Mesozoic geology of Eastern Australia plate conference. Geo Soc Aust, pp 450–454
33. Devison B (2004) The origin of precious opal – a new model. Aus Gemmologist 22:50–58
34. Brown LD, Thomas PS, Ray AS, Prince K (2006) A SIMS study of the transition metal element distribution between bands in banded Australian sedimentary opal from the lightning ridge locality. Neues Jb Miner Monat 182:193–199
35. Segnit ER, Stevens TJ, Jones JB (1965) The role of water in opal. J Geol Soc Aust 12:211–226
36. Langer K, Flörke OW (1974) Near infrared absorption spectra (4000–9000 $cm^{-1}$) of opals and the role of water in these $SiO_2.nH_2O$ minerals. Fortschr Mineral 52:17–51
37. Brown LD (2005) Characterisation of Australian opals. PhD Thesis, University of Technology, Sydney
38. Smallwood AG, Thomas PS, Ray AS (2008) Thermal characterisation of Australian sedimentary and volcanic precious opal. J Therm Anal Calorim 92:91–95
39. Smallwood AG, Thomas PS, Ray AS (2008) The thermophysical properties of Australian opal. Australian Institute of Mining and Mineralogy Publication Series No. 8, pp 557–560

40. Banerjee A, Wenzel T (1999) Black opal from honduras. Eur J Mineral 11:401–408
41. Caucia F, Ghisoli C, Adamo I, Boiocchi M (2008) Opal-C, Opal-CT and Opal-T from Acari. Peru Aust Gemmologist 23:266–271
42. Rondeau B, Fritsch E, Guiraud M, Renac C (2004) Opals from Slovakia ('Hungarian' opals): a reassessment of the conditions of formation. Eur J Mineral 16:789–799
43. Jones JB, Segnit ER (1971) The nature of opal. I. Nomenclature and constituent phases. J Geol Soc Aust 18:57
44. Williams LA, Crerar DA (1985) Silica diagenesis. II: general mechanisms. J Sediment Petrol 55:312–321
45. Landmesser M (1998) Mobility by metastability: applications. Chem Erde 58:1–22
46. Brown LD, Ray AS, Thomas PS, Guerbois JP (2002) Thermal characteristics of Australian sedimentary opals. J Therm Anal Calorim 68:31–36
47. Smallwood AG, Thomas PS, Ray AS, Šimon P (2009) A Fickian model for the diffusion of water in Australian sedimentary opal. J Therm Anal Calorim 97:685–688
48. Heide K, Woermann E, Ulmer G (2008) Volatiles in pillows of the Mid-Ocean-Ridge-Basalt (MORB) and vitreous basaltic rims. Chem Erde 68:353–368
49. Engelhardt Wv, Luft E, Arndt J, Schock H, Weiskirchner W (1987) Origin of moldavites. Geochim Cosmochim Acta 51:1425–1443

## Abbreviations

| | |
|---|---|
| DEGAS | High vacuum hot extraction gas analysis by mass spectroscopy |
| GAB | Great Artesian (Australian) Basin |
| LVT | Lunar volcanic theory |
| Opal-A | Amorphous opal |
| Opal-CT | Cristobalite-tridimite ordered opal |
| Opal-C | Cristobalite ordered opal |
| Opal-AG | Amorphous gel-like opal |
| Opal-AN | Amorphous network-like opal |
| TIT | Terrestrial impact theory |
| TMA | Thermomechanical analysis |
| XRD | X-ray diffraction |

# Chapter 20
# Hotness Manifold, Phenomenological Temperature and Other Related Concepts of Thermal Physics

Jiří J. Mareš

## 20.1 Introduction

Although the operative methods of temperature measurement are well-known and described in detail in various practical instructions [1, 2] and discussed in many textbooks [3–10], the systematic treatment of the central concept of thermal physics, the *temperature* itself, is paradoxically almost lacking in the current literature. The temperature, namely, is there at present defined mostly from the theoretical positions of statistical physics and not as a phenomenological quantity. Nevertheless, the predominant majority of practical measurements in physics, chemistry and technology or in thermal analysis and calorimetry particularly are performed by means of macroscopic devices (thermometers) yielding as a result the *phenomenological temperature*, and not by means of statistical analysis of properties of ensembles of particles and excitations. It is thus evident that prior to the identification of the temperature defined in the frame of statistical theory with the phenomenological temperature, the latter has to be satisfactorily defined first.

The purpose of this contribution is thus to re-examine a fundamental concept of thermal physics, *phenomenological temperature*, from the logical, epistemological and partially also from historical points of view. The mathematical structure of the precursor of temperature, *hotness manifold*, is, as far as we know, for the first time discussed here in terms of *elementary set theory*. In the exposition of the subject the emphasis is put on experiment, elucidation of allied concepts and on generalization of empiric data while the reader must confer with special references already given on mathematical proofs of some statements.

J.J. Mareš (✉)
Institute of Physics ASCR, v.v.i., Cukrovarnická 10, 162 00 Praha 6, Czech Republic
e-mail: maresjj@fzu.cz

J. Šesták et al. (eds.), *Glassy, Amorphous and Nano-Crystalline Materials*,
Hot Topics in Thermal Analysis and Calorimetry 8, DOI 10.1007/978-90-481-2882-2_20,

## 20.2 Early Thermal Measurements, Thermoscope

The first devices known as *thermoscopes* appeared during the later Renaissance in connection with the first edition of Latin translation of Hero's "*Pneumatica*" by F. Commandino Urbinate (1575) [11]. The influence of this book dealing with various unexplained natural phenomena and curious contrivances worked by air, water or steam was so general that it is almost impossible to tell whether a particular device was directly derived from Hero's descriptions or whether it is an original invention. Therefore one can find in the literature a long series of "inventors of thermometer", e.g. Cardano, Galileo, Sanctorius, Besson, de la Porta, Drebbel, Fludd, Leurechon, Ens, Harsdoerfer, Kirchner etc. [12, 13], but to decide about the priority of any one of them is a very difficult task. Quite early it was recognized that these devices, having various forms of fluid or air dilatometers, enabled the objectification of the subjective feelings of hot and cold. Almost simultaneously appeared an idea that the thermal states of bodies which were in common terms described by means of ordered series of terms cold, cool, tepid, warm, hot, could be characterized by the ordered series of thermoscope readings as well. The substitution of *thermoscope* for human *sensations* in experiments leads finally to the conviction that the thermal state of bodies or of environment can be characterized by thermoscope readings incomparably better than by means of sensation in itself. The thermoscope typically a glass tube provided with an arbitrary scale and bottom opened vessel filled with wine, the upper bulb of which contained a mixture of air with water and alcohol vapours, was thus promoted to the device apt to indicate with certain "absolute exactness" the thermal state of vicinal bodies. Such a (of course, very optimistic!) belief reflects also famous Fludd's quotation [14] "*Weather-glass* (i.e. thermoscope) *became a mighty weapon in the Herculean fight between Truth and Falsehood*". However, the principles behind the operation of thermoscope remain for a long time unclear and the conditions for its proper application were discovered only step-by-step by means of painful experimental work. Let us now summarize the fundamental concepts of thermometry in terms of modern language.

## 20.3 Introduction of Phenomenological Conjugate Variables and Thermal Equilibrium

The basic task of any mathematical theory of material systems is to establish general rules for the further treatment of *empirical constitutive relations* describing the state of a body in terms of suitably chosen parameters. We do not mean here the parameters specific for the description of thermal effects but just the parameters already introduced in other branches of physics and generally known as the *phenomenological variables* [15]. The existence of such variables, playing the roles of macroscopic conditions which are compatible with a huge number of parameters describing each microscopic component of the body, is clearly a matter of experience. Usually, there is an appreciable number of various phenomenological variables determining the state of the body but fixing a chosen one by external means, this number can ever be diminished by one. Continuing such a procedure, the number of significant phenomenological variables can be finally reduced to two. It is a fact worth to noticing that the two-parameter system is the simplest model of a real system because it enables one to construct a meaningful constitutive relation and, simultaneously, it is apt for straightforward generalization, e.g. by adding another pair of independent variables. In order to ensure easy perceptibility of mathematical description of two-parameter systems a special form of phenomenological variables was found to be desirable. It is a well known fact that the terms entering the energy balance equations in mechanics and electrodynamics have a canonical form which may be characterized by means of the following dimensional relation,

$$[\text{Energy}]=[\text{X}] \times [\text{Y}], \tag{20.1}$$

where square brackets mean the physical dimension of the quantities enclosed. As the energy is an extensive quantity, it is favourable to choose for the first phenomenological variable also an extensive quantity, say X. In such a case, however, the second parameter has to be inevitably an intensive quantity, Y [16]. Such a couple of quantities obeying relation 20.1 is then called a couple of *conjugate variables*. For example, extensive quantities are volume V, momentum G, electrical charge Q, mass M, and paired, conjugate, intensive quantities are pressure p, velocity v, electrical potential $\varphi$, and gravitational potential $\gamma$. The existence of the intensive and extensive "aspects" of heat which was already recognized by J. Black [17] is thus in this context the discovery of primary importance for the formalization of theory of heat and its compatibility with other branches of physics. His "intensity of heat" and "matter of heat" can be, namely, quite naturally assigned to a certain couple of conjugate variables, which may be tentatively called "temperature" and "heat". Formal compatibility of these two quantities with the system of quantities already introduced in other branches of physics is thus only a matter of proper choice of suitable operative definitions and units.

The idea to treat a real system in terms of conjugate variables enables one to introduce some fundamental concepts of thermal physics in a quite systematic way and, somewhat astonishingly, without a priori reference to the thermal phenomena per se, particularly to the quantities of temperature and heat [18]. The important role plays here the so called *correlation test*. It is the procedure frequently used in the practical thermometry which enables one to check whether the thermometer is in proper thermal contact with the measured body. Simultaneously, it provides the basis for the following operational definition of diathermic and adiabatic partitions (walls), viz: Let us have two systems characterized by couples of conjugate variables (X,Y) and (X‘,Y‘), respectively, and separated by a macroscopically firm material partition (wall) defining their common boundary. Such a partition is called *diathermic* if the changes of the variables (X,Y) induce the changes of the variables (X‘, Y‘) and vice versa, i.e. if the changes of both otherwise separated systems are correlated ("diathermic" originates from Greek $\delta\iota\acute{\alpha}=$ through, $\theta\varepsilon\rho\mu\acute{o}\varsigma=$ warm). A concept complementary to that of diathermic partition is the *adiabatic* partition (from Greek $\alpha$ = negation, $\delta\iota\acute{\alpha}$ = through, $\beta\alpha\acute{\iota}\nu\varepsilon\iota\nu$ = to go) which prevents the thermal contact of neighbouring bodies, i.e. ensures their thermal insulation. Obviously this concept is a limiting one, depending to an appreciable extent on the accuracy of the correlation test. The adiabatic partition is, namely, any partition for which the experimental proof of its ability to realize the diathermic contact by the said test failed.

Using then the language of the two-parameter description, the general definition of the equilibrium state as known from other branches of physics can be extrapolated also to the region of thermal phenomena. Let us recall first what the equilibrium state means e.g. in mechanics. Standard formulation for the two-parameter system reads: Any state of a body in which the conjugate coordinates X and Y remain constant in time so long as the external conditions are unchanged is called *equilibrium state* [19]. Combining then this definition with that of the diathermic partition, we can immediately define the concept of thermal equilibrium, which already belongs to the scope of thermal physics, namely:

> If two bodies being in diathermic contact are both in equilibrium state, they are in thermal equilibrium.

Let us then call *thermoscope* any two-parameter system in which one of the conjugate parameters, say Y, can be fixed, $Y=Y_0$. It is further assumed that thermoscope can be brought into a diathermic contact with other bodies and that it is sufficiently "small" in comparison with these bodies in order to not appreciably disturb their thermal equilibrium. The second conjugate real parameter X, which is called in this connection *thermoscopic variable*, is generally of quite a diverse physical nature and dimension. It may be length, volume, resistance, voltage, frequency and many others. In order to distinguish formally among various thermoscopic variables, differently constructed thermoscopes and physical conditions under which they operate, a small Latin index is used. Applying this convention, reading $X_k(P)$ of the k-th thermoscope which is in diathermic contact with a body under investigation defines the *thermoscopic state* P of the body. The corresponding set of the thermoscopic states which can be observed in this way is then marked as $\mathbf{H}_k$. Notice that the readings $X_k$ are related to the thermoscope while the indicated thermoscopic state such as $P \in \mathbf{H}_k$, already relates to the body. It is a matter of fact that the phenomenological parameters were introduced into classical mechanics and electrodynamics as continuous quantities covering certain closed intervals of real axis. Therefore, according to our definitions, such a property is transferred also to the thermoscopic variables $X_k$. We will thus assume that the numerical values of quantity $X_k$ also continuously cover a certain closed interval $\mathbf{I}_k$, operation range of the k-th thermoscope, which is a proper part of the set of real numbers, $\mathbf{E}_1$. In usual symbols we can thus write: $X_k \in \mathbf{I}_k \subset \mathbf{E}_1$. If it is further for every couple $P, Q \in \mathbf{H}_k$, $P \neq Q, \Leftrightarrow X_k(P) \neq X_k(Q)$, the set $\mathbf{H}_k$ can be ordered in accordance with the intrinsic order already existing in real interval $\mathbf{I}_k \subset \mathbf{E}_1$.

This provides basis for the construction of a primitive temperature scale. Indeed, unambiguous assignment of a certain value $X_k(P)$ to every state $P \in \mathbf{H}_k$ is nothing but realisation of *a local empirical temperature scale* in terms of thermoscopic variable $X_k$. It is, however, an important fact belonging rather to the scope of epistemology that although the empirical scales enable one to characterize the thermal states of bodies, for the revealing of the very nature of the physical quantity called temperature are almost useless and further development of more involved concepts is thus necessary.

There is another requirement ensuring the objectivity of the above conception. It is so called *Principle of indifference* [20] according to which different thermoscopes k, j operating in the common range of thermoscopic states should distinguish any two different states $P \neq Q$, $P, Q \in (\mathbf{H}_k \cap \mathbf{H}_j)$, regardless of their construction, thermometric substances, variables X and other physical conditions used. A procedure worked out by Dulong and Petit [21] is used in practical thermometry for the comparison of different empirical temperature scales. So called Dulong-Petit plot is a locus of readings $X_k$ of one thermoscope versus readings $X_j$ of another thermoscope both being in thermal contact with the same body (thermal bath). Evidently, in terms of such a plot the Principle of indifference may be formulated simply as follows: Two empirical temperature scales agree with the Principle of indifference just if their Dulong-Petit plot is monotonic. Interestingly, very similar method was used much earlier by savants of *Accademia del Cimento* who discovered in this way, just above the freezing point of water its anomaly, i.e. non-monotonic Dulong-Petit plot with respect to the other at that time known thermoscopes [14, 22], which excludes water from being in this range a suitable thermometric substance.

## 20.4 Fixed Thermometric Points, Mach's Postulates

A serious obstacle for the development of non-peripatetic thermal physics was an appreciable irreproducibility of early thermoscopes used. There were attempts to improve the situation by making exact copies of a standard instrument and by sending them to the various laboratories where they were intended to serve as secondary standards [23]. This, theoretically correct approach had nevertheless lot of practical limitations. It required a really high reproducibility of glass-blowing and preparation of glass and thermometric substances. Consequently, standard "thermometers" were very expensive and the transport of such delicate instruments over the long distances was quite risky. Therefore an important qualitative step toward the scientific thermometry was done when the so called *fixed thermometric points* were discovered and came into general use.

> The fixed point is called a body prepared by a definite prescription revealing by some observable qualitative property its physical state (e.g. boiling point of helium, melting point of water, melting point of platinum – all at normal atmospheric pressure) and which being in thermal equilibrium with other bodies defines unambiguously their thermoscopic state.

Fixed points thus may serve as a mean for the realization of fiducial points on empirical temperature scales corresponding to the different thermoscopes. Decisive steps in this direction were made by Danish astronomer O. Rømer, and especially by his immediate follower, German instrument maker G. D. Fahrenheit who used large deal of Rømer's know-how [24]. He devoted an enormous care to the purification of thermometric substance, improvements of glass-blowing procedure and exact specification of conditions for realization of fixed points. As a result, Fahrenheit's thermometers were much admired throughout the scientific community for their accuracy and extraordinary reproducibility and became thus for a long time a thermometric standard.

As was recognized very early the fixed points are of crucial significance for sewing together local thermometric scales. In order to cover much larger range of thermoscopic states it is, namely, necessary to combine the thermoscopes of different construction and sometimes of different physical nature of thermoscopic variables. It is then obvious that just the existence of a common fixed point incident with two different local empirical scales ensures that these scales really overlap and that they can be sewn at this point together.

Importance of fixed points for thermometry is, however, not confined only to the calibration of thermoscopes but as was recognized not before the end of the nineteenth century their theoretical significance is much more general. Quite interestingly, it has been originally taken for a self-evident empirical fact that it is always possible to find, in an operation range of any thermoscope, a sufficient number of fixed points enabling calibration of a local empirical scale. The very fact that such a liberty of choice can only be a consequence of the existence of enormous (if not infinite) number of fixed points falling into any interval of thermoscopic states remained for a long time quite unnoticed. Similar fate, i.e., being effectively undiscovered, has also the fact that the fixed points can be always found out of any interval of thermoscopic states.

These and other experimentally observed properties of fixed points have been generalized by means of the method of incomplete induction, the reasoning according to which the conclusion related even to the infinite number of cases is drawn from the knowledge of a finite number of cases provided that they, without exception, imply the same conclusion. Such a type of generalization of experience resulting into certain verities or postulates is quite analogous to that made e.g. prior to the axiomatic construction of Euclidean geometry. As the propositions given below received their first explicit formulation in the hands of E. Mach [21], we suggest calling them tacitly *Mach's postulates* (M1–M5).

First of all, as every fixed point defines unambiguously a certain thermoscopic state and because the set of thermoscopic states $\mathbf{H}_k$ is ordered by means of relation $(\prec, \succ)$, the *set of fixed points* $\mathbf{F}$ ($\mathbf{F}=\cup\mathbf{F}_k$, where $\mathbf{F}_k$ are sets of fixed points related to $\mathbf{H}_k$) can be also ordered just according to the same relation. Giving a physical meaning to such an idea, we can say that calibration of empirical scales by means of fixed points can be interpreted as an ordering of fixed points. This fact then enables one to postulate:

(M1) The set of fixed points **F** is ordered by means of binary relation ($\prec, \succ$).

The generalizations of experience with experimental establishment of new fixed points and of making their inventory list lead then to the following three postulates[1]:

(M2) To every fixed point P $\in$ **F** there exists at least one fixed point Q such that Q$\succ$P.

(M3) To every fixed point R $\in$ **F** there exists at least one fixed point S such that S$\prec$R.

(M4) For every couple of fixed points P$\prec$R there exists at least one interlaying point Q such that the relations P$\prec$Q and Q$\prec$R are simultaneously valid.

There is another remarkable empirical property of the set of thermoscopic states closely related to that of fixed points which can be formulated as follows:

(M5) Let A and B be two different fixed points such that A$\prec$B. Then if the body changes its thermoscopic state from the state corresponding to a fixed point A to that represented by a fixed point B, it must inevitably pass through all the interlaying thermoscopic states P, for which A$\prec$P$\prec$B.

## 20.5 Hotness Manifold and Definition of Temperature

It is a matter of historical fact that formulation of Mach' postulates (1896, [21]) and establishment of Cantor's set theory (1895, [25]) were practically contemporaneous events. That is probably why the mathematical structure of hotness manifold has not been fully appreciated and, as far as we know, has never been systematically analyzed from the point of view of the set theory. In terms of this theory (e.g. [26, 27]) Mach's postulates may be interpreted in the following way.

Taking first into account the fact that the realizations of fixed points are real bodies, their number must be either finite or equivalent to the set of natural

[1] Speaking for a while in terms of Kelvin's temperature which will be specified later, the temperatures observed range from ~$10^{-10}$ K (Low Temperature Lab, Helsinki University of Technology) up to ~$10^9$ K (supernova explosion) without any traces that the ultimate limits were actually reached. Speculative upper limit provides only the so called Planck temperature $T_P=\sqrt{(\hbar/G)}\times(c^2/k)\approx 1.417\times10^{32}$ K, hypothetically corresponding to the first instant of Big Bang and depending on the assumption that the constants c, G and k involved are really universal. Therefore the conjecture referred to as Mach's postulates M2 and M3, i.e. that the hotness manifold has no upper or lower bound, is obviously operating at least for all phenomena already known.

numbers, i.e. **F** must be countable. In the 4th Mach's postulate one can easily recognize the definition of dense sets belonging to Cantor's theory; from this we can immediately conclude that **F** is also dense. Postulates 2 and 3 then mean that the set **F** has no upper or lower bound. Any ordered countable dense set is, however, called rational series or a set of rational numbers. We can thus summarize, the set of fixed points **F** *is equivalent to an unbounded set of rational numbers.*

The mathematical structure of *hotness manifold* **H**, which is a union of all sets of thermoscopic states $\mathbf{H}_k$, $\mathbf{H} = \cup \mathbf{H}_k$, is not as simple as the structure of **F**. It is necessary first to make clear the operational method (i.e. a method related to the experimental procedures which can be really performed) enabling sewing-up the overlapping sets of thermal states and matching of corresponding empirical temperature scales. Let us assume that two sets of thermal states, $\mathbf{H}_k$, $\mathbf{H}_{k+1}$, overlap, i.e. that $\mathbf{H}_k \cap \mathbf{H}_{k+1} \neq \varnothing$. In order to realize this fact in experiment one has to find a fixed point $R \in \mathbf{F}$ belonging to both these sets, i.e. $R \in \mathbf{H}_k$, $R \in \mathbf{H}_{k+1}$. Theoretically the possibility of such an operation is ensured by 4th Mach's postulate M4. For the sake of definiteness and without loss of generality we can further construct the subsets $\mathbf{H}'_k \subset \mathbf{H}_k$, and $\mathbf{H}'_{k+1} \subset \mathbf{H}_{k+1}$ in such a way that $Q \prec R$ for every $Q \in \mathbf{H}'_k$ and $P \succ R$ for every $P \in \mathbf{H}'_{k+1}$. Evidently, the empirical temperature scale for thermal states from $\mathbf{H}_k \cup \mathbf{H}_{k+1} = \mathbf{H}'_k \cup \mathbf{H}'_{k+1}$ below R corresponds to the empirical scale in $\mathbf{H}_k$ and above R to that in $\mathbf{H}_{k+1}$. Moreover, in order to assign the same value of empirical temperature to the common point R, it is necessary to make formal changes at least in one of the empirical scales. Applying the procedure just described and simultaneously looking for new fixed points and for new physical effects enabling the construction of new kinds of thermoscopes, we can build a chain of $\mathbf{H}_k$'s more and more extending the region of accessible thermal states. We are obliged to Professor Mach for belief that such a procedure is limited only by our skills. Taking now into account the fact that every $\mathbf{H}_k$ is equivalent to a real interval $\mathbf{I}_k \subset \mathbf{E}_1$, it is obvious that $\mathbf{H}_k$ is a continuous set. Furthermore, fixed points, such as R, are then nothing but rational cuts in sets $\mathbf{H}_k$ and $\mathbf{H}_{k+1}$ [27]. Analyzing these circumstances, we can conclude that the properties of the hotness manifold $\mathbf{H} = \cup \mathbf{H}_k$ discussed above can be put in the form of two axioms already well-known from the set theory, namely

Dedekind's axiom: If $\mathbf{H}_1$ and $\mathbf{H}_2$ are any two non-empty parts of **H**, such that every element of **H** belongs either to $\mathbf{H}_1$ or to $\mathbf{H}_2$ and every element of $\mathbf{H}_1$ precedes every element of $\mathbf{H}_2$, then there is at least one element $R \in \mathbf{H}$ such that:

1. any element that precedes R belongs to $\mathbf{H}_1$,
2. any element that follows R belongs to $\mathbf{H}_2$.

Axiom of linearity: The hotness manifold **H** contains countable subset $\mathbf{F} \subset \mathbf{H}$ in such a way that between any two points $P \prec Q \in \mathbf{H}$ there is a point $R \in \mathbf{F}$ such that $P \prec R$ and $Q \succ R$.

As the second axiom ensures that the requirements of 5th Mach's postulate are automatically satisfied, the couple of Dedekind's axiom and axiom of linearity may be regarded as a concise reformulation of Mach's postulates. The simultaneous validity of both these axioms, however, defines in set theory the class of sets which are equivalent to the set of real numbers $\mathbf{E_1}$. The mathematical structure of hotness manifold may thus be summarized as follows:

Hotness manifold (a set of all accessible thermoscopic states) **H** is a set topologically equivalent to the set of all real numbers (real axis) $\mathbf{E_1}$. It contains a countable, dense and unbounded subset of all fixed points $\mathbf{F} \subset \mathbf{H}$, realizing the skeleton of **H**.

As we have seen above, the construction of this manifold is based on well-defined operational methods specifying conditions and procedures necessary for determination or reestablishment of a particular thermoscopic state. Manifold **H** is just the experimentally accessible entity enabling one to judge how hot or cold the bodies are. Therefore, it is this entity which is right to be regarded as an entity objectively existing in the Nature and representing the universal Platonic idea behind the usual concept of temperature, in philosophical jargon, the *temperature "an sich"*. Of course, as the set **H** has no intrinsic *metric properties*, it yields directly no physical quantity [16, 28]. Introduction of corresponding physical quantity, tacitly called temperature, thus requires special definition which obviously has to involve all the properties of hotness manifold.

As has been shown in previous paragraphs, the temperature and even the hotness manifold cannot be taken for primary concepts of thermal physics but are in fact the subjects of somewhat convoluted constructions. In the hierarchy of conceptual basis of thermal physics, however, the concept of temperature plays the role subordinate to that of hotness manifold **H**, which is characterized just only by its topological properties. On the other hand, it is quite clear that hotness manifold alone is not sufficient for the development of quantitative theory of thermal effects. For such a purpose, namely, a regular physical quantity [16] preserving simultaneously all essentials of hotness manifold, is necessary. Keeping the sufficient generality, such requirements may be satisfied by the following definition:

Temperature is any continuous one-to-one order preserving mapping of hotness manifold on a simple connected subset of real numbers.

## 20.6 Kelvin's Temperature Scale

It is quite obvious that the definition of temperature given in the preceding paragraph offers an enormous liberty for the construction of temperature scale. It is thus necessary to rationally choose the limitations which will be decisive for intelligibility of the future theory of thermal effects. Traditionally the safest guide for the introduction of new concepts into science is the so called *anthropomorphic principle* respecting the commonly accepted patterns of thinking and involving also practical and cultural aspects. The special mapping of the hotness manifold on an ordered subset of real numbers defining the operating temperature scale T was thus chosen on the more-or-less historical and practical grounds, in other words, on the basis of fully arbitrary anthropomorphic criteria. One of such restrictions having in fact no physical reason but which significantly simplifies the mathematical operations with temperature concerns the class of *absolute temperature scales* defined as follows [29]:

> Any temperature scale which is chosen in such a way that their functional values have the highest lower bound equal to zero (i.e. T is always positive) is called absolute temperature scale and the corresponding temperatures are called absolute temperatures[2].

Notice that the possible value $T = 0$ (equal to the highest lower bound) is already excluded by our definition of temperature because due to the absence of the lowest hotness level in the hotness manifold any continuous one-to-one order-preserving transformation on the set with lower bound $= 0$ has inevitably to map its improper point (e.g. $-\infty$) just on the point corresponding to absolute zero. *Nernst's law* of unattainability of absolute zero of temperature (the "Third Law of Thermodynamics") [30] is thus together with its consequences intrinsically involved in these definitions of temperature and absolute temperature and as such needs no additional, sometimes very curious, justifications or "proofs" [31].

It is a result of rather a complicated historical development that the present temperature scale (Kelvin's international temperature scale ITS [1]) is based on *two independent anthropomorphic idealizations*, namely, idealized substance, *ideal (perfect) gas* and idealized process, Carnot's *reversible cycle*. It is an important provable fact with large practical impact that both approaches define the identical scales which can thus be in particular cases used alternatively.

The first approach is based on the idealization of the most salient common features of the constitutive relations of real gases. The behaviour of majority of real gases is, namely, almost the same in cases where the gases have sufficiently low density. This fact was used for the definition of the perfect gas and later for the construction of the *ideal (perfect) gas temperature scale* T. The equation controlling the behaviour of the ideal gas, which is a hypothetical substance or concept rather than a real thing, reads:

$$T = pV/nR, \tag{20.2}$$

[2] Notice that this definition of absolute temperature scale differs from that due to Lord Kelvin who related the adjective "absolute" rather to the independence of temperature scale of thermometric substance than to the existence of lower bound of temperature values.

where p and V are respectively the pressure and the volume of the ideal gas which may both alternatively play the role of thermoscopic variables. As the hypothetical thermoscope a conventional gas thermometer [2] filled with n moles, $n > 0$, of ideal gas is considered. The constant R on the right side, has then a form of product $R = kN$ where k and N are Boltzmann's and Avogadro's constants, respectively (in SI system of units $k = 1.38 \times 10^{-23}$ J/K, $N = 6.02 \times 10^{23}$ mol$^{-1}$). The scale defined by means of Eq. 20.2 has some other remarkable properties. For example, as both quantities p and V have a natural lower bound = 0 (this very fact was already recognized by Amontons [21] and formulated as the hypothesis of *l'extrême froid*), the temperature T has also this lower bound and thus automatically belongs to the class of absolute temperatures. Moreover constitutive relation 20.2 reveals remarkable symmetry with respect to quantities p and V. We can thus exploit anyone of these two quantities as a thermoscopic variable keeping the other one constant. Comparing these two cases it must be inevitably:

$$T_p = T_V = T, \tag{20.3}$$

where $T_p$ and $T_V$ are temperatures of a body (e.g. corresponding to temperature of a certain fixed point) determined by means of constant pressure and constant volume method, respectively. The exact realization of condition 20.3 in experiments with real gases and with prescribed high accuracy (typically of the order of 0.1%) is very difficult if not impossible. However, Berthelot [32] devised a simple graphical method based on plausible assumptions which enables one to extrapolate experimental data obtained on real gases at finite pressures to the case corresponding to the ideal gas and finally determine also the value of T satisfying conditions 20.3. From these facts it is thus apparent that the ideal gas temperature scale can be in principle realized in the range where the gaseous phase of real gases and, of course, also the gas thermometer itself, can exist.

## 20.7 Carnot's Theorem and Kelvin's Proposition

Reasonably chosen temperature function which maps the hotness manifold on a subset of real numbers $\mathbf{E_1}$ should be, as was already mentioned, conformal with other terms entering the energy balance equation. In such a case temperature (intensive quantity T) and heat (extensive quantity $\varsigma$) will make up a couple of conjugate variables obeying dimensional equation 20.1, i.e.

$$[\text{Energy}] = [T] \times [\varsigma]. \tag{20.4}$$

The principal possibility to write down the thermal energy term just in this form was confirmed by early experiments on the development of mechanical work by means of heat engines. In spite of the fact that these experiments were backed by a rather primitive technique (e.g. temperatures were measured by roughly calibrated mercury thermometers and heat by the weight of burned coal), being thus of doubtful accuracy, their analysis enabled S. Carnot to introduce some new theoretical concepts and to draw out definite conclusions. In the present context among new Carnot's concepts the most important roles play two idealizations of real thermal process taking place in the heat engine, namely, the *cyclic process* and *reversible process*. By cyclic process (*cycle*) is meant any thermal process in which initial and final physical state of the heat engine are the same. The reversible process is then a thermal process in which the heat engine works without wastes of heat. For the heat engines utilizing the cyclic reversible process (so called *ideal heat engines*) Carnot was able to formulate a theorem which in its archaic version reads[3] [33]:

| |
|---|
| The motive power of heat is independent of the agents set at work to realize it; its quantity is fixed solely by the temperatures of the bodies between which, in the final result, the transfer of caloric is done. (S. Carnot, 1824) |

Of course, from the modern point of view Carnot's theorem is rather a desideratum than piece of scientific knowledge. (Remarkable is also a somewhat inconsequent use of heat and caloric as synonyms.) On the other hand, it has a form of the energy balance postulate we are searching for. Indeed, if we, namely, transform the theorem into mathematical symbols we can write it in terms of finite differences [34]

$$\Delta L = \varsigma F'(t)\, \Delta t, \tag{20.5}$$

where $\varsigma$ means the quantity of heat regardless of the method of its measurement, $\Delta L$ is the motive power (i.e. useful work done by heat engine) and $\Delta t$ is the difference between empirical temperatures of heater and cooler. The unknown function $F'(t)$ called *Carnot's function* should be for a concrete empirical scale determined by experiment [36, 37]. As the gained work $\Delta L$ has a dimension of energy and as this energy must be for reversible cycle equal per definition to the thermal energy of heat $\varsigma$ supplied to the ideal heat engine, we can conclude that the terms suitable for insertion into the energy balance equation have to have a form of products $\varsigma\Delta t$ properly modified by Carnot's function.

A revolutionary step toward the definition of the temperature scale independent of particular type of thermometer and thermometric substance was made in 1848 by Lord Kelvin [38]. He proposed to treat Carnot's theorem not as a heuristic statement

[3] It should be stressed here that there exist in the literature a lot of various arbitrarily changed forms of "Carnot's theorem" or "principle" which are not equivalent one to each other and which essentially differ in their very content from the original formulation. As was thus quite correctly pointed out by H. L. Callendar [35] distinguished researcher into the vapour turbines and president of Royal Society, the original oldest Carnot's formulation of his principle is at the same time the best one.

deduced from experiments of rather a limited accuracy but as a fundamental postulate of absolute validity. He further pointed out that the very purpose of Carnot's function is to modify or correct the difference of temperatures measured in a particular empirical temperature scale in such a way that it could serve as an exact proportionality factor between work, $\Delta L$, and heat, $\varsigma$. As this factor has to be according to Carnot's postulate the same for all substances and reversible cycles, Kelvin, inverting the logics of reasoning, suggested to define a universal (in his terminology "absolute", see footnote 2) temperature scale just by *prescribing* a proper analytical form of Carnot's function. For example, giving to Carnot's function the simplest permissible analytical form, namely, $F'({}^*T) = 1$ (so called "*caloric gauge*"), we are in fact defining a new temperature scale $^*T$ in terms of which Eq. 20.5 reads:

$$\Delta L \; = \varsigma \left({}^*T_2 - {}^*T_1\right). \tag{20.6}$$

It is immediately seen that using such a definition of the temperature scale the energy terms have the desired form of a product of two conjugate variables $\varsigma$ and $^*T$. Interestingly enough, Eq. 20.6 is simultaneously a fundamental relation of the caloric theory of heat (cf. [34, 39]). Accordingly, from the phenomenological point of view the heat is a kind of substance or fluid, caloric (*calorique, Wärmestoff, теплородЪ, teplík*), which being dissolved in all bodies is responsible for their thermal state. It is treated as an indestructible fluid (recall that the only method of how to get rid of heat is to convey it away), which is created in every irreversible processes such as rubbing, chemical reactions, burning, absorption of radiation and eating during which "something" simultaneously disappears for ever. The properties of so defined quantity are thus very near to the concept of heat in a common sense [40]. Taking further into account the structure of Eq. 20.6, we can also conclude that the development of moving force in an ideal heat engine is not connected with some actual consumption of heat as is claimed in thermodynamics but rather with its transfer from hotter body to a colder one (water-mill analogy [33]). At the same time, Eq. 20.6 defines an entropy-like unit of heat fully compatible with the SI system which may be, according to Callendar's suggestion, appropriately called "Carnot" (Abbreviation "Ct") [35]. One Carnot is then that quantity of heat which is in a reversible process capable of producing 1 J of work per 1 K temperature fall.

Nevertheless, in the present context another aspect of Eq. 20.6 is far more important. Accordingly, namely, the temperature difference $^*T_2 - {}^*T_1$ between two bodies used e.g. as "heater" and "cooler" of an ideal heat engine, is identical with the ratio $\Delta L/\varsigma$ where both of these quantities are measurable in principle; $\Delta L$ by means of standard methods well-known from mechanics and $\varsigma$ e.g. by the amount of fuel consumed by heating the heater or, if the cooler is kept at freezing point of water, by amount of ice melted during the cycle. It is quite obvious that such a technique of temperature measurement, although possible in principle, is rather a curiosity which would be very difficult to realize with sufficiently high accuracy in practice. The idea of this method is, however, of primary importance for theory. Obviously, due to Carnot's postulate, Eq. 20.6 has to be valid for any ideal heat engine regardless of its construction and working substance used. Analyzing thus one particular representative case of the ideal heat engine, general conclusions can be made. For example, if we imagine an ideal heat engine driven by perfect gas and working in cycles which consist of two isothermal and two isochoric reversible processes, the useful work can be easily computed, provided that the temperatures are measured in terms of perfect gas scale. The result of such a computation reads

$$\Delta L = nR \ln(V_2/V_1)(T_2 - T_1), \tag{20.7}$$

where $V_1$ and $V_2$ are the limits of volume between which the engine operates. It is apparent at first glance that the last equation is fully congruent with Eq. 20.6 with the proviso that the heat (measured in entropy units, e.g. Ct) transferred from heater to cooler per cycle is given by $\varsigma = nR \ln(V_2/V_1)$. The congruence of these equations means that the system of units can be always chosen in such a way that scales *T and T will be identical [16]. Expressing this fact more physically, we can say:

> The measurement of temperature by means of ideal gas thermometer is equivalent to the measurement of temperature by means of ideal heat engine.

The theoretical significance of this theorem is enormous because it enables one to relate without ambiguity the ideal gas (Kelvin) temperature scale to the temperatures defined by other types of ideal heat engines, e.g. "*gedanken*" reversible cycles, in systems controlled by electric, magnetic or electrochemical forces. Besides, it should be stressed that this theorem, although based on arbitrary assumptions, is by no means accidental. The idealization of the constitutive relation of real gases and the idealization involved in Carnot's postulate have the same anthropomorphic roots, namely, the feeling that the thermal dilation of bodies must be linearly dependent on their thermal state. Incidentally, in the range between 0°C and 100°C the air scale and the mercury temperature scale, prevailingly used in experiments related to establishment of Carnot's theorem, are almost identical.

For the sake of completeness we have to mention here also the so called *thermodynamic gauge* of Carnot's function. The general acceptance of this gauge in *classical thermodynamics* was, however, not a result of a free choice but a direct consequence of admittance of *Joule-Mayer's Principle* of equivalence of work (energy) and heat. This Principle which is till now in practically all modern textbooks on thermodynamics treated as an experimentally proved "truth" was, however, quite correctly from the very beginning criticized by M. Faraday [41] as an absolutely wrong "*strange conclusion*" which was "*deduced most illogically*" on the basis of fatal misinterpretation of Joule's paddle-wheel experiment. Accordingly, namely, the existence of exchange rate between two different quantities, mechanical work and heat, called mechanical equivalent of heat, $J \approx 4.2$ J/cal, is confused with the experimental proof of identity of these two entities. Thus in such a context Joule-Mayer's Principle has rather a character of arbitrary redundant postulate than that of experimental result [42]. Nevertheless, once this Principle is accepted, Carnot's function, in terms of ideal gas temperature scale T, has to have inevitably the form $F'(T) = J/T$ (*thermodynamic gauge*) [36], cf. also [43]. Such a gauge, however, enormously complicates the formalism of classical thermodynamics, because it requires introducing into the thermal term (4) of balance equation, instead of "quantity of heat" already associated with the energy, another somewhat artificial physical quantity, *entropy*, which has no clear phenomenological interpretation [40]. Typical claims which can thus be found in the current literature on thermodynamics sound "*Joule's experiment conclusively established that heat is a form of energy...*" It is, however, worth mentioning here some remarkable facts which undermine the credibility of such statements. For example, practically all measurements (i.e. more than about 30 serious extensive works Joule's works including from the second half of the nineteenth to the end of the twentieth centuries) of mechanical equivalent of heat were made at only *single* temperature. It is thus quite evident that the experimentalists tacitly assumed, prior making the experiment, the validity of Joule-Mayer's Principle, being convinced that the measured equivalent is nothing but a conversion factor between two different energy units, which has to be inevitably temperature independent. "Derivation" of Joule-Mayer's Principle from such an experimental data is obviously nothing but a case of circular reasoning. Moreover, the mechanical equivalent of heat was (with much smaller accuracy but in the correct way) determined by Carnot [33] more than 20 years before Joule within the frame of caloric theory, i.e. without any possible reference to Joule-Mayer's Principle.

Interestingly enough, the choice of particular gauge does not directly influence the properties of Kelvin's temperature scale itself but it is quite decisive for the mathematical behaviour and physical interpretation of corresponding conjugate extensive quantity.

## 20.8 Problem of Distant Measurements of Temperature

Under the term "*distant measurements*" in a restricted sense we mean the determination of a physical quantity belonging to a certain moving inertial frame by means of measurements made in the rest system. The operational methods for distant measurements of e.g. length, time, frequency and intensity of fields are generally known from the Special Theory of Relativity. In the case of temperature, however, due to its peculiar physical nature, we encounter serious difficulties which result into quite controversial solutions of the problem [44]. At first glance it may seem that the problem of distant measurement of temperature belongs to the scope of practical calorimetry and thermal analysis performed in laboratories only marginally, being of primary importance only for thermo-physical processes taking place on remote objects like stars or spacecrafts. It should be stressed, however, that the considerations dealing with relativistic transformations of thermal quantities reveal their physical structure and are thus quite crucial for a consistent interpretation of these quantities even under the terrestrial conditions.

The main difficulty in distant measurement of temperature is the principal impossibility of realization of correlation test and establishment of thermal equilibrium between two relatively moving inertial systems. Indeed, the relative movement of systems A and B prevents one from answering without ambiguity, on the basis of correlation test, the question of whether the common boundary is diathermic or not, which makes any judgment on the thermal equilibrium between A and B quite questionable. It is further clear that the boundary between two relatively moving systems has to move at least with respect to one of them. In such a case, however, the interaction between these systems can exist even if the boundary is non-diathermic (adiabatic). For example, the moving boundary can exert a pressure on one of the systems without changing the state of the other and/or a charged system A surrounded by a metallic envelope, regardless of the fact whether it is diathermic or adiabatic, can induce dissipative equalization currents in system B without affecting the charge distribution inside system A. In order to exclude such cases, the temperature of any body must be measured only by means of a thermometer which is in the rest with respect to the body, and this operation cannot be, in principle, performed by a relatively moving observer (cf. also [45]). Hence the temperature cannot be the subject of a direct distant measurement in principle. It can only be the result of local measurement and subsequent data transfer into another inertial system. (If possible, the digital mailing of the data would be the best choice.) The operational rules for distant measurement of temperature may then be formulated as follows:

1. Bring the body under investigation into diathermic contact with the thermometer placed in the same inertial frame
2. Reconstruct in another (e.g. rest) inertial system the reading of the thermometer applying transformation rules relevant to the thermoscopic variable used

Having already at our disposal the prescription defining the temperature in mathematical terms, it is in principle possible to perform the Lorentz transformation of the left-hand side of Eq. 20.2 and to obtain in this way the formula for the relativistic transformation of temperature T. However, in order to be able to analyze also other temperature measuring methods (using e.g. platinum resistance, blackbody radiation, thermoelectric voltage) in a sufficient generality a methodical approach is more relevant.

Fortunately, the properties of hotness manifold enable one to make the following fairly general considerations. First of all, it is evident that in order not to violate the Principle of Relativity the behaviour of bodies realizing fixed thermometric points has to be the same in all inertial frames (cf. [46]). For example, it would be absurd to admit an idea that the water violently boiling in its rest system can simultaneously[4] look calm if observed from another relatively moving inertial system. In other words, any fixed point has to correspond to the same hotness level regardless of the inertial frame used for the observation. Assigning, by means of some convention, to each body realizing the fixed point a certain "inventory entry", the resulting, by pure convention established list of numbers cannot be changed by a mere transfer from one inertial system to another.

For example, using thus as an operative rule for stocktaking of fixed points formula 20.2 (in SI units with $R = 8.3145$ J/K mol) and assigning to the triple point of water an inventory entry 273.16 K, we obtain an ordered table of fiducial points of ideal gas scale (similar to the ITS [1]) which must be valid in all inertial frames. As the set of fixed points provides a dense subset (skeleton) in continuous hotness manifold, such a Lorentz-invariant table can be extended and made finer as we like and consequently, any hotness level can be, by means of this table, approximated with arbitrary accuracy. Due to the continuity of prescription 20.2 the whole *ideal gas (Kelvin) scale* T *is then inevitably Lorentz invariant.*

The invariance of Kelvin scale has, however, a very interesting and far reaching consequence. Let us make the following thought experiment with two identically arranged gas thermometers both filled with one mole of ideal gas which are in two relatively moving inertial systems in diathermic contact with the same fixed point bath (for definiteness, with triple point of water) placed in their own frames. As the pressure in both devices is Lorentz invariant [47, 48], we can write:

$$p = p_0, \tag{20.8}$$

$$T = T_0, \tag{20.9}$$

where index 0 is related, as above, to the quantities measured in the a-priori chosen rest system. Taking now the well-known Lorentz transformation of volume into account, we obtain from 20.8 and 20.9 the following series of equations

$$pV = p_0 V_0 \sqrt{(1 - v^2/c^2)} = RT = R_0 T_0 \sqrt{(1 - v^2/c^2)}, \tag{20.10}$$

from which a somewhat astonishing relation immediately follows:

$$R = R_0 \sqrt{(1 - v^2/c^2)}. \tag{20.11}$$

The physical meaning of this formula is really far reaching. Taking into account, namely, that R is an entropy unit, Eq. 20.11 must simultaneously enter the transformation formulae for entropy in general. This is, however, in severe contradiction with *Planck's Ansatz* claiming that the entropy is Lorentz invariant. We have to

[4] Notice that we have to do here with the essentially time-independent stationary process where the Lorentz transformation of time plays no role. Let us also recall that the pressure, controlling e.g. boiling point of water, may be proved independently to be Lorentz invariant [47, 48]

recall here that this Ansatz, serving as a starting point of numerous considerations in relativistic thermodynamics, has never been proved with sufficient exactness but from the beginning it was mere an intuitive conjecture [49]. It was namely argued that the entropy has to be invariant, because it is the logarithm of a discrete number of states which is "naturally" Lorentz invariant. Nevertheless, such a seemingly transparent argument cannot be true in general as can be shown by means of the following consideration. Let us imagine first that a sample of paramagnetic salt is submerged in a bath of boiling helium kept at normal atmospheric pressure realizing the Lorentz invariant fixed point corresponding to the temperature of 4.2 K [8]. At zero magnetic fields certain entropy can be ascribed to such a state of paramagnetic salt independently of the fact in what inertial system the experiment is performed. Let us further assume that there is distributed static electric charge in the rest system not affecting the entropy of paramagnetic salt. As is well known from the special theory of relativity, however, the magnetic field is nothing but the electrostatic field observed from the relatively moving inertial system [50]. Therefore, the observer in the relatively moving inertial frame has to detect magnetic field, and the entropy of the said paramagnetic sample kept at 4.2 K must be smaller than that in the rest system. In other words, because the entropy in this particular case depends on the choice of inertial system of observation, it cannot be generally Lorentz invariant. If we thus once admit the relativistic invariance of temperature, we have to reject Planck's conjecture as unsound and particularly, we can also no more treat various entropy pre-factors, e.g. gas constant R and Boltzmann's constant k, as universal constants.

## 20.9 Summary

In conclusion, the central concept of thermal physics, *temperature*, is defined in terms of the set theory as an arbitrary one-to-one order preserving continuous mapping of the so-called *hotness manifold* (*set*) **H** on a certain simple connected open subset of real numbers. It has been shown that the hotness manifold representing all in the Nature existing thermoscopic (thermal) states is the only entity accessible to direct physical observation. This set which was further proved to be topologically equivalent to the set of all real numbers (real axis) $\mathbf{E_1}$, contains a countable, dense and unbounded subset of all *fixed points* $\mathbf{F} \subset \mathbf{H}$. Any fixed point is realized by means of a specially prepared body which defines just one thermal state. The properties of the set **F** and its relation to the manifold **H** are specified by means of Mach's postulates which are generalizations of empirical facts. As was further shown, the special mapping of **H** on the set of all positive real numbers known as the *International Kelvin Temperature Scale* T was chosen on the grounds of two essentially anthropomorphic idealizations providing a concordant result, namely, on ideal substance, *perfect gas* and on ideal process in heat engine, *reversible cycle*. Finally, on the basis of simple physical arguments taking into account the mathematical structure of the hotness manifold the Lorentz invariance of the temperature was proved. Consequently, the variable conjugate to temperature, i.e. entropy-like heat, cannot be Lorentz invariant in severe contradiction to Plank's Ansatz claiming the Lorentz invariance of entropy in general.

**Acknowledgments** This work was supported by Institutional Research Plan of Institute of Physics No AV0Z10100521.

## References

1. Preston-Thomas H (1990) The international temperature scale of 1990 (ITS 90). Metrologia 27:Z186–Z193
2. Nieto A, Paul F (1958) Mesure des températures. Éditions Radio, Paris
3. Poincaré H (1908) Thermodynamique, 2nd edn. Gauthier-Villars, Paris; Reprint: Éditions J Gabay, Paris (1995)
4. Cullen H (1960) Thermodynamics: an introduction to thermostatics. Wiley, New York
5. Roberts JK (1950) Heat and thermodynamics, 4th edn. Blackie, Glasgow
6. Epstein PS (1954) Textbook of thermodynamics. Wiley, New York
7. Truesdell C (1969) Rational thermodynamics. McGraw-Hill, New York
8. Waldram JR (1985) The theory of thermodynamics. Cambridge University Press, Cambridge
9. Bailyn M (2002) A survey of thermodynamics. AIP, New York
10. Blundell SJ, Blundell KM (2008) Concepts in thermal physics. Oxford University Press, New York
11. Boas M (1949) Hero's Pneumatica: a study of its transmission and influence. Isis 40:38–48
12. Hoppe E (1926) Geschichte der Physik. F. Vieweg und Sohn, Braunschweig
13. Mareš JJ (2000) On the development of the temperature concept. J Therm Anal Calorim 60:1081–1091
14. Meyer-Bjerrum K (1913) Die Entwicklung des Temperaturbegriffs im Laufe der Zeiten. Vieweg und Sohn, Braunschweig
15. Fürth R (1929) Allgemeine Grundlagen der Physik, Prinzipien der Statistik. In: Handbuch der Physik Bd. IV. Springer, Berlin
16. Palacios J (1964) Dimensional analysis. Macmillan, London
17. Black J (1803) Lectures on the elements of chemistry. W. Creech, Edinburgh; Germ transl.: Vorlesungen über Chemie. Crell, Hamburg (1804)
18. Balamuth L, Wolfe HC, Zemansky MW (1941) The temperature concept from the macroscopic point of view. Am J Phys 9:199–203
19. Keller FJ, Gettys WE, Skove MJ (1993) Physics. McGraw-Hill, New York
20. Barnett MK (1956) The development of thermometry and the temperature concept. Osiris 12:269–341
21. Mach E (1896) Die Prinzipien der Wärmelehre. Verlag von J. A. Barth, Leipzig
22. Burckhardt F (1867) Die Erfindung des Thermometers und seine Gestaltung im XVII Jahrhundert. Basel
23. Boyer CB (1942) Early principles in the calibration of thermometers. Am J Phys 10:176–180
24. von Oettingen AJ (1894) Abhandlungen über Thermometrie von Fahrenheit, Réaumur, Celsius. Ostwald's Klassiker No. 57, W. Engelmann, Leipzig
25. Cantor G (1895) Beiträge zur Begründung der transfiniten Mengenlehre. Math Ann 46:481–512
26. Frenkel A (1923) Einleitung in die Mengenlehre. Springer, Berlin
27. Huntington EV (1917) The continuum and other types of serial order. Harvard University Press, New York; Reprint: Dover Phoenix Editions, New York (2003)

28. Serrin J (1978) The concepts of thermodynamics. In: de la Penha GM, Medeiros LA (eds) Contemporary developments in continuum mechanics. North-Holland, Amsterdam
29. Truesdell C, Bharatha S (1977) Concepts and logic of classical thermodynamics as a theory of heat engines. Springer, New York
30. Nernst W (1926) The new heat theorem. Reprint: Dover, New York (1969)
31. Boas ML (1960) A point of logic. Am J Phys 28:675
32. Wensel HT (1940) Temperature and temperature scales. J Appl Phys 11:373–387
33. Carnot S (1824) Réflexions sur la puissance motrice du feu et sur les machines propres à développer cette puissance. Bachelier, Paris, Germ transl.: Ostwald's Klassiker, Nr. 37, Engelmann, Leipzig (1909)
34. Larmor J (1918) On the nature of heat, as directly deducible from the postulate of Carnot. Proc R Soc London A 94:326–339
35. Callendar HL (1911) The caloric theory of heat and Carnot's principle. Proc Phys Soc London 23:153–189
36. Cropper WH (1987) Carnot's function: origins of the thermodynamic concept of temperature. Am J Phys 55:120–129
37. Truesdell C (1979) Absolute temperatures as a consequence of Carnot's general axiom. Arch Hist Exact Sci 20:357–380
38. Thomson W (1848) (Lord Kelvin of Largs): On the absolute thermometric scale founded on Carnot's theory of the motive power of heat. Philos Mag 33:313–317
39. Mareš JJ, Hubík P, Šesták J, Špička V, Krištofik J, Stávek J (2008) Phenomenological approach to the caloric theory of heat. Thermochim Acta 474:16–24
40. Job G (1972) Neudarstellung der Wärmelehre – die Entropie als Wärme. Akad. Verlagsges, Frankfurt am Main
41. Smith CW (1976) Faraday as referee of Joule's Royal Society paper "On the Mechanical Equivalent of Heat". Isis 67:444–449
42. Job G, Lankau T (2003) How harmful is the first law? Ann NY Acad Sci 988:171–181
43. Helmholtz H (1889) Über die Erhaltung der Kraft. Ostwald's Klassiker, Nr. 1. Engelmann, Leipzig
44. Mareš JJ, Hubík P, Šesták J, Špička V, Krištofik J, Stávek J (2010) Relativistic transformation of temperature and Mosengeil-Ott's antinomy. Physica E 42:484–487
45. van Kampen NG (1968) Relativistic thermodynamics of moving systems. Phys Rev 173:295–301
46. Avramov I (2003) Relativity and temperature. Russ J Phys Chem 77:S179–S182
47. von Laue M (1911) Das Relativitätsprinzip. Vieweg und Sohn, Braunschweig
48. Møller C (1952) The theory of relativity. Oxford University Press, Oxford
49. Planck M (1908) Zur Dynamik Bewegter Systeme. Ann Phys (Leipzig) 331:1–34
50. Rosser WGV (1968) Classical electromagnetism via relativity. Butterworths, London

# Chapter 21
# Historical Roots and Development of Thermal Analysis and Calorimetry

**Jaroslav Šesták, Pavel Hubík and Jiří J. Mareš**

## 21.1 Historical Aspects of Thermal Studies, Origins of Caloric

Apparently, the first person which used a thought experiment of continuous heating and cooling of an illustrative body was curiously the Czech thinker and Bohemian educator [1], latter refugee *Johann Amos Comenius* (Jan Amos Komenský, 1592–1670) when trying to envisage the properties of substances. In his "*Physicae Synopsis*", which he finished in 1629 and published first in Leipzig in 1633, he showed the importance of hotness and coldness in all natural processes. Heat (or better fire) is considered as the cause of all motions of things. The expansion of substances and the increasing the space they occupy is caused by their dilution with heat. By the influence of cold the substance gains in density and shrinks: the condensation of vapor to liquid water is given as an example. *Comenius* also determined, though very inaccurately, the volume increase in the gas phase caused by the evaporation of a unit volume of liquid water. In Amsterdam in 1659 he published a focal but rather unfamiliar treatise on the principles of heat and cold [2], which was probably inspired by the works of the Italian philosopher *Bernardino Telesius*. The third chapter of this Comenius' book was devoted to the description of the influence of temperature changes on the properties of substances. The aim and principles of thermal analysis were literally given in the first paragraph of this chapter: citing the English translation [3–5]:

> In order to observe clearly the effects of heat and cold, we must take a visible object and observe its changes occurring during its heating and subsequent cooling so that the effects of heat and cold become apparent to our senses.

J. Šesták (✉)
New Technologies Research Centre, University of West Bohemia,
Univerzitní 8, CZ-30614, Plzeň, Czech Republic
e-mail: sestak@fzu.cz

P. Hubík and J.J. Mareš
Institute of Physics ASCR, v.v.i.
Cukrovarnická 10, 162 00, Praha 6, Czech Republic

J. Šesták et al. (eds.), *Glassy, Amorphous and Nano-Crystalline Materials*,
Hot Topics in Thermal Analysis and Calorimetry 8, DOI 10.1007/978-90-481-2882-2_21,

In the following 19 paragraphs of this chapter *Comenius* gave a rather systematic description (and also a partially correct interpretation) of the effects of continuous heating and cooling of water and air, and also stressed the reversibility of processes such as, for example, evaporation and condensation, etc., anticipating somehow the concept of latent heat. Comenius concludes this chapter as follows: All shows therefore that both heat and cold are a motion, which had to be proved. In the following chapter Comenius described and almost correctly explained the function of a thermoscope ('*vitrum caldarium*') and introduced his own qualitative scale with three degrees of heat above and three degrees of cold below the ambient temperature launching thus a concept of "*caloric*".

Nonetheless, it is difficult to trace [1, 3–6] and thus hard to say if it was possible (though likely) to disseminate the Comenius idea of caloric from Amsterdam (when he mostly lived and also died) to Scotland where a century later a new substance, or better a matter of fire, likewise called caloric (or caloricum), was thoroughly introduced by *Joseph Black* (1728–1799) [7] and his student *Irvine*. Unfortunately, Black published almost nothing in his own lifetime [5, 8] and his attitude was mostly reconstructed from contemporary comments and essays published after his death.

Caloric [1, 7, 9–11] was originally seen as an imponderable element with its own properties. It was assumed, e.g., that caloric creeps between the constituent parts of a substance causing its expansion. Black also supposed that heat (caloric) was absorbed by a body during melting or vaporization, simply because at the melting or boiling points sudden changes took place in the ability of the body to accumulate heat (~1761). In this connection, he introduced the term '*latent heat*' which meant the absorption of heat as the consequence of the change of state. Irvine accounted that the relative quantities of heat contained in equal weights of different substances at any given temperature (i.e., their '*absolute heats*') were proportional to their '*capacities*' at that temperature and it is worth noting that the term 'capacity' was used by both Black and later also Irvine to indicate specific heats [7, 9–11].

Black's elegant explanation of latent heat to the young *James Watts* (1736–1819) became the source of the invention of the businesslike steam engine as well as the inspiration for the first research in theory related to the novel domain of thermochemistry, which searched for general laws that linked heat, with changes of state. In 1822, *Jean-Baptiste Joseph Fourier* (1768–1830) published an influential book on the analytical theory of heat [12], in which he developed methods for integration of partial differential equations, describing diffusion of the heat substance. Based on the yet inconsistent law of conservation of caloric, *Siméon D. Poisson* (1823) derived a correct and experimentally verifiable equation describing the relationship between the pressure and volume of an ideal gas undergoing adiabatic changes. *Benjamin Thompson (Count Rumford,* 1753–1814) presented qualitative arguments for such a fluid theory of heat with which he succeeded to evaluate the mechanical equivalent of heat [11, 13]. This theory, however, was not accepted until the later approval by *Julius Robert Mayer* (1814–1878) and, in particular, by *James Prescott Joule* (1818–1889), who also applied Rumford's theory to the transformation of electrical work.

In the year 1826 *Nicolas Clement* (1779–1842) [11] coined the unit of heat as amount of caloric, necessary for heating 1 g of liquid water by 1°C. Though the expected temperature changes due to "thickened caloric" did not experimentally occur (cf. measurements in "Torricelli's vacuum" over mercury by *Gay-Lussac*) and in spite of that Thompson (1798) showed that the heat could be produced by friction ad infinitum, the caloric theory survived many defeats and its mathematical scheme is in fact applied for the description of heat flow until today. The above customary unit was called '*calorie*' (cal) or '*small calorie*', whereas a '*large calorie*' corresponded to the later '*kilocalorie*' (kcal). The word "calorie" was more widely introduced into the vocabulary of academic physicists and chemists by *Favre* and *Silbermann* [14] in 1852. The expression of 1 kcal as 427 kg m was given by *Mayer* in the year 1845.

We should add that caloric differed from the foregoing concept of *phlogiston* because, beside else, it could be measured with an apparatus called a *calorimeter,* however, it is not clear who was the first using such an instrument. If we follow the studies of *Brush* [8], *Mackenzie* [15] and *Thenard* [16] they assigned it to *Wilcke*. It, however, contradicts to the opinion presented in the study by *McKie* and *Heathcote* [17] who consider it just a legend and assume that the priority of familiarity of ice calorimeter belongs to *Laplace* who was most likely the acknowledged inventor and first true user of this instrument (likely as early as in 1782). In fact, *Lavoisier* and *Laplace* entitled the first chapter of their famous "*Mémoire sur la Chaleur*" (Paris 1783) as "*Presentation of a new means for measuring heat*" (without referring Black because of his poor paper evidence). Report of Black's employment of the calorimeter seems to appear firstly almost a century later in the *Jamin's Course of Physics* [1].

## 21.2 Underlying Features of Thermal Physics Interpreted Within the Caloric Theory

In the light of work of senior *Lazare Carnot* (1753–1823) on mechanical engines [11], *Sadi Carnot* (1796–1832) co-opted his ideas of equilibrium, infinitesimal changes and imaginatively replicated them for caloric (in the illustrative the case of water fall from a higher level to a lower one in a water mill). He was thinking about writing a book about the properties of heat engines applying caloric hypothesis generally accepted in that time within broad scientific circles [18–20]. Instead, he wrote a slim book of mere 118 pages, published in 200 copies only, which he entitled as the "*Reflections on the motive power of fire and on machines fitted to develop that power*" (1824) [21], which was based on his earlier outline dealing with the derivation of an equation suitable for the calculation of motive power performed by a water steam [11]. He discussed comprehensively under what conditions it is possible to obtain useful work ("*motive power*") from a heat reservoir and how it is possible to realize a reversible process accompanied with heat transfer. Sadi also explained that a reversibly working heat engine furnished with two different working agents had to have the same efficiency related to the temperature difference, only. Among other notable achievements [14, 22–27] there was the determination of the difference between the specific heats of gases measured at constant pressure and volume. He found that the difference was the same for all gases, anticipating thus the Mayer's relation for ideal gas: $c_p - c_v = R$. Sadi also introduced the "*Carnot's function*" the inverse of which was later (1850) identified by *Rudolph Clausius* (1822–1888) [28], within the classical thermodynamics, with the absolute temperature $T$. Finally, Sadi adjusted, on the basis of rather poor experimental data that for the production of 2.7 mechanical units of "motive power" it was necessary to destruct 1 cal unit of heat, which was in a fair correspondence with the later mechanical equivalent of heat: (4.1 J/cal). It is worth noting that already when writing his book he started to doubt the validity of caloric theory [11, 27] because several of experimental facts seemed to him almost inexplicable. Similarly to his father, Sadi's work remained unnoticed by contemporary physicists and permanently unjustly criticized for his principle of the conservation of caloric, which is, however, quite correct for any cyclic reversible thermal process.

Adhering to the way of Carnot's intuitive thinking [26, 27], the small amount of work done d$L$ (motive power in Carnot's terms) is performed by caloric $\varsigma$ literarly falling over an infinitesimal temperature difference d$T$ [11, 16, 26], $\mathrm{d}L = \varsigma\, F(T)\, \mathrm{d}T$. The function $F(T)$ here is the Carnot's function, which has to be determined experimentally, certainly, with respect to the operative definitions of quantities $\varsigma$ and $T$. Carnot assumed that caloric is not consumed (produced) by performing work but only loses (gains) its temperature (by d$T$). Therefore, the caloric has there an extensive character of some special substance while the intensive quantity of temperature plays the role of its (thermal) potential; the thermal energy may be thus defined as the product $\varsigma \times T$, in parallel with other potentials such pressure (choric potential) for volume, gravitational potential for mass and electrostatic potential for charge.

Taking into account that caloric is conserved during reversible operations, the quantity $\varsigma$ must be independent of temperature and, consequently, Carnot's function $F(T)$ has to be also constant. Putting the function equal identically 1 the unit of caloric fully compatible with the SI system is defined. Such a unit (*Callendar* [23]), can be appropriately called "Carnot" (abbreviated as "Cn" or "Ct"). One "Ct" unit is then such a quantity of caloric, which is during a reversible process capable of producing 1 J of work per 1 K temperature fall. Simultaneously, if such a system of units is used [26, 27], the relation $\mathrm{d}L = \varsigma\, \mathrm{d}T$ retains.

The caloric theory can be extended for irreversible processes by adding an idea of wasted (dissipated) motive power which reappears in the form of newly created caloric [26]. Analyzing Joule's paddle-wheel experiment from view of both this extended caloric theory and classical thermodynamics, it can be shown that the relation between caloric and heat in the form $\mathrm{d}\varsigma = J\, \mathrm{d}Q/T$ takes place, which, at first glance, resembles the famous formula for entropy, certainly if we measure the heat in energy units. This correspondence between *entropy and caloric*, may serve as a very effective heuristic tool for finding the properties of caloric by exploitation the results known hitherto from classical thermodynamics. From this point of view it is clear that the caloric theory is not at any odds with experimental facts, which are only anew explained ([26]). The factor $J$ historically determined by Joule ($J \sim 4.185$ J/cal) should have been rather related with the establishment of a particular system of units then with a general proof of the equivalence between heat and energy.

One of the central questions of the Carnot's theory of heat engines is the evaluation of engine efficiency. The amount of caloric $\varsigma$ which is entering the completely reversible and continuously working heat engine at temperature $T_1$ and leaving it at temperature $T_2$ will produce a motive power of amount $L$. Carnot's efficiency $\eta_C$ defined as a ratio $L/\varsigma$ is then given by a plain temperature drop $\Delta T = (T_1 - T_2)$ (as measured in the ideal gas temperature scale). Transforming the incoming caloric into thermal energy $T_1\varsigma$, we obtain immediately Kelvin's *dimensionless efficiency* $\eta_K$ of the ideal reversible heat engine, $\eta_K = \{1 - (T_2/T_1)\}$, which is well-known from textbooks of thermodynamics [3, 29].

However, $\eta_K$ is of little significance for the practical evaluation of the performance of real heat engines, which are optimized not with respect to their efficiency but rather with respect to their available output power. As a convenient model for such a case it may be taken an ideal heat engine impeded by a thermal resistance [26]. The effect of *thermal resistance* can be understood within the caloric theory in such a way that the original quantity of caloric $\varsigma$, taken from the boiler kept at temperature $T_1$, increases, by passing across a thermal resistance, to the new quantity equal to $\varsigma + \Delta\varsigma$, entering than the ideal heat engine at temperature $T < T_1$, and leaving it temperature $T_2$. If we relate the quantities $L$ and $\varsigma$ to an arbitrary time unit (we conveniently use for this purpose a superscript $u$), it follows $L^u = \lambda(T_1 - T)(T - T_2)/T$, where for the evaluation of temperature drop across the thermal resistance we can apply the *Fourier law* [12] $\varsigma^u T_1 = \lambda (T_1 - T)$, where $\lambda$ is a constant representing the inverse of thermal resistance. The condition for the optimum of the output power with respect to temperature $T$ then reads $\mathrm{d}L^u/\mathrm{d}T = 0$, from which we obtain $T = \surd(T_1 T_2)$ [26]. Consequently, the Carnot's true efficiency of such a system with optimized output power is thus given by a formula, $\eta_C = T_1 \{1 - \surd(T_2/T_1)\}$. Such a square root dependence, which is the direct consequence of linearity of Fourier's law, is also obviously repeated for the above mentioned dimensionless Kelvin's efficiency, $\eta_K$. Because of enormous effort of engineers to optimize the real output power of concrete heat engines, the above formula describes the *actual efficiencies* quite well as interestingly shown for authentic industrial cases by *Curzon* and *Ahlborn* [30].

## 21.3 Early Scientific and Societal Parentage of Thermal Analysis

Standard reference books [16, 19, 21, 29, 31] are rather coy about the history of thermometry and thermal analysis being the subject of specified papers and book chapters [1, 4–11, 15, 32–35], which goes back to historic times of *Isaac Newton* (1642–1727) who published his temperature scale in 1701 the significance of which lies both in its range of temperature and in its instrumentation presenting also the famous Newton's Law of Cooling [36]. First cornerstone of the theory of warmth propagation was provided by *J.-B. J. Fourier* who initiated the investigation of Fourier series and their application to problems of heat transfer [12]. The very roots of *thermal analysis* appear in the nineteenth century where *temperature* became an *observable* and experimentally decisive quantity, which thus turned into an experimentally monitorable parameter associated with an consequent underpinning of the field of thermodynamics [29, 34, 35]. The first characterization of thermometric measurements is identified in Uppsala in 1829 through the earliest documented experiment which nearly meets current criteria. It was *Fredrik Rudberg* (1800–1839) [15, 22] who recorded the *inverse cooling-rate* data for lead, tin, zinc and various alloys which were placed in a smaller vessel surrounded by a large double-walled iron vessel where the spaces between its two walls, as well as the top lid, were filled with snow to ensure that the inner walls were always kept at zero temperature. Once the experimental condition was set up, Rudberg noted and tabulated the times taken by the mercury in thermometer to fall through each 10° interval. The longest interval then included the freezing point.

One of important impacts came with the discovery of thermoelectric effect [37] by *Thomas J. Seebeck* (1770–1831) occurring in a circuit made from two dissimilar metals and the consequent development of a device called *thermocouple* [37, 38], suitable as a more accurate temperature-measuring tool, in which gas volume or pressure changes were replaced by a change of electric voltage (*Augustin G.A. Charpy* (1865–1945) [39]). *Henry L. Le Chatelier* (1850–1936) [38] was the first who deduced that varying thermocouple output could result from contamination of one wire by diffusion from the other one or from the non-uniformity of wires themselves. The better homogeneity of platinum-rhodium alloy led him to the standard platinum – platinum/rhodium couple so that almost 70 years after the observation of thermoelectricity, its use in thermometry was finally vindicated, which rapidly got a wider use. *Floris Osmond* (1849–1912) [15, 40] investigated the heating and cooling behavior of iron with a goal to elucidate the effects of carbon so that he factually introduced thermometric measurements to then most important field: metallurgy [40].

In 1891, Sir *William C. Roberts-Austen* (1843–1902) [41] was accredited to construct a device to give a continuous record of the output from thermocouple and he termed it as '*thermoelectric pyrometer*' (see Fig. 21.1) and in 1899, *Stanfield* published heating curves for gold and almost stumbled upon the nowadays idea of *differential thermal analysis (*DTA) when maintaining the thermocouple 'cold' junction at a constant elevated temperature measuring thus the entire differences between two high temperatures. Such an innovative system of measuring the temperature difference between the sample and a suitable reference material placed side-by-side in the same thermal environment, in fact initiated the consequent development of DTA instruments [42–44].

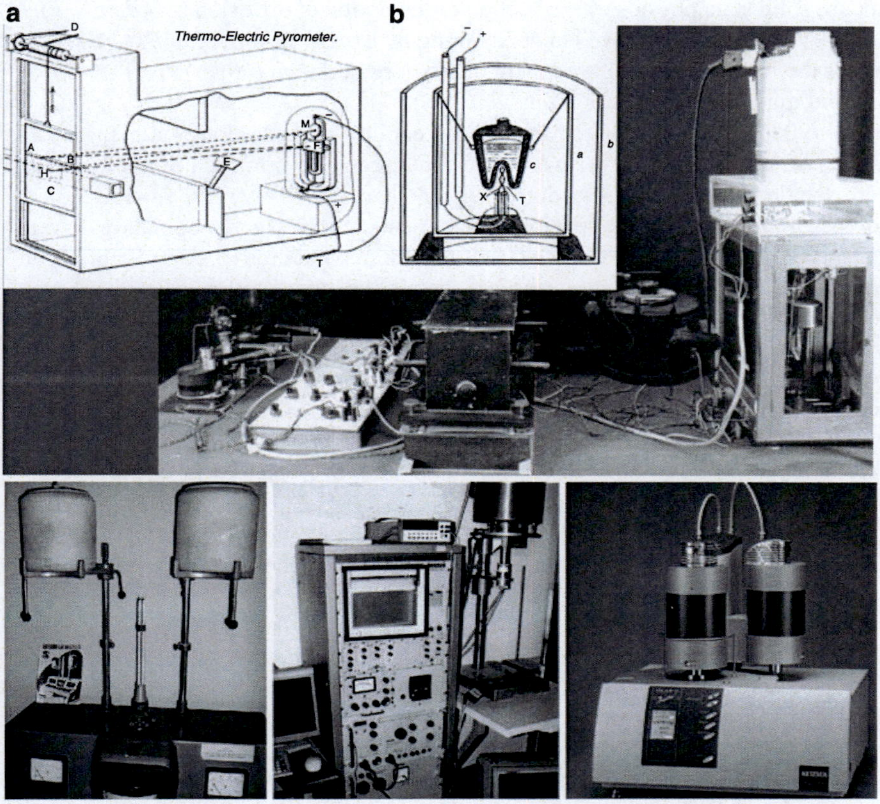


**Fig. 21.1** *Upper*: Thermo-electric pyrometer of Roberts-Austen (1881) showing the instrument (*left*) and its cooling arrangement (*right*) with particularity of the sample holder. *Middle*: Historical photo of the early set-up of Hungarian "Derivatograph" (designed by brothers Paulik), which was one of the most frequent instruments in the former Eastern bloc. *Below*: photo of one time very popular and widespread instruments for high-temperature DTA produced by the Netzsch Gerätebau GmbH (Selb, Germany) from its early version (*left*) presented to the market on 1950 up to the latest third-generation rendering new STA 449 F1 Jupiter (*right*). The middle type (yet based on then fashionable analogous temperature control) was particularly sold during 1970s and survived in many laboratories for a long period (being gradually subjected to enduring computerization and digital data processing)

In 1909 there was elaborated another reliable procedure of preserving the high-temperature state of samples down to laboratory temperature, in-fact *freezing-in* the high-temperature equilibrium as a suitably 'quenched' state for further investigation [34]. It helped in the consistent construction of phase diagrams when used in combination with other complementary analytical procedures, such as the early structural microanalysis (introduced by *Max von Laue* (1879–1960) and *Sir William L. Bragg* (1890–1971) when they detected the X-rays diffraction on crystals) along with the traditional metallographic observations. Another important step toward the modern solid state physics was induction of the notion of diffusion by *Adolf E. Fick* (1829–1901) and its improved understanding by *Ernest Kirkendall* (1914–2005) as well as the introduction of the concept of disorder by *Jakob I. Frenkel* (1894–1952) [45] and models of glasses by *Tammann* [46].

By 1908, knowledge of the heating or cooling curves, along with their rate derivatives and inverse curves were sufficient enough to warrant a first review and more detailed theoretical inspection given by *George K. Burgess* (1874–1932) [47]. Not less important was the development of heat sources where coal and gas were almost completely replaced by electricity as the only source of controllable heat. Already in 1895, *Charpy* described in detail the construction of wire-wound, electrical-resistance based, tube furnaces that virtually revolutionized heating and temperature regulation [39]. Control of heating rate had to be active to avoid possibility of irregularities; however, little attention was paid to it as long as the heat source delivered a smooth temperature-time curve. All early users mention temperature control by altering the current and many descriptions indicate that this was done by manual or clockwork based operation of a rheostat in series with the furnace winding, the system still in practical use up to late 1950s.

However, the first automatic control was published by *Carl Friedrich* in 1912, which used a resistance box with a specially shaped, clock-driven stepped cam-plate on top. As the cam rotated it displaced a pawl outwards at each step and this in turn caused the brush to move on to the next contact, thus reducing the resistance of furnace winding. Suitable choice of resistance and profiling of the cam achieved the desired heating profile. There came also the reduction of sample size from 25 g down to 2.5 g, which lowered the ambiguity in melting point determination from about >2 C down to ~0.5 C. Rates of about 20 K/min were fairly common during the early period later decreased to about quarter. Early in 1908, it was Burgess [47] who considered the significance of various experimental curves in detail concluding that the area of the inverse-rate curve is proportional to the quantity of heat generated divided by the rate of cooling.

The few papers published in the period up to 1920 gave, nonetheless, little experimental details so that *White* [48] was first to show more theoretically the desirability of smaller samples providing a more exhaustive study of the effect of experimental variables on the shape of heating curves as well as the influence of temperature gradients and heat fluxes taking place within both the furnace and the sample. It is obvious that DTA was initially more a qualitative empirical technique, though the experimentalists were generally aware of its quantitative potentialities. The early quantitative studies were treated semiempirically and based more on instinctive reasoning. *Andrews* (1925) was first to use *Newton's* law while *Berg* gave the early bases of DTA theory [49, 50], which was independently simplified by *Speil*. In 1939 *Norton* published his classical paper on differential thermal techniques where he made rather excessive claims for their value both in the identification and quantitative analysis exemplifying clay mixtures [51]. *Vold* (1948) [52] and *Smyth* (1951) [53] proposed a more advanced DTA theory, but the first detailed theories and applicability fashions, free from restrictions, became accessible by followers in 1950s [3, 50, 54–58], e.g., *Keer, Kulp, Evans, Blumberg, Erikson, Soule, Boersma, Borchard, Damiels, Deeg, Nagasawa, Tsuzuki, Barshad, Strum, Lukaszewski*, etc.

In general, the thermoanalytical methods gained theoretical description early 1960s [59–61]. The resulting thermal effects, explicitly temperature disparity ($\Delta T$), can be analyzed at four different but gradually escalating levels [3, 34, 62, 63]: fingerprinting *(identity), quality, quantity* (peak areas) and *kinetics* (peak shape) which were extensively applied to assessments of phase diagrams, transition temperatures, and chemical reactions, as well as to the qualitative analysis of metals, oxides, salts, ceramics, glasses, minerals, soils, and foods. Because of its easy accessibility DTA was used to study behavior of the constrain states of glasses [64–68], inherent processes conventionally viewed as a diagram of temperature (T) versus enthalpy (H) [66], which derivative resembles the entire DTA curve (informative for the analysis of glassforming processes [34]).

## 21.4 Theoretical Basis, Quantitative Thermometric and Calorimetric Measurements

In the beginning, DTA could not be classified as a calorimetric method since no heat was measured quantitatively [59–62]. Only the temperature was determined with the precision of the thermocouple. The quantitative heat effects were traditionally measured by *calorimetry*. Beside the above quoted ice-calorimeter pioneered by *Laplace* the early instrumentation for the determination of heat capacity was based on the classical adiabatic calorimeter and designed by *Walther H. Nernst* (1864–1941) [69, 70] for low temperature measurements [71] (in Germany 1911). Its original experimental arrangement involved the introduction of helium gas as a thermally conducting medium by which the specimen would rapidly reach the temperature required for the next measurement.

Although the measurements of heat changes is common to all calorimeters, they differ in how heat exchanges are actually detected, how the temperature changes during the process of making a measurement are determined, how the changes that cause heat effects to occur are initiated, what materials of construction are used, what temperature and pressure ranges of operation are employed, and so on. If the heat, Q, is liberated in the sample, a part of this heat accumulates in the calorimetric *sample-block* system and causes a quantifiable increase in the temperature. The remaining heat is conducted through the surrounding jacket into the thermostat. The two parts of the thermal energy are closely related. A mathematical description is given by the basic calorimetric equation, often called the *Tian* equation [72].

The calorimetry classification came independently from various sources, e.g. [3, 73–75]. The principal characteristics of a calorimeter are the calorimeter capacity, effective thermal conductivity, and the inherent heat flux, occurring at the interface between the sample-block, B, and the surrounding jacket, J. The temperature difference [3], $T_B - T_J$, is used to classify calorimeters, i.e., *diathermal* ($T_B \neq T_J$), *isodiathermal* $(T_B - T_J) = const.$ and $d(T_B - T_J) \rightarrow 0$, *adiabatic* ($T_B = T_J$), *isothermal* ($T_B = T_J = const.$) and *isoperibolic* $(T_B - T_J) \rightarrow 0$. The most common version of the instrument is the diathermal arrangement where the thermal changes in the sample are determined from the temperature difference between the sample-block and jacket. The chief condition is, however, the precise enough determination of temperatures. With an isodiathermal calorimeter, a constant difference of the block and jacket temperatures is maintained during the measurement, thus also ensuring a constant heat loss by introducing extra heat flux to the sample from an internally attached source (often called 'microheater'). The energy changes are then determined from the energy supplied to the source. For low values of heat, the heat loss can be decreased to minimum by a suitable instrumental set-ups and this version is called as adiathermal calorimeter. An adiabatic calorimeter suppresses heat losses by maintaining the block and jacket temperatures at the same temperature. Adiabatic conditions are more difficult to assure at both higher temperatures and faster heat exchanges so that it is preferably employed at low temperatures.

Eliminating the thermal gradients between the block and the jacket by using an electronic regulation requires, however, sophisticated circuits and more complex set-ups. For this reason, the calorimeters have become experimentally very multifaceted instruments. With compensation "quasiadiabatic" calorimeter, the block and jacket temperatures are kept identical and constant during the measurement as the thermal changes in the sample are suitably compensated, so that the block temperature remains the same. If the heat is compensated by phase transitions in the reseivoir in which the calorimetric block is contained, the instrument are often termed transformation calorimeter. *Quasi-isothermal* calorimeters are, in turn, instruments with thermal compensation provided by electric microheating and heat removal is accomplished by forced flow of a fluid, or by the well-established conduction through a system of thermocouple wires or even supplemented by *Peltier* cooling effect. The method in which the heat is transferred through a thermocouple system is often called *Tian-Calvet* calorimetry [76, 77]. A specific group is formed by isoperibolic calorimeters, which essentially operate adiabatically with an isothermal jacket.

Even in the 1950s, it was a doubtful prediction that classical DTA and adiabatic calorimetry would merge, producing a *differential scanning calorimeter* (DSC). The name DSC was first mentioned by *O'Neil* [78] for a differential calorimeter that possessed continuous power compensation (close-to-complete) between sample and reference. This development came about because the key concern of calorimetry is the reduction of, and certainly also correction for, heat losses and/or gains due to inadvertent temperature distribution in the surroundings of the calorimeter. The heat to be measured can never be perfectly insulated; even in a true adiabatic calorimeter certain heat-loss corrections have to be made and resulting adiabatic deviation must then be corrected through extensive calibration experiments. In order to cancel the heat losses between two symmetric calorimeters were used (e.g., twin calorimetry – one cell with the sample and the other identical, but empty or filled with a reference material), however presented control problems were not easy to handle [3].

True DSC is monitoring the difference between the counterweighing heat fluxes by two extra micro-heaters respectively attached to both the sample and reference in order to keep their temperature difference minimal, while the samples are maintained in the pre-selected temperature program. This technique was originally introduced by *Eyraund* in 1950s [84]. Such an experimental regime bears a quite different measuring principle when comparing with DTA because the temperature difference is not used for the observation itself but is exclusively employed for the regulation only. Certainly, it is the way for accomplishing the most precise measurements of heat capacity (close to adiabatic calorimetry) but technically restricted, to the temperature range up to about 700°C, where heat radiation become decisive making consequently the regulation and particularly compensation complicated.

Three major types of DSCs emerged that all are classified as scanning [79], *isoperibolic twin-calorimeters*. One type makes use of approximate power compensation between two separately heated calorimeters, and the other two merely rely on heat exchange of 2 calorimeters placed symmetrically inside a single heater, but differing in the positions of the controlling thermometers. Even the majority commercial DTA instruments can be classified as a *double non-stationary resembling calorimeter* in which the thermal behaviors of sample are compared with a correspondingly mounted, inert reference [3]. It implies control of heat flux from surroundings and heat itself is a kind of physico-chemical reagent, which, however, could not be directly measured but calculated on the basis of the measurable temperature gradients. We should remark that heat flow is mediated by massless phonons so that the inherent flux does not exhibit inertia as is the case for the flow of electrons. The thermal inertia of apparatus (as observed in DTA experiments) is thus caused by heating a real body and is affected by the entire properties of materials, which structure the sample under study.

The decisive theoretical analysis of a quantitative DTA was based on the calculation of heat flux balances introduced by *Factor* and *Hanks* [80], detailed in 1975 by *Grey* [81], which premises were completed in 1982 by the consistent theory made up by *Holba* and *Šesták* [3, 82, 83]. It was embedded within a 'caloric-like' framework centered on macroscopic heat flows encountered between large bodies (DTA cells, thermostats). Present DTA/DSC instruments marched to high sophistication, computerization and miniaturization, see, e.g., Fig. 21.1

All the equations derived to the description of theoretical basis of DTA/DSC methods can be summarized within the following schema [3, 34], which uses a general summation of inherent terms (each being responsible for the subsequent distinct function): *Enthalpy* + *Heating* + *Inertia* + *Transient* = *Measured Quantity*. It implies that the respective effects of enthalpy change, heating rate and heat transfer are reflected in the value of the measured quantity for all set-ups of the thermal methods commonly exercised. Worth noting is the *inertia term*, which is a particularity for DTA (as well as for *heat-flux* DSC) expressing a specific correction due to the sample mass thermal inertia owing to the inherent heat capacity of real materials. It can be visualized as the sample hindrance against immediate 'pouring' heat into its heat capacity 'reservoir' and it is apparent similarity to the definite time-period necessary for filling a bottle by liquid. Keep in mind, that the consequential compensation DSC calorimetry is of a different nature because it evaluates, instead of temperature difference ($\Delta T \Rightarrow 0$), compensating heat fluxes and thus the heat inertia term is absent [3, 34, 82]. The practice and basis of DSC has been treated numerously [85, 86].

In order to meet an experimental pre-requisition of the transient term (involving the instrumental constant characteristic of a particular DTA apparatus), the routine procedure of calibration is indispensable for a quantitative use of DTA. It is commonly guaranteed by a practice of an adequate incorporation of defined amounts of enthalpy changes by means of the selected test compounds (which widespread standardization, however, failed so that no ICTAC recommendation was issued). Nevertheless, in the laboratory scale, certain compounds (and their tabulated data) can be employed, but the results are questionable due to the various levels of the tabulated data accuracy. Thus it seems be recommendable to use the sets of solid solutions because they are likely to exhibit comparable degree of uncertainty (such as $Na_2CO_3$–$CaCO_3$ or $BaCO_3$–$SrCO_3$ or various sesquioxides mixtures like manganese spinels) [3]. However, the use of the Joule heat effect from a resistance element on passage of electric charge is a preferable method for achieving a more 'absolute' calorimetric calibration. It certainly requires special set-ups of the measuring head enabling the attachment of the micro-heater either on the crucible surface (similarly to DSC) and/or by direct immersing it into the mass of (often powdered) sample. By combination of both experimental methods (i.e., substance's enthalpies and electric pulses) rather beneficial results [87] may be obtained, particularly, when a pre-selected amount of Joule heat is electronically adjustable (e.g., simple selection of input voltage and current pairs) [3, 34]. It was only a pity that no commercial producer, neither an ICTAC committee, have ever became active in their wider application.

## 21.5 Modulated Temperature, Exploration of Constrained and Nano-Crystalline States, Perspectives

Yet another type of thermal measurement that had an early beginning, but initially did not see wide application, is the *alternating current* (AC) calorimetry [79, 88]. Advantage of this type of measurement lies in the application of a modulation to the sample temperature, followed by an analysis of responses. By eliminating any signal that does not correspond to the chosen operating frequency, many of the heat-loss effects can be abolished. Furthermore, it may be possible to probe reversibility and potential frequency-dependence of changes of the studied sample. The heat capacity Cs of the sample can be determined from the ratio of the heat-flow response of the sample, represented by its amplitude $A_{HF}$, to the product of the amplitude of the sinusoidal sample-temperature modulation $A_{Tt}$ and the modulation frequency $\omega = 2\pi/p$ ($p$ being the period). The next advancement in calorimetry occurred in 1992 with the amalgamation of DSC and temperature modulation to the *temperature-modulated* DSC

(TMDSC) [79, 89–91]. In this quasi-isothermal operation, sample temperature $T_S$ oscillates about the underlying temperature $T_0$ (constant/increasing) similarly as in an AC calorimeter (which bears an analogy modulus of a familiar isothermal dynamic mechanical analysis – DMA). The ensuing phase lag, ε, is taken relative to a reference oscillation, $T_S = T_0 + A_{Tt} \sin(\omega t - \varepsilon)$, and by deconvolution of the two signals; an average signal, practically identical to the standard DSC output and a reversing signal, related to the AC calorimetry. There, however, are additional factors necessary for consideration because of the peculiarity of twin calorimeter configuration, such as there is no thermal conductance between the sample and reference calorimeters, zero temperature gradients from the temperature sensors to the sample and the reference pans, and, also, zero temperature gradients within the contents of the pans. In other words, an infinite thermal conductance between temperature sensors and the corresponding calorimeters should be assumed. In summary, three directions of calorimetry were, thus, combined in the twentieth century, which dramatically changed the capabilities of thermal analysis of materials [79]: The high precision of adiabatic calorimetry, the speed of operation and small sample size of DSC, and the possibility to measure frequency dependence of thermal behavior in AC calorimetry.

Another reason for both the modulation mode and the high-resolution of temperature derivatives is the fight against 'noise' in the heat flow signal in temperature swinging modifications. Instead of applying a standard way of eliminating such noise (and other unwanted signal fluctuations) by a more appropriate tuning of an instrument, or by intermediary measurements of the signal in a preselected distinct window, the fluctuations can be forcefully incorporated in a controlled and regulated way of oscillation. Thus the temperature oscillations (often sinusoidal) are located to superimpose over the heating curve and thus incorporated in the entire experimentation (temperature-modulated DTA/DSC) [89]. This was, in fact, preceded by the method of so-called *periodic thermal analysis* introduced by *Proks* as early as in 1969 [92], which aimed at removing the kinetic problem of 'undercooling' by cycling temperature. Practically the temperature was alternated over its narrow range and the sample investigated was placed directly onto a thermocouple junction until the equilibrium temperature for the coexistence of two phases was attained.

Another way of a more clear-cut investigation was introduction of micro-analysis methods using very small samples and millisecond time scales [93, 94]. It involved another peculiarity of truthful temperature measurements of nano-scale crystalline samples [95] in the particle micro range with radius $r$. The measurement becomes size affected due to increasing role of the surface energy usually described by an universal equation: $T_r/T_\infty \cong (1 - C/r)^p$ where $\infty$ portrays standard state and $C$ and $p$ are empirical constants ($\approx$0.15 nm $< C <$ 0.45 nm and $p = 1$ or ½) [96–98].

Measurement in such extreme conditions brings extra difficulties such as measuring micro-porosity [99], quenching [94] and associated phenomena of the sample constrained states [64–68], variability of polymeric macromolecules [100, 101] together with non-equilibrating side effect or competition between the properties of the sample bulk and its entire surface [97] exposed to the contact with the cell holder [34]. Increasing instrumental sophistication and sensitivity provided possibility to look at the sample micro-locally [93, 101–103] giving a better chance to search more thoroughly toward the significance of baselines, which contains additional but hidden information on material structure and properties (inhomogeneities, local nonstoichiometry, interfaces between order–disorder zones [104]). Popular computer built-in smoothing of the noised experimental traces (chiefly baselines) can, however, become counterproductive.

In the future, we may expect certain refining trends possible returning to the original single-sample set-ups with recording mere heating/cooling curves. However, it will happen at the level of fully computerized thermal evidence involving self-evaluation of 'calibration' behavior of the sample thermal inertia and its subtraction from the entire thermal record in order to proliferate thermal effects possibly computing the DTA-like records. In addition, it may even incorporate the application of an arbitrary temperature variation enabling the use of self-heating course by simple placing the sample into the preheated thermostat and consequent computer evaluation of standardized effects or hitherto making possible to introduce fast temperature changes by shifting the sample within the temperature gradient of a furnace [3, 34], etc. Worth noting are special trends [105] particularly based on the modified thermophysical procedure of the rate controlled scope of thermal analysis (RCTA) [106] and/or on the diffusion structural diagnostics as a result of suitably labeled samples [107].

Upcoming prospect of thermal analysis scheme may go down to the quantum world [108] as well as may extend to the global dimension [109] touching even the remote aspects of temperature relativity [110], which, however, would become a special dimension of traditional understanding yet to come.

## 21.6 Some Issues of Socially Shared Activity, Thermoanalytical and Calorimetric Journals and Societies

The historical development and practical use of DTA in the middle European territory of former Czechoslovakia [33] was linked with the names *Otto Kallauner* (1886–1972) and *Joseph Matějka* (1892–1960) who introduced thermal analysis as the novel technique during the period of the so called "*rational analysis*" of ceramic raw materials [111] replacing the process of decomposition of clay minerals by digestion with sulphuric acid, which factually played in that time the role of the contemporary X-ray diffraction. They were strongly affected by the work of *H. Le Chatelier* [38] and their visits at the Royal Technical University of Wroclaw (*K. Friedrich, B. Wohlin*) where the thermal behavior of soils (bauxite) was investigated during heating and related thermal instrumentation was elaborated. Calorimetric proficiency was consequently gained from Polish *Wojciech Świętosławski* (1881–1968). Much credit for further development of modern thermal analysis was attributed with *Rudolf Bárta* (1897–1985) who stimulated thermal analysis activity at his coworkers (*Vladimír Šatava, Svante Procházka* or *Ivo Proks*) and his students (*Jaroslav Šesták*) at the Institute of Chemical Technology (domestic abbreviation VŠChT) in Prague.

In this aspect a special notice should be paid to the lengthy efforts, long journey and fruitful service of *International Confederation of Thermal Analysis* (ICTA and *Calorimetry* – ICTAC, instituted later in the year 1992 and facilitated by *G. Della Gatta*) as an important forerunner and developer in the field of thermal analysis, cf. Fig. 21.2. It has an important preceding history [6, 112, 113] connected with the former Czechoslovakia and thermoanalytical meetings organized by *R. Bárta* just mentioning the earliest first Conference on DTA, (Prague 1956), the Second (Prague 1958) and the Third Conference on Thermography (Prague 1961) and the Forth Conference on DTA (Bratislava 1966). *Robert C. Mackenzie* (1920–2000) from Scotland was an invited guest at the 1961 meeting and upon the previous communication with Russian *L.G. Berg* and US *P.D. Garn* as well as Hungarian *L. Erdey* an idea for the creation an international society was cultivated aiming to enable easier contacts between national sciences, particularly across the separating 'iron curtain', which in that time divided the East and West Europe [6]. The first international conference on thermal analysis was then held in the Northern Polytechnic in London, April 1965 and was organized by British scientists namely *B.R. Currell, D.A. Smith, J.P. Redfern, W. Gerrard, C.J. Keattch* and *D. Dollimore* with a help of *R.C. Mackenzie, B. Stone* and US professors *P.D. Garn* and *W.W. Wendlant*, Canadien *H.G. McAdie*, French *M. Harmelin*, Hungarian *L. Erdey*, Japanese T. *Sudo*, Swedish *G. Berggrenn* and Italian *G. Lombardi*. Some invited speakers from the East Europe were particularly asked to come to bridge then existing tough political control on physical, freedom and civil frontiers strongly restricting the human rights of the Easterners (dominated by Soviet Union until the late 1980s), such as *F. Paulik* (Hungary) and *J. Šesták* (Czechoslovakia). The consequent ICTA foundation in Aberdeen, September 1965, was thus established by these great progenitors of thermal analysis, Russian *Lev G. Berg* being the first ICTA presidents (with the councilors *J.P. Redfern, R.C. Mackenzie, R. Bárta, S.K. Bhattacharrya, C. Duval, L. Erdey, T. Sudo, D.J. Swaine, C.B. Murphy,* and *H.G. McAdie*).

The progress of thermal analysis was effectively supported by the allied foundation of international journal, which editorial board was recruited from the key-speaker of both 1965 TA conferences as well as from the renowned participants at the second ICTA in Worcester (USA 1968). In particular it was *Journal of Thermal Analysis*, which was brought into being by *Judit Simon* (1937-, who has been serving as the editor-in-chief until today) and launched under the supervision Hungarian Academy of Sciences (Akadémiai Kiadó) in Budapest 1969 (*L. Erdey, E. Buzagh, F. and J. Paulik brothers, G. Liptay, J.P. Redfern, R. Bárta, L.G. Berg, G. Lombardi, R.C. Mackenzie, C. Duval, P.D. Garn, S.K. Bhattacharyya, A.V. Nikolaev, T. Sudo, D.J. Swaine, C.B. Murphy, J.F. Johanson,* etc.) to aid preferably the worthwhile East European science suffering then under the egregious political and economic conditions. Secondly it came to pass *Thermochimica Acta* that appeared in the year 1970 by help of Elsevier [114] and, for a long time, edited by *Wesley W. Wendlandt* (1920–1997) assisted by wide-ranging international board (such as *B. R. Currell, T. Ozawa, L. Reich, J. Šesták, A. P. Gray, R. M. Izatt, M. Harmelin, H. G. McAdie, H. G. Wiedemann, E. M. Barrall, T. R. Ingraham, R. N. Rogers, J. Chiu, H. Dichtl, P. O. Lumme, R. C. Wilhoit,* etc.).

ICTAC

The field growth lead, naturally, to continuous series of the *US Calorimetry Conferences* (CalCon) [115–117], which supposedly evolved from a loosely knit group operating in the 1940s to a recent highly organized assembly working after the 1990s. Worth mentioning are *Hugh M. Huffman* (1899–1950) *and James J. Christensen* (1931–1987), whose names were recently used to shield the CalCon Awards presented annually for achievements in calorimetry. There is a number of other respectable cofounders, (let us point out *D.R. Stull, G. Waddington, G.S. Parks, S. Sunner, F.G. Brickwedde, E.F. Westrum, J.P. McCullough, D.W. Osborne, W.D. Good, P.A.G. O'Hare, P.R. Brown, W.N. Hubbart, R. Hultgren, R.M. Izatt, D. J. Eatough, J. Boerio-Goates, J.B. Ott*). It provided a good example how the democracy-respecting society changing their chairmanships every year, which, however, did not find a place in the statutes of later formed ICTA (with its 4-years period and its unfortunate consequence of the 15 years of personality cult having been in effect at the turn of twenty-first century). Consequentially, the *Journal of Chemical Thermodynamics* began publication in the year 1969 firstly edited by *L.M. McGlasham, E.F. Westrum, H.A. Skinner* and followed by others. More details about the history and state-of-art of thermal science and the associated field of thermal analysis were published elsewhere [3–6, 32–35, 79, 112, 113, 115–117].

A specific domain of thermal analysis worth of attention (but laying beyond this file) is the weight measurement under various thermal regimes, pioneered by Czech *Stanislav Škramovský* (1901–1983) who coined the term 'statmograph' (from Greek stathmos = weight) [1, 6, 35], which, however, was overcome by the generalized

**Fig. 21.2** Portraits show some influential personalities on the international scene, which are noteworthy for their contributions to the progress of the fields of thermal analysis (TA) and calorimetry including the founders of ICTA/ICTAC (around the inserted emblem), living persons limited to age above 75. *Upper from left*: *Cornelius B. Murphy* (1918–1994), USA (TA theory); *Robert C. Mackenzie* (1920–2000), Scotland (DTA, clay minerals, history); Sir *William C. Roberts-Austen* (1843–1902), England (thermoelectric pyrometer); *Gustav H.J. Tammann* (1861–1938), Germany (inventing the term thermal analysis) and *Nikolaj S. Kurnakov* (1860–1941), Russia (contriving the first usable DTA); *below*: *Lev G. Berg* (1896–1974), Russia (TA theory); *Rudolf Bárta* (1897–1985), Czechoslovakia (ceramics, cements); *Walther H. Nernst* (1864–1941), Germany (originating low-temperature calorimetry); *Edouard Calvet* (1895–1966), France (heat-flow caolrimetry) and *Henry L. Le Chatelier* (1850—1936), France (devising thermocouple); *yet below*: *David J. Dollimore* (1927–2000), England (later USA – theory, kinetics); *Hugh M. Huffman* (1899–1950), USA, founder of CalCon; *James J. Christensen* (1931–1987), USA (calorimtery); *Wojciech Świętosławski* (1881–1968), Poland (calorimetry); Čenk Strouhal (1850–1922), Czechoslovakia (thermics, Strouhal numbers); yet below: Hans-Joachim Seifert (1930–), Germany (phase diagrams); *Takeo Ozawa* (1932–), Japan (energetic materials, kinetics); *Eugene Segal* (1933–), Romania (kinetics); *Hiroshi Suga* (1930–), Japan (calorimetry) and *Giuseppe Della Gatta* (1935–), Italy (calorimetry); *yet below*: *Wesley W. Wendlandt* (1920–1997), USA (TA theory, instrumentation); *Bernhard Wunderlich* (1931–), USA (macromolecules, modulated TA); *Paul D. Garn* (1920–1999), USA (TA theory, kinetics); *Jean-Pierre E. Grolier* (1936-), France (calorimetry) and *Ole Toft Sǿrensen* (1933-), Denmark (CRTA, non-stoichiometry); *Bottom*: *Cyril J. Keattch* (1928–1999), England (thermogravimetry); *Hans G. Wiedemann* (1920–), Switzerland (TG apparatuses, instrumentation); *Shmuel Yariv* (1934–), Israel (earth minerals); *Joseph H. Flynn* (1922–), USA (DSC, kinetics) and *Patrick K. Gallagher* (1931–), USA (inorganic materials)

**Fig. 21.3** Recognized pioneers of thermal analysis, of Hungarian origin, who were accountable for the development of instruments (popular East-European TA apparatus "derivatograph") – *Ferenc Paulik* (1922–2005), right, and for initiation of fingerprint methodology (multivolume atlas of TA curves by Akademia Kiado) – *Geörge Liptay* (1931–), *left*

**Fig. 21.4** The photo from 28th conference of the Japanese Society on Calorimetry and Thermal Analysis (JSCTA) in Tokyo (Waseda University, 1992) shows *(from left) M. Taniguchi* (Japan), late *C.J. Keattch* (GB), late *R. Otsuka* (Japan), *S. St. J. Warne* (Australia, former ICTA president), *H. Suga* (Japan), *J. Šesták* (Czechoslovakia) and *H. Tanaka* (Japan). The regular JSCTA conferences started in Osaka 1964 under the organization of *S. Seki* who became the first president when the JSCTA was officially established in 1973. Since then, the JSCTA journal NETSU SOKUTEI has been published periodically

expression 'thermogravimetry' as early introduced by French *Clément Duval* (1902–1976) or Japanese *Kotaro Honda* (1870–1954) [33–35, 118–121]. Consequently, it yielded a very popular topic of simultaneous weight-to-caloric measurements under so called quasi-isothermal and quasi-isobaric conditions [35, 122] making use of the apparatus 'derivatograph', see Fig. 21.1, originated in Hungary in late 1950s [122], see Fig. 21.3. It apparently lunched an extended field of microbalance exploitation and their presentation in regular conferences [123].

## References


1. Šesták J, Mareš JJ (2007) From caloric to stathmograph and polarography. J Therm Anal Calorim 88:763–768
2. Comenius JA (1659) Disquisitiones de Caloris et Frigoris Natura. J. Jansson, Amsterdam
3. Šesták J (1984) Měření termofyzikálních vlastností pevných látek. Academia, Praha; English transl.: Thermophysical Properties of Solids. Elsevier, Amsterdam; Russian transl.: Teoreticheskyj termicheskyj analyz. Mir, Moscow (1988)
4. Šesták J, Mackenzie RC (2001) Heat/fire concept and its journey from prehistoric time into the third millennium. J Therm Anal Calorim 64:129–147
5. Mackenzie RC, Proks I (1985) Comenius and Black: progenitors of thermal analysis. Thermochim Acta 92:3–14
6. Šesták J (2005) Some historical aspects of thermal analysis: origins of Termanal, CalCon and ICTA. In: Klein E, Smrckova E, Simon P (eds) Proceedings of International Conference on Thermal Analysis "Termanal". Publishing House of the Slovak Technical University, Bratislava, p 3
7. Black J (1803) Lectures on the elements of chemistry. W. Creech, Edinburgh; Germ. transl.: Vorlesungen über Chemie. Crell, Hamburg (1804)
8. Brush SG (1976) The kind of motion we call heat, vol I and II. North Holand, Amsterdam
9. Fox R (1971) The caloric theory of gases: from Lavoisier to Regnault. Claredon, Oxford
10. Proks I (1991) Evaluation of the knowledge of phase equilibria. In: Chvoj Z, Šesták J, Tříska A (eds) Kinetic phase diagrams. Elsevier, Amsterdam
11. Proks I (2011) Celok je jednodušší než jeho části. (Whole is simpler than its parts). Publishing House of Slovak Academy of Sciences, Bratislava
12. Fourier J-BJ (1822) Theorie analytique de la chaleur. Firmin Didot, Paris
13. Barnett MK (1956) The development of thermometry and the temperature concept. Osiris 12:269–341
14. Favre PA, Silbermann JT (1852) Recherches sur les quantités de chaleur degagées dans les actions chimiques et moléculaires. Ann Chim Phys 34:357–450
15. Mackenzie RC (1984) History of thermal analysis. Special issue of Thermochim Acta 73, Elsevier, Amsterdam
16. Thenard L (1836) Treatise of chemistry, 6th edn. Crochard, Paris
17. McKie D, Heathcote NHV (1935) The discovery of specific and latent heats. Arnold, London
18. Socquet JM (1801) Essai sur le calorique. Desray, Paris
19. Kelland P (1837) Theory of heat. JJ Deighton, Cambridge
20. Lindsay RB (1975) Energy: historical development of the concept. Dowden, Stroudburg
21. Carnot S (1824) Réflexions sur la puissance motrice du feu et sur les machines propres à développer cette puissance. Bachelier, Paris, Germ transl.: Ostwald's Klassiker, Nr. 37, Engelmann, Leipzig (1909), English transl. by R.H. Thurston (ed) (1890) Reflections on the motive power of heat and on machines fitted to develop that power. Wiley, New York
22. Strouhal Č (1908) Thermika. (Thermics). JČMF, Praha
23. Callendar HL (1911) The caloric theory of heat and Carnot's principle. Proc Phys Soc London 23:153–189

24. Cropper WH (1987) Carnot's function: origins of the thermodynamic concept of temperature. Am J Phys 55:120–129
25. Mareš JJ (2000) On the development of temperature concept. J Therm Anal Calorim 60:1081–1091
26. Mareš JJ, Hubík P, Šesták J, Špička V, Krištofik J, Stávek J (2008) Phenomenological approach to the caloric theory of heat. Thermochim Acta 474:16–24
27. Šesták J, Mareš JJ, Hubík P, Proks I (2009) Contribution by Lazare and Sadi Carnot to the caloric theory of heat and its inspirative role in thermodynamics. J Therm Anal Calorim 97:679–683
28. Clausius R (1876) Mechanische Wärmetheorie. Vieweg und Sohn, Braunschweig
29. Muller I (2007) A history of thermodynamics. Springer, Berlin
30. Curzon FL, Ahlborn B (1975) Efficiency of a Carnot engine at maximum power output. Am J Phys 43:22–24
31. Bensande-Vincent B, Stenger I (1996) History of chemistry. Harvard, London
32. Cardillo P (2002) A history of thermochemistry through the tribulations of its devotees. J Therm Anal Calorim 72:7–22
33. Šesták J, Proks I, Šatava V, Habersberger K, Brandštetr J, Koráb O, Pekárek V, Rosický J, Vaniš M, Velíšek, J (1986) The history of thermoanalytical and related methods in the territory of present-day Czechoslovakia. Thermochim Acta 100:255–270; Šesták J (1999) The history and future of thermal analysis; thermochemical and thermodynamic background. J Min Metal 35:367–390
34. Šesták J (2004) Heat, thermal analysis and society. Nucleus, Hradec Králové; Science of heat and thermophysical studies: a generalized approach to thermal analysis. Elsevier, Amsterdam (2005)
35. Šesták J (2009) Some historical features focused back to the process of European education revealing some important scientists, roots of thermal analysis and the origin of glass research. In: Šesták J, Holeček M, Málek J (eds) Some thermodynamic, structural and behavioral aspects of materials accentuating non-crystalline states. OPS-ZČU, Plzeň
36. Grigull U (1984) Newton's temperature scale and the law of cooling. Heat Mass Transf 18:195–199
37. Hunt LB (1964) The early history of the thermocouple. Platinum Metals Rev 8:23–28; Drebushchak VA (2009) Universality of the EMF of thermocouples. ThermochimActa 496:50–53; Ott JB, Goates JR (1992) Temperature measurements. In: Rossiter BW, Beatzold RC (eds) Determination of thermodynamic properties. Wiley, New York, pp 451–572
38. Le Chatelier H (1887) De l'action de la chaleur sur les argiles. Bull Soc Franc Minéral 10:204–207
39. Mackezie RC (1982) The story of the platinum-wound electric resistance furnace. Platinum Metals Rev 26:175–183
40. Kayser FX, Patterson JW (1998) Sir William C. Roberts-Austen and his role in the development of binary diagrams and modern physical metallurgy. J Phase Equilib 19:11–18
41. Roberts-Austen WC (1899) Fifth report to the alloys research committee. Nature 59:566–567, and Proc. Inst. Mech. Eng. 1899, 35
42. Tammann G (1905) Über die Anwendung der thermische Analysen in abnormen Fällen. Z Anorg Chem 45:24–30
43. Kurnakov NS (1904) Eine neue Form des Registrierpyrometers. Z Anorg Chem 42:184–202; Raboty v oblasti cvetnoj metalurgii p 104, Gos Nautsh Tech Izd, Moskva (1954)
44. Kauffman B, Beck A (1962) Nikolai Semenovich Kurnakov. J Chem Educ 39:44–49
45. Frenkel J (1946) Kinetic theory of liquids. Oxford University Press, Oxford, and Dower, New York 1950
46. Tammann G (1897) Über die Grenzen des festen Zustandes. Annalen der Physik und Chemie (Wiedemann), Bd. 298/N.F.Bd. 62:280–299; Der Glaszustand. Leopold Voss, Leipzig (1933)
47. Burgess GK (1908) Methods of obtaining cooling curves. Bull Bur Stand 5(S99):199–225

48. White WP (1909) Melting point determination. Am J Sci 28:453–473
49. Berg LG (1952) Skorostnoj kolichestvennyj fazovyj analyz (Rapid quantitative phase analysis). Akad Nauk, Moscow (in Russian)
50. Piloyan FO (1964) Vedenije v termografiju. (Introduction to thermography). Nauka, Moskva (in Russian)
51. Norton FH (1939) Critical study of the differential thermal methods for the identification of the clay minerals. J Am Ceram Soc 22:54–64; Sykes C (1935) Methods for investigating thermal changes occurring during transformations in solids. Proc R Soc A 148:422–446
52. Vold MJ (1949) Differential thermal analysis – DTA. Anal Chem 21:683–688
53. Smyth HT (1951) Temperature distribution during mineral inversion and its significance in DTA. J Am Ceram Soc 34:221–224
54. Boersma SL (1955) Theory of DTA and new methods of measurement and interpretation. J Am Ceram Soc 38:281–284
55. Borchadt HJ (1956) Differential thermal analysis. J Chem Educ 33:103–109; Borchard HJ, Daniels F (1957) The application of DTA to the study of reaction kinetics. J Am Chem Soc 79:41–46
56. Mackenzie RC (ed) (1957) The differential thermal investigation of clays. Mineralogical Society, London
57. Smothers WJ, Chiang Y (1966) Handbook of DTA. CRC Press, New York
58. Schultze D (1969) Differentialthermoanalyse. VEB Deutscher Verlag der Wissenschaften, Berlin
59. Garn PD (1962) Thermoanalytical methods of investigation. Academic, New York
60. Wendlandt WW (1964) Thermal methods of analysis. Wiley, New York
61. Lombardi G (ed) (1977/1981) For better thermal analysis. ICTA Information Booklet, Rome University, Rome
62. Šesták J, Šatava V, Wendlandt WW (1973) The study of heterogeneous processes by thermal analysis. Special issue of Thermochim Acta 7(5), Elsevier, Amsterdam
63. Šesták J (1979) Thermodynamic basis for the theoretical description and correct interpretation of thermoanalytical experiments. Thermochim Acta 28:197–227
64. Vogel H (1921) Das Temperaturabhängigkeitsgesetz der Viskosität von Flüssigkeiten. Phys Zeitschr 22:645–646; Kauzmann W (1948) The nature of the glassy state and the behavior of liquids at low temperatures. Chem Rev 43:219–256
65. Hrubý A (1972) Evaluation of glass forming tendency by means of DTA. Czech J Phys B 22:1187–1193; Glass-forming tendency in the $GeS_x$ system. Czech J Phys B 23:1263–1272 (1973)
66. Šesták J (1996) Use of phenomenological enthalpy versus temperature diagram (and its derivative – DTA) for a better understanding of transition processes in glasses. In: Šesták J (ed) Vitrification, transformation and crystallization of glasses. Special issue of Thermochim Acta 280/281:175–191 Elsevier, Amsterdam
67. Kozmidis-Petrović AF (2010) Theoretical analysis of relative changes of the Hruby, Weinberg, and Lu–Liu glass stability parameters with application on some oxide and chalcogenide glasses. Thermochim Acta 499:54–60
68. Hutchinson JM (2009) Determination of the glass transition temperatures. J Therm Anal Calorim 98:579–589
69. Nernst W (1911) Der Energieinhalt fester Stoffe. Ann Phys 341:395–439
70. Nernst W (1918) Die theoretischen und experimentellen Grundlagen des neuen Wärmesatzes. W. Knapp, Halle
71. Gmelin E (1987) Low-temperature calorimetry: a particular branch of thermal analysis. Thermochim Acta 110:183–208; Gmelin E, Britt W (1987) Recent progress in low temperature microcalorimetry. Thermochim Acta 119:35–36
72. Tian A (1923) Utilisation de la méthode calorimétrique en dynamique chimique: emploi d'un microcalorimètre à compensation. Bull Soc Chim Fr 33:427–428; Recherches sur la calorimétrie – Généralisation de la méthode de compensation électrique – Microcalorimétrie. J Chim Phys 30:665–708 (1933)

73. Velíšek J (1978) Kalorimetrické měřící metody. (Calorimetric methods) Chemické listy 72:801–830; Vysokoteplotní kalorimetrie (High-temperature calorimetry). Čs čas fyz A 20:513 (1970)
74. Zielenkiwicz W, Margas E (2002) Theory of calorimetry. Kluwer, Dordrecht; Rouquerol J, Zielenkiewicz W (1986) Suggested practice for classification of calorimeters. Thermochim Acta 109:121–137; Randzio SL (2002) Recent development in calorimetry. Ann Rep Prog Chem, Sect. C 98:157–217
75. Hemminger W, Höhne GWH (1979) Grundlagen der Kalorimetrie. Verlag Chemie Weinheim; Calorimetry: fundamentals and practice. Verlag Chemie, Weinheim (1984)
76. Calvet E, Prat H (1963) Recent progress in microcalorimetry. Pergamon, New York; Oscarson JL, Izatt RM (1992) Calorimetry. In: Rossiter BW, Beatzold RC (eds) Determination of thermodynamic properties. Wiley, New York, pp 573–620
77. Attree RW, Cushing RL, Ladd JA, Pieroni JJ (1958) Differential calorimeter of the Tian-Calvet type, a review. Rev Sci Instrum 29:491–496; Gravelle PC (1972) Heat-flow microcalorimetry and its applications to heterogeneous catalysis. In: Eley DD (series editor) Advances in catalysis and related subjects. vol 22, p 191. Academic, New York
78. Watson ES, O'Neill MJ, Justin J, Brenner N (1964) A differential scanning calorimeter for quantitative differential thermal analysis. Anal Chem 36:1233–1238; O'Neill MJ (1964) The analysis of a temperature-controlled scanning calorimeter. Anal Chem 36:1238–1245
79. Wunderlich B (2000) Temperature-modulated calorimetry in the 21st century. Thermochim Acta 355:43–57
80. Faktor MM, Hanks R (1967) Quantitative application of dynamic differential calorimetry. Part 1–Theoretical and experimental evaluation. Trans Faraday Soc 63: 1122–1129; and Part 2 – Heats of formation of the group 3A arsenides. Trans Faraday Soc 63:1130 – 1135
81. Grey AP (1968) Simple generalized theory for analysis of dynamic thermal measurements. In: Porter RS, Johnson JF (eds) Analytical calorimetry. Plenum, New York, vol 1, p 209; In: Proceedings of the 4th ICTA, Thermal Analysis, Akademiai Kiado, Budapest, 1974
82. Holba P, Šesták J, Bárta R (1976) Teorie a praxe DTA/DSC (Theory and practice of DTA/DSC). Silikáty (Prague) 20:83–95; Šesták J, Holba P, Lombardi G (1977) Quantitative evaluation of thermal effects: theory and practice. Annali di Chimica (Roma) 67:73–87
83. Boerio-Goates J, Callen JE (1992) Differential thermal methods. In: Rossiter BW, Beatzold RC (eds) Determination of thermodynamic properties. Wiley, New York, pp 621–718
84. Eyraud MCh (1954) Appareil d'analyse enthalpique différentielle. CR Acad Sci 238:1511–1512
85. Flynn JH (1974) Theory of DSC. In: Porter RS, Johnson JF (eds) Analytical calorimetry. Plenum, New York, p 3 and 17; Thermodynamic properties from differential scanning calorimetry by calorimetric methods. Thermochim Acta 8:69–81
86. Heines PJ, Reading M, Wilburn FW (1998) Differential thermal analysis and differential scanning calorimetry. In: Brown ME (ed) Handbook of thermal analysis and calorimetry, vol 1. Elsevier, Amsterdam, pp 279–361
87. Svoboda H, Šesták J (1975) New system of DTA calibration by the predetermined amount of heat. In: Thermal analysis. Proceedings of the 4th ICTA, vol 3, p 725. Akademiai Kiado, Budapest (1975); In: "TERMANAL", Proceedings of the 9th SCTA, p PL-31. SVST, Bratislava (1976)
88. Sullivan PF, Seidel G (1968) Steady-state ac-temperature calorimetry. Phys Rev 173:679–685
89. Reading M (1993) Modulated differential scanning calorimetry: a new way forward in materials characterization. Trends Polym Sci 1:248–253
90. Reading M, Elliot D, Hill VL (1993) A new approach to the calorimetric investigation of physical and chemical transitives. J Therm Anal 40:949–955
91. Wunderlich B, Jin Y, Boller A (1994) Mathematical description of differential scanning calorimetry based on periodic temperature modulation. Thermochim Acta 238:277–293

92. Proks I, Zlatovský J (1969) Periodická termická analyza (Periodic thermal analysis). Chem Zvesti (Bratislava) 23:620 (in Slovak)
93. Minakov AA, Adamovsky SA, Schick C (2005) Non-adiabatic thin-film (chip) nanocalorimetry. Thermochim Acta 432:177–185
94. Adamovsky SA, Minakov AA, Schick C (2003) Scanning microcalorimetry at high cooling rate. Thermochim Acta 403:55–63; Ultra-fast isothermal calorimetry using thin film sensors. Thermochim Acta 415:1–7
95. Wunderlich B (2007) Calorimetry of nanophases of macromolecules. Int J Thermophys 28:958–967
96. Zhang Z, Li JC, Jiang Q (2000) Modelling for size-dependent and dimension-dependent melting of nanocrystals. J Phys D Appl Phys 33:2653–2656
97. Vanithakumari SC, Nanda KK (2008) Universal relation for the cohesive energy of nanoparticles. Phys Lett A 372:6930–6934
98. Guisbiers G, Buchaillot L (2009) Universal size/shape-dependent law for characteristic temperatures. Phys Lett A 374:305–308
99. Beurroies I, Denoyel R, Llewellyn P, Rouquerol J (2004) Melting solidification and capillary condensation: interpretation of thermoporometry data. Thermochim Acta 421: 11–18; Nedelec J-M, Grolier J-PE, Baba M (2006) Thermoporosimetry: a powerful tool to study the cross-linking in gels networks. J Sol-Gel Sci Tech 40:191–200
100. Wunderlich B (2005) Thermal analysis of polymeric materials. Springer, Berlin
101. Reading M, Price DM, Grandy D, Smith RM, Conroy M, Pollock HM (2001) Micro-thermal analysis of polymers: current capabilities and future prospects. Macromol Symp 167:45–55
102. Hammiche A, Reading M, Pollock HM, Song M, Hourston DJ (1996) Localized thermal analysis using a miniaturized resistive probe. Rev Sci Instrum 67:4268–4275
103. Price DM, Reading M, Hammiche A, Pollock HM (1999) Micro-thermal analysis: scanning thermal microscopy and localised thermal analysis. Int J Pharm 192:85–96
104. Hlaváček B, Šesták J, Koudelka L, Mošner P, Mareš JJ (2005) Vibration forms and structural changes in liquid/glassy state. J Therm Anal Calorim 80:271–283
105. Paulik F (1995) Special trends in thermal analysis. Wiley, Chichester; Balek V, Brown ME (1998) Less common techniques of thermal analysis. In: Brown ME (ed) Handbook of thermal analysis and calorimetry. Elsevier, Amsterdam, vol 1, pp 445–471
106. Sørensen OT (1999) Rate controlled thermal analysis used to study solid-state reactins. J Thermal Anal Calor 56:17–22; Sørensen OT, Rouquerol J (2003) Sample controlled thermal analysis. Kluwer, Dordrecht
107. Balek V (1984) Emanation thermal analysis. Thermochim Acta 22:1–156; Balek V Tölgyessy J (1984) Emanation thermal analysis and other radiometric emanation methods. Elsevier, Amsterdam
108. Mareš JJ, Šesták J (2005) An attempt at quantum thermal physics. J Therm Anal Calorim 82:681–686
109. Šesták J, Hubík P, Mareš, JJ (2010) Thermal analysis scheme aimed at better understanding of the Earth's climate changes due to the alternating irradiation. J Thermal Anal Calor, 101:567–575
110. Mareš JJ, Hubík P, Šesták J, Špička V, Krištofik J, Stávek J (2010) Relativistic transformation of temperature and Mosengeil-Ott's antinomy. Physica E 42:484–487
111. Kallauner O, Matějka J(1914) Beitrag zu der rationellen Analyse, Sprechsaal 47, 423; Matějka J (1919) Chemické proměny kaolinitu v žáru (Chemical changes of kaolinite on firing) Chemické listy 13, 164–166 and 182–185
112. Gallagher P, McAdie H (eds.) ICTAC celebrates its 40th anniversary. Special issue of ICTAC News – ISSN 1022–2510 (2006)
113. Mackenzie RC (1993) Origin and development of ICTA. J Therm Anal 40:5–28, and ICTA News 8(1), 3 (1975)
114. Wendlandt WW (1981) How thermochmica acta began: some reflections. Thermochim Acta 50:1–5

115. Izatt RM, Brown PR, Oscarson JL (1995) The history of the calorimetry conferences 1946-1995. J Chem Thermodyn 27:449–464
116. Sipowska JT, Izatt RM, Ott JB (2001) The history of the calorimetry conference. Netsu Sokutei (Japanese Journal of TA) 37:17
117. Stull DR (1971) Twenty-five years of calorimetry conferences. Bull Thermodyn Thermochem 14:431–441
118. Keattch CJ, Dollimore D (1993) Studies in the history and development of thermogravimetry. J Therm Anal Calorim 39:97–118
119. Duval C (1953) Inorganic thermogravimetric analysis. Elsevier, Amsterdam
120. Saito H (1962) Thermobalance analysis. Gijitsu Shin, Tokyo
121. Vallet P (1972) Thermogravimetry. Gauthier-Villars, Paris
122. Paulik F, Paulik J (1981) Simultaneous thermoanalytical examination by means of derivatograph. Elsevier, Amsterdam
123. Jayaweera SAA, Robens E (2003) The roots of vacuum microbalance techniques and the international conferences on this subject. J Therm Anal Calorim 71:7–11

# Index

**Q**



**R**

## U



## V